KU-141-192

KING'S SAFETY IN THE PROCESS INDUSTRIES

KING'S SAFETY IN THE PROCESS INDUSTRIES

Second Edition

Ralph King
BSc, CEng, FRAeS, FIChemE, FInstP
Consultant Chemical Engineer

Ronald Hirst
CEng, FIChemE, FIFireE, MIFS, AMRAeS
Fire Engineering Consultant

With contributions from:

Glynne Evans
CEng, MIMechE
Consultant Mechanical Engineer

A member of the Hodder Headline Group
LONDON • SYDNEY • AUCKLAND

Copublished in North America by
Wuerz Publishing Ltd, Winnipeg, Canada

First published in Great Britain in 1998 by Arnold,
a member of the Hodder Headline Group,
338 Euston Road, London NW1 3BH

Copublished in North America by
Wuerz Publishing Ltd,
895 McMillan Ave., Winnipeg,
Manitoba, Canada

© 1998 R W King & R Hirst

All rights reserved. No part of this publication may be reproduced or transmitted in any form or by any means, electronically or mechanically, including photocopying, recording or any information storage or retrieval system, without either prior permission in writing from the publisher or a licence permitting restricted copying. In the United Kingdom such licences are issued by the Copyright Licensing Agency: 90 Tottenham Court Road, London W1P 9HE.

Whilst the advice and information in this book is believed to be true and accurate at the date of going to press, neither the authors nor the publisher can accept any legal responsibility or liability for any errors or omissions that may be made.

British Library Cataloguing in Publication Data
A catalogue record for this book is available from the British Library

Library of Congress Cataloging-in-Publication Data
A catalog record for this book is available from the Library of Congress

ISBN 0 340 67786 4
ISBN 0 920063 75 6 (Wuerz)

Publisher: Matthew Flynn
Production Editor: James Rabson
Production Controller: Rose James
Cover designer: Stefan Brazzo

Typeset by Saxon Graphics Ltd, Derby
Printed and bound by J W Arrowsmith Ltd, Bristol

SOUTHAMPTON INSTITUTE
21
DATE 8·6·98

Contents

Foreword

I have been so busy with a very different project over the last two years that I have had little time to contribute to this second edition. Thus I must thank my co-authors Ron and Glynne for taking on the task of revising and partly rewriting it. I have spent only a small part of my professional career in the pursuit of safety. This really started with the investigations into the causes of the Flixborough disaster of 1974 and led to my getting the sack (not for the first time!). The reasons were mainly political. This has taught me that the honest pursuit of safety as a career has its dangers. This is largely due to the persistence of *Organisational Misconceptions* which can lead to disasters in companies and society. This valuable concept was discovered and pioneered by Professor Barry Turner, a sociologist. I have tried to develop it and apply it in the context of this book. Unfortunately it seems that the legal profession has much responsibility for the persistence and spread of *Organisational Misconceptions*. To give an example, there is the matter of the addictiveness of nicotine to the human system. Having fought long and hard to break my own smoking habit in which I finally succeeded while knowing many others who have tried hard and not yet succeeded, I can only express incredulity and disgust at the denials, supported by their lawyers on the part of the heads of tobacco companies, that nicotine is addictive.

The same kind of influences may have caused the Flixborough Court of Inquiry to describe the plant in its Report as 'well designed and constructed'. This was in spite of the fact that my co-author Glynne Evans had discovered a fatal flaw in the plant's design at an early stage in the Inquiry and which we identify as one of the most important causes of the disaster. My own puny attempts to have the official report withdrawn and a fresh investigation held have not (yet) succeeded.

Time and again we discover the harm done by *Organisational Misconceptions* which are held, often for obscure reasons, at high levels in an organisation or society and acquire the force and dogma of religion. To question an *Organisational Misconception* is then treated as disloyalty to the organisation concerned. This occurs in all walks of life.

The most important lesson for accident investigators was aptly put by Sherlock Holmes, the great fictional detective created by Conan Doyle:

'It is a capital mistake to theorise before one has data. Insensibly one begins to twist facts to suit theories instead of theories to suit data'.

Yet we all make it!

Ralph King

Preface to the Second Edition

When Ralph King realised that it was time to think about a second edition of this popular book he was, typically, heavily engaged in other projects. He decided to look for help, first in order to spread the load, and second so that some new thinking could be injected. Glynne Evans, who had been involved with him in the Flixborough inquiry, had recently retired from HSE, and was an obvious choice. Glynne had been the HSE pressure vessel expert and has his own unique approach to the problems of mechanical failure. He agreed to work with Ralph in writing a chapter on Flixborough [4]. He has also re-written much of Part III: Preventing Mechanical Failures [12–17], and has up-dated the information on legislation [2, 22]. My own role has been to co-ordinate this work, to revise the rest of the book, and to provide an input on fire and explosion hazards. It is unfortunate that most chemists and engineers receive little, if any, instruction in combustion technology. They can then find themselves responsible for the safety of a process or a plant without a proper understanding of either of these hazards. There may well be guidance in the operating manuals and the relevant codes and standards, but this may be no help in recognising potential hazards, nor in responding to an emergency. It is for this reason that Chapter 10 gets down to basics and discusses in some detail the causes and behaviour of premixed flames (explosions) and diffusion flames (fires). Ignition sources are also discussed, which is why area classification is included in this chapter.

The original work was based largely on Ralph's own extensive and varied experience in the process industries. His anecdotal material is of particular value (we can still learn from others' mistakes) and much of it has been retained. His use of 'I' and 'me' in these accounts has not been changed.

We are all grateful to our Senior Commissioning Editor at Edward Arnold, David Ross, for his advice and firm guidance in this complex project. The list of people who were thanked by Ralph for helping with the first edition is not repeated here, but clearly we are still indebted to them for some of the data in the book. We would welcome reports of any errors (for which we are jointly responsible) and also suggestions for additions and improvements in future editions.

Ron Hirst
Cumbria, December 1996

Introduction

Most human activities carry special risks. Steel erectors and roof workers are most at risk of falling while machine operatives are more at risk of lacerations. The risk profiles of particular industries change with time as certain hazards (such as boiler explosions) are conquered and new ones (such as gamma-rays) appear.

The main hazards of the process industries arise from the escape of process materials which may be inherently dangerous (e.g. flammable or toxic) and/or present at high pressure and high or low temperatures. Large and sudden escapes may cause explosions, toxic clouds and pollution whose effects extend far beyond the works perimeter. Such major accidents include the explosion of liquefied petroleum gas in Mexico City in 1984 which resulted in 650 deaths and several thousand injuries, followed two weeks later by the release of toxic methyl isocyanate gas in Bhopal, India, which caused over 2000 deaths and over 200 000 injuries. These have rightly attracted world attention. Small and persistent escapes may lead to chronic ill-health and environmental pollution. Their insidious effects which have taken longer to arouse the public have contributed to the present prominence of 'green' issues.

Hazards differ widely between processes. Their magnitude depends mainly on the process materials and their quantity. The probability of an accident depends more on the process conditions and their complexity.

To prevent repetition of past disasters, correct diagnosis and exposure of the relevant hazards is essential. The lessons then need to be incorporated in the training of managers and staff who may be faced with these hazards, in company rules, in codes of practice and sometimes in legislation. Diagnosis is often difficult and controversial and one seldom knows whether it is quite correct or complete. It first requires all known and possibly relevant facts to be disclosed, related and assessed. To do this a broad scientific and technical background is more important than a legal one. As Sir Geoffrey De Havilland and P. B. Walker pointed out after the early Comet disasters,[1] most accidents in the technical sphere are caused by combinations of (relatively straightforward) hazards. Unfortunately, the legal and political connotations of many accident inquiries may put investigators under pressure to give undue attention to explanations which would exonerate parties whom they represent. Yet even the most objective investigation may succeed only in identifying several possible causes, each of which must be treated to minimise the probability of future failures.

The process industries

I cannot better the definition issued by the journal *Process Engineering*:[2]

> The process industries are . . . involved in changing by chemical, physical or other means raw materials into intermediate or end products. They include gas, oil, metals, minerals, chemicals, pharmaceuticals, fibres, textiles, food, drinks, leather, paper, rubbers and plastics. In addition the important service areas of energy, water, plant contracting and construction are included.

From this we can visualise the process industries as an intermediate stage in the transformation of raw materials of every kind — animal, vegetable or mineral — into materials and finished goods. The process industries convert these diversified raw materials into standardised bulk products. Some are sold direct to the customer (e.g. motor fuel), some merely packaged before sale (e.g. milk and lubricating oil), and some (e.g. wood-pulp and polyethylene) supplied to factories which make finished products.

Clear dividing lines cannot always be drawn between process industries and those which precede or follow them. Those preceding include mineral dressing at mines, water treatment (e.g. for injection into oil wells to assist recovery) and milk pasteurisation and cereal treatment at the farm. Those following include thermal and mechanical forming and cooking processes such as the casting and cold drawing of metals, the spinning and weaving of fibres, the moulding of plastics and the baking of bread. Many typical hazards of the process industries discussed here are found again in the industries which supply them or purchase from them. This book is also addressed to those working in them.

Chemical hazards

Today there is widespread concern over the hazards of chemicals, not only to those who work with them but also to the environment and the general public. However well-designed a plant may be, it is very difficult to prevent some dangerous materials from escaping. The longevity and concentration in nature of chemicals such as chlorinated biphenyls and chlorofluorocarbons, whose hazards only became apparent after they had been in production for many years, has heightened this awareness. One major problem in dealing with hazardous chemicals is that there are so many of them. There are now about *five million* chemicals listed in *American Chemical Abstracts*, and over 100 000 compounds in NIOSH's *Registry of Toxic Effects of Chemical Substances*.[3]

Apart from the general problems of manufacture, special ones arise in bulk transport by road, rail and water, and when pregnant women are employed in manufacture and packing.[4] Much new legislation has followed this public concern. If its results are often disappointing, this is largely because of the wide gaps in understanding and experience between the legislators and those most at risk.[5]

Many readers will surely be familiar with the *Handbook of Reactive Chemical Hazards* by Bretherick,[6] which covers some 7000 chemicals, and *Dangerous Properties of Industrial Materials* edited by Sax[7] which refers to more than 19 000 such materials. A useful classification of hazardous chemicals for quick reference is that published by the National Fire Protection Association of America.[8] This provides a numerical rating of 0 to 4 for three regular hazards of every chemical — health, flammability and reactivity.

All those using, handling or making chemicals should have full information, which the supplier should provide, about their properties and possible hazards. Material safety data sheets (MSDS) giving this information should be brought to the special attention of persons and departments in need of it (e.g. fire, safety, medical, operations, maintenance, cleaning and transport). An EC directive and MSDS form was issued in 1991. Compliance with the directive will be judged on whether the user has sufficient information to work safely rather than on the provision of lists of specific data. OSHA[8] has issued an MSDS form for use in ship repairing, shipbuilding and shipbreaking. MSDSs for a wide range of chemicals are available from the on-line database OHS.MS produced by Occupational Health Services Inc.

In considering chemical hazards, we must think not only of chemicals in their restrictive sense but of all materials which may display hazards designated as chemical. They can include soils, minerals, metals, mineral waters, gases, food, drink, fuels, building materials, pharmaceuticals, photographic materials, textiles, fertilisers, pesticides, herbicides and lubricants. Each is composed of one or more of the 92-plus chemical elements and may well have hazards resulting from its composition.

Safety and technical competence

Safety in the process industries cannot be treated as a separate subject like design, production or maintenance, but is inextricably interwoven into these and other activities. It depends on both the technical competence and safety awareness of all staff and employees.

At least one company tries to solve this problem by assigning its key production and maintenance personnel for periods, usually of several months, during the early part of their careers, to work in the safety or loss-prevention department under a permanent safety manager. This gives them a new outlook and philosophy on safety which they do not easily lose when they return to face the myriad pressures of production. Furthermore, they are aware that their superiors in the management structure share their experience and outlook. It is hardly surprising that this company has an exceptional safety record.

Of the various specialists involved, the process engineer occupies a central position. While not always recognised in terms of his authority, he should by education and experience be able to appreciate, on the one hand, the chemistry of the process and the materials processed, and on the other, the factors involved in the mechanical design and construction, and in the materials of construction used. He should be thoroughly familiar with hazards inherent in the process and should be aware of those arising from the detailed engineering and other areas outside his direct concern.

About this book

Although many specialised books and papers have been written about specific facets of hazard control in the process industries, only a few have attempted to cover the whole field. One (published in the UK) is Lees's *Loss Prevention in the Process Industries*[9] which was written mainly as a reference book for students. It has a very comprehensive bibliography and gives quite a detailed mathematical treatment of reliability theory, gas dispersion and some protective systems. While it does not define the process industries, it is clearly slanted to the oil, gas, petrochemical and heavy chemical industries. Another is *Safety and Accident Prevention in Chemical Operations*,[10] with chapters by 28 specialists, edited by Fawcett and Wood and published in the USA. I have drawn extensively on both books and refer to them frequently. Like them, this book does not attempt to cover the special hazards of nuclear energy or biochemical engineering.

Having spent about a third of my working life in several different countries, I have tried to write from an international viewpoint. I have also tried to adopt a multi-disciplinary approach while using a minimum of mathematics. To avoid unnecessary repetition, each subject is treated as far as possible in a single appropriate place, with extensive cross-references to other chapters, sections and subsections. Here square brackets [ ] are used for '(see) chapter, section, appendix, etc.'. However, in order that each chapter can be complete in itself, a certain amount of repetition is unavoidable.

The 23 chapters of this book fall loosely into four parts.

Part I, 'Setting the scene', includes the first five chapters. These deal with history (mainly recent), the legal background and five major accidents, their causes and lessons.

Part II, 'Hazards — chemical, mechanical and physical', comprises the next six chapters. Five of these deal with the toxic, reactive, explosive, flammable and corrosive hazards of process materials.

Part III, 'Preventing mechanical failures', consists of the next six chapters. These include discussions of modern ideas about reliability, active and passive protection and control instruments.

Part IV, 'Management, production and related topics', comprises the last six chapters and includes permit-to-work systems, training, personal protection and hazards which arise in the transfer of modern technologies.

A glossary of abbreviations used is given at the end of the book.

References

1. De Havilland, G. and Walker, P. B., 'The Comet failure' in *Engineering Progress Through Trouble*, edited by Whyte, R. R., The Institution of Mechanical Engineers, London (1975)
2. Sales brochure, *Progress Engineering — The Market Leader*, Morgan-Grampian (Process Press Ltd), London (September 1984)
3. *NIOSH Registry of Toxic Effects of Chemical Substances* (revised annually), National Institute for Occupational Safety and Health, Rockville, Md. 20857
4. Brown, M. L., *Occupational Health Nursing — Principles and Practices*, Springer, New York (1980)
5. Ashford, N. A., *Crisis in the Workplace — Occupational Disease and Injury*, MIT Press, Cambridge, Mass. (1966)
6. Bretherick. L., *Bretherick's Handbook of Reactive Chemical Hazards*, 4th edn, Butterworths, London (1990)
7. Sax, N. I., *Dangerous Properties of Industrial Materials*, 6th edn, Van Nostrand, New York (1984)
8. National Research Council, *Evaluation of the Hazards of Bulk Water Transportation of Industrial Chemicals — A Tentative Guide*, National Academy of Sciences, Washington DC (January 1974)
9. Lees, F. P., *Loss Prevention in the Process Industries*, Butterworths, London (1996)
10. Austin, G. T., 'Hazards of commercial chemical operations' and 'Hazards of commercial chemical reactions', in *Safety and Accident Prevention in Chemical Operations*, 2nd edn, edited by Fawcett, H. H. and Wood, W. S., Wiley-Interscience, New York (1982)

Part I
Setting the scene

1 From past to present

History has several lessons for us about the hazards of the modern process industries. One is the toxicity of many useful metals and other substances won from deposits in the earth's crust. Although the dangers of extracting and using lead have been known from the earliest times, these seem later to have been forgotten. A second lesson, then, is that past lessons are sometimes forgotten after a lapse of a few years, when history has an unfortunate habit of repeating itself.

A third lesson is that there is often a time-lag between the initial manufacture of hazardous substances and general appreciation of the dangers. Today we are all aware of the hazards of asbestos, CFC refrigerants and aerosols, chlorinated hydrocarbon insecticides, benzene and the bulk storage of ammonium nitrate. Only recently all were considered to be safe and needing no special precautions. Similar time-lags occurred a few decades ago before the hazards of yellow phosphorus (matches) and benzidine (dyestuffs) were appreciated.

A fourth lesson is that it is usually the lowest and least articulate strata in society who bear the brunt of industry's hazards.

A fifth lesson is that the capacities of process plants and the magnitude of major losses involving them have increased continuously *and are still increasing*. Related to this and to the high capital:worker ratio of these plants is the high ratio of capital loss:human fatalities in most major fires and explosions. This does not, however, apply to poisoning and pollution incidents which spread well beyond the works boundary (cf. Seveso and Bhopal).

The recent world record of large losses in the process industries, especially oil and chemicals, is truly alarming and gives us no grounds for complacency. The numbers of losses in excess of $10 million (adjusted to 1988 values), and their total value, have increased greatly in each successive decade since 1958, particularly in the second decade as the following figures show:[1]

Period	Number of losses over $10 million at 1986 values	Total ($10 million)
1958–1967	13	442
1968–1977	33	1438
1978–1987	58	2086

Having persisted for so long, it would be very surprising if these themes did not continue into the future. If history teaches us nothing else, it should warn us to be sceptical of claims that the use of some new material in an industrial process is entirely safe. Usually only time will tell.

1.1 Origins of process hazards

Several typical hazards of the process industries have a very long history. This is because a number of the 92-plus chemical elements (particularly metals and semi-metals) of which all matter is compounded are poisonous, and are naturally concentrated here and there on and below the earth's surface.

As human prowess developed, the mastery of fire, and through it the invention of smelting to obtain bronzes and other metals, released fumes which affected the health of the craftsmen. Another early 'industrial' hazard was the making of flint tools, where abundant archaeological evidence of silicosis has been found.

As human occupations became more specialised, it was clear that some were more dangerous and less healthy than others. Thus Hephaistos, the Greek god of fire and patron of smiths and craftsmen, was lame and of unkempt appearance, while Vulcan, the Roman god of metal workers and fire, was also ugly and misshapen. It is now thought that the lameness of the smith-gods was the result of arsenic poisoning, since many of the ores from which copper and bronze articles were made contained arsenic, which improved the hardness of the resulting articles.

From the earliest times, there has been a strong prejudice among the articulate elite against such craftsmen. Socrates was reported to have passed the following judgement:

> What are called the mechanical arts, carry a social stigma and are rightly dishonoured in our cities. For these arts damage the bodies of those who work at them or who have charge of them, by compelling the workers to a sedentary indoor life, and in some cases spending the whole day by the fire. This physical degeneration results in degeneration of the soul as well.

The social cleavage illustrated by such attitudes to industrial hazards has persisted through human history. Despite the recent elimination of many of these hazards, our social and economic structure and the mental habits that go with it are slow to adjust to the possibilities of a golden age, free from occupational hazards and excessive working hours, in which all can enjoy our common heritage of knowledge, invention and accumulated technical progress. It is tragic, as the British miners' strike of 1984/1985 showed, that working people still feel compelled to fight for the right to continued employment in an occupation notorious for accidents and disease.

In more recent times, some occupational diseases were so common as to have acquired well-known names, such as those quoted by Hunter:[2]

> Brassfounders' ague, copper fever, foundry fever, iron puddlers' cataract, mule-spinners' cancer, nickel refiners' itch, silo-workers' asthma, weavers' deafness and zinc oxide chills.

1.2 Toxic hazards of ancient metals[2]

Several of these hazards have persisted to the present day, although their forms have changed. To say that any metal is poisonous is an over-simplification. Metals usually occur combined in nature and few are found in their free state. While several inorganic compounds of a metal display the same characteristic toxic features, the degree of toxicity of such compounds depends on their solubility in water and body fluids, as well as on the ionic and complex state of the metal. Insoluble elements and compounds are seldom toxic in themselves. The first lead ore worked at Broken Hill in Australia was the relatively soluble cerussite, $PbCO_3$, the dust of which caused much disease among the miners. Fortunately this was soon worked out, and the ore subsequently mined was galena, PbS, which is very insoluble, and has caused few cases of lead poisoning.

Besides the toxic inorganic compounds whose effects are typical of the metal present, there are many man-made organo-metallic compounds, which have different and more acute toxic effects. An example is the volatile tetraethyllead, used as a petrol additive, which produces cerebral symptoms. Nickel carbonyl, used in the purification of nickel, is another example. Even metals which exhibit no marked toxicity in their inorganic compounds, such as tin, can form highly toxic organo-metallic ones (tetramethyltin).

The hazards of two metals used since antiquity, lead and mercury, are next considered. They are discussed again [23] in the context of technology transfer to developing countries.

1.2.1 Lead

The symptoms of inorganic lead poisoning — constipation, colic, pallor and ocular disturbances — were recognised by Roman and earlier physicians. The symptoms of poisoning by organo-lead compounds include insomnia, hallucinations and mania. Lead ores have been smelted since early Egyptian times. Being soft, dense, easily worked and fairly resistant to corrosion, lead was long the favourite metal for water pipes, roof covering and small shot. With the invention of printing, it became the principal metal used for casting type. White lead (a basic carbonate) and red lead (an oxide) were long used as paint pigments and ingredients of glass and pottery glazes. The smelting of lead ores (Fig. 1.1) and the manufacture and use of lead compounds increased greatly during the Industrial Revolution, together with an increase in death and injury among workers exposed to them.

The health hazards to lead workers featured in Victorian factory legislation and it was eventually recognised by Sir Thomas Legge that:

> Practically all industrial lead poisoning is due to the inhalation of dust and fume; and if you stop their inhalation, you will stop the poisoning.

Although the conditions in established lead processes improved considerably after this, newer large-scale uses of lead, first, the manufacture of lead-acid car batteries, and second, the manufacture and use of volatile organo-lead compounds for incorporation into petrol to improve its performance, brought further hazards. Several multiple fatalities occurred during the cleaning of large tanks which had contained leaded petrol. The worst

Fig 1.1 Sixteenth century furnaces for smelting lead ore

happened at Abadan refinery during World War II while I was working there. There were then about 200 cases of lead poisoning with 40 deaths among Indian and Iranian workers.

As the hazards of lead are better appreciated today, its use has declined. One special hazard in its production is that many lead ores also contain arsenic, which is even more toxic [23.3.2].

1.2.2 Mercury

Mercury, the only liquid metal, is highly toxic and has an appreciable vapour pressure at room temperatures. Symptoms of poisoning from mercury vapour are salivation and tenderness of the gums, followed in chronic cases by a tremor. Another symptom is *erethism*, a condition in which the victim becomes both timid and quarrelsome (Fig. 1.2), easily upset and embarrassed, and neglects his or her work and family. Merely to be in an unventilated room where mercury is present and exposed to the atmosphere can, in time, lead to mercury poisoning.

Mercury occurs as its sulphide in the ore cinnabar, which has been mined in Spain since at least 415 BC, and mercury poisoning has long been prevalent among workers employed in such mines and reduction plants.

Mercury has long been used as such in the manufacture of thermometers and barometers, and more recently in the electrical industry for contact breakers, rectifiers and direct current meters. New compounds of mercury have been invented and commercialised, including mercury fulminate, used in detonators, and organo-mercury compounds used as antiseptics, seed disinfectants, fungicides and weedkillers. Mercury is used as cathode and solvent for metallic sodium in the Castner–Kelner process for the production of chlorine and caustic soda. This process has now been largely replaced by others which are free of the mercury hazard. In most of its industrial

Fig 1.2 The Mad Hatter, drawn by Tenniel

applications there are well-authenticated cases of poisoning by exposure to mercury vapour, or dusts containing its compounds. Mercury poisoning has been notorious in the felt-hat industry for centuries, where mercuric nitrate was used to treat rabbit and other furs to aid felting.

An infamous case of mass poisoning from organo-mercury compounds occurred among the fishermen and their families living along the shores of Minamata Bay in the south of Japan in the 1950s. This was ultimately traced to the discharge of spent mercury-containing catalyst into the bay from a nearby chemical factory which made vinyl chloride monomer. Organo-mercury compounds settled in the silt of the bay and were ingested by fish which were caught, sold and eaten by the local inhabitants and their cats. By July 1961 there had been 81 victims, of whom 35 died. The symptoms included numbness in the extremities, slurred speech, unsteady gait, deafness and disturbed vision. The mud of the bay remained loaded with mercury compounds for many years afterwards.

1.3 Changing attitudes to health and safety in chemical education

The last 50 years have shown great changes in attitudes to chemical safety in schools and colleges. This is clear from my own education in the 1930s. The first hazardous chemical to which I was exposed was mercury. This was in our school chemistry laboratory-cum-classroom (1932–1936). Our chemistry master, a middle-aged bachelor, had studied under Rutherford and had a penchant for research. For this he needed copious supplies of mercury, which he purified in his spare time in the school laboratory. Although I did not recognise his symptoms at the time, in retrospect the tremor of his hands, his high-pitched nervous twittering laugh, general shyness and odd mannerisms were typical of *erethism*. Globules of mercury, which was used for many juvenile pranks, were scattered on the laboratory benches and floor. Perhaps the fact that the laboratory was underheated saved me from serious mercurial poisoning.

At college (1936–1940), I was exposed to blue asbestos, from which we made mats for filter crucibles used in inorganic analysis, hydrogen sulphide, benzene, which was used as a common laboratory reagent and solvent, and again mercury, of which I used several kilograms for a research project. My most serious exposure was probably to a complex mixture of polynuclear aromatic compounds containing sulphur, which I was asked to prepare for a professor during a long vacation by bubbling acetylene through molten sulphur. The professor contracted cancer from his researches and died in middle age a few years later.

Chemical research was then held up to students as a vocation, demanding sacrifice of time and, where necessary, of health, in order to advance the frontiers of knowledge in the service of mankind. Madame Curie was quoted as a noble and inspiring example, whose work somehow justified the cancer which finally killed her. Other scientists such as J. B. S. Haldane and Dr C. H. Barlow who carried out dangerous and often painful experiments on their own bodies were also regarded as heroes.

It was only later that I realised that such sacrifices can only be justified if they improve the health of others. Often the reverse has been the case. A survey by Li *et al.*[3] of causes of death in members of the American Chemical Society between 1943 and 1967 showed that deaths from cancer of the pancreas and malignant lymphomas were significantly higher than among the general population. Scientists have tended to regard working conditions which they readily tolerate as quite good enough for their laboratory assistants. This is but a short step to expecting the same acceptance from industrial workers and the general public.

In spite of many exposures to harmful chemicals throughout my training and subsequent career, I am fortunate to be alive and in excellent health and still a keen squash player in my seventies. As most of my former colleagues are dead, I must be the exception which proves the rule!

The situation in schools today is very different from that in the 1930s. Safety policy has been greatly tightened over the past 20 years and most heads of science take their safety responsibilities very seriously indeed. A chemistry teacher was recently prosecuted and fined £500 for failing to take adequate safety precautions. The Association for Science Education has a Laboratory Safeguards Committee and its journal *Education in Science* carries regular updates on potential hazards. Local education authorities publish safety guidelines and individual schools are often required to have such guides to suit their particular situations. As examples of the changing situation, traditional asbestos bench mats were phased out during the 1970s, and the safe handling of chemicals and manipulation of apparatus is one of the features of pupils' practical chemistry work which is assessed for the GCSE examination.

References

1. Garrison, W. G., *100 Large Losses — A thirty-year review of property damage losses in the hydrocarbon-chemical industries*, 11th edn, Marsh and McLennan Protection Consultants, 222 South Riverside Plaza, Chicago, Illinois 60606 (1988)

1a. Garrison, W. G., *Large Property Damage Losses in the Hydrocarbon-Chemical Industries — A Thirty-Year Review*, 12th edn, Marsh and McLennan, Chicago (1989)

2. Hunter, D., *The Diseases of Occupations*, 6th edn, Hodder and Stoughton, London (1978)

3. Li, F. P., Fraumeni, J. F., Mantel, N. and Miller, R. W., 'Cancer mortality among chemists', *Journal of the National Cancer Institute*, **43**, 1159 (1969)

2 Law codes and standards

2.1 Early background and history

Since mankind formed itself into the most primitive societies it has been carrying out hazard and risk assessments. As societies became more sophisticated they laid down rules or standards by which they could work together. As they became more sophisticated they were able to collate and write their laws or rules down. One such law (Law 229 of HAMMURABI) dating from 1750 BC says:

> if a builder builds a house and does not make its construction firm, and the house that he has built collapses and causes the death of the owner of the house, that builder shall be put to death.

The Old Testament book of Deuteronomy Chapter 22 Verse 8 says:

> when you build a new house, you shall make a parapet for your roof, that you may not bring the guilt of blood upon your house, if anyone fall from it.

This was probably written down about 650 BC from preceding orally transmitted laws. This Old Testament safety legislation marks a transformation from a nomadic tent-dwelling population into an arable village-based society that needed laws to protect itself from shoddy workmanship. Obviously the parapet around the roof was needed to stop anyone falling from the roof and hence compliance with this law would make the roof access safer. But it also had another advantage in that the owner of the house, if he had put up a parapet in conformity with the law, would be exonerated should anybody fall from his roof because, for example, they were fooling around, had become drunk, were accidentally pushed or deliberately thrown from the roof. Health and safety law has always benefited both a potentially injured party, and at the same time limited the degree of liability of the responsible person.

The two illustrated examples of ancient law demonstrate different philosophies in law making. The example where a builder has to put a parapet around his roof is deemed to be *prescriptive* as it tells the responsible person what action to take. Much industrial law in Germany, UK, France and the Continent in general has until recently been prescriptive law. Such law reached its zenith within the German framework where designers, fabricators and installers were under legal obligations to comply with exactly prescribed standards. The standards themselves were not law, but the law defined the standard with which compliance was required. Manufacturers and suppliers welcomed such legislation because, provided they complied with the standard described in the law, then they complied with the law; the compliance with the standard by definition was easily measured and so they knew they were within the prescribed law. This gave them the advantage of being able to design and then fix a budget for an article or commodity and it was thought there was a common playing field for competition. Provided the code makers had undertaken the correct hazard and risk assessment and written into the code the necessary remedial action, then the users of articles, equipment etc. could expect to enjoy a uniformly acceptable standard of health and safety.

The disadvantage of prescribed particulars was that they stifled innovation. For example if a law prescribed a standard which called for a steel of a certain strength, and then due to research or advance someone could make that steel stronger, the standard had to be rewritten and approved which could take up to five years; this would lead to a trade disadvantage to both designers and users but with no added safety protection. There was a general realisation that *goal-setting* legislation was needed. This is illustrated by the HAMMURABI law which told the builder of a house that he had to construct it firmly but left it up to him how to carry out that obligation. Such goal-setting law necessitates an obligated person carrying out hazard and risk assessment, and then carrying out research and quality assurance so as to achieve the desired end result of producing an article which is safe as far as is reasonably practicable.

2.2 Historical background to the UK process industry

The petrochemical and food-processing industries naturally fall within the definition of process industries, but so too would glass manufacturing, iron, steel and other metal manufacturing as well as wood preserving, paper making and rubber manufacturing. Even some processes within the health care sector of industry can be included: central sterilising units undertaking process functions, laundries, dry cleaning and scrap metal reclamation come under the process industry heading.

The Industrial Revolution started early in Britain. After the black death of the 1300s the population had been decimated. There were no longer sufficient people to work the arable land and so landowners turned to sheep rearing. At first the surplus wool and hides were exported to the continent where they were processed, but gradually the small cottage industries in Britain expanded so as to wash, comb, spin and weave the wool and produce garments, the surplus of which was exported bringing in much larger revenue. Here can be seen the birth of the manufacturing and production industries. Further developments took place as mechanisms such as water-driven and then steam-driven mills, the spinning jenny and automatic weaving machines came into being.

The process industry had its foundations at about the same time. Originally wool had been to the fore in the industrial revolution but gradually linen, hemp and cotton textiles were produced. Linen, cotton and hemp all needed to be bleached. Originally bleaching had been done in 'bleaching fields' whereby the finished textiles were laid out on grass or bushes so as to be naturally bleached by the sun. But as quantity increased it was obvious that there was a considerable need for chemical bleaches and this coincided with the repeal of the UK Salt Laws.[1] There had been a high tax on the processing and supply of salt. The repeal of the Salt Laws meant that there was a cheap commodity available for the production of chlorine-based bleaches. The textiles industry also demanded a greatly increased supply of soaps, which had originally been made from weak lye, a caustic soda compound combined with animal fats. The first part of the nineteenth century saw the dawning of the detergent industry as well as a vast expansion in the production of dying chemicals. The increasing population put greater demands on arable farming leading to a requirement for artificial fertilisers and so the new chemical industry was born.

2.3 From the Industrial Revolution to the Robens Report

The revenue drawn to Government from the Industrial Revolution allowed tax changes and new concepts in law drafting and enforcement. By the beginning of the 1800s not everybody was benefiting from the new-found wealth. There had been a migration from rural Britain to the industrialised cities. Living accommodation was appalling. Life expectancy was curtailed both by poor sanitation and by working conditions in which violent accidents and lingering health problems were hazards. Even though Government had a casual attitude to such issues, there were influential people such as Lord Shaftebury who promoted the concepts of better working conditions, prison reforms and the end of slavery.

The first health and safety legislation in the UK was enacted in 1833. This was originally aimed at improving the morals of the workforce, improving living conditions in dormitories, reducing working hours and providing some level of education for children. Successive rounds of new legislation tackled long working hours for women, the abandonment of children and women in underground mines, and finally led on to the Factories Act.

The health and safety law of the period was pragmatic, and dealt only with the problems that were manifest at the time. As in much health and safety legislation, there were benefits to the people who could be injured, and also to the employers whose responsibilities were limited by the law. For example the introduction of the Factories Act meant that a factory owner or works manager could no longer be prosecuted for manslaughter if a worker died but would now only be susceptible to the maximum Factory Act fine, maybe £100.

The first Factory Act proper came into force in 1906, and this was accompanied by the Alkali Works Regulation which addressed the hazards and necessary remedial action in the chemical industries. There were from time to time amendments and new regulations such as the banning of phosphorus matches, because it had become known that phosphorus had caused bone deficiencies in workers. There are interesting small regulations such as the KIERS Regulations dealing with the dying of fabrics, but the legislation was so prescriptive and specific that many employers found useful loopholes and avoided complying with the law.

A main review of the health and safety legislation came about in 1961 but basically it re-enacted the 1906 Factories Act, and consolidating the useful and repealing the defunct regulations. At the time the Factories Act was very detailed, and specific. For example steam boilers and steam receivers as well as air receivers had specific legislation — that had come about because of the appalling explosions that occurred at the turn of the century — but at the time there was no specific legislation about other pressurised equipment that did not contain steam or air. In the late 1960s and early 1970s there was a general air of 'fair play', and a desire to make the world a better place. Society was ripe for change.

2.4 The Robens Report of 1972

The Government of the day set up under Lord Robens a committee to consider safety and health at work and make recommendations. At that time there were many Government departments each administering health and safety legislation in different sectors of society. For example there was separate and sometimes subtly distinct legislation applying to factories, offices, mines, fishing industries, local health authorities, railways etc. Legislation had come about piecemeal, and often had only been drafted in response to some previous disaster, or to public pressure. Much legislation was specific and totally prescriptive leaving little room for innovation or development.

The Robens Report responded to Government in a revolutionary way. It certainly did not advocate the strict prescribed system of law standards and direct information which had been developed on the Continent, but suggested a more pragmatic and open approach. The Robens Report recommended that there should be more voluntary standards and codes of practice developed by employers, employees and learned bodies. There was to be an emphasis on local safety committees which was based on a common concept of the early 1970s, that of a society which was much more cooperative and where unions were held in high esteem. There was to be a more open approach to disclosing information and the interests of the public and nearby neighbours of industrial complexes were to be recognised. The idea was that legislation would cover universally all aspects of work activities — including design, supply and manufacture, and also the off site effects of the process. In addition the hazards resulting from the use of the products would also be covered.

The Robens concepts were so far-reaching and the debate within industry and commerce so intense that many people feared that this new system of legislation would founder in a morass of debate. The Flixborough disaster changed all that. After the explosion at the Flixborough site in 1974 the Government of the day saw an opportunity to further its desire for industrial and commercial health and safety, and at the same time avoid devastating criticism of its inability to curb such disasters as had happened at Flixborough. Opposition parties feared the outcry if they did not support such legislation in the aftermath of the disaster and within one year of its occurrence the new Health and Safety at Work Act had been laid before Parliament and received the Royal assent.

2.5 The Health and Safety at Work (HSW) etc. Act 1974

The Act was intended to make further provision for securing the health, safety and welfare of persons at work, for the protection of others against risk to health or safety in connection with activities of a person at work, for controlling and keeping and preventing the unlawful use of dangerous substances and for controlling certain emissions into the atmosphere. It also made further provisions with respect to the Employment Medical Advisory Service.[2]

The Act is goal-setting. It lays obligations on designers, manufacturers, suppliers, employers, self-employed persons and employees in a non-prescriptive manner; it states what objectives should be achieved rather than how they should be achieved.

The HSW Act brought into being the Health and Safety Commission (HSC) which is responsible for executive functions. It also brought into being the Health and Safety Executive (HSE). The HSE was formed from the old inspectorates such as the nuclear inspectorate, factory inspectorate, offshore inspectorate, railway inspectorate, explosives inspectorate etc.

The HSC is controlled by the Secretary of State; it holds the budget which is allocated from Parliament, and has the ability to approve certain things as well as the power to direct investigations and inquiries, see HSW Section 14.[3]

The HSC via the HSE has an obligation to give advice on health and safety matters, can undertake research, and is empowered to 'police' the Acts and Regulations made under it.

The enforcing of the HSW Act is undertaken by HSE-authorised inspectors. They have warrants and have the authority to enter premises in pursuit of their responsibilities. Generally they undertake their enforcing duties alongside their advisory duties, and give verbal and written advice. Under the Act they can serve improvement notices, prohibition notices or undertake prosecutions.

Prosecutions at summary level, in a Magistrates Court, can be undertaken either by approved inspectors or by appointed solicitors, and the maximum fine is £20 000. For the more serious offences HSE representatives, accused parties, or magistrates can send cases to Crown Court on indictment. In Crown Court financial fines are unlimited. Within the HSE Act there is provision for custodial sentences of up to six months for individuals.

2.5.1 Legislation concerning the process industries

The HSW Act applies to all the various sectors in the process industries. But the modern legislation has been extended to consider specific parts of the process industry and to give specific obligations and duties by the making of regulations.

2.5.2 Notification of Installations Handling Hazardous Substances (NIHHS) Regulations 1982[4]

These regulations require that in all fixed installations including factories, warehouses, transport depots, ports, pipelines and vessels used for storage where more than specified quantities of 35 named substances or classes of substances are present, these should be notified to the HSE. The regulations do not apply to transport.

The specified substances only come within the regulations if they are of a specified quantity, the philosophy behind this being to try and collate the same hazards. For example it was thought that 25 tonnes of LPG would form approximately the same hazard as 25 tonnes of chlorine. Of course the one would give rise to a hazard by fire and explosions while the other would give rise to toxic hazards.

The objectives of these regulations were to provide the HSE with a recorded picture of the sites of the major hazard installations, to remind employers and employees of the hazards they were dealing with and to give the public, particularly those living within range of the hazards, an indication of what was being stored.

2.5.3 Control of Substances Hazardous to Health (COSHH) Regulations 1988[5]

These regulations were made with a view to protecting persons against risks to their health, whether immediate or delayed, arising from exposure to hazardous substances. They laid an obligation on employers to carry out assessments of the health risk and where appropriate to monitor the health of people exposed to danger. They prohibited certain substances and recommended the use of control measures and the maintenance of a system for testing these measures. It also deals with the monitoring of exposure at the workplace.

The regulations are prescriptive in form in that they give lists of substances that are banned and substances that have maximum exposure limits. They also lay down specific times when examinations and tests of local exhaust systems should be undertaken and specified substances where monitoring is required. Having said that, there is a definite element of goal-setting and the provision for employers to set their own remedial actions based upon their hazard and risk assessment gives them the flexibility for innovation and progress.

2.5.4 Control of Asbestos and Work Regulations 1987[6]

Asbestos can be made or processed within the process industries, but it is also a valuable asset used in the fire protection of installations where chemical processes are carried out.

Under these regulations employers shall not carry out any work that would expose any of their employees to asbestos unless they have ascertained what type of asbestos it is; they should assess the work that is being undertaken and carry out adequate training and give information to their employees. They should notify the HSE before that work is undertaken. There is an obligation to prevent or reduce exposure to asbestos and to use appropriate control measures and maintain those control measures; and also to provide and clean protective clothing. These regulations are very much goal-setting in that they allow the employer the flexibility to carry out work in a competitive world, and encourage health and safety in a pragmatic manner.

2.5.5 Control of Industrial Major Accident Hazards (CIMAH) Regulations 1984[7]

A major disaster in Seveso, North Italy, where a large quantity of toxic substances caused immediate health hazards and is suspected of causing long-term environmental damage, is considered to have spawned the CIMAH Regulations.

These regulations came from the European lobby. In order to restrict barriers to trade and produce a 'level playing field' it was considered that there should be European-wide legislation covering major hazard plant. Breaking away from the old UK position and the widely held European concept of safety achieved by prescriptive legislation backed with definitive standards and regulations the CIMAH Regulations were the first goal-setting legislation giving organisations the flexibility to achieve a common standard of safety by carrying out their own hazard and risk assessment and then undertaking remedial action if need be. This legislation covers both process and storage activities but it excludes major hazards at nuclear installations and those covered by the Ministry of Defence. It came fully into force on 1 April 1985 and covers very toxic substances, flammable substances and explosive substances. It is a natural follow-on from NIHHS Regulations. But in the UK the largest or the most hazardous installations have to prepare a Safety Report and submit that in writing to the HSE. The Safety Report has to be updated if any modifications are made to the industrial activity or in any case every three years. The manufacturer at notifiable installations, e.g. the largest or the most hazardous, also has to prepare an on-site emergency plan and this must be supplied to the local authority which has an obligation to prepare an off-site emergency plan.

It is considered that probably 200 large installations in the UK are subject to these requirements. The intention was to give more information and general awareness to local authorities, the HSE and above all to persons living or working close to major hazards. The heart of the CIMAH Regulations is Schedule 6, the information to be included in the report under Regulation 7. Here there is an obligation for the employer to name the dangerous substances used or stored on the site and give a map of the site and describe the management system and arrangements for training.

The key to the regulations can be found in Schedule 6 Part 5 which lays an obligation upon the employer to describe the potential sources of a major accident and the conditions or events which could be significant in bringing one about; this forces employers to accept that a potential disaster is possible and encourages them to think how one could materialise. Having done this it is sensible that a description of the measures taken to prevent, control or minimise the consequences of any major incident should be written down.

2.5.6 The Reporting of Injuries, Diseases, Dangerous Occurrences Regulations (RIDDOR) 1985[8]

These regulations require notification of all fatal accidents, major injuries, accidents causing more than three days off work, certain specified diseases and certain specified dangerous occurrences such as failures of pressurised storage pipework, major crane failures etc.

The objective of these regulations is to give the Government via the HSC and HSE an overall picture of where injuries and dangerous occurrences occur, so that future legislation, 'police work', encouragement of research or enforcement can be undertaken.

In the early days of the HSW Act there was an ambience of co-operation and a free exchange of information. With commercial pressures and competition ever to the fore and with the continuing fear of litigation that can arise after admissions have been made there is now a general air of secrecy, and it is only through independent sources, such as HSE, that general trends in health and safety can be judged and recommendations made.

2.5.7 Electricity at Work Regulations 1989[9]

The process industries, like all parts of industry, use electricity and are subjected to these regulations. Deaths from electrocution still occur within the UK and there are serious burns and shocks and therefore it is incumbent upon the employers and employees to make any necessary safeguards.

2.5.8 The Noise at Work Regulations 1989[10]

In recent years hearing loss sustained by employees has given rise to much litigation against employers and their insurance companies. The Noise Regulations, like much modern safety law, focus on carrying out assessments first and then taking remedial action so as to reduce the exposure to noise to levels below 85 dB(A). The objective is to prevent hearing loss at source but if all else fails then ear protection can be provided.

2.5.9 Pressure Systems and Transport of Gas Containers Regulations 1989[11]

At the turn of the century there were over 300 steam boiler explosions every year and this spawned the Factories Act 1906. By the mid-1960s it was realised that the number of steam boilers failing each year had diminished, but other pressurised systems and pipework failures were causing injury and financial hardship. It was realised that new health and safety laws were needed to deal with pressure systems, and to this end Lord Harman was invited to set up a Royal Commission which reported in 1968. The report was in the public domain and so highly debated that it was felt that very little progress would be made and consequently it was decided to shelve any concept of pressure system safety regulations pending the report of the Robens Committee.

The Robens Report[12] addressed pressure systems as well as storage under non-pressure conditions and made recommendations.

After the HSW Act was enacted it was thought by professionals within the health and safety world that sufficient responsibility was placed on employers and on the designers, makers and suppliers of pressurised equipment respectively by Sections 2 and 6 of the HSW Act. After the Flixborough disaster in 1974 there were rumbles from interested parties wishing to adopt pressure system regulations, but those applicants usually wanted prescribed particulars rather than the goal-setting of modern legislation. The debate went on for many years and finally in 1989 the Pressure Systems and Transport of Gas Container Regulations was enacted but did not come fully into force until 1 July 1994.

The Pressure Systems Regulations consider design, construction, repair and modification as well as installation, marking and the provision of information. The key sections of the regulations cover setting the safe operating limits, Regulation 7, and then operating the plant within those safe operating limits, Regulation 11. These regulations are clearly goal-setting and do not in any way prescribe how or what safe operating limits should be set; neither do they detailed instruction on what maintenance will be needed. The Pressure Systems Regulations came out almost 30 years after the Royal Commission, and 20 years after the Flixborough disaster.

Even now the regulations only deal with the pressure systems which are above half a bar pressure, and they still exclude storage where the pressure is low, even though the hazard may be very high. The Pressure Systems Regulations only apply where there is a hazard or danger from stored pressure energy. Therefore leakage of toxic or flammable substances would not be a breach of the regulations. For this reason some would argue that these regulations would not have applied to the Flixborough plant.

2.6 EC directives

One of the objectives of the Treaty of Rome was the elimination of barriers to trade within the European Community. It had been felt for some time that some countries and organisations used the health and safety laws of their own countries to inhibit free trade and it was realised that the harmonisation of standard codes of practice and legislation dealing with health and safety had become bogged down in a morass of self-interest.

The Treaty of Rome Article 100a introduced a new voting scheme which did away with the old veto system and streamlined the introduction of harmonised health and safety legislation across Europe. At first interested parties thought that there would be no reconciliation between the modern UK goal-setting type of legislation and the more traditional legislation that approved certain standards and codes of practice so as to promote health and safety. Those pessimistic predictions have not been fulfilled. Safety experts throughout Europe have worked together on many fronts. On the legislative field there has been the so-called 'six pack' which covers:

1. the Manual Handling Operations Regulations 1992[13] which came fully into force on 1 January 1993;
2. the Workplace Safety and Welfare Regulations 1992[14] which came into force 1 January 1993;
3. the Supply of Machinery (Safety) Regulations 1992[15] which came into force 1 January 1993;
4. the Personal Protective Equipment at Work Regulations 1992[16];

5. the Management of Health and Safety at Work Regulations 1992[17];
6. the Provision and Use of Work Equipment Regulations 1992[18] which came into force on 1 January 1993;
7. the Visual Display Regulations 1992[19] which came into force on 1 January 1993.

All of the above regulations are applicable throughout the EC and all of them have a commonality. Each requires a hazard and risk assessment to be undertaken, followed by remedial action to reduce that hazard and risk to an acceptable level. The regulations are not prescriptive; they set goals to be achieved without giving details on how, when and where. Within the Treaty of Rome Article 100a there is a concept that health and safety should be achieved using a hierarchy of measure, the principle being that the first remedy should be used and progress made to the next only if the first proves not to be reasonably practicable. Elimination should be tried first followed by substitution and then safeguarding. If none of these is acceptable then protective equipment may be used. If all else fails then safety must be achieved by the use of training, advice or management control.

2.7 Safe so far as reasonably practicable

All present-day UK legislation on health and safety contains wording such as 'safe so far as reasonably practicable'. The term does not readily translate into other European languages but because of the need to maintain working relationships with non-English speaking technical experts, translation difficulties have been overcome (while still maintaining the principle).

Within UK law the definition of reasonably practicable arose from the case of Edwards v National Coal Board 1949[20] in which it was stated that:

> reasonably practicable is a narrower term than physically possible implies — that a computation must be made in which the quantum of risk is placed in one scale and the sacrifice whether in money, time or trouble involved in the measure necessary to avert the risk is placed in the other; and if that it be shown that there is a gross disproportion between them the risk being insignificant in relation to the sacrifice the person upon who the duty is laid discharges the burden of proving that compliance was not reasonably practicable

and

> This computation falls to be made at the point of time anterior to the happening of an incident.

In layman's terms this means that an employer must carry out a hazard and risk assessment at a point of time before an accident or dangerous occurrence occurs. It is this risk assessment approach which is the keystone to all modern health and safety legislation. It will be discussed more fully in Chapter 14.

2.8 Standards (see also 2.11 and 2.13)

With the beginning of the Industrial Revolution many manufacturers and buyers wanted standardisation of articles, shapes, forms and contracts; to this end there was a standardisation of nuts, bolts, strengths of materials, sizes of clothes etc. It became the fashion within the UK and throughout the industrialised world to add sections dealing with health or safety in most standards. Rarely within the UK did safety legislation make a direct reference to a British standard, but on the Continent, German, Italian and French legislation often referred to a specific national standard when safety matters were being considered.

The aim of standards has always been to achieve a harmony in economic effort, expenditure and materials. This sometimes extends to giving consumer interests an opportunity of influencing the quality of goods and services.

At the same time that the Treaty of Rome was encouraging the harmonisation of legislation so as to reduce the barriers to trade there were also moves to harmonise national standards. Under the European Committee for Standardisation (CEN) there has been a process of replacing the old national definitive standards with ones which are consistent with the goal-setting principles of modern safety legislation.

Under CEN standards there is the concept of 'A standards' which look at design principles, 'B standards' which deal with safety devices such as photo-electric guarding, safety interlocks, hydraulic interlocks, pressure-relieving devices, temperature controls etc. and 'C standards' which are product-orientated, e.g. a standard of zinc alloy die casting machines.

All modern standards have an in-built quality assurance regime so that both the designer, the maker, the seller and the purchaser can be assured of the quality of the end product. Within the EC there is now a trend away from producing national standards and towards harmonised European standards, but with reference to experts in member states.

2.9 Codes of practice

Under the HSW Act originally it was considered that approved codes of practice could and probably would be made, whereby the HSE put forward to Government standards, codes and written information that could be approved and used in industry, including the process industries. It was soon acknowledged that this definitive information would be slow to modify and update as progress was made and very few approved codes of practice now exist. Nevertheless in line with the health and safety responsibility to produce help and guidance to industry on health and safety matters, HSE has produced many guidance documents, including one on each of the so-called six pack, the Pressure System Regulations, the LPG Regulations, CIMAH Regulations, COSHH Regulations, etc. All are available from HMSO. The

Royal Society for the Prevention of Accidents (RoSPA) has also produced much valuable information advising employers, employees, users and the general public on health and safety matters. Various trade associations have also given valuable help and advice, but neither the standards nor the guidance notes have official status.

There is a term 'deemed to satisfy' whereby it is meant that if the guidance within an HSE publication had been met in all respects then the legislation would be deemed to have been complied with. This 'deemed to satisfy' does not in anyway prohibit or restrict the implementation of any other methods of achieving equal safety.

Within the process industries it is being realised that a lead has been taken from the American petroleum industry which has produced much worthwhile information on the design, fabrication and maintenance of chemical and process plant. Much of this has been formed by committees, which are often slow to respond to advances in technology, and are at the same time trying to achieve a consensus, against a general background of vested interests. Learned bodies such as the Institution of Mechanical Engineers and the Institution of Chemical Engineers have often produced worthwhile guidance and advice particularly when considering process plant, but this again has no legal status, merely acting as one way of achieving health and safety and not precluding others.

2.10 The law and public inquiries into major accidents

We consider here the type of public inquiry which followed the Flixborough disaster [4], and will probably follow the next one in the UK. Like many others, although not always for the same reasons, the writer has serious misgivings about the effectiveness of such inquiries.

If an industrial disaster were to occur tomorrow, a public inquiry would probably be called for by the HSC under Section 14 of HSWA 1974 and would follow the Health and Safety Inquiries (Procedure) Regulations 1975.[21] The Commission would appoint a 'person' and assessors (who would constitute the court) to hold an inquiry, giving appropriate notice of its date, time and place to all persons entitled to appear.

The inquiry would have two main objectives: (a) to discover the causes of the disaster; and (b) to decide who was responsible for causing it.

2.10.1 The court of inquiry

The appointed person, who would be chairman of the court, would probably be a leading QC. The court's other members might be professors in the technical disciplines most involved, and a representative of the HSC. The court's main job would be to hear the evidence, reach its conclusions and write its report.

The appointed person would hold the inquiry in public except where:

- a Minister of the Crown rules that it would be against the interests of national security to allow the evidence to be given in public; or
- as a result of application made to him, he considers that the evidence to be given may disclose information relating to a trade secret.

2.10.2 'Interested parties', counsel and expert witnesses

A preliminary hearing would be held at which interested parties and their counsel (usually barristers), who would represent them in court, would be identified. The parties might be the operating company, its parent company, licensors, designers, contractors, trade unions, local authorities and Her Majesty's Factory Inspectorate (HMFI).

Both the court and the main parties would choose their own technical advisers (who might be rival firms of professional consultants). They, together with experts from HSE, would investigate the causes of the disaster before the court hearing started and later serve as expert witnesses in court.

2.10.3 Professor Ubbelohde's criticisms

Important criticisms of this type of inquiry were voiced by Professor Ubbelohde in 1974.[22] Some extracts follow:

> There are special features of every major accident . . . Promptitude in the scientific study of accidents is often hindered by established procedures of inquiry. One cause of delay is the age-old desire to find 'someone to blame' when things have gone wrong . . . Legal problems of allocating responsibility often overshadow scientific problems of how such accidents occur . . . Financial consequences of legal liability can become so oppressive that the whole tempo and management of accident inquiries (particularly while they are sub-judice) may acquire almost a nightmare quality. This concern with liability can distort and obscure basic scientific questions of how and why accidents occur . . . Scientists would stress that their inquiries must be concluded without any emotional or other pressures on them . . . British practice seems no better than elsewhere. It can even be termed wasteful, antiquarian and lopsided, since full detailed scientific study often cannot take place until the scent is cold.
>
> Whatever form of organisation is chosen, to be effective, an officially established scientific inquiry should run side by side with any judicial inquiry, not after it. It should aim to have all the information collected and presented in an organised way by a small team of experienced scientific assessors. The professional independence and scientific integrity of these assessors must be properly protected. Their task should be to collect facts scientifically and their submissions would be examined and freely discussed by a body of scientists.

2.10.4 Further criticisms

Counsel are at least as much concerned to prevent the party they represent from being blamed as to uncover hazards which contributed to the disaster.

They are so used to working in this way that they are unaware of their own blind spots and limitations. In technical matters they are largely dependent on the views of hand-picked experts whose 'proof of evidence' they have rehearsed and vetted. Who has ever heard of an expert witness giving evidence which might render his or her client responsible for the accident? The two objectives of the inquiry are thus in conflict.

Legal training and procedures (examination and cross-examination of witnesses) may have some value in establishing whether or not a witness is telling the truth. The procedures are, however, highly formal, and courts can only work on the evidence with which they are presented. They cannot crawl under the car and examine the under-chassis. They can only deal with what the mechanic or expert who has crawled under the car told them in evidence that he found.

Courts thus work in an artificial atmosphere of pre-selected facts and evidence over whose selection they may have rather limited control. They are seldom sufficiently familiar with the subject with which they are dealing to appreciate the relevance of every aspect.

Because of the heavy financial burden of daily legal representation at an inquiry, most parties want to see the whole thing over as soon as possible. In spite of this, some inquiries are very lengthy.

In the (usual) case of disagreement between experts, a court would probably prefer the views of its own advisers. This might be justified on the grounds that only its own advisers are genuinely impartial, but it is liable to lead to complaints from some parties that the court itself is biased.

Even the court's technical advisers may find themselves in a delicate position. The court has only a transitory life and will be out of business as soon as the inquiry is over, whereas interested parties are usually powerful multinational companies with more staying power.

Another weakness of public inquiries is the unaccountability of courts themselves. Once the court has produced its report, it becomes 'defunct', though its members are alive and well. The court cannot be sued if its findings are in error. There is usually no provision for re-opening such a public inquiry, and even if there were, most parties would be against it because of the cost.

2.10.5 Hazards of unrecognised causes

Thus there are many reasons why the court of a public inquiry may fail to unearth the real causes of a disaster. This can lead to serious dangers, since the party who has wittingly or unwittingly covered up a hazard is likely to be the one most affected if he or she has other installations where it remains dormant, and could cause a further similar accident.

- If the hazard has been consciously covered up, the party has to adopt a double standard, denying publicly that such a hazard could exist, yet alerting those within his or her own organisation to be prepared for it.

Thus one employee could be sacked for admitting the hazard's existence, and another for failing to take precautions against it!

- If it has not been recognised or if senior officials in the party's organisation have been persuaded by their own PR exercise that the hazard does not exist, it is likely that it will strike again.

2.10.6 What needs to be done?

An interesting paper by Dr Mecklenburgh[23] developed Professor Ubbelohde's thesis further. It suggested that the technical investigation of the causes of a disaster should be as complete as possible and reach definite conclusions before legal liability is dealt with. Thus there would be separate and consecutive inquiries to cover the technical and legal aspects, the first conducted by technical experts, the second by lawyers.

There are some like Professor Davidson[24] (a member of the Flixborough court) who consider that technical experts cannot be trusted to carry out a technical inquiry without the help of lawyers. Yet in the case of aircraft accidents where the regulations covering investigations are more advanced,[25] technical causes are investigated by technical experts unaided by lawyers. The views of Mr Wilkinson, former Chief Inspector of Accidents in the Department of Transport, are quoted below.[26]

> The UK approach to aircraft accident investigation has been tested and tried over many years and it works. This does not mean that it cannot be improved upon. The proof that the system has achieved general public acceptance is the fact that public inquiries and review boards are rare occurrences.
>
> I should now make it quite clear that, in my considered opinion and from bitter personal experience, there is no place whatever for lawyers in the investigation of aircraft accidents. Some superficially attractive arguments are regularly put forward, saying, in effect, that manufacturers, operators and airworthiness authorities, to name but a few, would, due to commercial pressures, not respond in a responsible manner to criticisms and recommendations without the immense pressures that are applied to them both collectively and individually after a major accident by lawyers acting for interested parties. This is nonsense, in fact, the reverse is much closer to reality. People who manufacture, operate and certify modern aircraft are generally responsible professionals who are concerned to give of their best.

There are, of course, important differences between disasters in air transport and those in the process industries. It is doubtful, for instance, whether it would be possible for a small country such as the UK to keep a permanent group of experts with sufficiently wide experience to be able to investigate the causes of all major accidents in all branches of the process industries. Nevertheless, HSE have gone a long way in developing such a capacity. They have investigated several serious accidents in the process industries (such as the explosion at King's Lynn in 1976) and published their findings,[27] with which there has been little criticism. A wider pool of expertise for such investigations might be organised on an international or regional basis, e.g. within the EC.

Should the UK again experience a process disaster on the scale of Flixborough, it is hoped that the HSE, with the help where necessary of outside experts, will be allowed to complete the technical investigations and report them to the public before questions of legal liability are examined.

2.10.7 Could a Parliamentary Commission of Inquiry be used?

Those not happy with the idea of leaving the investigation of the causes of major industrial disasters entirely to technical people (even when they are employed by the HSC) might like to know the origins of our public inquiry system. It started after the Marconi scandal of Edwardian times. The public inquiry was invented as a device to replace Parliamentary Commissions of Inquiry for handling contentious political issues on which MPs were so sharply divided (on party political lines) that consensus was impossible.[28]

This should not apply to the essentially technical issues of major accident causation. Members of Parliamentary Commissions, like lawyers, may not have the technical expertise to investigate the causes of process disasters, and they would thus be still largely dependent on outside help. Yet they seem to be more accountable to the public than the present courts of public inquiries. Parliamentary Commissions are certainly used in other countries to investigate the causes of industrial disasters. This was done in Italy after the Seveso disaster[29] [5.2], and no doubt could be done in the UK if this was agreed to be the best course.

2.11 The role of standards[30]

Modern industry operates with the aid of a complex network of standards (including codes of practice). Most of these have no legal status, although many form the bases of contracts, while non-compliance with important standards [2.14] would weaken the litigant's case in a lawsuit. The future safety of any plant depends on the choice and use of appropriate engineering standards which should be quoted in contracts between owners, designers and plant contractors. Those responsible for supervising design and construction on behalf of the owners must be aware of these requirements and be sufficiently familiar with the standards concerned to detect departures from them at an early stage. They must also have the necessary authority to be able to enforce compliance with the standards [23.5].

The aims of standardisation are defined as[31].

- overall economy in terms of human effort, materials, power, etc. in the production and exchange of goods;
- protection of consumer interests through adequate and consistent quality of goods and services;
- safety, health and protection of life;

- provision of a means of expression and communication among all interested parties.

These aims are interdependent, but where health and safety are concerned, it is seldom possible to adopt the most economic solution (except in the broadest sense to society as a whole). Although some standards, such as BS 2092, 'Specification for industrial eye protection', are primarily concerned with safety and health, many others are crucial to safety.

Standards may be classified according to their subject, level and aspect, examples of which are given in Table 2.1.

Some of the earliest standards were units of measurement. Most countries now use 'SI' units [3.1.1]. This simplifies the harmonisation of national standards and the creation of international ones, but due to the reluctance of the USA to abandon inches, pounds and gallons, many international standards are drawn up in two versions, one using SI units, the other American ones.

One group of standards reduces manufacturing costs by standardising the dimensions of manufactured items such as nuts, bolts and pipes. Another group covers material specifications and gives chemical compositions and other properties of materials specified in design standards.

Design standards form another group of engineering standards. These give detailed design procedures, e.g. for pressure vessels, with recommendations on manufacture, inspection and testing.

Application standards form another group. They give the main characteristics of the equipment, e.g. a pump or motor, specify the ancillary equipment needed, and the operating conditions for which the equipment is designed.

Some national standards bodies produce codes of practice on a range of subjects such as tower cranes and machine guards. These give general guidance, design parameters and modes of operation, and refer to other standards for further details.

Table 2.1 Classification of standards

Subject	*Level*	*Aspect*
Engineering	International	Terminology
Transport	Regional	Specification
Building	National	Sampling
Food	Industrial	Inspection
Agriculture	Company	Tests and Analyses
Textiles		Limitation of variety
Chemicals		Grading
Information		Code of practice
Science		Packaging
Education		Transport
Health		Safety

2.12 Levels of standards

Standards are next briefly discussed by level.

2.12.1 International standards

These, apart from those in the electrical field which started in 1908, are produced by the International Standards Organisation (ISO) which was set up in Geneva in 1946. It attempts to harmonise national standards and to assist developing countries to establish their own. Despite the growing number of international standards, their coverage is thinner than the national standards of major industrial countries. There are also often long delays before international standards can be agreed. Since an international standard on any subject cannot be established without the agreement of the countries which have already produced their own standards on it, the standard usually has to wait until some countries give way and bring their national standards into line with others.

2.12.2 Regional standards

Regional standards committees have been established in various areas, often characterised by the use of a common language. They aim to encourage trade in the area by establishing common national standards within it.

2.12.3 National standards

These are produced by national standards bodies which have technical divisions responsible for different areas of technology. They work through specialist committees which draft new national standards and revise old ones.

Some national standards can restrict competition from countries with different ones. The UK, however, does not insist that all goods and equipment, whether made in the UK or imported, should comply with a British standard. Critical plant items such as pressure vessels which do not comply with the appropriate British standard may be imported so long as they conform to a reputable foreign standard carrying an equivalent assurance of safety and reliability.

2.12.4 Industrial standards

These cover particular industries and are produced mainly by professional bodies within those industries. They are specially important in the USA, many such standards being used internationally. The bodies producing these standards are agents of the American National Standards Institute, and include:

- ASTM American Society for Testing and Materials
- ASME American Society of Mechanical Engineers

- API American Petroleum Institute
- FPA Fire Protection Association
- ACGIH American Conference of Governmental Industrial Hygienists.

In the UK, there has been an unfortunate tendency for different organisations with a common interest in any subject each to produce and publish their own standard, all differing in matters of detail. As an example, codes covering the handling and use of liquid petroleum gases have been published by the HSE, the Home Office, the Institute of Petroleum, the Liquid Petroleum Gas Industry Technical Association, the Fire Protection Association, and by ICI in conjunction with RoSPA. Not all of these are periodically amended to keep them up to date. Such organisations should be encouraged to pool their efforts to produce a joint standard or code which is regularly updated.

When modern industrial technologies are transferred to other countries, particularly Third World ones which have little industry, appropriate standards are often lacking in the recipient country. The safety problems which result are discussed in Chapter 23.

2.12.5 Company Standards[32]

Company standards supplement national and industrial standards and incorporate the company's own experience. A company which pioneers a new technology is obliged to develop its own standards for it. These are often the forerunners of standards at industrial, national and international levels. Most companies have their own standards and an organisation for creating them and keeping them up to date. Most company standards are of three types:

- buying standards, e.g. for raw materials, plant and equipment which it buys, and staff and labour whom it hires;
- internal standards for design, operation, safety and general practice within the company;
- standards and specifications for its own products.

Company standards are an important management tool, and play a key role in determining the company's competitiveness and safety record. If the standard contains valuable commercial information, it will be treated as confidential.

Company buying standards usually refer first to any accepted international, national or industrial standards on which they are based (e.g. ASME, BS, DIN or ISO), and then detail the deviations and additions to the standard which the company requires.

Some company standards are important for the safety of process plant. It is easier to develop effective company standards in large companies than in small ones. Some large companies have developed extensive codes of practice in safety matters for their own use.

2.13 Safety standards and codes of practice

While most standards have some significance for safety, some deal specifically with safety topics, e.g. protective headgear, scaffolding, insulation and colour coding of electrical wiring. Many of these are codes of practice. Several have been published by the ILO dealing with safe working practices in entire industrial fields.

Under HSWA 1974, standards issued by the British Standards Institution and professional and industrial bodies are treated as 'guidance literature'.

In 1982 HSC issued a guidance note on standards relating to safety[33]. This proposed that HSE should become more involved in BSI technical committees in cases where it expects to make use of the resulting standards. It also discussed legal backing for some British standards, which might take any of three forms (listed in order of diminishing status).

1. 'Application of a standard by regulation', i.e. the standard becomes mandatory in all circumstances;
2. 'Approval of a standard under Section 16 of HSWA';
3. 'Reference to standards in the course of guidance' (i.e. in HSE Guidance Notes).

HSE have issued a list[34] of some 625 British Standards which are significant to health and safety. Despite HSE's list, there are still no adequate British standards for some important features of process plant, e.g. for the design of pressure relief systems for oil and petrochemical plant. For these we usually rely on appropriate foreign standards, particularly American ones.

It is impossible to list here all relevant British, American and European standards. Lees[35] gives extensive lists of standards in use in 1996, although some of these (such as those issued by ICI in conjunction with RoSPA) are no longer published for general use.

Despite the proliferation of standards, it is essential for professional people to be familiar with and up to date in the standards in their own fields. There is a need for special courses to update professionals in the standards they have to use.

2.14 Offshore legislation

A substantial amount of the safety legislation which appeared during the twentieth century was written as a result of disasters: indeed it is sometimes called ghoul law. When the exploitation of North Sea oil and gas began, the necessary legislation was already in place. Comprehensive regulations on the construction of installations and on their fire protection were provided in two Statutory Instruments, each of which was amplified in detailed guidance notes. The regulations were published by the Department of Energy (DEn) under the authority of two enabling Acts: the Mineral Workings (Offshore Installations) Act 1971, and the Petroleum and Submarine Pipelines Act 1975. Fire protection was covered by S.I. 611 1978 Offshore Installations (Firefighting Equipment) Regulations,[36] and the guidance notes were contained in a 90-page book, *Offshore Installations: Guidance on Firefighting Equipment*[37] (known as the *Red Book*). When a disaster did occur in the North Sea, on the Piper Alpha platform (July 1988), the regulations were scrutinised by Lord Cullen, who conducted the public inquiry.[38] He considered that when the regulations were written, and particularly when their amendment was discussed, no consideration was given to the great advances that had been made in the preparation of onshore legislation, particularly in the transition from prescriptive to goal-setting requirements. Although comprehensive and detailed, the regulations were also restrictive and inflexible, and were not necessarily applicable to every installation. It was necessary that the hazards existing on each platform should be identified, and that appropriate protection should be provided. Lord Cullen made a total of 106 recommendations, all of which have been accepted by the Government.

An account of the disaster is given in 5.4, where it is explained that a massive increase in the intensity of the fire resulted from the rupture of three gas risers which connected Piper to other platforms. The failures happened on the riser side of the isolating valves. This and other happenings were considered by a separate technical inquiry which was conducted by Jim Petrie of DEn prior to the main inquiry. A statutory instrument was issued, the Emergency Pipeline Valves Regulations, 1989,[39] which require that all hydrocarbon risers are provided with emergency shutdown valves located as near as possible to sea level. This action was endorsed by Lord Cullen, who rightly accepted that some prescriptive legislation would be necessary in addition to the new goal-setting approach.

All of the platforms installed prior to 1990 will have been equipped in compliance with S.I. 611 (although many operators provided protection which was in excess of these requirements). A brief indication of the regulations is therefore given here. Production could not begin on a platform until the firefighting equipment had been inspected and approved, and a Certificate of Examination had been issued. A new certificate was required every two years. The more important of the *Red Book* requirements are these:

- automatic self-monitoring fire detection throughout all working spaces, providing an indication of the location of the fire at a continuously manned control room;
- a similar provision for gas detection wherever flammable gases may accidentally accumulate;

- a fire alarm system;
- a fire main to supply hydrants, monitors and water-deluge systems, connected to no less than two pump units situated in different parts of the platform;
- a water-deluge system to protect any equipment in which petroleum is stored, conveyed or processed, and to protect critical structural steelwork;
- a sprinkler system within the accommodation areas;
- a total flooding system in the control room, and in any areas containing internal combustion engines or open-flame heaters;
- a foam system to protect the helideck, together with suitable extinguishers;
- suitable portable and non-portable (wheeled) extinguishers in all the areas in which extinguishing systems are needed.

These provisions are quite consistent with a realistic fire-protection philosophy. Rapid detection, particularly in areas which are infrequently visited, is essential if fires are to be attacked while they are still small. Hand extinguishers and wheeled units provide the essential first line of defence against these incipient fires. Larger fires can be tackled by a trained team using monitors and foam-making equipment. It must be accepted however that fires can occur offshore which are beyond the capabilities of this team because they are inaccessible, or too large, or the fuel is escaping at too high a pressure. Two actions are then vital in preventing a disastrous escalation: the amount of fuel feeding the fire must be limited by isolating and depressurising equipment; and during the time needed to consume the released fuel, or for the fire to be diminished to a manageable size, all of the surrounding plant and structural steelwork must be kept cool by the application of water. It is the purpose of the deluge systems (and monitors) to deliver this water. Without adequate cooling, structural steelwork could collapse, and additional fuel could be released into the fire from ruptured equipment. In an intense fire, unprotected steel could fail in as little as ten minutes. It is unfortunate that, as is explained in 5.4, none of these actions could be taken on Piper.

There can be no doubt that compliance with the regulations ensured a good standard of fire protection. However, it has already been noted that these requirements were inflexible (although an inspector might agree to accept that certain conditions could be deemed to comply with a particular regulation). They were also restrictive in that there was little allowance for new developments. An example is the statement that flame detectors 'should generally be of the ultra-violet type'. This was written at a time when some infra-red devices had proved to be unreliable offshore. Currently available detectors have a good record and are in fact better suited to the offshore environment. Compliance with the regulations has also led to unsatisfactory decisions on the design of platforms. One example results from the requirements for the water-deluge systems, which are based on a 'reference area' concept. The *Red Book* says in effect that water must be applied to any equipment containing oil or gas, and to local principal load-bearing structural members. The quantity of water applied to the equipment should be not less than that which would be required for an application rate of 12.2 litres/minute to each square metre of the protected area. If the platform is equipped with two fire pumps (the minimum requirement) then each must be capable of supplying the largest deluged area and at least two hydrants. Thus, if an operator wants to avoid the need for large fire pumps, then large process areas can be divided into a number of smaller areas, using fire resisting bulkheads. This has two disadvantages: a reduction in the amount of water available for firefighting, and an increase in the probability that an explosion will cause critical structural damage.

There is no mention of explosion protection in either of the S.I.s or their guidance notes. Nor is there any discussion of the vital importance of good safety management: adequate training, good housekeeping, correct attitudes, safe working practices and an effective permit-to-work system. DEn were aware of many of these deficiencies and had commissioned a number of reviews of their guidance notes. They also had a considerable input of advice from many organisations, and in particular from UKOOA (The UK Offshore Operators Association). It became apparent, however, that some of the suggested amendments would necessitate changes in the regulations themselves.

Much of this new thinking appeared in a draft second edition of the *Red Book*, which was issued just after the completion of the Cullen Inquiry. However, both this and the regulations themselves were withdrawn as a result of the recommendations in the Inquiry report.

The first of Lord Cullen's 106 recommendations says that:

> the operator should be required by regulations to submit to the regulatory body a Safety Case in respect of each of its installations.

The safety case should show that a number of objectives have been met, including:

- that the safety management system of both the company and the installation are adequate to ensure that the design and operation of the installation are safe;
- that the potential major hazards have been identified and appropriate controls provided;
- that provision has been made for a temporary safe refuge (TSR) for personnel on the installation, and for their safe and full evacuation, escape and rescue. The TSR is expected to be the accommodation module itself, or possibly a citadel within it. Other matters to be covered in the safety case include:

- a demonstration that the hazards resulting from hydrocarbons in plant and pipelines have been minimised;
- a demonstration by quantified risk assessment that the integrity of the TSR, the escape routes, the embarkation points and the lifeboats is adequate;
- a fire risk analysis;
- a specification of the events in which, and the times for which, the integrity of the TSR, the escape routes, the embarkation points and the lifeboats can be maintained. A safety case should be prepared for all existing installations and should be updated regularly.

The fire risk analysis very rightly covers both fire and explosion risks. The recommendations say that the approach to fire and explosion protection should be integrated as between active and passive fire protection, the different types of passive protection, and the protection provided for fires and explosions. Safety assessments rather than regulations should be used to determine the location and resistance of fire and blast walls, and the design of water-deluge systems. The use of fire scenarios is to be preferred to the reference-area method in determining fire protection requirements. The ability of the deluge systems and the fire pumps to survive severe accident conditions should be demonstrated. In another part of the recommendations it is suggested that a study should be made of:

> the feasibility of dumping in an emergency large oil inventories . . . in a safe and environmentally acceptable manner . . .

The safety case must be submitted to the 'Regulatory Body', which is also responsible for the legislation. At the time of the Piper disaster this function was carried out by DEn. One possible criticism of this arrangement is that the Petroleum Engineering Division of DEn contained both the Safety Directorate and the department concerned with oil and gas production: there could have been a clash of interests. Also, although the Safety Directorate had comprehensive knowledge of offshore equipment and operations, Lord Cullen found that they had failed to apply the new concepts of goal-setting regulations. On the other hand, HSE had considerable experience of the new-look legislation but had little if any contact with the offshore industry. Perhaps surprisingly Lord Cullen opted for a new division exclusively concerned with offshore safety within the HSE. Within very few years after the establishment of this division, a great amount of new guidance and regulations had been published in accordance with Lord Cullen's recommendations. Considerable use was also made of documents produced by UKOOA. In 1990 for example UKOOA produced a Procedure on Formal Safety Assessment which established a common framework for the preparation of safety cases.

Mention has already been made of the Offshore Installations (Constructional and Survey) Regulations, S.I. 289 of 1974.[40] By 1993 the fourth edition of *Offshore Installations: Guidance on Design, Construction and Certification*[41] was published. This contained a vast amount of information in 99 sections and 11 appendices, but at that time additional material was still required before a new S.I. could be issued. The regulations in S.I. 611 covering fire protection, with which we are particularly concerned, were withdrawn and replaced by the Offshore Installations (Prevention of Fire and Explosion, and Emergency Response) Regulations 1995.[42] One improvement is immediately obvious: there is no mention of explosions in S.I. 611.

These new regulations are unusually and very acceptably presented in a publication, issued in April 1995, which contains also an approved code of practice (ACOP), and guidance notes. Each regulation is quoted in full and is followed by the guidance, which explains the requirements of the regulations and the meaning of particular phrases. This in turn is followed by the ACOP, which has a special legal status. A court will find an owner guilty of a breach of the law if it is proved that he has not followed the provisions of the ACOP. It is not compulsory to follow the guidance notes and it is quite permissible to take other appropriate action. However, if the guidance is followed this will normally be sufficient to comply with the law. Inspectors will regard the guidance as an example of good practice.

The first three of the regulations are concerned with citation, interpretation and geographical application. It is in Regulation 4 that the goal-setting format, in accordance with Lord Cullen's recommendations, becomes apparent. This is concerned with the general duty of the 'duty holder': on a fixed installation this is the supervisor appointed by the operator, and is normally the OIM (although this title is not used in these regulations). The duty holder must take appropriate measures to protect everyone on the platform from fires and explosions, and for securing an effective emergency response. The guidance explains that 'protection' implies design, prevention, detection, control and mitigation, and that 'measures' can include plant and equipment (hardware) and management systems (software). It is also necessary to take into account the assessments required by Regulation 5 (discussed below), and compliance with other regulations. The ACOP makes it clear that the main purpose of the regulation is to promote a risk-based systematic approach to managing the hazards and to emergency response. The general principles to be followed are set out in the ACOP of the Management of Health and Safety at Work Regulations 1992.[17]

Regulation 5 requires the duty holder to carry out 'an assessment'. This will identify the events which can cause a major accident, or the need for evacuation, escape or rescue. The likelihood and consequence of these events will be evaluated. Standards of performance needed to cope with the events will be established, and appropriate measures will be selected. Records must be kept of these decisions. The guidance points out that the assessment will be used in determining the actions to be taken in complying with other regulations, and can also be fed into the safety case (discussed below). Attention is drawn to the Safety Case Regulations and also to the excellent guidance that has been provided by UKOOA. The ACOP provides detailed instructions on the way in which the assessment is to be made.

Regulation 8 covers the emergency response plan, and again the guidance mentions the relevant UKOOA publication. Regulation 9 is concerned with the prevention of fires and explosions. As might be expected, much of the area covered by the old *Red Book* is described, but without the detailed specifications. Mention is also made of the need for area classification and the choice of suitable electrical equipment, both of which are discussed in 10.5.

Regulation 10 covers fire and gas detection, which should be based on the assessment which was made under Regulation 5. Again much of the *Red Book* discussion is repeated, but without the specifications. Communication is covered in Regulation 11, and the Control of Emergencies in 12. Regulation 13 is concerned with the mitigation of fires and explosions. It requires measures to protect people and also measures to ensure that the protection will be effective throughout an emergency. The guidance makes it clear that the measures include deluge systems, fixed extinguishing systems, fire-resistant coatings, manual firefighting, ventilation control and fire and blast walls. Again the plant and the areas to which this protection is to be applied are not specified, but the choice must be in accordance with the assessment. Regulation 14 covers Muster Areas, and complies with Lord Cullen's recommendations concerning a TSR. His advice on evacuation, escape and rescue are covered in Regulations 15, 16 and 17. The requirements for life-saving appliances are detailed in Regulation 20.

The first of the 106 recommendations made in Lord Cullen's report says that the operator should be required to submit a Safety Case to the regulating body for each of its installations. The regulation requiring this should be analogous to Regulation 7 of the CIMAH Regulations 1988.[7] The Offshore Installations (Safety Case) Regulations[43] were issued in 1992 in a publication which contained guidance but did not include an ACOP. Three different safety cases may have to be submitted during the lifetime of a platform. The design safety case, prepared before the installation is built, must show how the design will provide inherent safety, how the quality of the detailed design and construction will be checked, and how safe construction and commissioning will be achieved. The pre-operational safety case must be submitted six months before drilling is attempted. It must describe the installation, activities and systems, and the limits for safe operation. Before the plant is decommissioned, the wells are closed, and the structure is removed, an abandonment safety case must be prepared. Additional cases will be needed to cover combined operations with another installation, major modifications and periodic updating.

The principal function of a safety case is to show: that the management system is adequate to ensure compliance with regulations; that adequate arrangements have been made for auditing and for the preparation of audit reports; and that all hazards which have a potential to cause a major accident have been identified, their risks have been evaluated, and adequate measures have been taken to ensure that the risk to people is as low as is reasonably practicable. The Safety Case Regulations and the associated guidance provide detailed information on the preparation, content and submission of safety cases covering a variety of situations.

Lord Cullen was concerned with the need to involve the entire workforce in safety matters by the use of safety committees and safety representatives. He said that there should be a feedback from the workforce to supervisors and management on safety matters, and the safety representatives should receive adequate training at the expense of the operator. His recommendations resulted in the publication of the Offshore Installations (Safety Representatives and Safety Committees) Regulations 1989[44] and Guidance Notes for these regulations in 1992.[45]

2.15 Health and safety legislation in the USA

Until the second half of the twentieth century the USA appeared to be lagging behind the UK, and most of Europe, in safety legislation. This can be partly explained by the strong American tradition of the freedom of the individual: neither a factory owner nor his workforce would tolerate interference by the state. The situation was changed dramatically in 1970 by the publication of the Occupational Safety and Health Act (OSHA). During the furore which followed its introduction one (apparently) sympathetic UK observer said that we experienced exactly the same problems after the publication of our fifth Factories Act, in 1833. OSHA was only one of many new Acts (some are discussed below) which resulted from the growing public awareness of environmental, safety and health problems. Legislation is enacted by both State and Federal Governments, but much of the growth in health and safety controls resulted from Federal Acts, like OSHA.

To some extent the USA is still lagging behind Europe in that USA safety legislation is mainly prescriptive, and there is little interest in the use of goal-setting regulations. Much of the detailed prescriptive legislation in the UK has now been replaced by these new-look regulations. For example, the owner of a process plant is required to identify the potential hazards and then to demonstrate that he is taking adequate steps to protect both his employees and the public. This procedure is to some extent self-regulatory, but the owner's decisions will be influenced by non-mandatory guidance which is published in conjunction with the regulations. The US legislation which is discussed here does not make use of these methods. However, we in Europe could with advantage follow the American practice of citing codes and standards in their regulations. Considerable use is made of the standards produced by the American National Standards Institute (ANSI) and the National Fire Protection Association (NFPA). The National Fire Codes which are published by NFPA are written by unpaid committees drawn from a wide cross-section of academic institutions, learned societies, government departments, trade unions, end-users and trade organisations (but not from individual manufacturers). The codes can be updated if necessary every year, but nothing is accepted for publication until it has received the democratic support of the NFPA membership at an annual meeting. This interesting use of standards is characteristic of both Federal and State legislation. It clearly has some advantage over the UK practice by which both regulations and guidance are provided by the same Government department.

Safety legislation will have little impact unless it is well supported and can be rigorously enforced. Ever since the 1833 Act, the Factory Inspectorate in the UK has had the right to walk into any factory without warning at any time of the day or night. Similar rules in OSHA allowed inspections to be made at any reasonable time without search warrants, but in 1977 a Federal Court found that this violated the Fourth Amendment of the US Constitution. The Court also said that if admission to a factory was refused, the inspectors had no authority to apply for a court order to gain entry. To some extent this problem was overcome by encouraging employees to report any violation of safety and health regulations which had been made by their employer. This method of enforcement (whistle-blowing) gained public support, but was condemned by the Environmental Protection Agency (EPA). However the EPA's attitude was subsequently modified.

Legislation in the USA which is concerned with the protection of the environment is enforced by the EPA, which was created in 1969 by the National Environmental Policy Act. Safety in the workplace is the concern of the OSHA, which was established by the Occupational Safety and Health Act (also OSHA) in 1970. Unfortunately the areas of responsibility of the two agencies are not clearly defined. The discussions below on Federal legislation show that there is a considerable overlap of responsibilities, particularly in the Acts which are concerned with toxic materials and with accidental releases. The situation may be further complicated by the existence of similar State legislation.

2.15.1 The Occupational Safety and Health Act 1970

OSHA, like the UK Factories Acts, is an enabling measure. It authorises the Secretary of Labor to issue standards, to inspect places of work, and to assess penalties for breaches of the law. The Secretary is primarily responsible for the administration and enforcement of the Act, and the Secretary of Health, Education and Welfare is required to produce the necessary safety standards. The Occupational Health and Safety Agency carries out the enforcement duties and is also responsible for publishing the standards. The National Institute for Occupational Safety and Health (NIOSH) was established by the Act, and one of its functions is the development of new standards. However, as we have seen, much use is made of existing or new standards written by non-governmental organisations, mainly ANSI and NFPA.

Particular mention is made in the Act of toxic materials. Standards must be set to ensure:

> that no employee will suffer material impairment of health or functional capacity even if such employee has regular exposure to the hazard . . . for the period of his working life.

The standards concerning toxicity must be:

> based upon research, demonstrations, experiments, and such information as may be appropriate. In addition to the attainment of the highest degree of health and safety protection for the employee, other considerations shall be the latest available scientific data in the field.

There is also mention in the Act of adequate labelling and the use of protective equipment. Regular medical examinations must also be available, and must be provided by the employer.

It should be noted that toxic materials are also covered by the Toxic Substances Control Act 1976 and by Title III of the Superfund Amendments and Reauthorization Act 1986 (SARA). OSHA also issued an additional regulation on the Process Safety Management of Highly Hazardous Chemicals in 1990. All of these measures are discussed below.

Section 5(b) of the Act says that:

> each employee shall comply with occupational safety and health standards and all rules, regulations and orders issued pursuant of this Act . . .

There are similar requirements in UK legislation, under which an employee can be prosecuted for failing to comply with the regulations. However, the section of OSHA dealing with procedure and enforcement places the onus for compliance mainly upon the employer rather than the employee. If a violation of the Act has occurred, the employer will be issued with a citation which must 'describe with particularity the nature of the violation' and will fix a reasonable time within which the violation must be corrected. Following this, the employer may also receive an assessment of the fine which he is required to pay. In the UK the fine would be imposed by a Court, but under OSHA the Department of Labor has this authority. However, the employer can appeal against either the citation or the assessment, asking for them to be modified or set aside, at a Court of Appeals.

Compliance with the OSHA regulations has resulted in US industry having to complete a vast number of official forms. OSHA was the first piece of legislation to be investigated by the Federal Paperwork Commission, in 1976.

2.15.2 Siting of hazardous installations

A number of different agencies became involved in decisions concerning the location of hazardous plant (and some appear to operate on an *ad hoc* basis). For example, the EPA has powers under the Clean Air Act 1970 to specify the purity of the air in a particular area. It may therefore be found that a plant from which some emissions are inevitable is planned in an area in which no change in air quality will be permitted. If the plant is to be located near to the coast, then the US Coast Guard may also be involved through the Coastal Zone Management Act 1972. If a liquefied natural gas (LNG) plant is planned, then the Federal Power Commission (FPC) will be involved and also, under the Natural Gas Pipeline Safety Act 1968, the Office of Pipeline Safety Operations (OPSO). The FPC is concerned only with existing regulations and is not involved in policy making. It has no national policy and does not publish guidelines, so its decisions are difficult to predict, and there is no guarantee that they will be in agreement with those of OPSO. As with other hazardous plant, permission will also be needed for any new construction from the local and State authorities.

2.15.3 Pollution and noise

There has been growing public concern in Europe and the USA concerning pollution by substances and by noise. An early piece of US legislation was the Water Quality Act 1965. This was followed by the National Environmental Policy Act 1969 (NEPA), which established the EPA. This Act requires a developer to submit an Environmental Impact Statement (EIS) to an appropriate Federal agency before permission can be granted for this construction of a new project. (The information to be provided in an EIS was specified by a regulation of the President's Commission on Environmental Quality in 1978.) The Clean Air Act 1970 (CAA) gives the EPA the authority to produce air quality standards and to enforce them. The Agency publishes lists of the maximum permitted concentrations in air of various gases. The CAA has been strengthened by amendments published in 1977, 1981 and 1990. The Federal Water Pollution Control Act 1972 (FWPCA) gives the EPA similar powers to control water pollution. They have published standards and guidelines on the maximum concentrations of materials which can be present in water. The EPA also becomes involved under the Safe Drinking Water Act 1974. They also enforce the Noise Control Act 1972. The EPA is responsible for the control of hazardous wastes under the Resource Conservation and Recovery Act 1976 (RCRA). The Agency issues rules covering the storage of waste materials in tanks, including underground storage. The control of hazardous waste sites through the EPA is also strengthened by the Comprehensive Environmental Response, Compensation and Liability Act 1980 (CERCLA), which established the Superfund. This Act enables the EPA to enforce the clean-up of waste sites. If the person responsible for the site cannot be identified, then the cost can be paid out of the Superfund. It is necessary to renew the Act (and the fund) at regular intervals, and this was done in 1986 by SARA (already mentioned). Title III of SARA extended the application of CERCLA to accidental toxic releases, which are discussed below.

2.15.4 Transportation

Again, increased public concern, particularly as a result of fatal accidents, has led to a great deal of legislation covering the transport of hazardous materials. Early legislation includes the Rivers and Harbors Act 1899; the Explosives Transportation Acts 1906, 1909 and 1921; the Natural Gas Act 1938, and the Federal Aviation Act 1958. The Natural Gas Pipelines Act 1968 has already been mentioned.

The movement of hazardous materials over State boundaries is clearly a Federal matter and was at one time the concern of the Interstate Commerce Commission (ICC). This is now covered by legislation enforced by the DOT. The legislation includes, in addition to the Acts already mentioned, the Tank Vessels Act 1936, the Dangerous Cargo Act 1970, the Hazardous Materials Transportation Act 1970, the Coastal Zone Management Act 1972 (noted above because of the Coast Guard involvement), and the Ports and Waterways Safety Act 1972.

Within the DOT, the National Transportation Safety Board (NTSB) is responsible for investigating accidents. Regulations are produced by the Hazardous Materials Regulation Board, which is advised by the Office of Hazardous Materials. The regulations are applied by the Federal Highway Administration, the Federal Railway Administration, the Federal Aviation Administration and the Coast Guard. The Coast Guard has produced a particularly useful classification of hazardous materials, including a guide to the compatibility of goods loaded into adjacent holds in a ship. This is used by the other organisations, and indeed by other countries. It was developed for the Coast Guard by the National Academy of Sciences.

Road transport regulations, like other similar legislation, make considerable use of standards. Both State and Federal regulations are frequently based on the NFPA National Fire Codes, and in particular: NFC 30, Tank Vehicles; 495, Explosive Materials; 498, Explosives Terminals; 512, Truck Fire Protection; and 513, Motor Freight Terminals. Both rail and canal transport are covered by similar legislation, canals being the responsibility of the Coast Guard. The Coast Guard is also the regulating authority for harbours and all marine operations. However, the international transportation of hazardous materials is controlled by the Intergovernmental Maritime Consultative Organisation (IMCO). This organisation has published a wide range of codes and regulations covering the construction and equipment of ships, their fire protection, the handling of dangerous cargoes and pollution. Air transportation is similarly regulated by the International Air Transport Association (IATA).

North America has a very extensive pipeline network, which is also regulated by the DOT (through the Natural Gas Pipeline Safety Act 1968). Other materials which are distributed extensively by pipeline include chlorine, ammonia, ethylene and other petroleum products. The DOT produces an annual report on pipeline failures.

2.15.5 Accidental releases

The disasters at Flixborough, Seveso and, in particular, Bhopal, resulted in the development of new legislation in Europe to deal with major hazards.

This was followed in the USA by the Emergency Planning and Community Right to Know Act 1986 (ECPRA). This is Title III of SARA, which has already been mentioned above. Title III, which became law in 1988, is again enforced by the EPA. Its main requirements cover the need for emergency planning; the notification of an emergency; the provision of an inventory of toxic chemicals; and the community's right to know about the presence and possible effects of the chemicals. Additional regulations covering accidental releases were issued in 1990 by OSHA: Process Safety Management of Highly Hazardous Chemicals. These rules require the establishment of management systems to identify, evaluate and control highly hazardous processes.

The State of New Jersey passed its own Toxic Catastrophe Prevention Act in 1985. This requires that where more than a specified quantity of certain toxic materials is present on an industrial site, a risk management programme must be prepared. California also has a Hazardous Materials Planning Program, which was based on four existing laws within their Health and Safety Program.

2.15.6 Lobbying

The USA has an extensive and effective lobbying industry. A UK manufacturer became aware of its activities during the marketing in the States of a high-performance chemical extinguishant. This was being sold in direct competition with a similar agent manufactured in the USA. The US material had an advantage over the UK material of being less toxic, although neither created a hazard in normal use. Soon after the launch of the UK extinguishant, a law was passed through a State legislature banning the sale of any extinguishing agent which had a toxicity in excess of a specified value. Not surprisingly, the US agent had a toxicity below, and the UK agent above, this value. The passing of similar laws in other States would have resulted in the loss to the UK firm of a very attractive market. Fortunately their prompt action secured the repeal of this particular piece of legislation.

References

1. Hardie, D. W. F., Davidson, J. and Ratt, P., *A History of the Modern Chemical Industry*, Pergamon, Oxford (1966)
2. HSE, *A Guide to the Pressure Systems and Transportable Gas Containers Regulations*, HMSO, London (1989)
3. HM Government 1974 C. 37, *Health and Safety at Work etc. Act* (1974)
4. S.I. 1982, No. 1357, *Notification of Installations Handling Hazardous Substances 1982*, HMSO, London
5. S.I. 1988, No. 1657, *Control of Substances Hazardous to Health Regulations 1988*, HMSO, London
6. S.I. 1987, No. 2115, S.I. 1990, No. 556, S.I. 1992, No. 3068, *Control of Asbestos and Work Regulations 1984*, HMSO, London
7. S.I. 1984, No. 1902, *Control of Industrial Major Accident Hazard Regulations 1984*, HMSO, London

8. S.I. 1985, No. 2023, *Reporting of Injuries, Diseases, Dangerous Occurrences Regulations 1985*, HMSO, London
9. S.I. 1987, No. 635, *Electricity at Work Regulations 1989*, HMSO, London
10. S.I. 1989, No. 1780, *Noise at Work Regulations 1989*, HMSO, London
11. S.I. 1989, No. 2169, *Pressure Systems and Transport of Gas Containers Regulations 1989*, HMSO, London
12. Lord Robens Committee, *Safety and Health at Work Report 1972*, HMSO, London
13. S.I. [1992], No. [2793], *Manual Handling Regulations 1992*, HMSO, London
14. S.I. [1992], No. [3004], *Workplace Safety and Welfare Regulations 1992*, HMSO, London
15. S.I. [1992], No. [3073], *Supply of Machinery (Safety) Regulations 1992*, HMSO, London
16. S.I. [1992] No. [2966], *Personal Protective Equipment at Work Regulations 1992*, HMSO, London
17. S.I. [1992], No. [2051], *Management of Health and Safety at Work Regulations 1992*, HMSO, London
18. S.I. [1992], No. [2932], *Provision and Use of Work Equipment Regulations 1992*, HMSO, London
19. S.I. [1992], No. [2379], *Visual Display Regulations 1992*, HMSO, London
20. Edwards v National Coal Board 1949, 1 KB 704
21. S.I. 1975, No. 335, as amended by S.I. 1976, No. 1246, *Health and Safety Inquiries (Procedure) Regulations 1975*, HMSO, London
22. Ubbelohde, A. R., letter to *Financial Times*, 4 December (1974)
23. Mecklenburgh, J. C., 'The investigation of major process disasters', *The Chemical Engineer*, August (1977)
24. Davidson, J., 'On Public Inquiries', *The Chemical Engineer*, June (1984)
25. S.I. 1983, No. 551, *The Civil Aviation (Investigation of Accidents) Regulations 1983*, HMSO, London
26. Wilkinson, G. C., 'UK aircraft accident investigation procedures', *Air Law*, Kluwer, Deventer, The Netherlands, IX, No. 1 (1984)
27. HSE, *The Explosion at Dow Chemical Factory, King's Lynn, 27 June 1976*, HMSO, London
28. *Report of the Royal Commission on Tribunals of Inquiry*. (chairman, Lord Justice Salmon), HMSO, London (1966)
29. *The official report of the Italian parliamentary commission of inquiry into the Seveso disaster, translated by HSE*, HSE, Bootle (about 1984)
30. King, R. W., 'The role of standards in the safe transfer of technology to developing countries', paper given at ILO symposium on *Safety, Health and Working Conditions in the Transfer of Technology to Developing Countries*, ILO, Geneva (1981)
31. ISO, *The Aims and Principles of Standardisation*, Geneva (1972)
32. British Standards Institution, *Guide to the Preparation of a Company Standards Manual*, London (1979)
33. HSC Consultative Document, *Reference to Standards in Safety at Work*, HMSO, London (1982)
34. HSE, *Standards Significant to Health and Safety at Work*, HSE (1985)
35. Lees, F. P., *Loss Prevention in the Process Industries*, Butterworths, London (1980)
36. S.I. 1978, No. 611, *Offshore Installations (Firefighting) Equipment Regulations 1978*, HMSO, London
37. DEn, *Offshore Installations: Guidance on Firefighting Equipment*, HMSO, London
38. Lord Cullen, *The Public Inquiry into the Piper Alpha Disaster*, Cm 1310, HMSO, London (1990)
39. S.I. 1989, No. 1029, *Emergency Pipeline Valves Regulations 1989*, HMSO, London
40. S.I. 1974, No. 289, *Offshore Installations (Constructional and Survey) Regulations 1974*, HMSO, London
41. HSE, *Offshore Installations, Guidance on Design, Construction and Certification*, 4th edition, HMSO, London (1993)
42. S.I. 1995, No. 743, *Offshore Installations (Prevention of Fire and Explosion and Emergency Response) Regulations 1995*, HMSO, London

43. HSE, *A Guide to the Offshore Installations (Safety Case) Regulations 1992*, HMSO, London (1992)
44. S.I. 1989, No. 971, *Offshore Regulations (Safety Representatives and Safety Committees) Regulations 1989*, HMSO, London
45. HSE, *Safety Representatives and Safety Committees on Offshore Installations, Guidance Notes*, HMSO, London (1992)

3 Meanings and misconceptions

People's reactions to particular words and symbols differ widely, and are not always what their users intended. This problem, which can be serious enough when only English is used, becomes more acute when two or more languages are involved. Mistranslations may be amusing [23.4.1] but they lead to misunderstandings which can cause accidents. (That's why the Tower of Babel was never finished!)

Here units and general nomenclature are discussed first [3.1], followed by health and safety terms [3.2]. Section 3.3 discusses misconceptions, particularly those which sometimes permeate an entire organisation and lead to disasters.

3.1 Units and nomenclature

3.1.1 Units

Most of us use different systems of units at various times. This, with the conversions between them, can be a source of misunderstandings and possible accidents. Writers seldom define their units completely, and assume that their readers will automatically know which system they are using. Although this is true for most readers, a minority usually get it wrong. Examples are Imperial and American gallons (6 American gallons = 5 Imperial gallons), long tons (2240 pounds) and short tons (2000 pounds) and ounces of different kinds (avoirdupois, US fluid and troy). Furthermore, engineers and others who are alert to discrepancies when using familiar units easily lose this critical faculty when working in unfamiliar ones.

While the UK has been in the process of changing from Imperial units to metric ones for the past 30 years, this has coincided with the introduction (starting in 1960) of a special version of the metric system, *Le Système*

International d'Unités (SI units), in which several hitherto widely used metric units have no place.[1] In it calories have given away to joules (J) and kg/cm^2 to pascals (Pa). The wholesale adoption of the SI system (as indeed that of metrification generally) within industry and commerce has still a long way to go. The *Handbook of Chemistry and Physics* continues to give most thermal data in calories.

An advantage claimed for the SI is that it has only one unit for each kind of quantity.[2] This, however, has proved so restrictive that several non-SI units (e.g. minutes, hours, days, degrees Celsius, litres, tonnes, bars) (but not calories!) are used with it to make the system workable[1] in various fields. The prefixes and their symbols (representing various powers of ten) which must be added as factors to base SI units to convert them to larger or smaller units can also be sources of error, as BS 5555[1] admits:

> 4.4 Errors in calculations can be avoided more easily if all quantities are expressed in SI units, prefixes being replaced by powers of 10.

In this book a minimum of formulae and mathematics are used. Most numerical quantities which are taken from other sources are quoted and where necessary calculated in the units quoted. With the occasional exception of joules (for which the writer still prefers calories and kilocalories (cals and kcals) for sensible and latent heats) SI units and others allowed in conjunction with them[1] are used in examples given here for the first time.

Two pitfalls in quoting pressures should be noted.

1. While many British and Americans use psi to mean gauge pressures, only writing psia for absolute pressures, many Europeans quote pressures in bars or pascals (1 bar = 10^5 Pa) to mean absolute pressures, in conformity with the historical origin of these units. Thus when we quote gauge pressures in bars, we are well advised to write 'bar g.'. We should also check what others mean when they quote bars or pascals without qualifying them.
   It is curious that the Flixborough report [4] quoted plant pressures before the disaster as kg/cm^2 only, when some company personnel understood this to mean gauge pressure while others understood it to mean absolute.

2. A second pitfall is that some units have approximately the same value, yet the differences are large enough to cause hazards if they are taken as identical in critical cases, e.g. the setting of pressure relief valves. Thus one standard atmosphere = 1.01325 bar = 1.0333 kg/cm^2.

3.1.2 Nomenclature

This section is mainly intended for non-UK readers. The abbreviation HS is frequently used for 'health and safety' in a general sense and persons responsible on a full-time basis for specified aspects of safety are referred to as safety professionals, despite wide differences in their training, backgrounds and responsibilities. Most large companies and 'works' engaged in the process industries employ several safety professionals, with a safety manager in charge. The overall responsibility for HS should lie with a named director of the company[3] [19.1].

Some technical terms used here where British and American usage sometimes differs are given in Table 3.1.

Other preferred terms are 'supervisor' rather than foreman or chargehand, 'non-return valve' rather than check valve, 'bund' rather than dike, 'flametrap' rather than flame arrester, 'generator' rather than dynamo, electrical 'earth' rather than ground, 'branch' (on a vessel) rather than nozzle, 'bursting disc' rather than rupture disc.

3.2 Meanings of health and safety terms used

The definitions given here are mainly derived from the writer's earlier book[4] and from a short IChemE guide.[5]

3.2.1 Hazard

Hazard is a very important concept. Put simply by Heinrich:[6]

> A hazard is a condition with the potential of causing injury or damage.

The pursuit of safety is largely a matter of identifying hazards, eliminating them where possible or otherwise protecting against their consequences.

Table 3.1 Some technical terms

Term	*Description*
Mixer	Used for complete mixing equipment of various kinds
Stirrer	Only used for rotary shaft stirrers
Spectacle plate	An 8-shaped plate permanently installed between flanged joints to isolate equipment
Line blind	Circular plate with stub temporarily inserted between flanged joints to isolate equipment
Mild steel	Used for carbon steel except when speaking of high, low or medium carbon steel
'P & V' valve (pressure and vent)	Weight-loaded valve used to maintain a low positive pressure in tanks and equipment by controlling flow of inert gas entering and venting (referred to in the USA as 'conservation vent')
Interlock	A device which prevents a valve, switch, etc. from being actuated while another valve, switch, etc. is in a certain mode
Trip	An automatic device, such as a combination of a pressure-sensitive switch and a three-way solenoid in the air supply to a pneumatic valve controlling a process stream, which opens or closes the valve when the temperature, pressure, level or some other variable reaches a pre-set value

The IChemE guide[5] defines **chemical hazards, major hazards** and **hazardous substances**.

Often two hazards need to be present simultaneously to cause a major accident. The law has not always recognised this, and much time has been wasted arguing which of two hazards was the 'proximate cause' of an accident.[7]

3.2.2 Accident

The many meanings of this common word are the subject of two reviews.[8,9] The writer prefers the definition given by Heinrich[6] with the word 'sudden' added. This definition includes 'near misses':

> An accident is a [sudden] unplanned event which has a probability of causing personal injury or property damage.

The inclusion of the word 'sudden' differentiates accidents from slower forms of deterioration such as corrosion or exposure to low levels of airborne asbestos. But sudden total failure following a long period of deterioration might be classed as an accident.

The word is also commonly used to mean an accidental injury,[10,11] when it is often qualified by adjectives such as 'fatal', 'notifiable', 'reportable'.

3.2.3 Injury

The word injury as used here, without qualification, refers only to physical injury to a person caused by an industrial accident, which by the above definition is sudden. Other types of injury are qualified.

Sudden injuries caused by accidents need to be distinguished from disabilities caused by industrial diseases over long periods.

3.2.4 Industrial disease

Following Hunter,[12] the term 'industrial disease' or 'occupational disease' is used here for any type of ill-health arising from conditions at work.

The border-line between an accident such as 'gassing' caused by brief exposure to a high concentration of a toxic gas and a disease caused by chronic exposure to low concentrations of the same gas is often difficult to draw.

A number of industrial diseases are notifiable in Britain, with an obligation to report all cases of them to the appropriate health authority.[13] Many are furthermore 'prescribed'. This means that those suffering from them who have worked in particular occupations for a minimum period are entitled to claim compensation from the state.

3.2.5 Damage

Accidental damage applies to things rather than people and is usually measured in money terms. Damage may be to plant and equipment (part of the fixed capital) and/or to materials in process (part of the working capital).

Damage control consists of recording all accidents, including near-misses, analysing their causes (hazards) and working out and implementing steps to remove or reduce them.[14] These activities help to reduce injuries, since many accidents which damage things also injure people.

3.2.6 Loss

This term is used in insurance and refers to the costs of injuries and damage.

Consequential loss is additional to direct financial loss and refers to that caused by the interruption of production during the period of repair following an accident.

3.2.7 Disaster

This word generally refers to a major accident or natural event and/or a serious epidemic of an occupational or natural disease which results in the death of a number of people. The word is a subjective one, and refers more to the degree of shock and suffering rather than the size of the event. Western has produced a classification of natural and man-made disasters.[15] Turner's study[16] of man-made disasters is discussed in 3.3.

3.2.8 Probability

> Probability is a mathematical term having a value between 0 and 1, where 0 represents complete impossibility and 1 represents absolute certainty.[17]

Probability may also be quoted as a probable frequency (the number of specified events occurring in unit time). It is vital to the study of **risk** and **reliability** [14].

3.2.9 Risk

This word is often used with different meanings, occasionally being used to mean hazard. I prefer the definition given by the Chartered Insurance Institute:[7]

> Risk is the mathematical probability of a specified undesired event occurring, in specified circumstances or within a specified period.

Other meanings of risk are sometimes used in insurance, e.g. the subject insured (a dancer's legs) or the eventuality insured against (having twins).

Within our chosen meaning of risk as a probability, further contrasting types of risk have to be considered:

- **pure risk** or **speculative risk**[17]
- **individual risk** or **societal risk**.[17]

A **pure risk** is one where the only possibility is loss or breaking even, whereas a **speculative risk** is one where there is a possibility of gain or loss.[7]

Many risks found in industry which at first sight appear as pure risks, are found when we look closer to be speculative ones. The risk of head injuries to workers not wearing protective hats in an area where overhead construction work is going on might appear to be a pure risk, until we realise that by taking the risk someone has saved the cost of the hats. Estimates of the value of human life are based on the consideration of speculative risks which people are prepared to take with their own lives and those of others.[3]

Total elimination of risk is never possible. The questions of what degree of risk is acceptable, how much money should be spent in reducing them and how different risks should be balanced have led to two professional disciplines, **risk analysis** and **reliability engineering**.[17]

3.2.10 Safe

The best definition[18] the writer has found is:

> A thing is provisionally categorised as safe if its risks are deemed known and in the light of that knowledge judged to be acceptable.

Safety case has a special meaning in connection with the CIMAH Regulations.

3.2.11 Other terms

Several other specialised terms relating to explosions, fires, toxicity, release and dispersion of hazardous substances, hazard studies, reliability and risk criteria are defined in the IChemE guide.[5] These and others are explained as needed in the text.

3.3 Misconceptions and disasters

This section develops some of Dr Barry Turner's ideas[16] and shows how they can be applied to safety in the process industries. Examples are given in Chapters 4 and 5.

3.3.1 Learning from past accidents

Do we learn as much as we should from past accidents? Is the necessary information available? Or are we swamped with so much information that we cannot find the right bits? Technology can help. The proverbial needle in a haystack is easier to find if we have a metal detector. Similarly, a computer can search information for us quickly when it is stored in a well-indexed data bank. Yet the 'information explosion' can overburden our minds and hinder learning. As one writer put it, 'The more an organism learns, the more it has to learn to keep itself going'.[19]

Nevertheless we expect to be able to learn from disasters, for as Robert Stephenson said in 1856:[20]

> Nothing was so instructive to the younger Members of the Profession, as records of accidents in large works . . . A faithful account of those accidents, and of the means by which the consequences were met, was really more valuable than a description of the most successful works. The older Engineers derived their most useful store of experience from the observation of those casualties which had occurred to their own and other works, and it was most important that they should be faithfully recorded in the archives of the Institution.

In spite of this, there are often obstacles which prevent us gaining more than a superficial knowledge of the causes of major accidents in the process industries. Major accidents tend to destroy or obscure much of the physical evidence. But perhaps more important is the fear in some directors' minds of publicising information which might reveal some negligence on the part of their companies. Commercial confidentiality may be invoked as a (legal) reason for concealing it.[21] Thus sometimes only the more obvious hazards of major accidents are revealed.

More fruitful lessons in hazard identification can often be learnt from the investigation of lesser accidents, and even 'near-misses', where there is less concern about being blamed and a greater readiness to examine and admit possible causes. This is a good argument for the system known as 'damage control'[14] [3.2.5].

3.3.2 The ingredients of disasters

Turner[16] concludes from a study of several man-made disasters that:

> Large scale disasters need time, resources and organisation if they are to occur . . . Since these conditions are most unlikely to be met solely as a result of a concatenation of random events, we can almost suggest that simple accidents can be readily arranged, but that disasters require much more organising ability!

Usually a combination of two or more different types of hazard is necessary. This was well expressed by Sir Geoffrey De Havilland and Mr P. B. Walker:[22]

> There is a modern trend which is steadily changing the overall character of investigations, though without affecting the basic principle. Accidents are becoming less and less attributable to a single cause, more to a number of contributory factors. This is due to the skill of the designers in anticipating trouble, but it means that when trouble does occur, it is inevitably complicated. . . .This trend can make nonsense of public inquiries and of lawsuits where allocation of responsibility is attempted.

Turner proposed the following general formula or equation for man-made disasters:

> 'Disaster equals energy plus misinformation.'

Energy here may be the kinetic energy of a train, the potential energy of water stored in a dam or the chemical energy of combustible materials and explosives in a process plant or in storage.

Examination of several major process plant disasters [4, 5] shows clear evidence of multiple causation. Two types of hazard are usually involved.

1. The hazardous material or agent itself, its nature and its quantity, which determine the type and maximum extent of the damage it can do, if it is unleashed. I call this the **inherent hazard**. It corresponds to Turner's 'energy', although we need to add 'toxic substances' as an alternative to energy in his equation.
2. The mechanism by which the inherent hazard becomes unleashed. This generally consists of one or more inadequacies in the means of control and containment. These I term **initiating hazards**. Since the means of control and containment should be known, the initiating hazards generally correspond to misinformation in Turner's equation.

3.3.3 Misinformation

Misinformation may have several causes.

- The information is completely unknown at the time (e.g. the carcinogenic properties of vinyl chloride in the 1950s).
- The information is known to some people, but not known to or properly appreciated by those in need of it at the time. The dissemination of this information may be inhibited if the senior management have misconceptions about it. These are then termed **organisational misconceptions**.
- The information needed is about fast-moving events as they unfold. Here misinformation may arise through failure of people to respond quickly enough, or through inadequacies in the means of perception or the channels of communication. An example is an air collision or near-miss, the avoidance of which depends on rapid communication between the pilots of both aircraft and a ground flight controller.
- Information which is misunderstood, e.g. due to language problems, or because it is too complicated, or because it contains apparent contradictions or illogicalities.

Even, however, when all these types of misinformation are included, there still appears to be something missing from the right-hand side of Turner's equation. In its crudest form, this consists of deliberate actions in defiance of information known at the time. In many cases this defiance of information involves taking a 'speculative risk' [3.2.9]. A person may be praised or advanced if he 'gets away with it' and perish or be pilloried if he fails. If the pressures, economic, social or emotional, are strong enough he may feel that he has no option but to take the risk of ignoring information on which even his life may depend. In other cases it may involve a death wish on the part of the individual, or it may be an act of wilful sabotage. A person may also act in defiance of information which he knows to be right, because of some uncontrollable desire (e.g. for sex, nicotine, alcohol or drugs).

In yet other cases, although the information lies in the person's unconscious mind, he cannot recall it to his consciousness when it is needed, and behaves as though he were misinformed. This happens to all of us when our minds are overloaded, or because of some 'hang-up' or mental blockage (often caused by an inner conflict).

Defiance of information may thus be due to economic, sociological or psychological causes, and a number of disasters had such origins. Behind many of them there is a human drama. As the American humorist Artemas Ward put it, 'It ain't so much the things we don't know that gets us into trouble. It's the things we know that ain't so'. I would therefore modify Turner's equation to read:

> 'Disaster equals energy and/or toxic substances plus misinformation or rejection of information.'

The **initiating hazard** is then often the result of misinformation, or the rejection of correct information.

3.3.4 The hazard of over-centralised management control

The need for standardised procedures tends to lead to rather centralised control, with hierarchical management structures. In the case of a transnational enterprise, with subsidiary companies operating similar plants in different countries, the procedures may be worked out by the parent company, which passes them on as established wisdom to its foreign subsidiaries. Personnel from the parent company who visit the subsidiary company to 'advise' on this or that are often imbued with status symbols reminiscent of religious orders:

> In holy orders, social distinction and the symbols of such distinction become so interwoven that the vestments themselves of the priest inspire awe and wonder. Who knows whether the priest is a holy man? And who knows whether the majestic scholar, wending his way in mortarboard, hood and gown to the chapel on Convocation Day, is a wise man? What we do see is a drama of hierarchy wherein rank is infused with a principle of hierarchical order.[23]

Thus while there is a real need for strictly standardised procedures, the type of organisation to which this gives rise may make it more difficult to correct organisational misconceptions within itself.

Four of the five major accidents discussed in Chapters 4 and 5 occurred within the foreign operating subsidiaries of transnational companies. Organisational misconceptions appear less likely to occur on the home pitch of a transnational company, because of the greater knowledge, experience and self-reliance of its technical staff compared with those of its subsidiary companies, who are less expected to think for themselves.

3.3.5 The incubation of disasters

Turner has shown that man-made disasters do not come 'out of the blue' in an organised situation, be it a railway system, a mine or a chemical plant, but usually after a lapse of time which may range from a few hours to several years after one or more departures from normality have occurred. He reached this conclusion after studying a number of official reports of disasters investigated by UK authorities. This time lapse represents the **incubation period**, during which action could have been taken to prevent the disaster, had the danger been recognised by the organisation concerned.

This may be a very sensitive matter and it often needs a character of independent mind within the organisation to perceive and warn of the hazard — in short a whistle-blower.[24] To respond properly to such warnings and deal fairly with the whistle-blower is a severe test for any management [19].

Turner's concept of the incubation period is given in the following passage:

> We began by imagining a situation where a group or community possessed of sufficiently accurate information about their surroundings to enable them to construct precautionary measures which successfully warded off known dangers, to provide us with a 'notationally normal' starting point for the development of disasters. From this starting point, for each disaster or large-scale accident which emerges, we have suggested there is an 'incubation period' before the disaster which begins when the first of the ambiguous or unnoticed events which will eventually accumulate to provoke the disaster occurs, moving the community away from the notationally normal starting point. Large-scale disasters rarely develop instantaneously, and the incubation period provides time for the resources of energy, materials and manpower which are to produce the disaster to be covertly and inadvertently assembled.

The incubation period for a disaster on a safely designed, built and managed process plant may begin when some misguided action is taken which goes against the original norms and reduces the integrity of the plant, although by itself it could hardly cause a disaster.

The mistake might then be recognised and corrected, and the integrity of the plant restored, so that resulting accidents are averted. If not, some minor accident may follow without the mistake being recognised. When the damage has been repaired and the plant is started again, its integrity remains impaired, since the hazard resulting from the mistake is still present.

Now a second mistake may be made, which is also not recognised. A further hazard is introduced which reduces the integrity of the plant still further. One or two further accidents due to one or both of these hazards may follow. After each the damage is repaired and the plant restarted, although the mistakes and the hazards which resulted from them remain unrecognised and uncorrected. Finally, when a certain combination of conditions is reached, both hazards are activated, and disaster occurs. The incubation period has lasted from the time of the first mistake through a succession of minor accidents until the disaster.

It is possible that the first mistake was made in the initial design of the plant. Here one could date the incubation period either from the time the mistake was made or from the time the plant was first started up.

Having discussed Turner's general theory of the origin of a disaster and of the opportunities available (during its incubation period) for preventing it, some examples are examined in the next two chapters. They discuss the technical causes of some past disasters in the process industries, the incubation periods which preceded the disasters, and the organisational misconceptions which led to them. They also extend the lessons drawn from the reports of official inquiries. The mistakes or departures from norms which occurred were, in hindsight, quite elementary ones and not caused by ignorance of abstruse information.

References

1. BS 5555: 1981/ISO 1000–1981, *Specification for SI units and recommendations for the use of their multiples and of certain other units*
2. Weast, R. C. (editor in chief), *Handbook of Chemistry and Physics*, 1st student edition, CRC Press, Boca Raton, Florida (1988)
3. Lord Robens Committee, *Health and Safety at Work*, HMSO, London (1972) and 'The Health and Safety at Work etc. Act 1974'
4. King, R. W. and Majid, J., *Industrial Hazard and Safety Handbook*, Butterworths, London (1980)
5. The Institution of Chemical Engineers, *Nomenclature for Hazard and Risk Assessment in the Process Industries,* Rugby (1985)
6. Heinrich, W. R., *Industrial Accident Prevention*, 4th edn, McGraw-Hill, New York (1968)
7. CII Tuition Service, *Elements of Insurance*, Chartered Insurance Institute, London (1974)
8. Crane, N. C., 'Just what is an accident?', *Industrial Safety*, **23**, No. 3, 10 (1977)
9. The National Institute of Industrial Psychology, *A Review of the Industrial Accident Research Literature,* HMSO, London (1972)
10. The Factories Act 1961, Section 80, Part V, HMSO, London
11. The International Labour Office, *International Recommendations on Labout Statistics*, Geneva (1976)
12. Hunter, D., *The Diseases of Occupations*, English Universities Press, London (1957)
13. S.I. 1980, No. 377, *The Social Security (Industrial Injuries) (Prescribed Diseases) Regulations 1980*, HMSO, London
14. Bird, F. E. and Germain, G. L., *Damage Control*, American Management Association, New York (1966)
15. Western, K. A., *The Epidemiology of Natural and Man-made Disasters: the present state of the art*, Academic Diploma in Tropical Public Health, London University (1972)
16. Turner, B. A., *Man-made Disasters*, Taylor and Francis, London (1978)
17. McCormick, N. J., *Reliability and Risk Analysis*, Academic Press, New York (1981)
18. The Council for Science and Society, *The Acceptability of Risks*, Barry Rose (Publishers) Ltd, London (1977)
19. Dewey, J., *The Night Country*, Garstone Press, New York (1974)
20. Stephenson, R. quoted in *Engineering Progress through Trouble*, edited by Whyte, R. R., The Institution of Mechanical Engineers, London (1975)

21. HSC Discussion Document, *Access to Health and Safety Information by Members of the Public*, HMSO, London (1985)
22. De Havilland, G. and Walker, P. B., 'The Comet failure' in *Engineering Progress through Trouble*, edited by Whyte, R. R., The Institution of Mechanical Engineers, London (1975)
23. Duncan, H. D., *Communication and Social Order*, Oxford University Press, Oxford (1968)
24. The Council for Science and Society, (report of a working party), *Superstar Technologies*, Barry Rose (Publishers) Ltd, London (1976)

4 The Flixborough disaster

4.1 Introduction

As the prime cause of this disaster lay in a mistake in the mechanical design of the plant, we start this chapter with the plant's design and deal with subsequent events in the order in which they happened. For so long as the plant continued to operate without substantial modification to rectify this mistake, it was bound to fail sooner or later, whether disastrously or not.

After the disaster on 1 June 1974, a public inquiry was set up under Section 84 of the Factories Act 1961:

> to establish the causes and circumstances of the disaster and to point out any lessons which we might consider were immediately to be learned therefrom.

4.2 Design features of the cyclohexane oxidation plant

Design of the cylohexane oxidation plant started in 1968 when Nypro was then owned 45% by DSM (Dutch State Mines), 45% by the NCB (National Coal Board) and 10% by Fisons Ltd. The process had been developed in Holland by DSM where another broadly similar plant had been built. The process design was DSM's responsibility and principally done by their engineering subsidiary Stamicarbon (NV). The plant was constructed by Sim-Chem Ltd, chemical plant contractors, and completed in 1972 when DSM bought out Fisons and now owned 55% of Nypro. World demand for caprolactam was growing and several other cyclohexane oxidation plants were designed by DSM for other countries. These plants all had a chain of stirred reactors operating at about 9 bar g. and 155°C. These were supported in the plant structure at progressively lower levels. Each reactor had two large flanged openings on opposite sides so that liquid cyclohexane could flow through them in cascade. Compressed air passed in parallel flow through the liquid in each reactor which it entered through fixed distributors. Each reactor had a powerful stirrer above the distributor (Fig. 4.1). The spent air saturated with cyclohexane vapour left the top of each reactor to join a common manifold.

As the reactors had to be progressively heated up from cold to the operating temperature of 155°C during start-up, means had to be provided to prevent excessive stresses due to differential thermal expansion between the flanged openings of adjacent reactors. Sometimes engineering practice was to use large 'U' bends in the pipework to take up these stresses. This was known to have been done in at least one other DSM-designed cylohexane oxidation plant built in the former USSR. The alternative arrangement adopted for Flixborough was to employ large stainless steel expansion bellows between reactors.

Such flanged openings in pressure vessels are reinforced, generally by steel pads, which are welded to them during their construction. The sizes of these reinforcing pads are calculated by the pipework and vessel designers taking all forces to which the vessel will be subjected into account. The reactors of the cyclohexane oxidation plant at Flixborough were designed for nominal thrusts on their sides of 9 tonnes whereas the thrusts caused by the pressurised bellows were at least 38 tonnes!

4.3 Before the big bang

4.3.1 Start-up and operation of the plant in 1973

The plant was started up early in 1973, apparently unaffected by the design error described in 4.2. Figure 4.1 shows the plant as it was designed and built. During operation 250–300 m^3 per hour of cold cyclohexane were pumped into column C2521 where it scrubbed and cooled the gases leaving C2544. Liquid from C2521 was pumped through a water separator S2522 to the top of column C2544 (containing 40 m^3 of 2-inch ceramic rings), where it contacted the same gas leaving the reactors, cooling it and becoming heated to about 155°C. The operating temperature in C2544 was kept high enough to prevent water vapour in the gas from condensing in it.

Liquid leaving C2544 at 155–160°C and 8–8.5 bar g. flowed in cascade through six stirred reactors each holding 33 m^3 of liquid which flowed between them through 28-inch stainless steel bellows connected by flanged joints to branches on opposite sides of each reactor. Compressed air entered the base of each reactor and bubbled through the liquid, a small percentage of its oxygen being converted via organic peroxides to cyclohexanol and cyclohexanone with smaller amounts of organic acids. The gas leaving the reactors was mainly nitrogen with some unconverted oxygen, oxides of carbon, cyclohexane and water vapour.

Liquid leaving No. 6 reactor passed first through the after-reactor R2529, which provided further time for peroxides to decompose. It was then

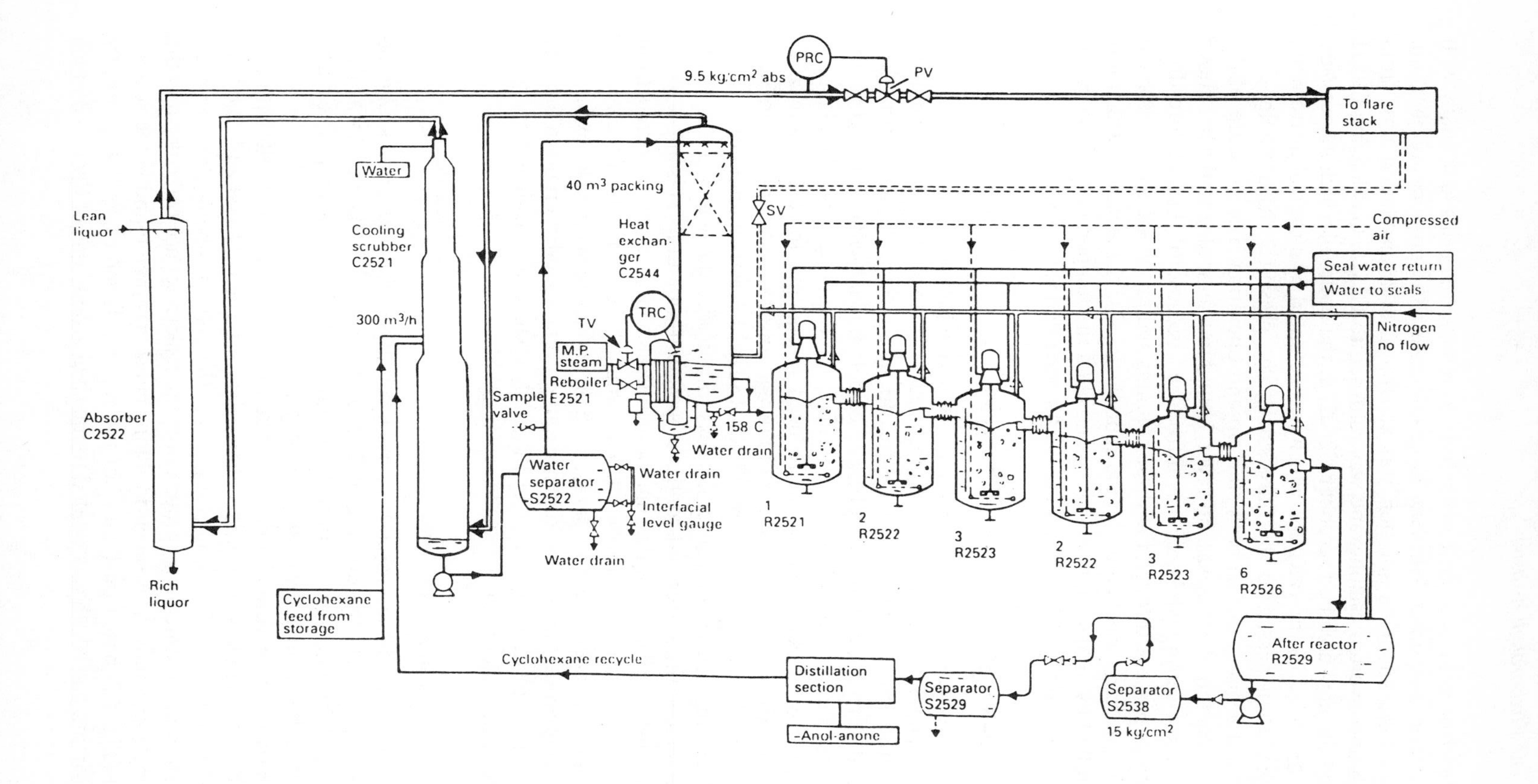

Figure 4.1 Cyclohexane oxidation plant as built, in operation, 1973

washed first with aqueous caustic soda to remove the acids and finally with water. The washed cyclohexane layer containing the product cyclohexanol and cyclohexanone in solution was separated in a distillation unit from which unconverted cyclohexane was returned to C2521. Gas leaving C2521 passed through a chilled scrubber C2522, at the outlet of which its pressure was controlled by the valve PV which was actuated by a control instrument. The gas then passed to the flare stack. The pressure was normally controlled at 8 bar g. during both start-up and operation. The pressure of the gas leaving the reactors was, however, bound to rise by at least 0.5 bar above the control setting when air was introduced because of the pressure drop during its passage through the three columns and pipework. During start-up the cyclohexane was heated by the reboiler E2521, the steam to which was controlled by an instrument which controlled the temperature of the cyclohexane leaving E2521 by actuating the air-operated valve TV.

4.3.2 Removal of stirrer from No. 4 reactor

The plant operated normally with all stirrers running until 13 November 1973 when a 'state of emergency' was declared in Britain following the miners' overtime ban. This included measures to save electricity[1] and it was then decided to try running the cyclohexane oxidation plant without using the stirrers since the air was well dispersed as it entered the reactors. The plant ran satisfactorily in this condition at reduced throughput for a few weeks. In January 1974 with the restoration of normal power supplies the stirrers were restarted. The drive of the stirrer of No. 4 reactor was then found to be damaged. Having operated for several weeks with unstirred reactors, it was decided to remove the stirrer and continue operation without it. This appeared to work satisfactorily, and at the time of the disaster the removal of stirrers from all reactors was being considered. This would have been safe if the reactor pressures had been above 11 bar g., i.e. high enough to keep cyclohexane plus any water present in the liquid phase and prevent a sudden eruption through the release of 'latent superheat' by the sudden boiling and mixing of dry cyclohexane with water in the unstirred reactor [4.5]

4.3.3 The split in No. 5 reactor

Most of the events discussed here are given in detail in paragraphs 52 to 56 of the report.[2] On the evening of 27 March 1974, while the plant was running, cyclohexane was found to be leaking from No. 5 reactor. The plant was shut down hurriedly to avoid leakage.

All six reactors were identical and had half-inch mild steel shells with one eighth-inch stainless steel liners. They were externally lagged and clad with aluminium for weather protection. While the plant was being depressurised and shut down, water was sprayed over the leak to reduce fire risks, and the lagging was removed. A seven-foot long crack was then found in the reactor shell (Fig. 4.2, based on Plate 7 of the report). It ran close to the annular reinforcing pad which was welded to the shell around one of the two 28-inch

Figure 4.2 Damaged reactor R2525

flanged openings. There was a shorter crack in the stainless steel lining. The reactors had been designed and tested individually (without the bellows) to a pressure of 16.2 bar g. There was no apparent reason for one to fail at half this pressure.

Specimens of the shell through which the crack passed were cut out for metallurgical examination by DSM in Holland. These were cut from the ends of the crack rather than from near the middle, where it might have been expected to start. The metallurgical report, dated 3 May 1974, did not reach Nypro till several days after the disaster. It concluded that the crack was caused by (intergranular) nitrate stress-corrosion cracking of the carbon steel shell. It noted that the scale on the shell near the crack contained 0.75% of nitrate. This it attributed to cooling water which had been sprinkled over the reactor casing for some weeks in November 1973 for fire prevention when a valve above the reactor was leaking.

There was widespread surprise in court about this report. Any nitrates found in the specimens were more likely to have come from water sprayed over the reactor while it was being depressurised than from water applied earlier before the crack started. According to a subsequent report by the HSE, nitrate stress corrosion is only found in highly stressed metal near its yield point[3]. Although this applied to the ends of the crack, there was no apparent reason for such high stresses where the crack started, nor was there any evidence of nitrate stress corrosion at this point. The court had taken evidence on the cause of the crack from several engineers and executives of DSM, the NCB and Nypro during days 26 to 28 of the inquiry but this was far from conclusive. At the close of the inquiry, there was still no agreement on the cause of the crack. Nevertheless, the court accepted DSM's assertion without seeking or hearing other explanations.[2]

> §212 The cracked Reactor R2525 initiated the sequence of events which led to disaster. Examination of the crack by expert metallurgists showed that the crack had been caused by nitrate stress corrosion.

Important information about the design fault in the reactors had, however, been uncovered during the inquiry by HMFI's pressure vessel inspector Glynne Evans, now retired and co-author of this book. He did not give evidence on the cause of the crack in court, because it was considered to be outside the court's terms of reference. Nevertheless several other people involved in the inquiry also knew this information. Having practically no personal contact with the inquiry, I only heard about it from Glynne later when he informed me

> that the vessel drawing specified a maximum thrust on the 28-inch nozzle stubs of 9 tonnes, whereas by calculation and experiment it was seen that there was a thrust of 38 tonnes at normal operating pressure. The crack ran longitudinally down the vessel, right at the toe of the weld at the doubling plate. No other so-called stress-corrosion cracks were found in this vessel that was sprayed by nitrate contaminated water, but cracks were discovered in identical places at other vessels that had never been sprayed with contaminated water.

I was not surprised by this since I never believed that nitrate stress corrosion was the sole cause of the crack although it might have played a part in its propagation. The reactors were simply not designed for the thrust of the bellows, whose effect on a reactor was, as Evans put it, 'like driving a two-inch nail into an inflated truck tyre'.

There was however another possible contributory cause to the failure of No. 5 reactor, and the fact that it was No. 5 which failed and not another increases this probability. The failure of No. 5 reactor occurred after the stirrer had been removed from No. 4, thus providing a place where water could have settled during start-up just as it would have done on the day of the disaster. The same mechanism which I believe contributed to the failure of the by-pass assembly could also have contributed to the failure of No. 5 reactor on 27 March.

We have to realise that when the 'latent superheat' in the two (liquid) phase system (dry cyclohexane and water) is released in an unstirred vessel, this will not happen instantaneously. Vaporisation will first occur at the interface between the two liquids in the unstirred No. 4 reactor, and it may take many seconds while the rest of the water layer becomes heated to the interfacial boiling point (about 148°C). The initial boiling in No. 4 reactor would first eject slugs of liquid through the bellows into No. 5 reactor where the stirrer was running. This would have rapidly broken up the slug into fine droplets which would instantly have vaporised as they dispersed in hot cyclohexane. Thus we might expect that the most vigorous stage of the eruption to have been in the stirred No. 5 reactor rather than in the unstirred No. 4 reactor where it started.

I am not suggesting that this was the sole cause of the failure of No. 5 reactor, since the very rapid boiling which I have postulated would probably not have damaged a well-designed reactor with normal stresses. But it would have been far more likely to initiate a crack in a grossly under-designed and over-stressed reactor. Another unsolved problem is why the drive mechanism of No. 4 reactor failed in the first place.

Following the failure of No. 5 reactor, the vessel was removed and a 20-inch pipe was made and installed in its place. This pipe and the two 28-inch bellows to which it was attached is subsequently referred to as the 'by-pass assembly'. Since the opposing branches on the remaining reactors were at different levels the pipe was made in the shape of a dog's leg with two mitred joints (Fig. 4.3).

4.4 The disaster on 1 June 1974 and the inquiries which followed

The events discussed earlier led to the largest peacetime explosion in Britain (equivalent to over 30 t of TNT) at 16.53 hours on Saturday 1 June 1974 above the chemical works of Nypro (UK) Ltd, near Scunthorpe. The disaster followed the rupture of the 'by-pass assembly' and the escape at sonic velocity of about 80 t of hot liquid cyclohexane at up to 155°C and 8 bar g. This formed an enormous vapour cloud as big as a football pitch which exploded within a minute of the rupture. It wrecked the works and main office block (fortunately unoccupied), and damaged property within a radius of 5 km. Twenty-eight men working on the site, nearly all in the control room, were killed. Hundreds of others, mostly outside the site, were injured.

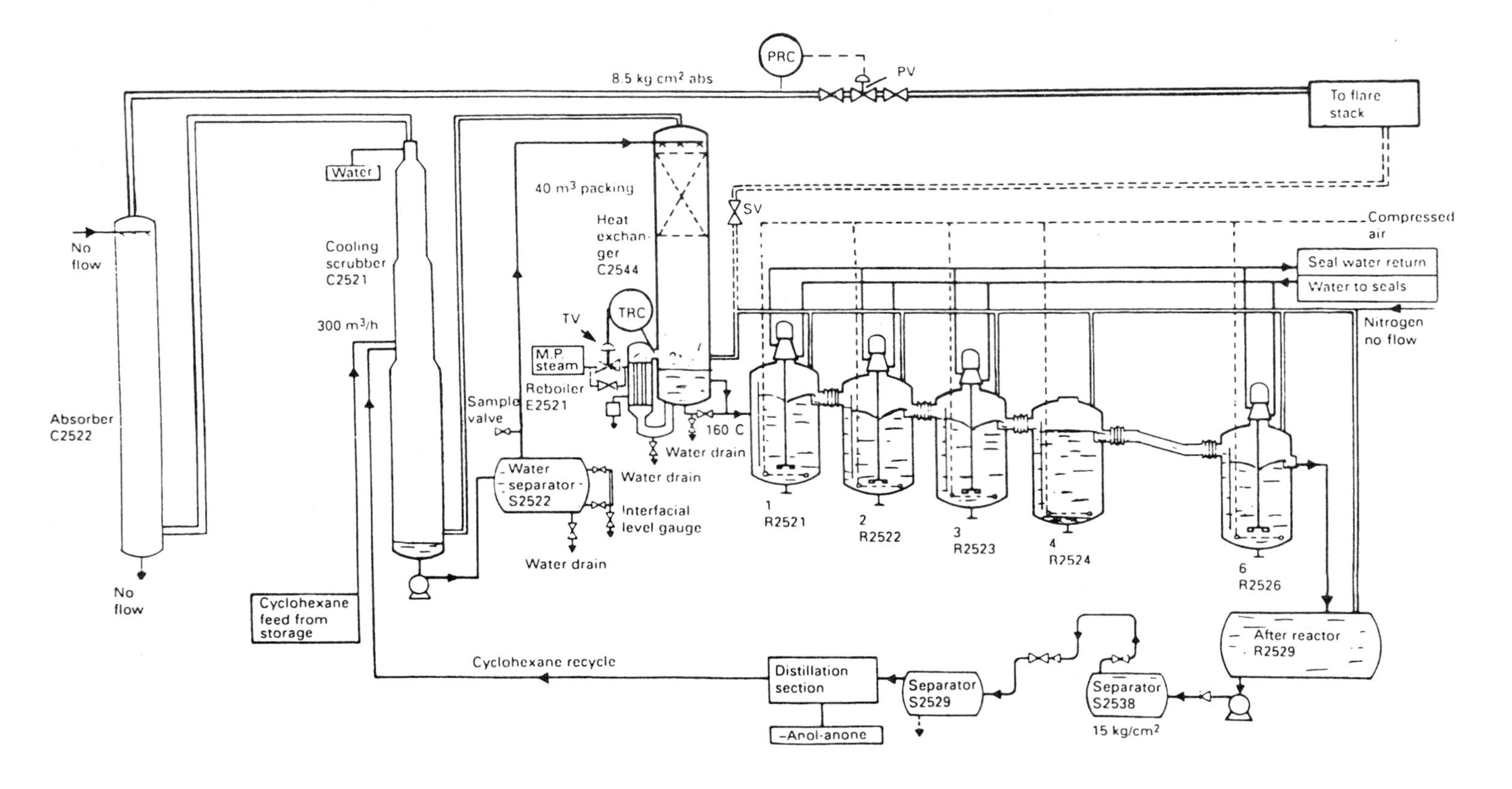

Figure 4.3 Cyclohexane oxidation plant as modified, in hot circulation, 1 June 1974

Within minutes the fire brigade and ambulance service reached the site and within an hour the first factory inspector arrived. Throughout that evening and the following Sunday TV and radio broadcasts were interrupted to bring updates to the nation. The Prime Minister Harold Wilson visited the site on the Sunday. A statement was made to the Houses of Parliament on the Monday indicating that there would be a public inquiry and formal investigation into the accident.

The first concern was to rescue anyone trapped within the wreckage and to reduce the risk of further explosions to a minimum. To this end HM Factory Inspectorate 'took over' the site, the local fire brigade were put on standby, and the mines rescue team attempted to enter the debris to rescue injured workers and recover the bodies of the dead.

4.4.1 Conditions during start-up on 1 June 1974[2,4]

Normal start-up of the plant as built was described earlier [4.3.1]. The subsequent changes involving the removal of the stirrer from No. 4 reactor and the replacement of No. 5 reactor by the by-pass assembly were described in Sections 4.3.2 and 4.3.3. The by-pass assembly was inherently weak because the compressive forces caused by the bellows produced a strong bending moment on both mitred joints. On top of this each bellows was now only anchored at one end, thus allowing the whole assembly to squirm, so that the bending moment could vary continuously.

The plant was restarted at the beginning of April, and ran with the by-pass in position till the end of May. It was then shut down to repair a leak on the sight glass of S2522, from which the cyclohexane was first displaced by water into C2544 and the reactors. Most of the water left in C2544 was subsequently drained, but any water entering No. 1 reactor could not be drained and would have stayed there until the plant was restarted. After repairing the leak the plant was pressurised to 4 bar g. with nitrogen.

At 05.00 on Saturday 1 June, circulation of cyclohexane was restarted and steam was reapplied to E2521. The plant pressure rose abnormally rapidly. By 06.00 it had reached 8.5 bar g. when the temperature in No. 1 reactor had only reached 110°C and the other reactors were successively cooler. Another leak was then found. Circulation was stopped and the steam was shut off. The leak was cured after the shift change at 07.00. The pressure had then fallen to about 4.5 bar g. without venting.

Circulation was restarted at 09.00 and steam was reapplied to E2521, with the temperature controller set to maintain the outlet temperature of E2521 at 160°C.

By 11.30 to 12.00, the pressure had reached 8.8 to 9 bar (0.3 to 0.4 above that required but still considerably[5] below its safety valve setting of 1 kg/cm^2). The block valve on the off-gas line (which had been closed earlier to conserve nitrogen) was opened slightly until the pressure fell to 8.6 bar, and was then closed again. The plant warm-up continued without venting to the end of the shift at 15.00.

According to David Hewitt,[6] the supervisor of the morning shift, the pressure and temperature had steadied out by about 13.30 and the conditions by 15.00 at the shift change when he handed over to Richard Simpson (who was killed on the following shift) were:

- cyclohexane circulation rate: 125 m^3/hour;
- system pressure: 7.8 bar with off-gas block valve shut;
- reactor temperatures: about 155°C with steam by-pass valve shut.

Both the court's technical advisers' (Cremer and Warner) staff and I were informed independently (in my case in writing) by Nypro staff that the plant pressure had been lowered by one bar after the installation of the by-pass assembly. This was, however, later denied on behalf of DSM. But whether the pressure was 7.8 bar g. or abs. it makes little difference to our explanations of the cause of the failure (although it makes more difference to the explanation accepted by the court).

It is interesting to note a curious anomaly in the court's reporting of plant pressures and in the questioning of witnesses by court and counsel on this point. Pressures given in the report and by witnesses during questioning were simply given in kg/cm^2, without 'g.' for gauge or 'abs.' for absolute. On the other hand, pressures quoted in Appendix 1 of the report for the various laboratory tests were quite specifically given as kg/cm^2 gauge. This might suggest that the court knew for certain in one case and not in the other.

At 16.52 the by-pass failed with the consequences outlined in Section 4.4.

4.4.2 Early HMFI investigations

The first HM Factories Inspector arrived before 6.30 on Saturday 1 June 1974 and specialist inspectors were on the scene by 11.30 a.m. on Sunday 2 June. They were mostly chemists and chemical engineers because it was then thought that the escape of flammables and the subsequent explosion probably had a chemical cause. When no chemical cause came to light mechanical and civil engineers from HMFI were called in.

The initial investigation showed that there had been a very large explosion. This had crushed and broken several pressure vessels and many pipes but there was no indication that there had been an explosion inside the plant and pipework. The investigation focused on two gaping holes facing each other in the ruptured 28-inch stainless steel bellows attached to Nos 4 and 6 reactors and the remains of the by-pass assembly which lay on the ground below them, 'jack-knifed' at one of the mitred joints. The bridge-pipe had been subjected to a large bending moment at the mitred joints caused by the thrust forces of the pressurised bellows whose flexibility allowed it to 'squirm'.

By 23 June HMFI had completed their investigations and produced their report. This contained stress calculations which indicated that the bridge probably failed as a result of 'progressive deterioration' such as one gets when a paper clip is repeatedly bent until it breaks. Glynne Evans, who was an HMFI pipe and pressure vessel expert and closely involved in this work, still stands by this conclusion.

4.4.3 The start of the public inquiry

On 27 June 1974 Michael Foot, then Secretary of State for Employment, announced that there would be a public inquiry into the disaster under the chairmanship of Roger Parker, QC. He would be assisted by Dr A. J. Pope, formerly Professor of Mechanical Engineering, Dr J. E. Davidson, Professor of Chemical Engineering at Cambridge University, and Mr W. J. Simpson, head of the newly appointed Health and Safety Commission which had just been formed on the recommendations of the Robens Committee. The inquiry was to establish:

> the causes and circumstances of the disaster and to point out any lessons which they considered were immediately to be learned therefrom.

In consequence the inquiry did not consider more general matters such as the proper policy with regard to safety, site layout, and construction of plants such as that at Flixborough. On the same day Michael Foot stated that a:

> Major Hazards Committee of experts would be formed to examine the hazards of large scale plants and ways in which people working in them and living nearby could be safeguarded.

This committee started work considerably later and produced three reports over as many years.

The court first met on 2 July when it was agreed that HMFI would hand over its investigational work, preliminary report and conclusions to the court and that HMFI would be legally represented at the inquiry because in Parker's words:

> if in the course of evidence one found that various people were banding together to say that the Factory Inspectorate was in one way at fault, they ought to have someone there to answer the charges that are made.

It was also agreed that that all evidence to be given in court should be funnelled through the Treasury Solicitor. Dr Davidson was concerned to know what investigations Nypro would be making and wanted to be involved in them from the beginning, but this was vetoed by Parker. Parker suggested that the court needed its own technical advisers and Davidson suggested the firm of Cremer and Warner who were appointed soon after. Within days Manderstam and Partners Ltd were appointed as technical advisers to Nypro and its owners and John Cox (originally Peter Grimshaw) and I were nominated for this role.

A preliminary public hearing was held on 24 July at which the Chairman explained the terms of reference of the court, outlined the procedures to be followed and heard applications for legal representation. Representation was granted to several parties on the grounds that they or their members could be 'prejudicially affected' (by statements made) during the course of the inquiry. The parties legally represented at the inquiry were DSM, the NCB, Nypro (UK) Ltd, SimChem the contractor which built the plant, HMFI, the court itself, representatives of those killed by the explosion and various trade unions and local authorities. The hearings began in Church House, Westminster, and a few were held near the site for the benefit of local witnesses before finishing in the *Rembrandt* Hotel, Kensington.

4.4.4 The court's views on the immediate cause of the disaster

The report concluded that the disaster was caused by the failure of the inadequate by-pass installed in March 1974. It also stated (without explanation):

> §225 (vi) on 1 June the assembly was subjected to conditions of temperature and pressure more severe than any which had previously prevailed.

> §88(a) The unusually fast rise in pressure during the early hours of 1st June remained unexplained.

and

> §88(f) During the course of our investigations the possibility of a sudden rise in pressure during the final shift due to some internal incident was considered. We were able to exclude all of the possible internal incidents suggested.

4.4.5 Other possible immediate causes

Early in the inquiry DSM's counsel A. P. Leggat, QC, produced a list of possible causes[7] which were quoted in the first edition of this book. For the sake of completeness I repeat them here. Leggat was employed to protect DSM's interests at the inquiry so it is not so surprising that progressive deterioration of the by-pass assembly is not included in the list, which is as follows:

1. failure of the 20-inch pipe resulting from a hypothetical but relatively small pressure rise;
2. failure of a different (8-inch) pipe (followed by an initial fire or explosion) before the failure of the 20-inch pipe;
3. prior failure of some other part of the system;
4. explosion in the air-line to the reactors.

Leggat also suggested the following possible causes for the pressure rise (1):

(a) entry of high-pressure nitrogen into the system due to some instrument malfunction;

(b) entry of a slug of water into the system;

(c) temperature rise in the system due to the use of excessive steam on the reboiler of C2544;

(d) a leaking tube in the reboiler of C2544 causing both actions (b) and (c);

(e) explosion of peroxides formed in the process;

(f) air in the reactor system which might cause a local explosion.

Of these I was satisfied that all except (1b) (and progressive deterioration of the by-pass assembly) were rendered unlikely by the investigations carried out after the disaster, and I have heard no suggestions to the contrary.

4.5 The pressure rises and 'latent superheat'

The pressure rises (referred to in the court's report) stem from the fact (well known in physical chemistry) that a physical mixture of water and a hydrocarbon liquid such as cyclohexane or petrol boils at a considerably lower temperature than the boiling point of either liquid alone. Thus, far from being unexpected, they would in the circumstances have required minor miracles to prevent them from occurring.

My paper published in September 1975[5] showed that 100 litres or so of water left in the packing of C2644 after the shut-down at the end of May 1974 would have been vaporised together with cyclohexane during hot-circulation, leading to the fast pressure rise observed at about 06.00 which the report noted but could not explain. This was surprising, for there was a professor of chemical engineering in the court who should have understood its cause. There was also nothing unusual in it to Nypro's staff who told me they had experienced such pressure rises before.

The same phenomenon may also have caused another and more serious pressure rise at the time of the disaster. Without repeating the calculations given in my paper, I try to explain it here in simple terms for the benefit of mechanical engineers and others who find it difficult to understand.[5,8] I use the term *latent superheat*[9] for the energy stored up between a large layer of hot and nearly dry cyclohexane (or similar hydrocarbon) and a small layer of water below it, as in the unstirred No. 4 reactor during hot circulation at start-up. This energy only exists when the total pressure in the system is less than the sum of the vapour pressures of cyclohexane and water at the temperature of the cyclohexane, which is assumed to be higher than that of the water. Had the pressure in the plant exceeded 10 bar g. and the temperature been 155°C, no latent heat could have been stored up and nothing would have happened. The lower the pressure in the plant the more latent superheat could be stored up. I really think that our HSE should publish a Technical Data Note on the subject and set up a simple demonstration unit for public display.

The reason for this energy is that the boiling point of a mixture of cyclohexane and water is substantially lower than the boiling point of dry cyclohexane at the system pressure. Thus when the two liquids at higher temperatures than the boiling point of the mixture come together rapid mixing and spontaneous boiling always occurs accompanied by cooling of the remaining liquid cyclohexane to the boiling point of the mixture. It is not so much the pressure rise itself that is dangerous as the rapidity with which it occurs and its effects on the liquid. The pressure rise would not be sufficient to open the pressure relief valve, particularly when this is some distance from the reactors. The more important and dangerous effects would be similar to those of a depth charge under the liquid, when slugs of liquid will be flung upward hitting the top and sides of the reactor with considerable force. A useful analogy which may help the reader to appreciate the phenomenon on a small scale is found in many chemistry laboratories when a clean liquid (such as dilute sodium silicate) is heated in a clean test tube. The temperature of the liquid may then rise to several degrees above its boiling point before suddenly boiling and ejecting itself like a champagne cork out of the tube. A tell-tale circle on the ceiling is often left.

Later, on 1 June when hot-circulation was continued, the water in the packing of C2644, plus any present in the first three (stirred) reactors, would have been carried in suspension in cyclohexane into the fourth reactor whose stirrer had been removed, and would have settled as a pool at the bottom. Continued circulation of hot cyclohexane at 155°C through the reactors would have heated the interface between the pool of water and the cyclohexane above it until it reached a temperature of about 145°C. At this point boiling would have started at the interface which was unstable above that temperature, followed by rapid mixing of water and cyclohexane. A more detailed scenario of how slugs of water were probably ejected through the by-pass assembly into the stirred No. 6 reactor and almost exploded there was given in Section 4.3.3.

Four conditions (all present in the plant at the time) were necessary for this to happen:

- an unstirred reactor in which water can settle;
- a system pressure lower than the sums of the vapour pressures of water and cyclohexane at the mean temperature in the unstirred reactor;
- enough water in the system to form a pool in the unstirred reactor;
- continued circulation of hot cyclohexane prior to oxidation for long enough to allow the water–cyclohexane interface to reach the temperature at which it becomes unstable.

The hazard is discussed more generally in Section 6.6 in a later paper[9] and more generally in an oil company's safety booklet.[10] The main concern for anyone who may be faced with a similar hazard is to ensure that the first and second of these conditions cannot arise simultaneously. A quantitative analysis of the consequences when all four conditions are met requires knowledge of several factors which include:

- the minimum amount of water present;
- the latent and specific heats of water and cyclohexane (or other hydrocarbon);
- the mutual solubilities of water and the hydrocarbon present;
- the volumes of the system filled with gas and liquid.

The amount of superheat suddenly released for an initial plant pressure of 8.5 bar g. and an initial temperature of 158 °C as calculated in my original paper[5] was 270 000 kilo-Joules. This may not have ruptured a well-designed and constructed plant, but was probably quite enough to destroy the faulty by-pass assembly. I was in fact surprised that Warner, who had earlier alerted me both to the possible roles of water and the unstirred reactor, had not anticipated me in unravelling this phenomenon.[11] I know that the possible role of water had been discussed by the management committee of the inquiry and I have read the Cremer and Warner report[11] which came quite near the mark. Had the court challenged DSM's repeated statement that over 6 tonnes of water were needed in the reactors to give a pressure rise of just one bar, the inquiry might have taken a very different turn. I know of little or no experimental research which has been done to evaluate the effects of the release of latent superheat in pressure vessels, e.g. in oil and chemical plants. I believe that this would be useful, perhaps as a university research project.

4.6 Glynne Evans' views on the inquiry

In early September the Court requested HM Factory Inspectorate to assume control of the safety of the dismantling of the devastated site. As the Court of Enquiry progressed, HM Factory Inspectorate found themselves with no adverse criticism or prejudicial effects, and their role to some extent changed whereby they aided the Court's technical experts Messrs Cremer and Warner.

When the actual hearings commenced the Court had already received the so-called HM Factory Inspectorate report, but there were many other theories and technical matters being put forward, by some of the parties who could be prejudicially affected during the course of the enquiry. The technical experts were requested to agree on as many technical matters as possible before the hearings commenced. To this end a so-called 'management committee' was formed comprising technical experts, their advisers and the parties represented. Initially technical experts were accompanied by their legal representatives but as the enquiry progressed the technical experts called in by representatives who could be prejudicially affected were allowed to meet together. It was Cremer and Warner, the Court's technical experts' responsibility to report the management committee decisions back to the official enquiry. This management committee was often the ground for bitter debate, but many of the technical problems were clarified and the Court was saved from harsh wrangling.

The Court met in public for seventy days and completed its published findings on 11 April 1975. For most of those involved with the detailed investigation, management committee meetings, and court procedures, the final report was proclaimed as both just and balanced. Inevitably some of the so-called experts had their egos dented, and they attempted to 'stir up' debate so as to rubbish the report.

From the very beginning it was realised that the plant was protected by a pressure relieving device set at 11 kg/cm^2; it was therefore considered that any pressure up to that safety valve setting could not be regarded as over pressure. Indeed there has never been any postulation during the enquiry or since that gave the impression that the pressure could have risen beyond safety valve settings. From the onset of the mechanical engineering investigations theoretical calculations showed that the mitred pipe would experience large thrusts (38 tonnes) and this would tend to buckle the pipe at the mitred joint. (See Flixborough Report paragraph 62.) The report criticises the plant operator, Nypro, for not having anyone in their employment who appreciated that fact, but in reality the Court's advisers, and many of the so-called experts who were usually chemical engineers or chemists were incapable of the expertise of recognising what is in essence a simple mechanical engineering problem.

The report, particularly at Appendix 1, Paragraph 9, acknowledges that the tests carried out on a reconstructed pipe showed progressive deterioration. Evidence submitted under oath by Glynne Evans shows that there was considerable progressive deterioration in stresses near to the mitred joint. Although the Court of Enquiry never took evidence they somehow dismissed progressive deterioration, not realising that the stresses involved were compressive stresses which by their nature lead to the buckling typical of the jack-knifed mitred pipe.

After the report was published the HM Factory Inspectorate published technical articles on creep cavitation, nitrate stress corrosion cracking of mild steel and zinc embrittlement of stainless steel thereby putting any discrepancies in the Flixborough Report to rights. The major hazard committee went on to publish three outstanding reports which considered any residual discrepancies which may have lingered after the Flixborough Disaster Report. For example in their report, number one, they readily acknowledge at page 64 'that the integrity of pressurised systems is of the most highest importance' and at paragraph 70 recommend that 'existing installations should be brought up to acceptable standards'. In its third and final report page 77 it acknowledged evidence that errors in design and construction may be a more significant cause of serious incidents than has so far been anticipated. The UK now has regulations which have and will do much to update existing and new process plant so as to diminish the possibility of another Flixborough Disaster.

4.7 Ralph King's comments on the inquiry and its report

I generally agree with Glynne's views. I also agree with the court's rejection of the '8-inch pipe hypothesis' as extremely unlikely. However, I am surprised that it spent so much of its time and 15 out of 38 pages of its report in considering it whilst ignoring more likely causes and failing to draw the resulting lessons. Of these, the failure of the plant designers to appreciate the thrust forces on the reactors caused by the 28-inch bellows (which Glynne first recognised) was the most vital. This sealed the fate of the plant from day one.

Another but less important lesson which the inquiry missed, and which I could only point out after the inquiry, was that all the conditions in the plant were ripe for the sudden impulsive boiling of a mixture of water and cyclohexane at the time of the disaster. This incidentally would not have caused the pressure relief valve to lift as its effects though violent were transient. I have no proof or evidence that this actually happened but in my view the final disastrous failure of the plant most probably resulted from a combination of both these causes. I have to admit that in some respects the inquiry was thorough whilst in others it seemed to me to be a little casual. Some of the parties to it and their legal representatives seemed so obsessed with protecting themselves from blame or criticism that they obscured the main object of revealing the cause(s) of the disaster.

4.8 After the report

When the report came out in April 1975 it ran into a little criticism of certain points in the technical press (particularly *The Engineer* and *Process Engineering*), and from The Council for Science and Society. Typical of these was an article by Brigadier Allen in *The Acceptability of Risks*.[12] One thrust of this criticism lay in the narrow view which the court took of its remit. An astonishing proportion of its time was spent in hearing and refuting the heavily funded arguments (ultimately paid for by the taxpayer) in favour of the eight-inch pipe theory.

An article in *The Engineer*[4] which reviewed the inquiry concluded thus:

> Perhaps a good way of starting along the new path might be to declare the conclusion as reached in the Flixborough inquiry 'non-proven' and either leave it at that or re-open the inquiry . . . The probable outcome of a re-opened inquiry is that it would come to an altogether different conclusion.

This article has itself been criticised by my co-author Glynne Evans in that it ignores the simplest of all explanations for the direct cause of the disaster, progressive deterioration of a highly stressed weld under compression. Whilst I am not a mechanical engineer, I now suspect that he may well be right.

4.9 Organisational misconceptions

Here, following Turner's ideas as developed in 3.3, I consider the organisational misconceptions (OMs) which made the disaster possible, and the incubation period leading up to it during which it might have been averted. To avoid misunderstandings I must emphasise that the misconceptions quoted are intended to convey in words the erroneous ideas in human minds which alone or in combination led ultimately to catastrophe. *They are not in any sense to be taken as recommendations, and are the exact opposite. Neither I nor the publisher can take any responsibility for them being misread in this way.*

The first organisational misconception (OM1) occurred during the design of the plant. Crudely stated it might read thus:

> Unrestrained pressurised bellows pieces impose no significant thrust forces on the pipes or equipment to which they are connected.

A second organisational misconception (OM2) occurred in November 1973 when it was decided to operate the plant with the stirrers stopped. It might be stated simply thus:

> There is no risk of a sudden eruption if you heat a hydrocarbon liquid and pass it through an unstirred vessel which contains some water.

The hazard arising from this misconception was 'frozen into the plant' in January 1974 when the stirrer of No. 4 reactor was removed.

The failure of No. 5 reactor in March 1974 followed from OM1. This was repeated in the 'design' (or lack of design) of the by-pass assembly. The disaster was caused by OM1, probably acting in combination with OM2.

A chronological chart of the incubation period showing the OMs and significant accidents within this period is shown in Fig. 4.4. The disaster could have been averted at any time during this period had the OMs been appreciated and the appropriate (but expensive!) action taken to eliminate the faults caused by them from the plant.

OM1 seems to have originated in the design organisation. It may be that the reactor specification was made with the intention of connecting the reactors with large pipe U bends, before a decision was taken to use bellows instead in order to give a more compact layout. Had the cost of constructing reactors which would adequately withstand the thrust from the bellows been considered, these might never have been used. The error seems to have passed unnoticed and uncorrected by the engineers of several companies involved in the plant design and construction. Following this high-level mistake in the design organisation, No. 5 reactor failed in service in April 1974. Another mistake — the inadequately supported by-pass, also resulting from OM1 — then occurred at a lower organisational level (i.e. in Nypro's maintenance department). This happened under stressful circumstances and led to the disaster. This illustrates the point made in 3.3.4 that once an OM exists at a high level in the organisation (e.g. in the parent company), it is

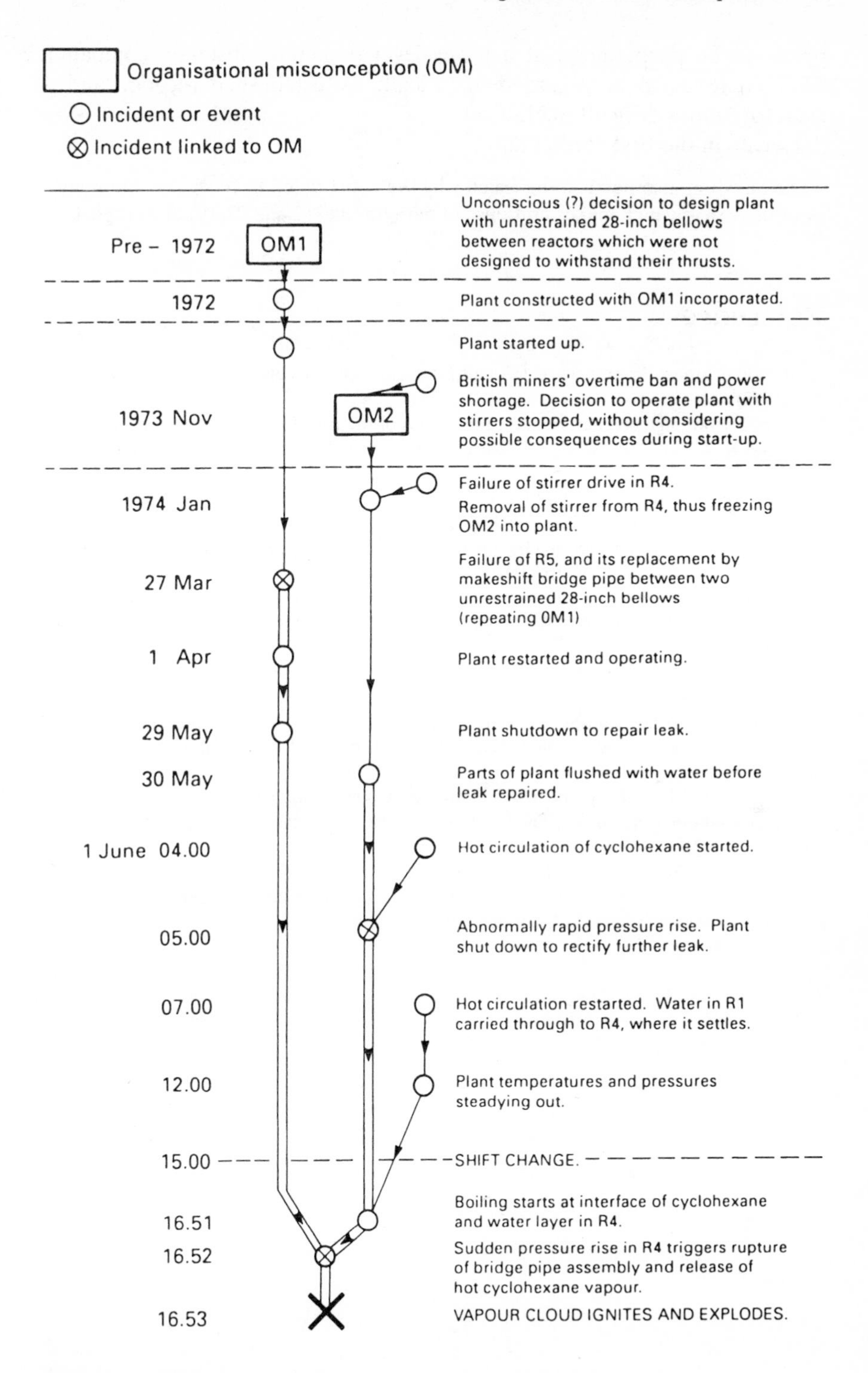

Figure 4.4 Chronological chart of organisational misconceptions and incidents leading to disaster (time scale is non-uniform)

prone to be perpetuated at a lower level (e.g. the subsidiary company), where it tends to be regarded uncritically as 'established wisdom' and is therefore more difficult to eradicate.

I wrote in the first edition that:

> The question also needs to be asked why the court failed to recognise and point out this error, which makes nonsense of paragraphs 212 and 225(v) of its report.

References

1. Leading article, 'Electricity chiefs call for further power cuts', *The Times*, 7 December (1973)
2. 'The Flixborough disaster, Report of the Court of Inquiry', HMSO, London (1975)
3. HSE Technical Data Note 53/2, 'Nitrate stress corrosion of mild steel', HMSO, London (1976)
4. '*The Engineer* Report — The lessons to be learned from the Flixborough enquiry', *The Engineer*, Morgan Grampian, London, 11 December (1975)
5. King, R. W., 'Flixborough — The role of water re-examined', *Process Engineering*, Morgan Grampian, London, pp. 69 to 73, September (1975)
6. Hewitt, D. W., during examination by Jupp, K. G. (QC), *Daily transcripts, Flixborough Court of Inquiry*, day 17, p. 47, by courtesy of W. B. Gurney & Sons, Official Shorthand Writers (1974)
7. Leggat, A. P. (QC), *Daily transcripts, Flixborough Court of Inquiry*, day 8, pp. 74 and 75, *ibid*.
8. King, R. W., 'A mechanism for a transient pressure rise', paper at Nottingham symposium, *The Technical Lessons of Flixborough*, The Institution of Chemical Engineers, Rugby (1975)
9. King, R. W., 'Latent superheat — a hazard of two-phase liquid systems', paper in *I. Chem. E. Symposium Series No. 49*, The Institution of Chemical Engineers, Rugby (1977)
10. AMOCO, Booklet No. 1, *Hazard of water*, 5th edn, AMOCO, Chicago (1971)
11. Cremer and Warner Report No. 2 to the Flixborough court of inquiry, 'General' (1974)
12. The Council for Science and Society (report of a working party), *Superstar Technologies*, Barry Rose (Publishers) Ltd, London (1976)

5 Four other major accidents

This chapter examines four other major accidents in recent years in the process industries:

1. the explosion in Shell's Pernis oil refinery in Holland on 20 January 1968;[1]
2. the unintended formation and release of a few kg of the super-toxin 2,3,7,8-tetrachlorodibenzoparadioxin (TCDD) at Icmesa Chemical Company's works at Seveso on 10 July 1976, causing widespread illness and lasting environmental damage;[2]
3. the release of some 30–40 t of toxic methyl isocyanate vapour at Union Carbide India Ltd's pesticide factory at Bhopal during the night of 2/3 December 1984, causing over 2000 deaths and affecting over 50 000 people.[3]
4. the Piper Alpha disaster, 6 July 1988.[4]

A common theme runs through all four accidents. All resulted from a combination of technically simple hazards which had been created by conscious decisions of technically trained staff, usually those responsible for production operations. In few cases would they have lacked the knowledge to appreciate the consequences of their decisions had they thought them through. One must then conclude that generally they failed to project their decisions through to their likely consequences, and face up to them. Why was this so?

- Were the persons too busy and overstretched in their jobs?
- Did they lack the ability to think their decisions through?
- Were they inhibited by mental blockages caused by preoccupation with other matters, e.g. production targets and the odium of failing to meet them?

Such failures were collectively termed 'organisational misconceptions' in 3.3. Between the time of their occurrence and the disaster, several hours, days, weeks or months, referred to in 3.3.5 as the 'incubation period',[5] had elapsed. During this time the consequences of the decision might have been appreciated and averted had someone had the guts and authority to do so.

The general lesson is that every decision should be projected forward. Every decision maker should be conscious of the fact that the present is possibly some stage in the incubation period of a disaster, and that the fatal decision(s) might already have been taken. A special lesson for managements is that they must ensure that their production and engineering staff have the ability, knowledge, time and support to think their decisions through, and that they are not subject to (usually management created) pressures which inhibit them from doing so, or from acting on their conclusions. In plants where the potential for disasters is high, there is a need for competent staff with no production responsibilities to act as 'technical longstops'. Their task should be to monitor the consequences of technical decisions taken by line managers and veto them where necessary, even when this leads to lost production.

5.1 The explosion at Shell's Pernis refinery in 1968[1]

A large aerial explosion[6] occurred above Shell's Pernis oil refinery near Rotterdam at 04.23 on Saturday 20 January 1968. Human fatalities and injuries were relatively light (two dead, nine hospital cases and 76 persons slightly injured), but the damage to the world's largest refinery was severe and estimated at US $28 million. Much neighbouring property including buildings in nearby Rotterdam was also damaged.

5.1.1 The inquiry and its findings

An inquiry was held by a group consisting of four senior police officers, seven officers of the factory inspectorate and the head of Shell's safety inspection division. Its sole task was to establish the technical cause of the explosion. Its report[1] was published within two months of the disaster. Its summary stated:

> The inquiry showed that the cause is to be found in a storage tank for oil slops which had steam heating near the bottom. Owing to the cold weather during the previous two weeks, the steam heating was in operation. As a result of difficulties in breaking down an emulsion during the preceding days in a desalting installation for crude oil there had been a large amount of waste oil, partly in the form of a water-in-oil emulsion with a high water content, which was among the liquids present in the tank concerned. There was a vigorous boiling effect in the tank, as a result of which a large quantity of hydrocarbons were expelled into the air within a short period of time. The presence of a very light, variable wind caused a large cloud to form, consisting of an explosive mixture of air and mist from these

hydrocarbons. This explosive cloud was ignited by a source which cannot be established with certainty and exploded violently.

An Annex to the report concluded that 50 to 100 t of hydrocarbons must have been involved in the explosion.

The term 'slops' in an oil refinery refers to oil, often emulsified, which has been recovered, mainly in oil–water separators, from refinery effluents. The slops tank, no. 402, was 9 m high with a diameter of 15.2 m and contained four steam coils in its base with a heating surface area of 36 m^2. Steam had been connected to the coils on 9 January.

The slops at that time contained unusually large quantities of a crude oil–water emulsion from a crude-oil desalting unit. Most crude oils contain inorganic chlorides which have to be removed before distillation to prevent severe corrosion in the distillation units [13.9.4]. This was done in a continuous unit by mixing clean water into the crude oil to dissolve the salts present, thus forming a water-in-oil emulsion. The emulsion was separated into oil and water layers by high voltage, the oil going to the crude distillation units and the water layer to the effluent oil separators.

Because of the stability of the emulsion formed with the Bachaquero crude oil then in use, the desalting unit could not separate it properly, and discharged a considerable amount of water-in-oil emulsion with a high water content to three settling tanks. As these in turn failed to break the emulsion, it passed into tank 402. Here the desperate remedy of heating the tank contents was being tried to an attempt to break the emulsion. The hazard of doing this has long been recognised by oil companies.[7]

The only alternatives to this were to process untreated crude oil in the distillation units, with risk of severe corrosion and other hazards, or to shut down one or more crude distillation units and reduce refinery throughput.

Tank 402 had been full for two days and contained about 1500 m^3 of slops. The steam to the coils was controlled by a valve which was found to be partly open after the explosion. The slops had a specific gravity of 0.82, and were estimated to have an initial boiling point of 60°C and to contain about 30% of light oil fractions.

The report concluded:

> The explosion was in all probability the result of the ignition of a large cloud consisting of air and a large quantity of hydrocarbon mist. The formation of this cloud was due to the weather conditions, namely a very light variable wind, after there had been an eruption of hydrocarbons from tank 402 in a very short space of time.
>
> 1. The contents of tank 402 consisted mainly of a mixture of hydrocarbons with a relatively low initial boiling point. In addition there was a relatively large quantity of a stable water-in-oil emulsion in the tank.
> 2. As the tank was heated by means of a steam coil, the emulsion at the bottom of the tank was able to reach a relatively high temperature, while the oil layer above it remained considerably cooler.

3. At a temperature of approximately 100°C the emulsion layer split into a water layer and a relatively heavy oil which therefore also had a temperature of approximately 100°C.
4. The moment the temperature of the light oil layer reached the initial boiling point, vapour developed as a result of vapour formation at the unstable interface between the hot and the cooler oil layer, and this caused rapid mixing of the hot oil layer with the rest of the contents of the tank. The vapour thus formed led to the overflowing of the tank and the subsequent eruption of hydrocarbon mist.

Little can be said with certainty about the way the explosive cloud ignited. The front of the explosive mist-cloud had moved approximately 100 m from its source when the explosion occurred.

5.1.2 Comment

It is clear that a considerable temperature difference had developed between the lighter and more volatile oil in the top of the tank and the heavier water-containing layer in the bottom while the tank contents were being heated. The contents of the tank as a whole were then in a thermodynamically unstable condition. The sensible heat of the hot layer in the lower part of the tank was sufficient to vaporise a considerable proportion of lighter and cooler hydrocarbon above it when the contents mixed, causing a physical explosion [6.2.3] inside the tank. When its contents then

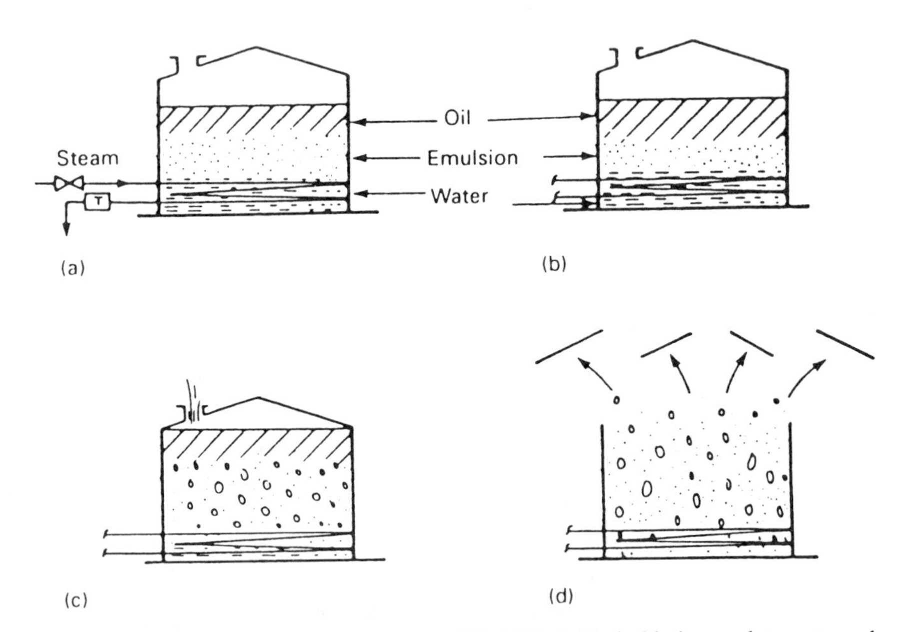

Fig 5.1 Sequence of events in Pernis disaster, 1968: (a) Tank filled with slops and steam turned on to heating coil; (b) emulsion partly separates as tank contents warm up; (c) water layer reaches its boiling point with cooler oil layer above it; (d) water–oil mixture boils explosively and erupts.

overflowed, the hydrostatic pressure inside it would have been reduced, thereby accelerating the boiling process. The sequence of events is illustrated diagrammatically in Figs 5.1(a) to (d).

The phenomenon, discussed elsewhere in general terms,[8] was similar to the 'missing link' [4.5] in the explanation of the Flixborough disaster.

The accident followed from the decision to heat the lower part of an unstirred tank containing an oil with volatile components to a temperature above its initial boiling point. The hazard was accentuated by the presence of a considerable amount of partly emulsified water in the tank.

The die was set for disaster when the steam supply was connected to tank 402 on 9 January, thus giving an incubation period or lead time of 10 days. Had the mistake been recognised then and the steam supply to the tank disconnected, the disaster would have been averted. The resulting loss of refinery production would have cost Shell far less than the disaster.

5.2 The 'Dioxin' release at Seveso on 10 July 1976[2,9]

This accident has influenced European thinking on the control of chemical hazards and led to the 'Seveso' Directive (82/501/EEC) which is implemented in the UK by the CIMAH Regulations [2.5.5]. This account is based on *The Superpoison*, by Margerison, Wallace and Hallenstein,[2] on the report of the Italian Parliamentary Commission of Inquiry,[9] and on a more recent paper by Cardillo and Girelli.[10]

5.2.1 The accident

At 12.37 hours on Saturday 9 July 1976 a bursting disc fitted in the vent line of the reactor producing 2,3,5 trichlorophenol (TCP) at the chemical factory of Icmesa ruptured because of internal overpressure. Icmesa was an Italian company which, since 1969, was owned by Givaudan, which itself was part of the (Swiss) Hoffmann La Roche group. The factory was at Meda, on the outskirts of Seveso, a town of about 17 000 inhabitants about 20 km from Milan.

The bursting disc discharged a cloud of about two tonnes of hot chemicals directly into the open air, on a hot summer's day with a light northerly breeze. Heavy rain fell soon after the discharge, bringing the contents of the cloud down to earth over a largely urban area.

Unfortunately the cloud contained a small quantity, later estimated at 2 kg, of one of the most toxic compounds known, 2,3,7,8 tetrachlorodibenzoparadioxin, commonly known as TCDD or simply dioxin.

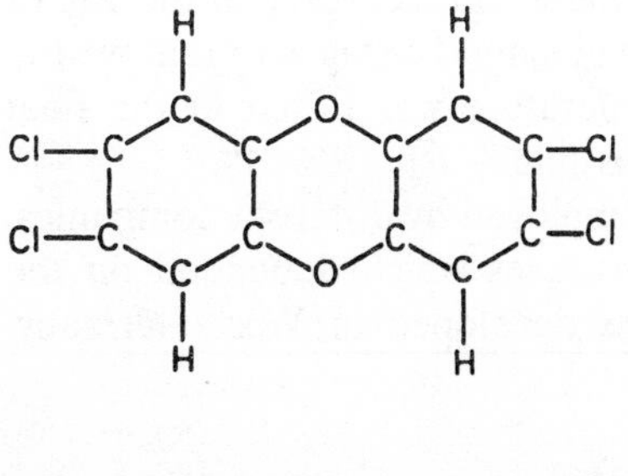

TCDD or 'dioxin'

An area of about ten square miles (which included some 40 factories) was contaminated, and by August at least 730 people had been evacuated.

The fatal dose of TCDD for the average man is less than 0.1 mg. It can enter the human body by ingestion, inhalation and through the skin, leading in mild cases to chloracne. Its symptoms include cysts and pustules, grey or brown staining of the skin, and purple urine. TCDD causes liver and kidney damage, and birth defects in children born to mothers exposed during pregnancy. It is a stable slow-acting poison, insoluble in water, and very difficult to destroy.

Over 700 local inhabitants were affected by the poison and many animals died. A considerable area of agricultural land was rendered unusable for many years.

5.2.2 TCP production and dioxin formation

TCP is a chemical intermediate in the manufacture of hexachlorophene, an anti-bactericidal agent made by Givaudan, and a hormone type weed-killer known as 2,4,5-T.

TCP is generally made by reacting tetrachlorobenzene (TCB) with an excess of caustic soda in the presence of a solvent, in the temperature range 160–200°C, preferably 170–180°C. This reaction, which is strongly exothermic, produces the sodium salt of TCP and sodium chloride and water as by-products.

$$C_6H_2Cl_4 + 2\,NaOH \longrightarrow C_6H_2Cl_3ONa + NaCl + H_2O$$

Tetrachlorobenzene (TCB)   Sodium trichlorophenate

The reaction is generally carried out in a stirred vessel, with a jacket or internal coil for heating or cooling and an overhead condenser for condensing solvent vapour and water formed (Fig. 5.2).

Condensed solvent could either be returned to the reactor to cool its contents during the reaction or collected separately and recovered at the end of it. While the reaction needed cooling during its middle stages to remove the large amount of heat generated, the temperature was raised in the final stage to drive it as near to completion as possible.

Several variations on the process were employed by different companies, using different solvents and operating pressures which depended on the boiling point of the solvent. In a process developed in West Germany,

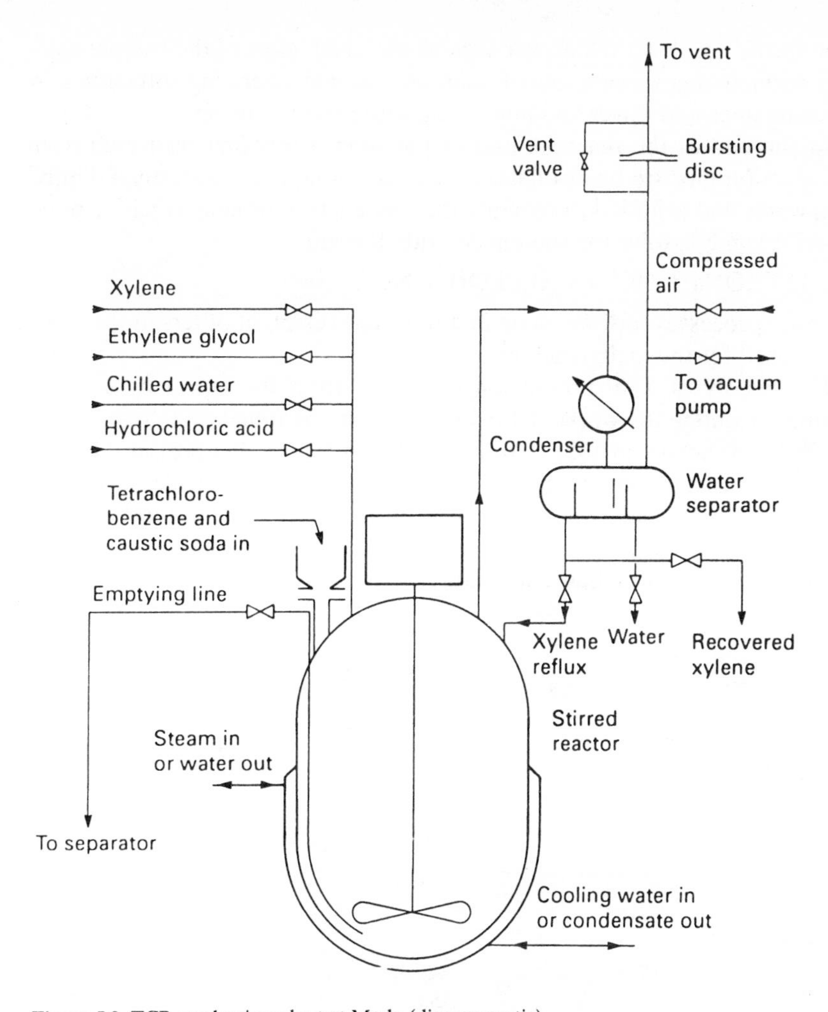

Figure 5.2 TCP production plant at Meda (diagrammatic)

methanol was used as solvent. While this was cheap, the process required an operating pressure of about 20 bar g. When a higher-boiling solvent such as ethylene glycol was used, the reaction could be carried out at atmospheric pressure, but vacuum was needed to recover the solvent by distillation after the reaction.

When a solvent in which water was insoluble was used it was easy to separate the condensed water and remove it during the reaction, so that only solvent was returned to the reaction mixture. This removal of water favoured the reaction and enabled it to proceed nearer to completion. But the solvent had to be one (such as an alcohol) in which sodium hydroxide is soluble. As a compromise, Givaudan used a mixture of two solvents, ethylene glycol (BP 197°C) in which sodium hydroxide is soluble, and orthodichlorobenzene (BP 181°C) or xylene (BP 142°C) in which water is insoluble. A mixture of ethylene glycol and xylene was in use at Meda at the time of the accident. Whichever solvent was used, most of the sodium chloride formed was thrown out of solution, causing operating difficulties by blocking lines and valves and interfering with reactor stirring.

At the end of the reaction, part of the solvent was first recovered from the reaction mixture by distillation. The reactor contents were then diluted with water and acidified to separate the molten TCP product (melting point 60–68°C) and dissolve the sodium chloride formed:

$$C_6H_2Cl_3ONa + HCl \rightarrow C_6H_2Cl_3OH + NaCl$$

In some processes this was done in a different vessel, in others in the same vessel used for the main reaction.

The history of TCP production was beset from the outset by cases of chloracne among the workers, for which dioxin was found to be responsible by Dr K. H. Schulz in 1957.[11] Traces of dioxin (about 25 ppm) in the TCP are formed as a by-product by the condensation of two molecules of sodium trichlorophenate and elimination of sodium chloride. This reaction, like that of TCP production itself, is exothermic and the quantity of dioxin formed increases at higher temperatures.

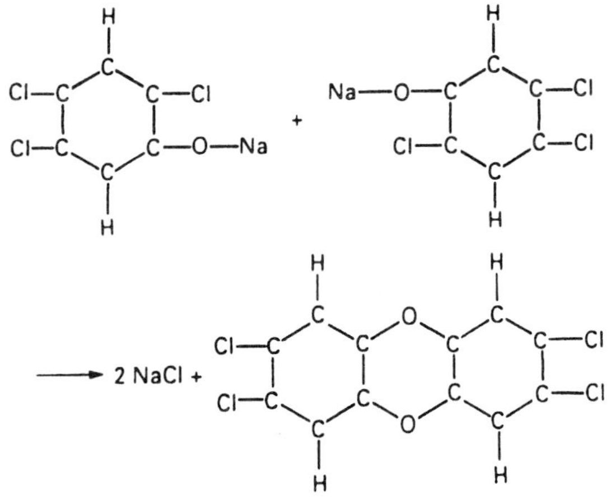

5.2.3 Previous incidents at TCP plants and remedial measures

A list of known incidents involving escapes of dioxin from TCP plants which resulted in illness among workers is given in Table 5.1. Several of these escapes resulted from runaway reactions which led to high temperatures and pressures, the opening of relief valves, and sometimes the failure of joint gaskets. After such escapes it proved extremely difficult to decontaminate the process buildings and workers' clothing.

Dioxin present in TCP also caused trouble in the final products (e.g. disabilities caused by exposure to dioxin in the 2,4,5-T used as a chemical defoliant during the Vietnam war). To overcome this problem Dow and some other chemical companies reduced the dioxin content of their TCP to less than 1 ppm by a special purification process, as well as adopting special hygiene measures in TCP production.

Table 5.1 Incidents causing human exposure to dioxin before 1976[9]

Year	*Location*	*Company*	*Number of cases*
1949	Nitro, W. Virginia, USA	Monsanto	117
1952/1953	Hamburg, W. Germany	Boehringer	37
1953	Ludwigshagen, W. Germany	BASF	55
1953 to 1971	Grenoble, France	Rhone-Poulenc	97
1963	Amsterdam, Holland	Philips-Duphar	30
1964	Midland, Michigan, USA	Dow Chemical	60
1964	Neuratovice, Czechoslovakia	Spolana	72
1968	Bolsover, Derbyshire, UK	Coalite	79
1972	Linz, Austria	Chemie-Werk	50

5.2.4 TCP production at Meda

The TCP plant at Meda used a mixture of ethylene glycol and xylene as solvent, and operated at atmospheric pressure except for distillation of glycol, which was done under vacuum. Trials in which several blockage problems were encountered were carried out in 1970/1971. There was little production for several years, but in 1974/1975 the plant was modified and production was expanded to meet Givaudan's needs for its US and Swiss factories. These needs had become urgent because many manufacturers had abandoned TCP production on account of its health hazards.

The reactor was heated by medium-pressure steam which had a maximum temperature of 190°C. After distilling off the xylene at the end of a batch, as much as possible of the glycol was distilled directly from the reactor under vacuum. The contents of the reactor were then quenched with cold water and acidified in the reactor itself. They were then transferred by compressed air into another vessel where they were separated into a water layer containing dissolved salt and glycol, and a molten TCP layer (MP *ca* 68°C). The unit was protected against overpressure during product transfer by a bursting disc fitted in the vent line from the condenser. This relieved directly to the atmosphere and was designed to rupture at 3.6 bar g.

5.2.5 Operation of the Givaudan plant in 1976

The TCP plant was operated for five days a week on three shifts, with two men per shift. In principle, a new batch of TCP was started every morning at 06.00 when the new shift arrived, and was completed before the night shift left at the same time the following morning. Because of minor operating problems and breakdowns, each batch tended to start progressively later during the week, until on Friday there was not enough time to complete the fifth batch of the week. It was then often decided to start a batch later on Friday and interrupt it part-way through, sometimes after distilling off the solvents, and before adding water and neutralising and washing the crude TCP product. The last shift then had instructions to shut off the steam to the reactor jacket and apply cooling water to it, and leave the water running until work restarted on Monday morning. This caused considerable delays on Monday morning as the contents of the reactor were now solid and had to be melted by applying steam to the jacket for several hours before it was safe to start the stirrer and continue work on the batch.

Several months before the accident, the Friday night shift workers were instructed that if a batch were unfinished at 06.00 on Saturday morning they were to shut off the steam heating but not to turn on the cooling water, so that the reactor cooled only slowly over the weekend. The contents were then sufficiently fluid on Monday morning for work to be started quickly, and time and money to be saved. Thirty-four batches had been interrupted over a weekend since production started in 1975, half of these after the decision was taken not to leave the cooling water running. On all but four of these occasions, the reaction had been completed and quenched with water before the night shift left.

5.2.6 The final batch

On Friday 9 July the fifth run of the week could not be started until late in the afternoon. By the time the night shift started, the reactor had been loaded with its charge, which consisted of:

Ethylene glycol	3235 kg
Xylene	609 kg
Tetrachlorobenzene	2000 kg
Flake caustic soda	1100 kg

The contents were heated to reaction temperature with steam at about 12 bar g. (190°C). The reaction was apparently completed during the night, and all of the xylene and about 500 kg of glycol were distilled off. At 04.45 the steam valve to the heating coil was closed and the stirring was continued for 15 minutes before the stirrer was stopped. The vacuum was turned off and the night shift then went home.

During Saturday morning, an exothermic runaway reaction developed unnoticed in the reactor, accompanied by considerable rises in temperature and pressure. This caused the bursting disc to rupture at 12.37 hours, with the discharge of a cloud of vapour and fine droplets high into the air. Some calculations indicated that a temperature of 400°C was needed in the reactor to produce a pressure high enough to rupture the bursting disc. A sample of material taken from the top of the reactor showed a dioxin content of 3500 ppm.

5.2.7 The cause of the runaway reaction

At the inquiry various theories were advanced to explain the runaway reaction. The most probable was that the reaction between tetrachlorobenzene and caustic soda had not finished when the reactor was shut down, and continued slowly in the unstirred and uncooled reactor, liberating heat with accelerating rises in temperature and pressure. This is consistent with an experimental study made some time after the disaster by Cardillo and Girelli in which they prepared a mixture of the same composition as that left in the reactor at 04.45 on 10 July, and subjected it to sensitive thermal analysis.[10] They found that 'a slow moderate exothermic process is noticeable at 180°C after about 4 hours'. This was much lower than the temperature of 230°C which was previously reported to be necessary for such a runaway reaction to start.

Other theories included:

- the exothermic condensation of glycol with elimination of water;
- that air entering the reactor reacted with material in it thus developing a hot spot which promoted other exothermic reactions;
- that hydrochloric acid was either accidentally or deliberately added to the reactor on Saturday morning.

5.2.8 Mistakes which led to the disaster

Two crucial mistakes whose possible consequences had not been appreciated (i.e. organisational misconceptions [3.3]) clearly played key roles in the disaster.

1. *The reliance on a pressure relief device (bursting disc) which discharged directly to the atmosphere to cope with any runaway reaction which might develop.* This caused both the pressure and temperature in the reactor to rise considerably when a runaway reaction occurred, leading to a very large increase in dioxin formation.

   It furthermore ensured that when the pressure built up and the bursting disc broke, much of the contents of the reactor were rapidly discharged high into the air and spread over a much larger area than would have happened if a vent valve on the reaction system had been open.

   This reliance was probably not a deliberate decision but rather one taken by default. The bursting disc was a safety device to protect the reaction system from overpressure from whatever cause and was probably installed to meet some code requirement. Had the consequences of a runaway reaction been considered properly, any gas/vapour passing the bursting disc (and a vent valve which by-passed it) should have been piped to a 'blow-down' vessel and condenser large enough to condense and contain all vapour released by the reaction.

   It appears that this mistake had persisted for nearly a year, from the time the TCP plant was modified in 1974/1975.

2. *The act of allowing a mixture of (hot) chemicals which can react further in a closed, unstirred and uncooled vessel, virtually unattended over a weekend.* This again may not have been a deliberate act, but it resulted from the

dislocation of a batch-production programme which was geared to a particular shift system. The possible consequences of leaving an unfinished and still reactive batch unattended had not been foreseen.

This mistake seemed to have persisted for several weeks, since four interrupted batches had already been left uncooled over a weekend before water and acid were added to quench the reaction and neutralise the caustic soda present.

Had these mistakes been recognised and their consequences thought through in time, they could surely have been rectified. Thus the disaster was preventable.

5.3 The Bhopal disaster in December 1984[3]

This account of the world's worst industrial disaster is based on a detailed article in *India Today*[3] and on information researched by Granada Television for an award-winning documentary[12] which its director Laurie Flynn kindly made available to me.

Some 2500 Indians were killed and 100 000 injured by the release of 26 t of highly toxic methyl isocyanate (MIC) vapour into the midnight air of Bhopal, a city of more than 700 000 inhabitants which lies in the centre of India. The plant was located in the north-east of the city where it was surrounded by commercial premises and workers' quarters (Fig. 5.3).

The following problems discussed elsewhere in this book featured in the disaster:

- the siting of major hazard installations [15]
- toxic vapours [7]
- spontaneous exothermic polymerisation [8.6.4]
- process protective systems [17]
- staff training [21] and selection [19]
- emergency procedures [20.3]
- metal corrosion [13]
- safety in technology transfer to Third World countries [23.5].

5.3.1 Background[3]

Most insecticides, while considered vital to food production, are toxic to humans to some degree. Their sales are often short-lived as more effective and safer ones are invented.

Union Carbide India Ltd (UCIL) started its Agricultural Products Division in 1966/1967 in Bombay. This moved in 1968 to Bhopal, where a formulation plant was set up for the insecticide carbaryl, under the trade

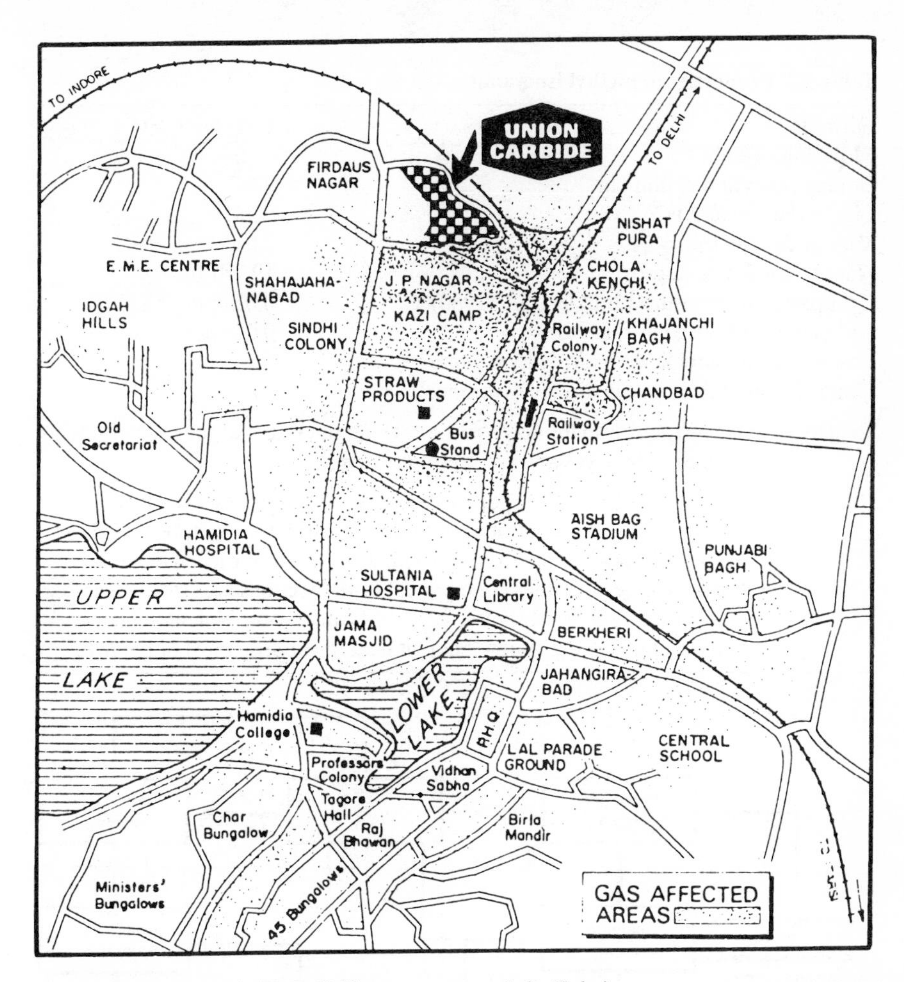

Fig. 5.3 Map of Bhopal. (By B. K. Sharma, courtesy *India Today*)

name Sevin. At first the concentrate was imported from Union Carbide Corporation (UCC) in America, only grinding, blending and packing being done at Bhopal.

Sevin, Temik and MIC, their common intermediate, were made by UCC at a plant at Institute, West Virginia. MIC, itself highly toxic, was made by a route involving several other toxic gases and volatile liquids. Some properties of MIC are given in Table 5.2.[13]

UCC made plans for the manufacture of 5000 t/yr of Sevin and Temik at Bhopal in two phases. In the first phase, which came into operation in 1977, the MIC was imported from the USA in stainless steel drums, and reacted with locally produced chemicals to make Sevin and Temik. In the second phase, which went into production in February 1980, MIC was also made at Bhopal.

The MIC plant, which started production in 1980, included the following stages (Fig. 5.4).

Table 5.2 Properties of methyl isocyanate

Formula	$CH_3-N=C=O$
Molecular weight	57.05
Boiling point at 760 mm Hg	40°C
Liquid density at 20°C	0.96
Vapour density (air = 1)	2.0
Heat of combustion, gross	19 770 J/g
Occupational Exposure Limit in USA (1980)[14]	0.02 ppm or
and in many other countries[15]	0.05 mg/m^3
Lower flammability limit in air, % by volume	5.3
Upper flammability limit in air, % by volume	26
Toxicity	The vapour is extremely toxic, attacking the skin and the mucous membranes of the eyes and respiratory system.
Reactivity	MIC is very reactive. It polymerises readily in the presence of various catalysts, including chlorides of iron, to form the cyclic trimer, 1,3,5-trimethyl isocyanurate. It reacts with water to form carbon dioxide, methylamine and other compounds. Both reactions are strongly exothermic, proceeding slowly below 20°C, but can become violent at higher temperatures and generate enough heat to vaporise most of the MIC

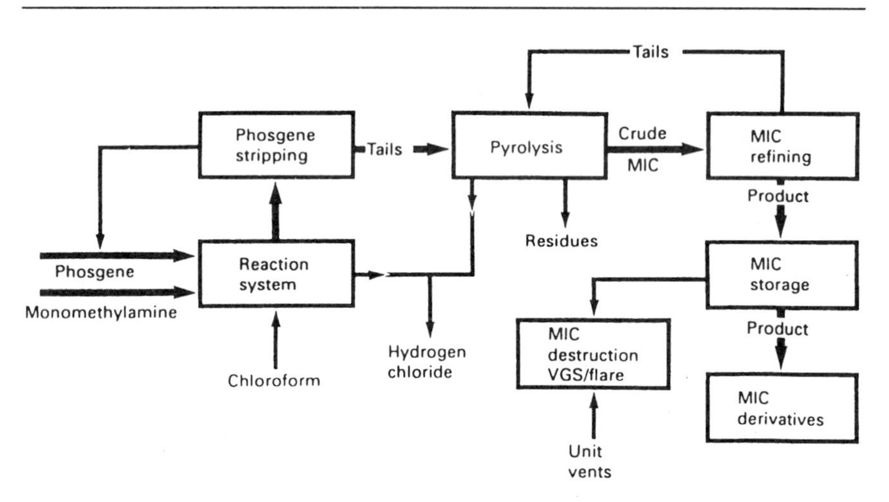

Fig. 5.4 Methyl isocyanate (MIC) process

- Production of phosgene from chlorine, and carbon monoxide produced on site:
  $CO + Cl_2 \rightarrow COCl_2$
- Production of methylcarbamoyl chloride from phosgene and monomethylamine, using chloroform as reaction solvent:
  $COCl_2 + CH_3NH_2 \rightarrow CH_3NHCOCl + HCl$
- Pyrolysis of methylcarbamoyl chloride:
  $CH_3NHCOCl + heat \rightarrow CH_3NCO + HCl$

- Fractional distillation of crude MIC in a 45-plate column at high reflux ratio, to remove chloroform and unconverted methylcarbamoyl chloride which was recycled to the pyrolysis stage.

Chlorine, monomethylamine and chloroform were brought by road or rail from other parts of India and stored on site.

UCC's specification for MIC included a maximum chloroform content of 0.5%, but because of difficulties in operating the MIC column it was sometimes higher.

5.3.2 Storage of MIC and safety features

MIC was stored in two horizontal stainless steel pressure vessels referred to as 610, 611. These were each of 57 m^3 nominal capacity, designed for full vacuum to 2.8 bar g. at 121°C (Fig. 5.5). The grade 604 stainless steel employed, while resistant to rusting and oxidising acids, would be rapidly attacked by hydrochloric acid, to form chlorides of iron and other metals [13.8.1]. These vessels were covered with earth mounds with concrete decks to protect against accidental impact, external fire and for thermal insulation. This insulation, while useful in preventing heat from entering the vessels when the contents were refrigerated, also prevented escape of heat generated by polymerisation of MIC if the refrigeration system was out of action [8.2, 8.6.4].

A third similar pressure vessel, 619, was installed to receive off-specification material which could either be returned for reprocessing or destroyed in the vent gas scrubber (VGS).

A 30-ton refrigeration system with a heat exchanger through which MIC was circulated was provided to maintain its storage temperature at 0°C or lower, in order to retard, control and minimise polymerisation.

Instrumentation of each MIC storage vessel included:

- a temperature indicator and high-temperature alarm;
- a pressure indicator/controller to regulate the pressure by admitting nitrogen or venting vapour to the VGS or flare;
- a liquid level indicator/alarm for high and low level.

The pressure in the vessels was controlled between the limits 0.14 and 1.7 bar g. This minimised nitrogen consumption and loss of MIC vapour. A pressure relief system fitted with a bursting disc and a relief valve in series was set to open if the pressure in a tank reached 2.8 bar g., discharging the vapour to the relief vent header.

The VGS was a 1.68 m diameter packed column in which gases entering were contacted with circulating caustic soda solution, the strength of which could be maintained by adding 50% caustic from storage tanks. Gas leaving the top of the VGS discharged through a stack to the air at a height of 33 m. Instrumentation included a caustic solution flow indicator/alarm which would automatically start the spare caustic circulation pump in case of low

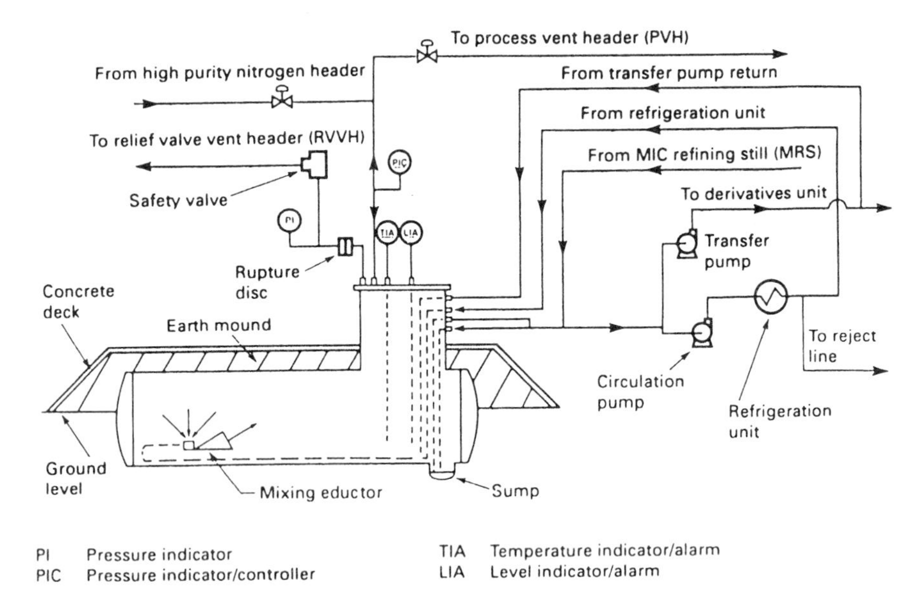

Fig. 5.5 MIC storage vessel 610. (Courtesy Laurie Flynn and Granada Television.)

flow. Under emergency conditions, the VGS could neutralise 4 t of MIC in the first 30 minutes, provided the circulating caustic solution was fresh. After that its capacity would be reduced to below 2 t/h because there were no means of cooling the circulating soda.

A 30 m high flare tower equipped with a flame-front generator and wind shield was primarily intended to burn vent gases from the carbon monoxide unit and methylamine from a relief valve. It was situated some distance from the MIC storage tanks.

Both the process vent header and the relief vent header were connected to the VGS and the flare line and could be routed to either. Fixed fire-water monitors were provided which could be used to provide a water curtain round sensitive areas and knock down some toxic gases and vapours.

An artist's impression of these safety features with notes on their condition at the time of the disaster is given in Fig. 5.6.

5.3.3 The period before the disaster

Following the start-up of the MIC plant, there were several accidents and minor escapes of toxic vapours. A maintenance fitter was fatally injured at the end of 1981 through exposure to a leak of phosgene. Six weeks later 16 workers were injured by another leak of toxic gas. Further accidents occurred in October and December 1982 and in February 1983, and morale fell.

The plant was surveyed in May 1982 by a team of Union Carbide's American safety experts who were concerned by the extent of corrosion and warned of a possible escape of toxic gases.

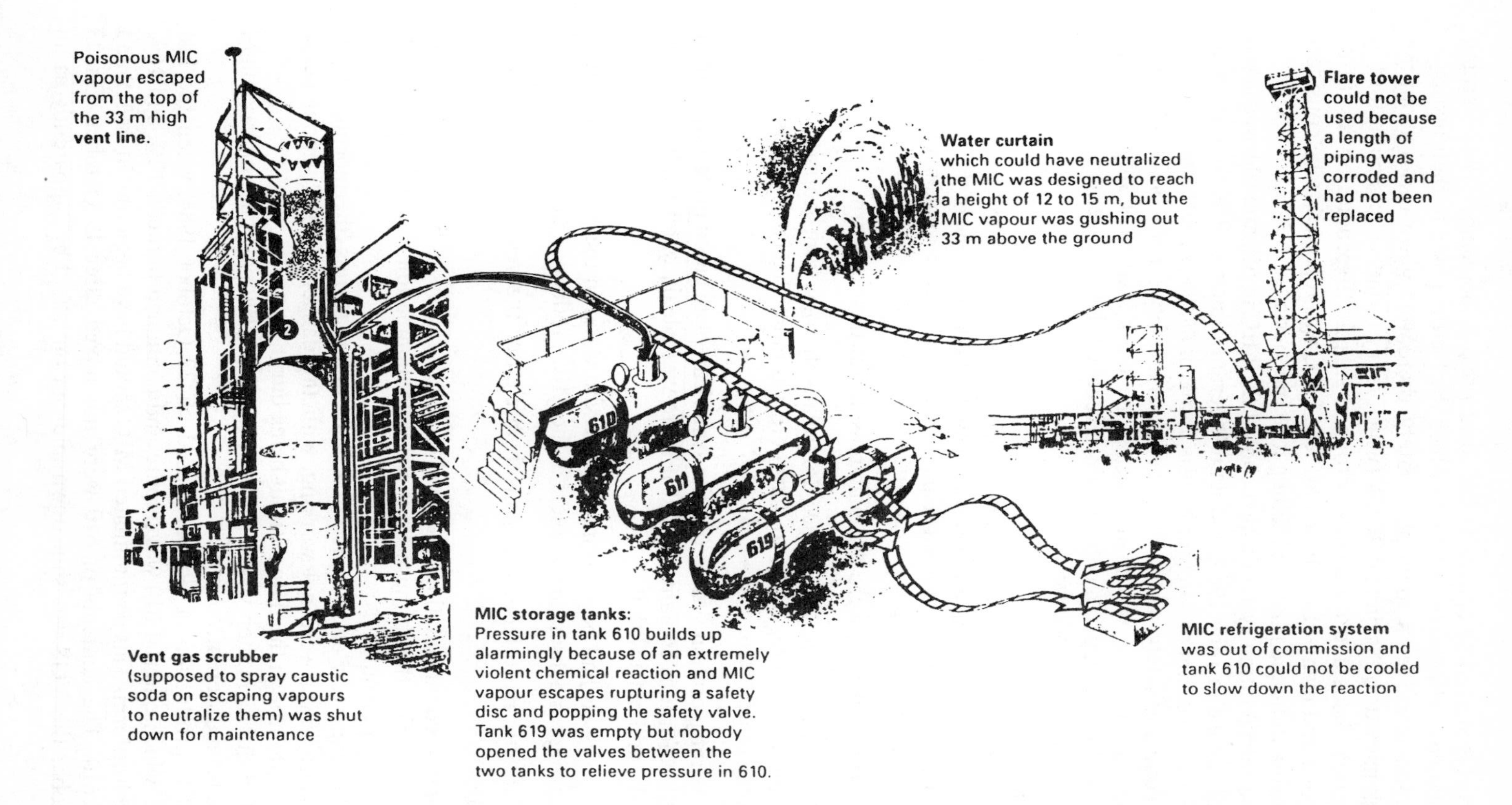

Fig. 5.6 State of safety features of MIC plant at Bhopal at time of disaster. (Drawing by Itu Chaudhuri, courtesy *India Today*.)

Articles started to appear in the local press in September 1982 warning the people of Bhopal of the threat which the factory posed to their lives.

Besides these fears about plant safety, there were now worries about the health hazards of Sevin and Temik (which many farmers could not afford in any case). Sales and production dropped sharply. Expenditure was cut to the bone and many of the best staff left. UCC was reported to be considering dismantling the plant and selling it to Indonesia.

The refrigeration system which cooled the MIC in 610 and 611 was taken out of service in June 1984, and its refrigerant removed. It is not clear who made that decision or whether he was aware of its likely consequences.

A company safety survey by six health and safety experts on UCC's MIC II Unit at Institute was made during the week starting 9 July 1984. An extract of their report, a copy of which reached the management at Bhopal on 19 September, follows:

> *MAJOR CONCERNS*
> SM1 *MIC Storage Tank Runaway Reaction*
> There is a concern that a runaway reaction could occur in one of the MIC Unit Storage tanks [vessels] and that response to such a situation would not be timely or effective enough to prevent catastrophic failure of the tank.

These (in my words) were the main reasons for their concern.

1. The vessels were being used for relatively long-term storage, and not being sampled regularly, so that there was more chance of contamination going undetected.
2. The refrigeration system (on the American plant) used brine rather than chloroform as coolant, so that if there was a leak, water could enter the MIC and react with it.
3. The vessels had been contaminated with water in the past as a result of leaking condensers and coolers in the MIC production unit.
4. Catalytic materials could conceivably enter the vessels from the flare system.

Their recommendations included one that the vessels be sampled at least daily, and that the seriousness of water contamination be emphasised to all operating personnel.

The team were also worried that flame failure alarms (thermocouples) on the flare tower were not working, with the danger of MIC vapour being released through the flare without being burnt.

Had the team then visited and inspected the Bhopal plant, history might have been different.

MIC production at Bhopal ceased on 22 October 1984, when the last MIC which entered 610 was impure and contained between 12% and 16% of chloroform. This contaminated MIC should have been routed to 619 and not 610. The contents of 610 were not mixed after 19 October. On 23 October the VGS was shut down and put on 'stand-by', apparently as an economy measure. At this time or soon after, the flare tower went out of service because of corrosion.

By Sunday 2 December, the day of the disaster, 610 contained 41 t of (impure) MIC. This had not been stirred or sampled for nearly six weeks, nor had its temperature been logged. The high-temperature alarm had been deactivated when the refrigeration system was shut down. The pressure on Saturday 1 December was apparently recorded as 0.14 bar g. (2 psig), and the same pressure was reported at 22.20 hours the next day.

5.3.4 The disaster

At 23.00 hours on Sunday 2 December 1985 a rather high pressure (0.7 bar g.) was noticed on the pressure indicator of 610 by the control room operator. One and a quarter hours later this had risen rapidly to over the top of the scale (3.8 bar g.). The vessel was hot, the concrete above it was cracking and MIC vapour was screeching through the relief valve. An attempt was made to start caustic circulation through the VGS. At 01.00 hours on Monday the toxic gas alarm was sounded by an operator in the derivatives area and fire-water monitors were turned on and directed to the VGS stack to try to knock down escaping MIC vapour. At 02.00 hours the relief valve of 610 (set at 2.8 bar g.) reseated and MIC emission ceased. 610 and its associated pipework were later found to be intact and pressure tight. Four hours later the temperature in the caustic accumulator of the VGS was about 60°C, thus indicating that some circulation of caustic had occurred.

5.3.5 Quantity of MIC vapour released

Estimates made from mass and energy balances indicated that about 36 t of material had escaped from 610, of which about 25 t were MIC vapour, the rest being entrained liquids and solids formed from MIC. The average pressure in 610 during the two-hour period of the MIC escape was estimated to have been 12.2 bar g. As this was well above the hydrostatic test pressure of 4.1 bar g., it is fortunate that the vessel did not rupture. A peak temperature of 200°C in 610 appears to have been reached after the relief valve closed.

5.3.6 Initiation and nature of the reaction in 610

Two possible explanations for the initiation of the reaction soon emerged.

1. A gradually accelerating runaway polymerisation reaction started in 610 some time after the refrigeration system was decommissioned, and its contents had warmed up to the ambient temperature. Such a runaway reaction [8.6.4] was involved in the accident at Seveso [5.2]. The critical temperature for any material above which one can start falls as the mass, thermal insulation and surrounding temperature increase. In this case the runaway reaction might well have started at 15–20°C.

The interval of five months between the shut-down of the refrigeration system and the disaster is not, however, easy to explain. Unfortunately we have no record of the temperature of 610 since June. At 0°C and below, both the polymerisation and hydrolysis rates of MIC are probably very low, even with metal chlorides present. Little reaction would have occurred until sufficient heat had entered the vessel from the ground to raise its temperature to about 15°C. Once the temperature in 610 reached that of the ground (15–20°C), the reaction, though still very slow, probably generated heat faster than it could be lost, and accelerated as the temperature rose. The vessel also contained metal chlorides (resulting from HCl corrosion of the stainless steel equipment), which catalyse the polymerisation. These could have lain dormant for a considerable period in the base of the vessel before becoming activated.

2. The polymerisation was initiated by the entry of water into 610, either through a leaking valve or as a result of sabotage, shortly before the disaster.

This explanation has received wide publicity.

Chemical analysis of the solids left in 610 might be expected to throw light on the question. A summary of the analysis by UCC of such a sample is given in Table 5.3.

The major constituent, MIC trimer, would be expected from the metal chloride catalysed polymerisation of MIC. The small amount of chloroform resulted from chloroform impurity in the MIC.

The formation of methylamine hydrochlorides required hydrogen chloride. The UCC investigators thought the hydrogen chloride was formed by hydrolysis of chloroform in the MIC, although it could have been formed from methylcarbamoyl chloride or even phosgene had they been present in the MIC in 610. (Sir Frederick Warner and Dr Luxon, health and safety director of the Royal Institute of Chemistry, were reported as saying that 'the victims suffered from phosgene poisoning'.)[15]

The total amount of iron, chromium and nickel in the salts found

Table 5.3 Analysis of core sample of solids in 610

Constituent	*% weight*
1,3,5-trimethyl isocyanurate (MIC trimer)	40–55
1,3-dimethyl isocyanurate	13–20
Methylamine hydrochlorides	7–10
Methylbiurets	7–14
Dihydro-1,3,5-trimethyl-1,3,5-triazine-2,4-dione	5–7
Methylureas	3–6
Iron, chromium and nickel salts	0.18–0.26
Chloroform	0.4–1.5
Water[a] up to	2

[a] The water may have entered the sample after the disaster

amounted to approximately 9 kg. Their ratio was approximately the same as in 304 stainless steel, indicating chloride corrosion of 610 (and/or other equipment) as their source. Some of the metal salts were probably present before December, and some formed during the runaway reaction in 610.

The analysis shows a number of organic compounds which might have been formed directly from MIC, or indirectly by 'cooking' of its trimer after the relief valve had closed. The UCC team concluded from their analyses of the residues that:

> 1000 to 2000 lb of water and 1500 to 3000 lb of chloroform were required.

Far less water would, however, have been needed if the hydrochlorides had been formed from methylcarbamoyl chloride rather than chloroform.

Dr S. Varadarajan, a senior Indian government scientist, stated[15] that:

> Just half a kilogram of water entered the underground methyl isocyanate tank, triggering a runaway reaction.

A pipe-washing operation was suggested as the source of water. UCC team's comment on this reads:

> Water could have been introduced inadvertently or deliberately directly into the tank through the process vent line, nitrogen line or other piping. Records indicate that the safety valve discharge piping to the relief valve vent header from four MIC process filters was being washed shortly before the incident. Oral discussions indicate that a slip blind was not used to isolate the piping being washed. However, entry of water into Tank 610 from this washing in the MIC unit would have required simultaneous leaks through several reportedly closed valves, which is highly improbable.

Whether or not water entered 610 after the MIC plant was shut down, the shutting down of the refrigeration unit in June was clearly a major factor in enabling the runaway reaction to occur.

5.3.7 Emergency measures

According to Wally Schaffer, chairman of the UCC workers' safety committee at Institute, there was no emergency or evacuation plan for people outside the plant at Bhopal, as there was at Institute. The toxic gas alarm at Bhopal was not activated until MIC had been escaping for at least an hour, and half of it had escaped. Even then it was turned down after a few minutes, when it could only be heard by UCIL workers. Criticisms have been raised[3] that more might have been done to introduce fresh caustic into the VGS so as to have made it more effective. There were, however, few personnel present at the time of the disaster as most plants were shut down. They were working in high concentrations of MIC vapour with apparently only 'half-hour masks' for protection.[16]

One desperate measure might have averted the worst effects of the disaster once the MIC emission had started. This would have been to set fire to the vapour escaping from the vent stack. Rockets or a machine gun firing incendiary bullets might have done this! Such ignition might, however, have caused a vapour cloud explosion [10.2.4] leading to an escape of stored chlorine or phosgene.

5.3.8 Mistakes or organisational misconceptions

Two major mistakes, expressed here as 'organisational misconceptions', lay at the root of the disaster.

1. That it was safe to build a plant storing large quantities of very toxic gases and volatile liquids close to a densely populated area. This seems to have arisen about ten years before the disaster.
2. That it was safe to shut down the refrigeration system designed to allow MIC to be stored safety. This happened 5 to 6 months before the disaster.

Union Carbide have been criticised on several other points which, though serious in themselves, appear less important than these two. They include:

- a safer route for the manufacture of MIC or of the insecticides themselves could have been chosen;
- the MIC storage capacity should have been much smaller;
- the storage vessels should have been made strong enough to contain a runaway reaction if it occurred;
- the VGS should have been large enough to cope with the maximum rate at which MIC vapour could have been released in a runaway reaction;
- the pressure relief system should have been designed for two-phase flow of vapour and liquid and not for vapour only.

Most of these criticisms lie in the grey area between what is 'practicable' and what is 'reasonably practicable' [2]. Compliance with them would have added significantly to the project cost, and might well have rendered it a non-starter in the first place. Such questions are now studied under 'risk-benefit analysis'. Many other Third World chemical companies would have followed the same practices as Union Carbide on these points, and would probably have got away with them.

Had someone deliberately tried to organise the disaster, he could hardly have done better! The words of Turner[5] quoted in 3.3.2 are particularly apt.

The first mistake probably had a profound effect on the morale, quality and competence of the technical management at UCIL's Bhopal works. 'Good company men' would obviously be expected to subscribe to the misconception. The effect on their morale when they could no longer accept it would have been shattering.

This was well expressed in the TV programme by engineer Pareek, who was responsible for safety for one and a half years at Bhopal:

Interviewer: 'Were inexperienced people being brought in?'
Pareek: 'Yes, er, you know in the industry the news gets around that a plant is not healthy, so nobody who was worth anything would like to come and join that plant'.

The lead time or 'incubation period' [3.3.5] during which the first mistake persisted thus lasted about ten years from 1974, when the siting of the MIC plant was decided. From then on, although many came to recognise this as a mistake, it became increasingly difficult and expensive to rectify. One might as well have tried to move the city as the plant. One hopes a better site would have been found for the plant in Indonesia!

5.4 The Piper Alpha disaster on 6 July 1988

The explosions and disastrous fire on the Piper Alpha offshore platform on 6 July 1988 resulted in the deaths of 165 men and the loss of the complete installation. Only 61 men survived. Lord Cullen, who conducted the public inquiry, recommended substantial changes in the legislation covering offshore safety, and a transfer of the responsibility for its enforcement from DEn to HSE.[4] All of his 106 recommendations were accepted.

5.4.1 General background

The reservoir fluids emerging from the wellheads comprised oil, gas, water and sand. These were separated on the platform. The water and sand were cleaned and dumped; the oil and gas were pumped down separate sub-sea pipelines to the Flotta terminal. The gas leaving an oil/gas separator contains a high proportion of methane, but also ethane, propane, butane and small amounts of pentane, hexane and higher alkanes. These are likely to condense out of the gas during processing, and are collectively called condensate. Their presence is undesirable in the exported gas because condensed liquid could collect in dips in the pipeline, and also because the gas would not comply with the customer's specification. On Piper, much of the condensate was removed when the gas pressure was dropped from 100 bar to 43 bar across a JT valve: this caused substantial cooling by the Joule–Thomson effect. The condensate collected in the JT flash drum and was then passed through a condensate booster pump, which raised its pressure to 47 bar. A condensate injection pump then further increased the pressure to 75 bar before it was spiked into the main oil export line. There were in fact two booster pumps and two injector pumps, one of each pair being on duty and the other on standby.

Gas produced on the neighbouring Tartan platform was pumped through a 19 km pipeline to Piper, from which it was exported, together with Piper's own gas, to the Frigg field pipeline. The gas from Tartan came onto the platform through a riser, which was connected to a number of valves. Gas was also exported to the Claymore platform, which produced only oil, for use in its gas turbines.

5.4.2 Events leading to the explosion and fire

The condensate injection pumps were protected by pressure safety valves (PSV). The one on pump A was removed by a fitter on the day shift and taken to the workshop for a regular maintenance inspection. The flange to which it had been attached was then covered by a blank plate. The shut-down of the pump and the rest of the procedure were correctly covered by the issue of a permit to work. Unfortunately neither the permit, nor the information that the valve had been removed, were passed on to the night shift. During the evening, pump B, which was then on duty, tripped out. This may well have been due to the presence of hydrate in the condensate. This is a crystalline solid which can be formed by a reaction between a hydrocarbon and water when the conditions of temperature, pressure and moisture content are favourable. Its presence in a reciprocating pump could well overload the motor. Methanol was injected at a number of points in the condensate system in order to inhibit hydrate formation, but owing to the need to repair a leak, none had entered the JT valve for about four hours prior to the pump failure. When pump B could not be restarted, the decision was taken to change over to pump A. This could only be done at the pumps themselves, but unfortunately the PSVs were located in a module which was above the one containing the pumps, and the men were therefore not aware that a valve was missing. When the pump was started up, a loud screeching noise was heard, presumably caused by escaping condensate: the blank plate did not provide a leak-tight seal at the flange. The violent ejection of a very volatile liquid produced a substantial premixed vapour/air volume. When this encountered an ignition source there was a violent explosion. The resultant damage to the plant caused the release of a large quantity of crude oil, which burned in a devastating fire.

In this situation, the only hope of containing the fire would be the application of cooling water to the affected steelwork, the isolation of the leaks, and, possibly, the application of foam. The water was available from a deluge system and from hydrants. These were connected to the pressurised fire main, and the pressure drop resulting from a flow of water from the main should have actuated the automatic start-up of the diesel fire pumps. It is unfortunate that, because divers were working beneath the platform, the pumps had been set to manual operation only. This unnecessary precaution (which was peculiar to Piper) seriously affected the firefighting capability of the platform. An attempt was made to reach the pumps by men wearing breathing apparatus, but they could not penetrate the smoke and flames. The fire continued unabated until the riser from Tartan ruptured on the riser side of the emergency shut-down valve (which had been closed). The escaping gas caused a second major explosion and a massive intensification of the fire (production on Tartan had not been stopped, nor had the gas pipeline been depressurised). Later, the fire was even further intensified by the failure of the gas risers on the pipelines to Frigg and Claymore.

The Occidental company, the owners of Piper, had provided an Emergency Procedures Manual which gave detailed instructions on the actions to be taken in an emergency by people both on and off the platform. Apart from the transmission of Mayday messages on the international distress frequency, these procedures were largely ignored. Although no instructions were given, most of the people on the platform assembled in the accommodation module. The OIM (Oil Installation Manager) told them that the whole world knew of their problems, and that helicopters and shipping were coming to their resue. No other information or instructions were given. At that time it was obvious that helicopters would be unable to land on the platform. A number of men therefore decided to get out of the accommodation and attempt to find an escape route down to the sea. Of the unknown number who did this, 28 survived. Everyone who remained in the accommodation, including the OIM, was killed.

5.4.3 Organisational errors which led to the disaster

The basic error was the failure to hand over either the permit or the relevant information at the shift change. Also, although the isolating valve for pump A was shut, it was not locked off or tagged. Exactly the same errors had resulted in a death 10 months before the disaster. Adequate procedures were provided in Occidental's Work Permit Booklet issued in 1985, and in a Safety Procedures Manual of 1987. The inquiry found:

> that the operating staff had no commitment to working to the written procedure: and that the procedure was knowingly and flagrantly disregarded.

When the disaster happened, the requirements of the Emergency Procedures Manual were also disregarded. There was clearly a failure of management to provide adequate initial and continuing training, and proper supervision. It is unfortunate also that inspections which were carried out by DEn in June 1987 and in June 1988 (after the fatal accident) failed to discover the inadequacy of the safety precautions. The design fault which resulted in the rupture of the three gas risers was quickly identified by a separate technical inquiry led by Jim Petrie of DEn. Two reports, giving guidance to the offshore industry on this and many other matters, were issued before the start of Lord Cullen's inquiry. The new legislation which was written as a result of the inquiry is discussed in Section 2.14.

References

1. Ministry of Social Affairs and Public Health, *Report concerning an inquiry into the causes of the explosion on 20th January, 1968 at the premises of Shell Nederland Raffinaderij in Pernis.* State Publishing House, The Hague, Holland (1968)
2. Margerison, T., Wallace, M. and Hallenstein, D., *The Superpoison* Macmillan, London (1981)
3. 'Bhopal — city of death', *India Today*, pp. 6 to 25, 31 December (1984)

4. Lord Cullen, *The Public Inquiry into the Piper Alpha Disaster,* Cm 1310, HMSO, London (1990)
5. Turner, B. A., *Man-made Disasters*, Taylor and Francis, London (1978)
6. Davenport, J. A., 'A study of vapour cloud incidents', Paper 24, *Eleventh Loss Prevention Symposium*, 83rd National Meeting, Am. Inst. Chem. Engrs, New York (March 1997)
7. AMOCO, Booklet No.1, *Hazard of Water*, 5th edn, AMOCO, Chicago (1964)
8. King, R. W., 'Latent superheat — A hazard of two phase liquid systems', I. Chem. E. Symposium Series No. 49, *Chemical Process Hazards with Special Reference to Plant Design VI*, Rugby (1977). Translated by J. P. Appleton, Plenum Press, New York (1969)
9. *The official report of the Italian parliamentary commission of inquiry into the Seveso disaster, translated by HSE*, HSE, Bootle (about 1984)
10. Cardillo, P., and Girelli, A., 'The Seveso runaway reaction: A thermoanalytical study', paper in I. Chem. E. Symposium Series No. 68, *Runaway Reactions, Unstable Products and Combustible Powders*, Rugby (1981)
11. Schulz, K. H., *Arch. Klin. Exp. Derm.*, **206**, 540 (1957)
12. Flynn, L., and Smithson, J., *The Betrayal of Bhopal,* 'World in Action' TV documentary, available as video cassette from Granada Television, London
13. Morel, C., Gendre, M., Limasset, J. C. and Cavigneaux, A., *Isocyanate de méthyle–Fiche toxicologique no. 162*, INRS, Paris, (1982)
14. The International Labour Office, *Occupational Exposure Limits for airborne toxic substances — a tabular compilation of values from selected countries*, 2nd edn, Geneva (1980)
15. 'Water leak caused fatal chain reaction says Bhopal expert', *The Guardian*, Saturday 5 January (1985)
16. Bhatia, S. and McKie, R., 'Time bomb at Bhopal', *The Observer*, 9 December (1984)

Part 2
Hazards – chemical, mechanical and physical

6 Electrical and other physical hazards

This chapter discusses electrical and other physical hazards of a non-mechanical nature which can to a large extent be controlled during design.

The ignition of flammable atmospheres by sparks, static discharges and lightning is discussed in [10.5.1]. The classification of hazardous areas and the choice of suitable electrical equipment is covered in [10.5.3].

6.1 Electricity

6.1.1 Physiological effects of electricity

The human body can receive an electric shock when the two hands are holding metal objects which have a potential difference (a voltage) between them. More usually, one hand (or another part of the body) touches a metal conductor which is at a potential above that of the surface on which the person is standing. In either case the current flows by the shortest route: between the two hands, or from one hand to the feet. The current density is greatest near to the points of contact, and this is where the most physical damage (trauma) will occur. The body behaves as a large volume of electrolyte, with a very low resistance, surrounded by a high-resistance container (the skin).[1] An indication of the resistance of the skin can be obtained by measuring the current which flows between two electrodes which are in contact with it. The measured resistance will vary with the area of contact, but typical values are between 3600 ohm for very dry hands and as little as 500 ohm for wet, or sweaty, skin. Much higher values are obtained at low voltages because there is no chemical breakdown of the skin's insulating properties. Electrical safety can therefore be greatly enhanced by a reduction in the voltage of the equipment.

The current flowing through the body can be calculated by using Ohm's Law, but remembering that the resistance is higher at low voltages. The damage which can be caused depends on the energy which is released, and this can be calculated from

$$I^2Rt \text{ or } \left(\frac{E^2}{R}\right)t$$

where I is the current,
R is the resistance,
E is the voltage drop between the conductor and the skin,
t is the time.

If the skin is very dry, a high voltage may cause a severe burn but there may be no other damage. On the other hand, a lower voltage applied to wet skin could cause death, particularly if the current passed through the heart, but there might then be no sign of burning. When death is caused by contact with the normal 240 V mains supply, the only visible injury may be a slight skin discoloration at the point of contact. Table 6.1[2] provides an indication of the effects caused by different current flows within the body. The appropriate voltages needed to produce these flows are also shown, based on a skin resistance of about 3000 ohm at low voltages and 2000 ohm at the higher values. The release current threshold of 76 mA is important. At current flows above this it may not be possible for a person to let go of a conductor because of muscle contraction. This effect is more serious with DC than with AC circuits. The values quoted in the table are surprisingly low and it is fortunate that our skin does have a high resistance. If it can be penetrated by a conductor, then death can be caused by only a few volts. However, the effect that an electric current has on different parts of the body depends on a number of factors, including the voltage, the frequency and the duration of the shock, and the location of the organ.[3] The important message is that severe shock, injury and death can be caused by contact with normal electrical supplies. People get used to the safe operation of electrical equipment and tend to forget the dangers which are always present.

Four different kinds of damage can result from the passage of an electric current through the body. The first has already been mentioned: burning close to the contact point, particularly at high voltages. However, an apparently superficial skin burn may be located above an area where a substantial

Table 6.1 Physiological effects of electricity

Current (mA)	*Voltage (V)*	*Effect*
5.2	15	Slight tingling
9	25	Shock: not painful and muscular control not lost
62	125	Painful shock, but muscular control not lost
76	150	Painful shock: release current threshold
90	180	Painful and severe shock, muscular contraction, breathing difficult
120	240	Serious burns, shock, death from heart or breathing failure

amount of tissue damage has occurred. Continuing medical attention is then needed because the presence of dead tissue, an infarct, may be life-threatening. Quite substantial burns, which are slow to heal, can also be caused by the arc which passes when a fuse is removed or a switch contact is opened, but in this case it is unusual for current to flow through the body.

The second effect is that noted in the table at 90 mA. At current flows above this, breathing becomes increasingly difficult, and the release current threshold has been exceeded. The effect is the same as that of suffocation.

The third and fourth types of trauma directly concern the heart, and may rapidly become fatal. First, the passage of a current through the heart may cause cardiac arrest. Unless the heart can be re-started within a few minutes, the brain will be irreversibly damaged by lack of oxygen, and death will quickly follow. The other effect which can be caused by an electric current is ventricular fibrillation. Instead of the synchronous contractions of the heart's auricles and ventricles, the muscles, particularly those in the ventricles, become unco-ordinated. The individual muscles fail to contract in unison and start to twitch violently. No blood is then pumped by the heart, and again the brain is starved of oxygen.

If someone cannot let go of a conductor, because the release current threshold has been exceeded, then they should be dragged away by their clothing. Alternatively it may be possible to push them using a piece of wood. If artificial respiration or heart message are needed they must be applied immediately and continued, if necessary, for at least an hour. If no burning has occurred, people who have survived an electric shock usually recover completely. There may be some numbness and pain for a long time, and occasionally there may be after-effects like cataracts, angina and nervous disorders.

6.1.2 Prevention

Electricians are necessarily very aware of the hazards involved in their work. Self-discipline (backed by management checks) is essential to ensure that proper procedures are always followed and that the correct equipment (including insulating gloves and mats) is used. Much of their work will be covered by a permit-to-work procedure, which should also provide protection for the plant operators. Other workers may become exposed to electrical hazards through the use of powered hand tools. In the worst situation, a positive lead comes into contact with a metal casing which is not earthed, at a time when the man holding it has good contact with earth. One way in which this can happen is the inadvertent cutting of the supply cable by the tool. All electrical circuits are protected against overloading by fuses or circuit breakers. However, the small current values in Table 6.1 show that serious injury or death could be caused by a current increase which is quite incapable of actuating these devices. Instead, there are two very positive ways of protecting people against receiving a shock from faulty equipment. The first is to monitor the integrity of the earth wire connected to the tool, and to cut the supply as soon as it is lost. The second method is to detect the small current that is passing to earth through the affected person, and again to cut the supply. Various names have been used for these detectors in the past, but by international agreement they are now called residual current devices (RCD).[4] The device comprises a transformer in which the core is in the shape of a ring. Current flowing in both the live and the neutral wires connected to the hand tool passes through two identical windings on the core. In normal conditions the two currents are equal and opposite, and no magnetic flux is induced in the core. When the currents are unbalanced, a flux will be present, and this will induce a voltage in a third winding. A fast-acting trip is actuated by this coil. A typical RCD will respond to a leakage current as low as 5 mA, and will trip within 0.2 s. This is still time enough for the person to receive an unpleasant shock, but the 0.2 s delay is less than the time for one heartbeat, and no fibrillation will be caused. Even an RCD which is designed to trip at 30 mA will provide a high level of protection against a lethal shock.[1]

An alternative method of protection is provided by the use of low-voltage tools. These are supplied from a transformer in which the low-voltage winding is completely isolated from the mains. It is sometimes thought that arc welding is safe because a similar arrangement is used. This is not so, because the voltage can be as high as 100 V. Suppose for example that the welder's hands are wet with perspiration, and he changes the electrode without putting on gloves and without switching off the supply. If he is in good contact with earth, and his skin resistance is down to about 800 ohm, then at 100 V the current passing through his body could be 125 mA. Table 6.1 shows that this could be fatal. Clearly this is an unsafe practice. Some welding equipment is so designed that the voltage is reduced as soon as an open circuit occurs. The full welding voltage is then restored as soon as the electrode again makes contact with earthed metal. This will clearly provide some protection, but should be regarded as a useful addition to the vital practice of following proper procedures. It should not be forgotten that a welder is exposed to a number of additional hazards: sparks which can cause burns; toxic fumes; intense UV and IR radiation, which can damage the eyes; and hot surfaces which can cause dangerous burns.

One electrical hazard which does occasionally happen results from the contact of large mobile machines, like a crane, with an overhead high voltage line. The machine is insulated from earth by its rubber tyres, and anyone who touched it could be electrocuted. It is important that the driver should not step out of the vehicle, but should jump clear.

6.2 Physical hazards involving liquids

The following hazards are discussed in this section:

1. the thermal expansion of liquids in closed systems;
2. the freezing of liquids in pipes, valves, etc.;

3. physical explosions which involve two phases, one or both of which are liquids and which are initially at different temperatures;
4. 'turnover' of liquids at their boiling point in tanks and vessels (a serious hazard in the cryogenic storage of liquefied gases);
5. 'boilover' which sometimes occurs during oil tank fires;
6. irregular boiling of superheated liquids characterised as 'bumping'.

Other physical hazards are discussed in Chapter 15. While the hazards described here are deceptively simple, their elimination can be difficult and some 'solutions' cause other hazards as bad or worse than the one eliminated.

6.2.1 Thermal expansion of liquids

The problem arises when the temperature of a closed liquid-filled system increases. It is found:

1. in hot water systems;
2. in pipes and pipelines containing toxic and flammable liquids and liquefied gases;
3. in storage and transport containers filled with similar liquids and liquefied gases.

The coefficient of thermal expansion by volume $\times 10^{-3}$ of common pipe materials at 20°C ranges from about 0.03 for steels and cast iron to 0.07 for aluminium, and from 0.2 to 0.4 for polyethylenes. Figures for liquids on the same basis range from 0.21 for water to 1.12 for methanol, 1.6 for pentane and 2 or more for butane and propane. Since liquids have a very limited compressibility, very high pressures can be reached in closed liquid-filled systems.

1. *Hot water systems*
When water or any liquid is heated in a closed system which is not connected to an elevated vented expansion tank provision must be made for its thermal expansion. This is commonly done by connecting an expansion chamber containing an impervious rubber bag filled with air or nitrogen to the pipework. Some means are then needed of checking that the bag contains sufficient gas and is not punctured.

2. *Pipes and pipework containing hazardous liquids and liquefied gases*
Many pipes and pipelines carrying liquids and liquefied gases have to be provided with valves at both ends. If these are both closed when the pipe is full of liquid, and its temperature later rises, the pressure generated may distort or rupture the pipe, or blow a gasket in a flanged joint or the packing of a valve. This can occur if the pipe is exposed to hot sunlight, or if the temperature of the liquid in the pipe when the valves are closed is below ambient.

The usual solution[5] is to fit small pressure relief valves to pipe sections which are liable to be 'boxed in'. Provided the valve closes once the over-pressure is relieved, the quantity of liquid released is small. But if the pipe contains scale which lodges in the valve and prevents it from reseating, its entire contents may be released.

In the case of LPGs, for relatively short pipes within the battery limits of a plant, the relief valves are often arranged to discharge onto the ground where the LPG soon evaporates. If the relief valve fails to reseat, this is revealed by ice on the downstream side of the valve. The pipe then has to be drained of liquid and the relief valve cleaned or replaced. But if the pipe is a large one, say 8 inches or more in diameter and 100 or more metres long, several tonnes of LPG may escape if the relief valve fails to reseat. In such cases the relief valve should discharge to a properly designed relief system, with an adequately sized liquid catch pot and a tall vent or flare stack.

Another solution is to route the discharge from the relief valve to part of the system at a lower pressure, e.g. a pressure storage vessel. In this case it is difficult to detect whether the relief valve has reseated or not, while an alternate means of disposal becomes necessary if the receiving vessel has to be emptied and isolated while the rest of the system is still in use.

Another solution is to establish a strict operating procedure whereby the line is partially drained of liquid before the second valve is closed. This needs close liaison between those responsible at each end of the line.

3. Storage and transport containers for liquids and liquefied gases
It is vital to leave a certain percentage of the volume of such containers free of liquid. This applies to drums and bottles containing liquids and especially to cylinders and road and rail tankers filled with liquefied gases.

Filling ratios for liquefied gas containers of different sizes, for different liquefied gases, and for use under different climatic conditions are given in BS 5355.[6] Cylinder filling plants and filling stations for road and rail tankers must be provided with positive means for ensuring that the specified filling ratio is not exceeded. Clear operating procedures must be established and those responsible for filling must be carefully trained and supervised. The Spanish camping site disaster in July 1978 where there was heavy loss of life was attributed to overfilling of a road tanker with liquid propylene.[7]

6.2.2 Pipes and equipment containing liquids liable to solidify

The unintended formation of solids in pipes and equipment containing fluids is always troublesome and can be very hazardous, particularly when the fluid is water which expands when it freezes. Perhaps the worst case was the Fézin disaster of 1966.[7] This resulted from the freezing of water in a drain valve under an LPG storage sphere when both LPG and water were passing through the valve, thus preventing it from being closed [10.1.1]. Since then, the design and operating procedures for the pressure storage of liquefied flammable gases for bulk LPG storage installations have received much attention.[8] Although different authorities have different views[9] on the safest way to drain settled water, all agree that it is essential to have two valves in series. The first valve nearest the storage vessel should be the larger of the two and of a type which can be quickly closed if the downstream valve becomes frozen in the open position. When draining water the first valve is opened wide and the flow is controlled by the second valve. Some authorities now insist on having a separate drain pot below the sphere, with a valve between it and the sphere which has to be closed whenever water is drained from the pot. In a cold climate there can then be a hazard of water freezing in the pot and bursting it.

Many hydrocarbon gases form hydrates which solidify at temperatures above 0°C. This tendency increases with the pressure, and can cause blockages in gas pipelines and valves[10] [5.4]. Such pipelines must be thoroughly dried before hydrocarbon gas is introduced under pressure, and the gas must be dried to a low dew-point. Some hydrocarbon liquids solidify at temperatures above 0°C (benzene at 5.5°C, cyclohexane at 6.5°C). Many substances which are solid at the ambient temperature are moved by pipe in their molten state. For pipes carrying water and other non-flammable liquids, a common solution to the freezing problem is to 'trace' exposed pipes, valves and fittings with electrical heating tape and lag them. For some fuel oil pipes, steam tracing is used. The use of hot tracing, however, accentuates the problem of thermal expansion in 'boxed-in' pipes [6.2.1], while only flameproof electrical equipment may be used in zones where flammable gases and vapours may be present. Care must also be taken to determine the correct quantity and resistance of the tape to be used, and to safeguard against short circuits and shocks. For water pipes in which water is usually flowing and where only short lengths are exposed, it is sufficient in many climates to ensure that water is either kept continuously flowing in cold weather or that the pipe is completely drained when frost is forecast.

A serious hazard arises when part of a fire water system is frozen. I saw a large warehouse opposite a hotel where I was staying in Amsterdam burn down at breakfast-time before the fire brigade could pump water from a frozen canal.

Many accidents have been caused by trying to clear blocked pipes by partly dismantling them and/or rodding them through while the contents were still under pressure. Specific instruction and warnings are needed to suit different situations. Special techniques are sometimes needed.

6.2.3 Physical explosions or eruptions

I use the term 'physical explosion' for the sudden vaporisation of a liquid in a closed or partly closed system. The term 'eruption' is sometimes more appropriate when a blast wave is not involved. The phenomenon occurs when a liquid (usually water) suddenly comes into contact with a hot solid or another liquid, or when the energy contained in a superheated liquid system is suddenly released.

Marshall[11] uses the term physical explosion for what I call a 'mechanical explosion', namely the blast wave which follows the sudden rupture of a boiler or pressure vessel. Others might include electrical and even nuclear explosions as physical explosions. An important point to be noted is that the pressure rise resulting from a physical explosion is usually too fast for any standard pressure relief device to handle.

A serious physical explosion in which seven men were killed and several injured occurred on 4 November 1975 at a steel plant at Scunthorpe.[12] This happened when molten steel was being poured into a special transport container into which water had entered through a leak in the cooling system. Molten steel was scattered over a wide area. The largest recorded physical explosion occurred during the Krakatoa eruption in 1883, when the sea poured into a submarine crater containing millions of tons of molten lava.

Physical explosions have frequently occurred during the start-up of distillation columns (especially vacuum columns) of oil refineries, when hot oil has come into contact with a pocket of water lying in a low point of the column or its associated pipework.[13] It is important to eliminate such low points as far as possible during design, and to ensure that the remaining ones are provided with drain valves at the lowest point of the pocket to remove water during start-up. An example of this problem is shown in Fig. 6.1.

It is also possible for a physical explosion to occur in an unstirred tank or vessel containing water and a hydrocarbon layer above it, when one or other of these layers is heated. The phenomenon starts with boiling at the

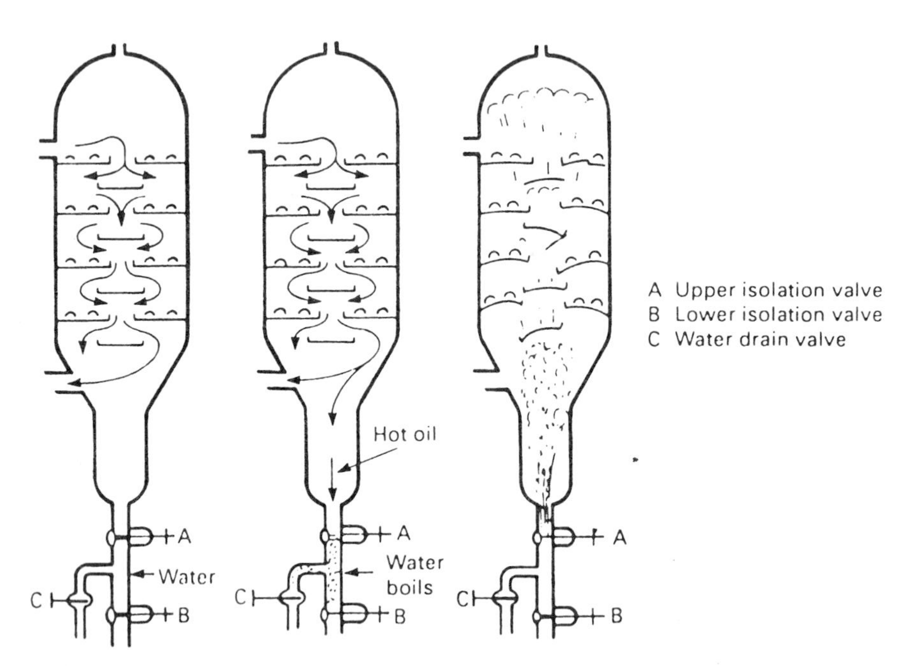

Fig 6.1 Physical explosion caused by mixing water and hot oil inside oil distillation column. The connection from valve C to the pipe joining A and B is too far above B, so that too much water is left undrained.

interface between the two liquid layers. This occurs at a temperature lower than the boiling point of the lower boiling liquid. The initial boiling rapidly mixes the two liquids which boil together at this lower temperature, thereby releasing a good deal of stored energy. I have shown elsewhere[14] how this phenomenon probably triggered a pressure rise causing the rupture (a mechanical explosion) of the by-pass assembly at Flixborough, which in turn was followed by a large vapour cloud explosion. The Pernis disaster discussed in 5.1 was another example of a physical explosion (or eruption) which led to a vapour cloud explosion. Another fire and explosion which apparently resulted from an internal physical explosion occurred on a plant in New Zealand in September 1974.[15]

The number of accidents of this type suggests that the hazard is still not widely recognised. The most important lesson is to avoid heating a tank or vessel containing two immiscible liquids unless one is sure that they are well stirred.

Figure 6.2 is a generalised temperature-composition or phase diagram for two nearly immiscible liquids at constant pressure. When in contact with each other these boil at a lower temperature than that of either liquid alone.

6.2.4 Rollover of LNG in cryogenic storage tanks

Rollover of liquids which are on the point of boiling is accompanied by a sudden increase in pressure and evolution of vapour and can occur in fields as different as jam making and the storage of liquefied natural gas. Its hazards are only discussed here in the last context[15], although the same principles apply generally. At the start of a rollover the entire contents of a tank or vessel may be at their boiling point, which is higher at the bottom of the tank because of the head of liquid above it. This situation is only possible when the composition of the tank contents is non-uniform and in such a way

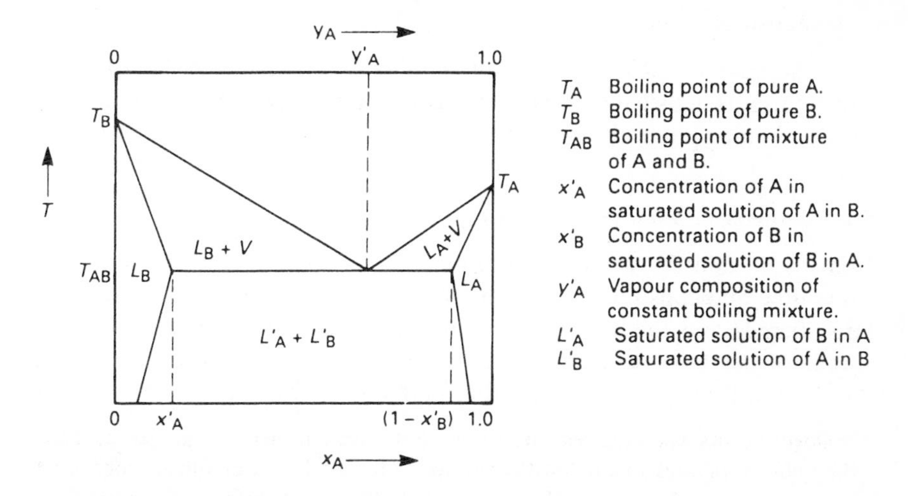

Fig 6.2 Liquid–vapour phase diagram at constant pressure for two liquids A and B of limited mutual solubility.

that the density of the liquid at the bottom would be higher than that at the top when both were measured at the same temperature. Due to its higher temperature, the density of the liquid at the bottom may then become less than that of the liquid at the top, causing the entire contents to roll over. As the hotter liquid rises, its hydrostatic pressure and boiling point are reduced so that it boils violently as sensible heat is converted to latent heat.

Gases are frequently stored as refrigerated liquids in insulated storage tanks or vessels at pressures only slightly above atmospheric. The word 'cryogenic' is used when the storage temperature is below –110°C. The surface of the liquid is in equilibrium with its own vapour. Refrigeration is often achieved by compressing and condensing the gas and returning it as liquid to the tank. If liquid of slightly higher density enters the tank and forms a layer in the bottom, its temperature can rise substantially before its density falls sufficiently for it to rise through the colder liquid above it. This is a rollover. It is accompanied by rapid turnover of the liquid contents, evolution of vapour and a sudden rise in pressure, the extent of which depends partly on the depth of the liquid. Refrigerated liquid storage installations must therefore be designed to withstand the maximum pressure rise which could result from a rollover. Other features which should be incorporated to reduce this hazard include the following.[16]

- *Top (splash) filling.* This relieves superheat in the incoming liquid. If the incoming liquid were superheated and of higher density than the bulk liquid, it would retain its superheat if it entered the tank at the bottom, where the hydrostatic pressure would prevent it from boiling.
- *Provision for circulating liquid from the bottom of the tank to the top.*
- *Provision of means for detecting a superheated state in the tank and predicting the incidence of a rollover.* This requires temperature monitoring at several levels in the tank and may also need the liquid density and/or composition to be monitored at several levels.

In the case of LNG which is predominantly methane the constituents which raise its density are ethane, propane and nitrogen. While the first two raise its boiling point, the last lowers it.

6.2.5 Boilover or slopover in oil tank fires[2]

In many oil tank fires a sudden eruption occurs when the fire has burnt for some time and consumed a considerable proportion of the oil. These cause large flames and the scattering of burning oil over a wide area, with the spread of fire to other tanks, injury to firemen, and the loss of firefighting equipment. Boilover is explained thus in the NFPA Manual:[17]

> Boilover occurs when the residues from the surface burning become denser than the unburnt oil and sink below the surface to form a hot layer which progresses downward much faster than the regression of the liquid surface. When this hot layer, called a 'heat wave', reaches water or a water-in-oil emulsion in the bottom

of the tank, the water is superheated and subsequently boils almost explosively, overflowing the tank. Oils subject to boilover must have components having a wide range of boiling points, including both light ends and viscous residues. These characteristics are present in most crude oils and can be produced in synthetic mixtures.

According to this explanation, a boilover is the result of a physical explosion [6.2.3]. But since a 'heat wave' depends on the density of the hot surface layer exceeding that of the cooler oil below it, boilovers have features in common with rollovers [6.2.4].

This prompts the question 'Is water in an oil tank essential to a boilover?' At the start of an oil tank fire the surface layer loses light ends while its density falls initially because of the increase in its temperature. As the middle oil fractions evaporate its density rises again. The temperature of the lower layer meanwhile rises slowly by conduction while its density falls. By the time the densities of the two layers are again equal, the hot upper layer consists mainly of heavy ends, while the lower layer still retains most of its light ends. Mixing of the two layers could then lead to a boilover caused by the sudden vaporisation of some of these light ends.

Because of the risk of water causing a boilover, special care is needed when using water jets on oil tank fires. Jets should be used to cool the walls of tanks and other objects exposed to flames and radiation, but foam and water fog should be used for fighting the fire. According to Lees,[7] the use of vertical strips of a temperature-sensitive paint on the wall of the tank can warn a trained fireman of the imminence of a boilover.

6.2.6 Irregular boiling of superheated liquids

'Bumping' and eruptions similar to physical explosions and rollovers sometimes occur when organic liquids or alkaline solutions are distilled or evaporated in clean industrial equipment and laboratory glassware. This is pronounced under vacuum. In the absence of bubbles, liquids are capable of sustaining several degrees of superheat before boiling commences, which then occurs with considerable violence, sometimes ejecting liquid from the container, sometimes rupturing it. The problem is cured by providing a source of bubbles in the liquid, e.g. pieces of a porous solid, or by arranging a small flow of inert gas through fine holes at the bottom of the boiling liquid. Laboratory workers should never point the mouth of a test tube which is being heated at anyone. The presence of circular stains on the ceilings of many a chemistry laboratory provides mute evidence of this hazard!

References

1. Adams, J. M., *Electrical safety — a guide to the causes and prevention of electrical hazards*, Institute of Electrical Engineers, London (1994)
2. Hirst, R., *Underdown's Practical Fire Precautions*, 3rd edn, Gower Technical, Aldershot (1989)

3. Geddes, L. A., *Handbook of Electrical Hazards and Accidents*, CRC Press, New York (1995)
4. Regulation 471–16 of Requirements for Electrical Installations, *IEE Wiring Regulations*, 16th edn (BS 7671), Institute of Electrical Engineers, Stevenage (1992)
5. API RP:520 *Recommended practice for the design and installation of pressure relieving devices in refineries*, American Petroleum Institute, New York (Part 1, 3rd edn, 1967, Part 2, 2nd edn, 1963)
6. BS 5355: 1976, *Specification for filling ratios and developed pressures for liquefiable and permanent gases*
7. Lees, F. P., *Loss Prevention in the Process Industries*, Butterworths, London (1980)
8. HSE, *The Storage of Liquefied Petroleum Gas at Factories*, HMSO, London (1973)
9. Anon., 'Safety in design of plants handling liquefied light hydrocarbons', *Loss Prevention Bulletin 042*, The Institution of Chemical Engineers, Rugby, (1981)
10. King, R. W. and Majid, J., *Industrial Hazard and Safety Handbook*, Butterworths,
11. Marshall, V. C., 'Dust explosions and fire balls', paper given at Major Hazards Summer School, Cambridge, organised by IBC Technical Services Ltd, London (1986)
12. HSE, *The Explosion at Appleby-Frodingham Steel Works, Scunthorpe on 4th November 1975*, HMSO, London
13. The American Oil Company, *Hazard of Water*, 5th edn, AMOCO, Chicago (1964)
14. King, R. W., 'Flixborough — The role of water re-examined', *Process Engineering*, Morgan Grampian, London, pp. 69 to 73, September (1975)
15. *Report of the commission of inquiry into the explosion and fire which occurred at the factory of Chemical Manufacturing Company Ltd on 26th September 1974*, Government Printer, Wellington, New Zealand (1975)
16. Drake, E. M., 'LNG rollover — update', *Hydrocarbon Processing*, January 1976, pp. 119–122
17. NFPA 30, *The Flammable and Combustible Liquids Code*, National Fire Protection Association, Boston, Mass. (1984)

7 Health hazards of industrial substances

Substances are considered toxic if they have some adverse effect on the human body, although as Paracelsus (1493–1541) said, this all depends on the size of the dose — anything is a poison if you take enough of it. Thus while very small quantities of elements such as manganese, fluorine and iodine are essential to health, they are toxic in larger doses. Calciferol or Vitamin D_2, which is essential to development in all mammals, is lethal in higher doses and is used in formulation with warfarin as rat poison.[1]

Bitter experience has shown that persistent exposure of workers to relatively low levels of many industrial substances, particularly chemicals, can produce chronic disease, leading to serious disability or premature death. Many chronic health hazards of the last century, e.g. silica, yellow phosphorus and lead used in the manufacture of cutlery, matches and paint, respectively, have been overcome by the use of safer materials. In recent decades their place has been taken by other hazards, some new ones such as vinyl chloride, others, old ones in new guises, such as asbestos and chromates. It seems inevitable that new problems will continue to appear as old ones are solved.

Most readers, particularly those who have worked in production or research in the chemical and allied industries, know some of these problems only too well. I knew at least four former colleagues who worked with known carcinogens and died in youth or middle age from cancers — a professor of organic chemistry specialising in higher aromatics, a research chemist in a dyestuffs company, another working for a company making asbestos brake linings, and a physicist working on nuclear-powered submarines. I was also in Abadan in the 1940s when 40 refinery workers died from TEL poisoning [1.2.1]. Those like me who are in their seventies and enjoy good health are the lucky ones.

Thanks to the many man-made chemicals in use today, toxicology has become an advanced scientific discipline. In spite of continuous research for several decades, our present knowledge on the extent of ill-health caused by exposure to harmful chemicals and their vapours is still very limited. Most of us rarely see more than the tip of the iceberg. Disease usually follows low-level occupational exposure to a variety of harmful substances over many years. Their effects are chronic rather than acute, and may be confused with those arising from other causes inherent in the worker's general health or life-style.

Several occupations in specific industries where workers have been exposed to particular hazardous process materials (mainly carcinogens) show appallingly high fatal incident frequency rates. A selection of the worst ones reported by Kletz[2] is given in Table 7.1. The work of Hamilton and Hardy[3] in the USA and Hunter[4] in Britain gives a clear and balanced picture.

While toxicology aims to study the effects of substances on the human body, living animals, which are less articulate than humans and enjoy fewer legal rights, are mainly used for experiment. Records of previous chemical exposure of workers also provide clues.

A very large amount of biological testing and health screening is now necessary in many countries before a new chemical is manufactured. In the USA these requirements are quite specific under the Toxic Substances Control Act.[5] In spite of this, a study made in 1984 showed that there was no toxicity data for nearly 80% of the chemicals used in commercial products and processes in the USA.[6] In the UK the obligation of employers to examine and protect workers against the health hazards of substances used in industry is implied under 'General duties' of Sections 2, 5, and 6 of HSWA 1974.

Current thinking in the EC is reflected by Council Directive No. 80/1107/EEC. This has led in the UK to the Control of Substances Hazardous to Health Regulations 1988[7] (COSHH) [2.5.3]. Other EC coun-

Table 7.1 Fatal incident frequency rates in particular occupations (FIFR is defined as fatalities per 10^8 working hours, which approximately equals the number of deaths in a group of 1000 employees during their working lives) (courtesy Kletz[2])

Occupation	*Cause of fatality*	*FIFR*
Coal carbonising	Bronchitis and cancer of the bronchus	140
Viscose spinners (aged 45–64)	Coronary heart disease (excess)	150
Asbestos workers		
Males (smokers)	Cancer of the lung	115
Females (smokers)	Cancer of the lung	205
Rubber mill workers	Cancer of the bladder	325
Mustard gas manufacture (Japan 1929–1945)	Cancer of the bronchus	520
Cadmium workers	Cancer of the prostate	700
Amosite abestos factory	Asbestosis	205
Nickel workers (employed before 1925)	Cancer of the nasal sinus	330
	Cancer of the lung	775
β-Naphthylamine manufacture	Cancer of the bladder	1200

tries now have similar legislation. The measures called for under COSHH are discussed in 7.8.

While COSHH revokes most of the previous piecemeal legislation for the protection of workers' health in particular industries, it follows, extends and complements other recent regulations such as the Classification, Packaging and Labelling of Dangerous Substances Regulations[8] (CPLR). Thus to the questions, 'How do we know if a substance is harmful to health, and in what way can it be harmful?' the first answer is: 'Read the label!' (Whether this always gives adequate warning, e.g. as in the case of tobacco, is debatable.) Besides protecting the health of workers, COSHH aims to extend our knowledge of how it is affected by the substances to which they are exposed at work.

Added to the fivefold classification of substances hazardous to health given in the supply provisions of CPLR, three overlapping classes of very dangerous substances listed in Part 1 of Schedule 1 of the first draft of the COSHH regulations[7a] are given in Table 7.2. Special precautions are needed in handling substances possessing these properties.[9]

Carcinogenicity concerns the ability of the material to produce cancer in human beings. Studies have suggested that well over 50% of human cancer has its origin in occupational and other man-made environments. According to data available in 1980, about 2000 chemicals were then suspected carcinogens.[9] Among the methods of testing used, the use of bacteria has proved to correlate well with the results of human exposure, and to be speedy and economic. Up-to-date lists of chemicals for which there is substantial evidence of carcinogenicity, and specific information about them, are prepared by the Carcinogen Assessment Group of the (US) Environmental Protection Agency.

Teratology is concerned with the birth of abnormal offspring when the material is administered to pregnant females. Teratogenic effects are produced by a variety of mechanisms. Their study is based mainly on administering the test material to the pregnant animal during a critical period of pregnancy.

Mutagenicity is concerned with the ability of the material to produce genetic changes. It is closely related to carcinogenicity and teratology. While

Table 7.2 Three classes and characteristic properties of substances with extreme health hazards

Class	*Characteristic properties*
Carcinogenic	A substance which if it is inhaled or ingested or if it penetrates the skin may induce cancer in man or increase its incidence
Teratogenic	A substance which if it is inhaled or ingested or if it penetrates the skin may involve a risk of subsequent non-hereditable birth defects in offspring
Mutagenic	A substance which if it is inhaled or ingested or if it penetrates the skin may involve a risk of hereditable genetic defects

difficult to investigate, it is of immense importance to the whole human race, since genetic changes once made cannot be undone, and are passed on from one generation to the next. Since our genetic make-up is the finely balanced result of long aeons of natural selection, any accidental genetic change resulting from chemical exposure is likely to be for the worse.

Before discussing the health hazards of various substances further, we first consider the roles of the professionals who deal with them.

7.1 Occupational health professionals

Professionals trained in several disciplines are employed to deal with health problems arising from exposure to harmful substances at work. Some are in government service, some are employed by industry, some work in universities and research institutions and some as consultants. While the professionals listed below have their own special fields of responsibility, because of their limited numbers there is a good deal of overlap in their work. Four types of professional are most involved with day-to-day problems in industry:

- the occupational physician
- the occupational hygienist
- the occupational nurse
- the dermatologist.

Professionals involved in more basic research include:

- the toxicologist
- the epidemiologist.

Ergonomists and psychologists are also sometimes involved.

7.1.1 The occupational physician[10]

Prior to 1973, there were some 1300 appointed factory doctors in the UK. Most were general practitioners who carried out various statutory duties such as:

- examination of young people first entering employment;
- regular examination and in some cases certification and medical supervision of workers in processes covered by regulations;
- investigation of gassing incidents and industrial diseases.

This system changed in 1973 with the formation of the Employment Medical Advisory Service (EMAS), under the Department of Employment. This is staffed by doctors with appropriate qualifications. Their services are provided free except for statutory medical examinations. They have powers to investigate and advise on occupational health problems on their own initiative and at the request of employers, employees, trade unions, factory inspectors and others.

Doctors employed within industry are known as occupational physicians. Their recommended duties include:[10]

1. responsibility for the health of the whole enterprise;
2. concern with the working environment and its health hazards;
3. accessibility to workers for individual consultations and investigation of work-related hazards arising from them;
4. education and advice for management and employees on work-related health problems, especially those arising from new processes, materials and equipment;
5. in a comprehensive health team, the occupational physician should be the leader and co-ordinator, while recognising the mastery of other specialists such as occupational hygienists in their own fields.

The occupational physician is often employed on a part-time basis, making regular visits to the establishment, which often has a well-equipped medical centre with a full-time trained nurse in charge of it. He is involved to a major extent with medical examination of workers and to a lesser extent with treatment. While he should not duplicate or take over the role of the general practitioner, skin diseases, other ailments and minor injuries arising from the working environment can often be best treated in the works medical centre. Under the COSHH Regulations he will normally be the 'appointed doctor'. He will be expected to have overall control of occupational health records and health surveillance procedures called for under the regulations. He will also be responsible for medical surveillance of employees liable to be exposed to substances listed in Appendix 1 of the general ACOP published with the regulations. The duties involved in medical surveillance are discussed in 7.8.1.

Medical examinations should be based on practical needs as well as statutory requirements. They fall into two groups, pre-employment examination and routine examination. Medical examinations are most needed for those whose work carries a specific health hazard.

7.1.2 The occupational hygienist

The British Occupational Hygiene Society, which is open to anyone with an interest in the subject, has about 1500 members, most of whom are professionally involved in it. The Institute of Occupational Hygienists, which has an entrance examination, has a somewhat smaller membership. Several universities and technical colleges have courses in occupational hygiene. About 1000 British occupational hygienists are understood to be employed in industry. Some large companies have well-established occupational hygiene departments whose duties include the monitoring of hazardous substances in the working atmosphere [7.7]. In medium-size and smaller companies with no occupational hygienist, this work may be done by a staff chemist or safety professional, or by one of several consulting firms which offer industrial hygiene services on a fee-paying basis.

In the USA the profession received considerable stimulus with the passing of the Occupational Safety and Health Act in 1970. The COSHH Regulations should provide a similar stimulus in the UK. The American Industrial Hygiene Association has given clear definitions of the profession and its practitioners, including their training[11] which should equip them:

1. to recognise the environmental factors and stresses associated with work and work operations and to understand their effect on Man and his well-being;
2. to evaluate, on the basis of experience and with the aid of quantitative measurement techniques, the magnitude of these stresses in terms of ability to impair Man's health and well-being;
3. to prescribe methods to eliminate, control or reduce such stresses where necessary to alleviate their effects.

The stresses which concern the industrial hygienist are chemical, physical, biological and ergonomic, and thus basically include all the health hazards considered in this book.

To evaluate health hazards in the working environment first requires detailed standards on tolerable levels of such hazards, and second, means of measuring the levels at which they are present. The standards with which we are mainly concerned in this chapter are the Occupational Exposure Limits for airborne substances given in HSE's Guidance Note 40[12] and discussed further in 7.5. Strategies for such monitoring are suggested in HSE's Guidance Note 42[13] and discussed further in 7.7. This work of the occupational hygienist should complement that of the occupational physician in monitoring the health of the workers.

As well as this monitoring task, the occupational hygienist is trained to recommend remedies to problems which are revealed in the course of it. Such remedies are basically of three types:

- finding a less hazardous substitute for the harmful substance which causes the problem;
- employing suitable means for reducing the intensity of the hazard in the working area. Typical examples are the use of local exhaust ventilation, and the substitution of a canned or diaphragm type pump for one with a packed gland when handling highly toxic materials;
- providing personal protective devices such as dust masks and respirators and ensuring that these are worn. This is a remedy of last resort.

7.1.3 The occupational nurse[14]

Trained nurses are employed in many manufacturing establishments. They take on many of the duties of the occupational physician, both in medical examinations (e.g. of sight and hearing) and in the treatment of some injuries and diseases, particularly those of the skin. Most manufacturing enterprises have a first-aid centre or medical department with a full-time occupational nurse in charge during day working hours. Emergency first-aid treatment needed during shift work when the nurse is not present is given by trained voluntary first-aid workers, as required by the Health and Safety (First-Aid) Regulations.[15] When nurses are available to treat persons who are injured or suffering from exposure to harmful substances, they are usually in the best position to handle reports required under RIDDOR[16] and other regulations, and to ensure that all relevant details provided by the injured or affected person, especially in regard to causation, are correct. In doing this they often become involved in occupational hygiene work. Many of the responsibilities of medical record keeping, which are more demanding under COSHH, fall on the shoulders of the occupational nurse.

7.1.4 The dermatologist[17]

The dermatologist is primarily concerned with diagnosing, treating and preventing skin diseases, many of which are caused by allergic reactions to particular substances. Many skin disorders have psychological causes and a good dermatologist has to be a bit of a psychologist as well. In diagnosing allergens that cause skin complaints in particular workers, the dermatologist can be assisted by observations made by other workers, supervisors, nurses, physicians and safety professionals.

7.1.5 The toxicologist[18]

The toxicologist is generally a research scientist specialising in the poisonous effects of substances on the human body. HSC's Advisory Committee On Toxic Substances (ACTS), which represents the views of British toxicologists, advises HSE on the subject. Methods used by toxicologists for assessing the toxicity of substances used or produced in the process industries include:

- epidemiological studies [7.1.6]
- experiments on animals
- experiments with micro-organisms.

Rats are mainly used for testing the toxicity of chemicals, despite protests from animal welfare organisations. Strict standardisation of several factors in such tests has been established by the EC to give statistically valid results when a group of animals is exposed to a chemical. These factors include route of admission (oral, dermal, inhalation), strain of rat, sex, age, temperature, diet, housing, season and time of observation. The toxicity of different substances and their effects on the human body are discussed in 7.3.

7.1.6 The epidemiologist[10]

The epidemiologist tries to diagnose the causes of outbreaks of diseases from the activities and environments of the population exposed to them. He is well versed in the use of statistics and computers. He is involved in the preparing of questionnaires for surveys involving large numbers of people, and in the analysis of the data obtained from them. While every occupational physician and most occupational nurses like to don an epidemiologist's hat from time to time, full-time epidemiologists are more concerned with the frequency and distribution of diseases in defined populations than with day-to-day medical practice.

Epidemiologists have played an important role in bringing to light the health hazards of several materials which were previously thought to present little danger.

7.2 How harmful substances attack us

To produce a harmful effect a substance has either to attack an external surface of the body and/or to enter it in some way before attacking an internal organ. Three modes of entry need to be considered in the industrial context: by swallowing, through the skin, and by breathing. (Other modes used medically are by injection into a vein, muscle, or under the skin.) Once a toxic substance has entered the blood-stream it is transported to most internal organs, one or more of which may be attacked.

7.2.1 External attack

Some corrosive liquids and solids, particularly acids and alkalis and strong oxidising agents, attack the skin directly, causing inflammation and sometimes destroying living cells. Many organic liquids, especially solvents, dissolve and remove natural oils and fats and so render the skin more vulnerable to injury. Inflammation is a biological response to tissue injury in which the flow of blood increases and particular blood cells migrate to the affected area. It is one of our immune mechanisms for the repair of tissue damage.[19] This response can become hypersensitive after repeated exposure to small traces of many substances, e.g. ethylene diamine. The result is allergic dermatitis.[17] Sometimes the action of strong light on the skin exposed to the chemical is needed to complete the process, known then as photo-sensitisation. Creosote is a common example of this.

The skin is a sensitive indicator not only of symptoms of external attack but also of emotional reactions, which may result in pallor or blushing.

Some allergies respond to psychiatric treatment. The development of allergic symptoms in one or two workers out of a number sharing the same environment can create special problems. The first step is to identify the material responsible (the allergen). This may involve special 'patch' tests under the supervision of a dermatologist.[17] If total protection against it cannot be guaranteed, it is advisable to transfer the worker to another job where he or she will no longer be exposed to the allergen.

The eyes are even more sensitive to harmful substances than the skin, and are much affected by sensory irritants in the atmosphere — fine dusts, and gases such as ammonia and sulphur dioxide. Because the eyes are so vital, there are many situations where eye protection [22.6.5] is essential. Gloves [22.6.2] and special clothes [22.6.1] to protect the skin are also needed in many of these situations.

A large number of substances can enter the body and attack it internally without affecting the eyes and skin.

7.2.2 Ingestion (swallowing)

Of the three methods of entering the body, swallowing is the most easily controlled, since eating and drinking are deliberate acts which should not be necessary at the workplace. Entry of many toxic materials into the body by ingestion is also less dangerous than by other means, since oral toxicity is in many cases lower than toxicity from inhalation or skin penetration.[6] The provision of adequate lockers, changing rooms, washrooms, canteens and restrooms and the enforcement of strict standards of hygiene and housekeeping are, however, essential in places where toxic materials are handled or processed. A total ban on eating, chewing, sucking, drinking and smoking in the workplace may be necessary when toxic or harmful materials are present. The dangers of putting fingers, pencils, etc. on which toxic dust has settled into the mouth should be pointed out, as such actions are often done unwittingly. The sucking by mouth on the ends of flexible tubes to start syphoning liquids from carboys and other containers should be banned, and proper equipment for the safe and easy transfer of liquids should be provided. Labels, small brushes and sewing thread should not have to be licked, especially when toxic substances are around. In laboratories, the use of mouth pipettes for measuring and dispensing solutions and of laboratory beakers for tea and other drinks should be prohibited and proper alternatives provided.

7.2.3 The dermal route (entry through the skin)

Several toxic liquids and their vapours pass readily through the skin into the bloodstream. These include tetraethyl lead, carbon disulphide, hydrogen cyanide, aromatic hydrocarbons such as benzene and toluene, nitrocompounds such as nitrobenzene, trinitrotoluene, and nitropropane, aromatic amines such as aniline, polychlorinated hydrocarbons such as trichloroethylene, organic phosphates, nicotine, and several insecticides and herbicides.

Outer clothing of PVC-impregnated fabrics with a nylon or terylene base offers good protection against most accidental chemical splashes, with the exception of some organic solvents and hydrofluoric acid. Supposedly impervious outer clothing only aggravates the problem, however, if the liquid penetrates it and is absorbed on underclothing. Where this is suspected, particularly in a worker who has collapsed, one of the first actions should be to remove his clothing, wash the affected area of his body with soap and warm water, and dress or wrap him in clean underclothing or pyjamas. If he is taken to hospital by ambulance while still wearing contaminated clothing, he may be dead on arrival. It is important to inform suppliers of protective clothing of the chemicals and harmful substances against which protection is needed, and secure their help and advice.

Health and safety precautions should include listing all liquids used which are absorbed through the skin, warning workers of the dangers of their contact with skin and clothing, and providing showers or suitable washing facilities and clean clothing for anyone affected.

7.2.4 Inhalation (entry by breathing)

The main route taken by toxic substances into the human body is by inhaling. The human respiratory system consists of an upper part, the nose, mouth and larynx, and a lower part, the trachea, bronchus, bronchioles and alveoli. The last two are the gas-exchange components of the lungs, and have a light, porous and spongy structure. The alveoli, of which there are a very large number, are blind cup-like pouches with thin elastic walls through which gases and vapours can pass readily into and out of the bloodstream. The system, which has a capacity of 5–50 1/min of air, is an easy route for contaminants to enter the body. Most gases and vapours reach the alveoli, but ones like ammonia and hydrogen chloride which are very soluble in water are absorbed before they get so far.

Whether particles will enter the respiratory system, how far they are likely to penetrate, and whether they will be trapped depends largely on their size. Particles whose mean diameters are greater than 50 μm do not usually enter it. Those with diameters greater than 10 μm are deposited in the upper respiratory tract. Those in the range 2–10 μm are deposited progressively in the trachea, bronchi and bronchioles. Only those smaller than 1–2 μm reach the alveoli. These are generally invisible to the naked eye, and are the most dangerous. Fibres with diameters of 3 μm and less and lengths of up to 50 μm can enter the lung; those longer than 10 μm tend to be trapped there. Fibres with diameters less than 1.5 μm and lengths up to 8 μm have the greatest biological activity. Asbestos was once the worst offender. Now that glass fibres with the same physical characteristics are becoming common, they may present an equal hazard.

7.3 Effects on body organs

Organs frequently affected are the respiratory system itself, the blood and bone marrow, the liver, the kidneys, and the brain and central nervous system. Some chemicals have specific effects on organs such as the bladder, the pleural cavity, the gastrointestinal tract, the prostate, the nasal cavities and larynx.

7.3.1 The respiratory system

The natural protection of the respiratory system includes:

- a wet filter formed by the fine hairs (cilia) of the nose and trachea. Particles trapped there are washed out by mucus which can be eliminated by sneezing, blowing the nose, spitting or swallowing;
- involuntary nervous contraction of the bronchioles which is triggered by various air contaminants and restricts breathing. This results in a struggle for breath and escape into cleaner air. Some subjects become hypersensitive to pollens and chemicals such as diisocyanates and fine spray containing platinum salts, which produce symptoms of asthma and hay fever.

Contaminants which primarily affect the respiratory system fall into five groups.

- Reactive gases such as ammonia, chlorine and sulphur dioxide; these irritate the upper respiratory system, swell the walls of the airways and cause involuntary flight reactions.
- Substances such as isocyanates, chromates and some wood dusts which cause bronchial restriction, and to which sensitive persons may become allergic.
- Most mineral dusts; some such as iron oxide appear merely to block the nasal passages and reduce breathing capacity; others such as silica and asbestos cause harmful and irreversible changes in the lung structure.
- Some contaminants which include blue asbestos, chromium and arsenic compounds and some complex hydrocarbons found in tars lead in time to the growth of malignant tumours in the respiratory system.
- Asphyxiant gases such as carbon dioxide and nitrogen which are all but inert; when present in high concentrations these reduce the oxygen concentration of the atmosphere to levels below those needed for normal work (about 18%) or life itself (about 10%).

7.3.2 The blood and bone marrow

The blood which is the internal transport medium for the whole body and the bone marrow which produces the red and white blood cells as well as platelets which initiate blood clotting following a wound are attacked by carbon monoxide, lead compounds, benzene and TNT, among other compounds.

7.3.3 The liver

The liver is the main detoxifying organ in the body. Most chemicals entering the body reach it and are converted there to other compounds which may be more or less toxic than the original ones. Most are more soluble in water than the original compounds, and end up in bile which is removed in the urine or faeces. Although the liver is fairly resistant to toxins, massive doses will inflame it, causing hepatitis and jaundice. Chlorinated hydrocarbons including insecticides such as DDT damage the liver, and vinyl chloride monomer causes a malignant liver tumour known as angiosarcoma.

7.3.4 The kidneys

The kidneys maintain the required levels of water, salt and pH in the body. They are damaged by compounds of mercury, cadmium and lead and by some organic compounds which include carbon tetrachloride.

7.3.5 The brain and central nervous system

The brain and central nervous system are affected by a number of organic compounds which include ethanol, nicotine and other alkaloids, diethyl ether and other anaesthetics, and a range of volatile solvents which are mild narcotics. The vapours of mercury and many volatile organo-metallic compounds also affect the central nervous system. The effects of many compounds on the central nervous system are fairly specific, producing hallucinations, exuberance, depression, tension, sedation, etc. Most of these symptoms can also have psychological causes.

7.4 Units and classes of toxicity

7.4.1 Units of toxicity

The main unit is the lethal dose for rats which is determined under standard laboratory conditions [7.1.5]. It is usually expressed as LD_{50}, defined as that dose administered orally or by skin absorption which will cause the death of 50% of the test group within a 14-day observation period.[20] It is generally expressed as mg or μg of substance per kg of body weight of the animal (mg or μg/kg). All test conditions must be standardised, with proper controls. A number of experimental animals of as uniform a population as possible must be sacrificed in the test, since individual responses vary considerably. The results must be subjected to rigorous statistical analysis. The LD_{50} unit was developed in 1927 for the standardisation of drugs. It is now used for virtually all chemicals and has legislative backing.

Since LD_{50} is based on oral or dermal entry into the body, another unit, the lethal concentration LC_{50}, is used for airborne materials which are inhaled. This is defined as the concentration of airborne material, the four-hour inhalation of which results in the death of 50% of the test group within a 14-day observation period.[20]

7.4.2 Classes of toxicity

The classifications 'very toxic', 'toxic' and 'harmful' are used in CPLR. More extensive toxicity ratings of substances were given by Hodge and Sterner,[21] who placed them in six classes, according to their probable lethal human dose taken orally (Table 7.3).

Chemicals in the super-toxic class of this table are included in the 'very toxic' CPLR class. The CIMAH Regulations[22] also recognise such 'super-toxins', and list a number of them.

7.4.3 The NFPA health hazard rating[23]

The NFPA provides health hazard ratings of chemicals in addition to those for reactivity and flammability. The main purpose of these ratings is as a guide for firefighting, where exposures are short (from a few seconds up to an hour). The resulting health hazards arise both from the properties of the chemicals themselves and from those of their combustion products. The hazard rating for any chemical indicates the nature and degree of protection required by those exposed to its spillages and fires.

7.5 Occupational exposure limits (OELs)

While LD_{50} and LC_{50}, values are derived from experiment, OELs (a more arbitrary set of values) have been established and published in various countries for the control of several hundred airborne substances in the working atmosphere. They are the maximum concentrations in air of the substances

Table 7.3 Toxicity classes (Hodge and Sterner[21])

Class	*Toxicity*		*Probable oral lethal dose (human) for 70 kg person*
6	Super-toxic	<5 mg/kg	A taste (less than 7 drops)
5	Extremely toxic	5–50 mg/kg	Between 7 drops and 1 tsp
4	Very toxic	50–500 mg/kg	Between 1 tsp and 1 oz
3	Moderately toxic	500–5000 mg/kg	Between 1 oz and 1 pint
2	Slightly toxic	5–15 g/kg	Between one pint and one quart
1	Practically non-toxic	15 g/kg	More than one quart

which should not be exceeded in the breathing zone of workers. OELs are given for gases, vapours, liquid droplets and solid particles, but not for substances which are hazardous solely through their radioactive or pathogenic properties.

Published OELs apply to single substances only. In the absence of other data the effect of two or more toxic gases/vapours is estimated by adding the fraction:

$$\frac{\text{(concentration ppm)}}{\text{OEL (ppm)}}$$

for each of them. If the sum of these fractions exceeds unity, the OEL for the mixture is assumed to be exceeded. But since some toxic/irritant substances tend to neutralise each other while others interact to produce more serious results than the sum of their individual effects (synergism), expert advice is often needed when dealing with two or more toxic/irritant airborne substances present at the same time.

As more has been learnt about the toxicities of some substances, their OELs have had to be reduced. The OEL for vinyl chloride which was set at 500 ppm in 1962 has subsequently been reduced to 3 ppm on an annual basis, since it was discovered in 1974 to be the cause of a rare liver cancer. Other sharp reductions have been found necessary for benzene, acrylonitrile, carbon tetrachloride and many other compounds, and more are expected to follow as knowledge increases. Compliance with a published OEL is thus no guarantee that the atmosphere is harmless, and the concentrations of all unnatural airborne contaminants should be reduced as far as is reasonably practicable.

Limits for gases and vapours are specified both as ppm by volume and as mg per m^3 in air, but limits for solids and liquid droplets can only be given as mg per m^3.

7.5.1 OELs for the UK

OELs for the UK are set, published and revised annually by the HSE.[12] Until 1984 they were based on those recommended and published in the USA by the American Conference of Governmental Industrial Hygienists and known as threshold limit values (TLVs). OELs in the UK fall into two categories:

1. *Control limits*. These are given in regulations, ACOPs, EC directives or have been agreed by HSC. They are judged to be 'reasonably practicable' for all work activities in the UK. Failure to comply with control limits may result in enforcement action.

2. *Recommended limits*. These are limits recommended by HSE, usually on advice from HSC's Advisory Committee on Toxic Substances. They provide realistic criteria for plant design and engineering and control of exposure, and assist in the selection and use of personal protective equipment.

'Maximum exposure limit' (MEL) and 'occupational exposure standard' (OES), both used in the COSHH Regulations, are new terms for 'control limit' and 'recommended limit' respectively. The latest ideas on toxic substances which may be reflected in future Guidance Notes EH 40 are given in the *Toxic Substances Bulletin*, published by HSE.

The number of substances and groups of similar compounds for which both categories of limits are given has risen steadily and Guidance Note EH 40/89 contains over 700 entries. Table 7.4 lists the substances assigned maximum exposure limits in Schedule 1 of COSHH. This corresponds to the control limits listed in EH 40/89. Asbestos, coal dust and lead which are subject to other regulations are not included in Table 7.4.

OELs for toxic substances which feature in Part 2 of Schedule 3 of the CIMAH Regulations are given in Table 7.8.

HSE Guidance Note EH 40/89 gives two sets of values for many substances — for long-term (usually 8-hour) and short-term (normally 10-minute) exposures. Specific short-term exposure limits are given for substances for which brief exposure may cause acute effects. Both are expressed as time-weighted averaged concentrations (TWAs) over the period specified. For substances for which no short-term limit is listed, HSE recommends that a figure of three times the long-term limit, averaged over 10 minutes, be used as a guideline for short-term excursions.

With the exception of the annual control limit for vinyl chloride monomer, and the recommended limit for cotton dust, all the limits relate to personal exposure measurements [7.7.7] rather than background levels.

The airborne substances listed enter the body mainly through normal respiration, but some (marked 'Sk' in the table) can also enter through the skin. There are no sharp dividing lines between 'safe' and 'dangerous' concentrations, and for some substances there are no apparent thresholds below which adverse effects do not occur.

The substances listed in Table 7.4 show long-term limits ranging from 0.005 to 1900 mg/m^3. The stringency of the precautions needed in handling these different substances thus varies considerably. While the limits for one substance may be met without difficulty by general dilution ventilation, even the best local exhaust ventilation will not meet the more stringent limits of some others, which may require handling by remote control in totally enclosed systems (e.g. 'power fluidics'). While some airborne vapours may be safely discharged from building and ventilation vents to the outside atmosphere, others must be removed by adsorption, neutralisation or incineration to prevent unacceptable environmental pollution.

Dusts require special consideration[24,25] since only the smaller particles are liable to be inhaled. As an approximation it is often assumed that half of an airborne dust is respirable. Some substances present in minute concentrations which are well below the published OELs can cause allergic reactions in persons who have become sensitised to them.[10] The absence of a substance from the list does not imply that it is harmless, since the toxicities of

Table 7.4 List of substances assigned maximum exposure limits

Substance	*Long-term max. exposure limit (8-hour TWA value)*		*Short-term max. exposure limit (10-minute TWA value)*		*Notes*
	ppm	mg/m³	ppm	mg/m³	
Acrylonitrile	2	4	–	–	Sk
Arsenic and compounds, except arsine and lead arsenate (as As)	–	0.2	–	–	
1,3 Butadiene	10	–	–	–	
Cadmium and cadmium compounds, except cadmium oxide fume and cadmium sulphide pigments (as Cd)	–	0.05	–	–	
Cadmium oxide fume (as Cd)		0.05	–	0.05	
Cadmium sulphide pigments respirable dusts (as Cd)		0.04	–	–	
Carbon disulphide	10	30	–	–	Sk
Dichloromethane	100	350	–	–	
2,2′-Dichloro-4,4′ methylene dianiline (MbOCA)	–	0.005	–	–	Sk
2-Ethoxy ethanol	10	37	–	–	Sk
2-Ethoxy ethyl acetate	10	54	–	–	Sk
Ethylene dibromide	1	8	–	–	
Ethylene oxide	5	10	–	–	
Formaldehyde	2	2.5	2	2.5	
Hydrogen cyanide	–	–	10	10	Sk
Isocyanates, all (as–NCO)	–	0.02	–	0.07	
Man-made mineral fibre	–	5	–	–	
1-Methoxypropan-2-ol	100	360	–	–	
2-Methoxyethanol	5	16	–	–	Sk
2-Methoxyethyl acetate	5	24	–	–	Sk
Rubber fume[a]	–	0.75	–	–	
Rubber process dust	–	8		–	
Styrene	100	420	250	1050	
1,1,1-Trichloroethane	350	1900	450	2450	
Trichloroethylene	100	535	150	802	Sk
Vinyl chloride[b]	7	–	–	–	
Vinylidene chloride	10	40	–	–	
Wood dust (hard wood)	–	5	–	–	

[a] Limit relates to cyclohexane-soluble material.

[b] Also subject to an overriding annual maximum exposure limit of 3 ppm.

a high proportion of the substances found in the process industries have not yet been fully assessed. All substances should therefore be handled with care.

7.5.2 OELs in other countries

In 1980 the ILO published the second edition of its report on the OELs for airborne toxic substances adopted in 18 countries.[26] This contains a review of the methodologies adopted by these countries, a list of 1116 substances with the OELs adopted for them, and sections on particulate matter and carcinogens showing how they are dealt with by the 18 countries. The report provides data for the following countries: Australia, Belgium, Bulgaria, Czechoslovakia, Finland, East Germany, West Germany, Hungary, Italy, Japan, Netherlands, Poland, Romania, Sweden, Switzerland, the USSR, the USA and Yugoslavia as well as the Council of Europe. UK figures were not shown separately as the US figures then applied in the UK. While there are considerable differences between the OELs adopted by various countries, the limits for any substance at any one time in most industrialised non-communist countries usually lie within a fairly narrow band. The figures for Sweden are generally lower, i.e. a quarter to a half of these figures, while those for the USSR are even lower, particularly for organic compounds. Some of these differences have been discussed by Fawcett.[9]

7.6 Sources of exposure to airborne substances hazardous to health

Sources of exposure are classed by Carson and Mumford[6] as:

- *Periodic emissions* which arise from the need to open or enter the 'system' occasionally, for example, during sampling, cleaning, batch additions, bulk tank-car loading, line breaking, etc. Periodic emissions tend to be large and include both anticipated events and unplanned releases, in which human error may be a factor.
- *Fugitive emissions* are small but continuous escapes from normally closed sources; 15–20% of total volatile organic chemical emissions are fugitive. They occur from dynamic seals such as valve stems and pump or agitator shafts and from static seals such as flange gaskets.

The consequences of these two types of emission may be different. A periodic emission which results in a high local airborne concentration of a toxic substance may cause acute effects on an exposed person, which qualify as an accidental injury [3.2.3]. Fugitive emissions of toxic substances may produce airborne concentrations high enough to eventually cause occupational diseases [3.2.4] in those chronically exposed to them, but are unlikely to produce the acute symptoms associated with an accidental injury.

7.6.1 Periodic emissions

Several causes of emissions, mostly of this type listed by Carson and Mumford, are given below under the headings of spillage, leakage, unintended venting, failure of item at normal working pressure and failure of item due to excessive pressure. Some present fire as well as toxic hazards and several are discussed elsewhere in this book. The list should be of value in keeping accident records and identifying and controlling the hazards responsible for them. Causes which could appear under more than one heading are given only under the first one. The list includes contributory causes of all the major accidents discussed in Chapters 4 and 5.

Causes of spillages

Overflow, backing-up, blowback, air-lock, vapour-lock;
Excess pressure, wrong routing, loss of vacuum;
Vessel damaged, tilted, collapsed, vibrated, overstirred;
Overloading of open channel/conveyor;
Poor isolation, drains or doors open, flanges uncovered;
Failure of control or major service;
Surging, priming, foaming, puking, spitting;
Condensed products in vapour, change in normal discharge;
Malicious intent, vandalism.

Causes of leakage

Broken, damaged or badly fitted pipe, vessel, instrument, glass, gasket, gland, seal, flange, joint or seam-weld;
Internal leaks, overpressure of pipe or vessel;
Deterioration of bursting disc (pinholing).

Causes of unintended venting

Evaporation through open line, drain, cover;
Relief valves leaking, bursting discs blown, lutes blown;
Valve stuck, scrubber overloaded, ejector failure;
Dust formation, escape and accumulation;
Equipment failed/out of service (e.g. scrubbers, flares).

Causes of item failures at normal working pressure

Inadequate design, materials, construction, support, operation, inspection or maintenance;
Deterioration due to corrosion, erosion or fatigue;
Mechanical impact.

Causes of failure due to excessive pressure

Overfilling, overpressurising or drawing a vacuum;;
Overheating or undercooling;
Internal release of chemical energy;
Exposure to fire or other source of external heating (e.g. radiation).

Most periodic emissions can be avoided by careful design and operation. This means providing drains and vents which lead to safe disposal systems from equipment which may have to be opened, together with inert gas or steam purges and sometimes solvents to remove residual material in the equipment before it is opened. This also applies to hose connections to road and rail tankers, etc. Reliable and quick-closing valves can be provided as close as possible to the flanges or connections that have to be broken.

Batch additions of liquids should be made from calibrated vessels through pipes and valves which drain completely into the receiving vessels. Batch additions of solids should also be made as far as possible from enclosed hoppers through enclosed chutes and suitable valves, although this often presents problems. Some low-melting, soft and putty-like materials (such as sodium metal) can be added to stirred batch reactors by extrusion through a die, which is simpler than melting them in a separate melter and then adding them as liquids. The operating procedures must ensure that the design features provided are properly used, and in cases where they present problems, that they are modified until the problem is overcome. Operators, especially those new to the work, must also be fully informed by management of the toxic and other hazards of the materials with which they are dealing [21.1.3]. Awareness is easy in the case of a pyrophoric liquid which bursts into flames as soon as it enters the air. Appreciation is far more difficult for toxic, odourless and relatively inert substances which produce no immediate symptoms in persons exposed to them.

In spite of all these precautions, occasions sometimes arise, usually during maintenance or cleaning operations, when there are risks of personal exposure from periodic emissions of hazardous substances. For such operations, it is essential that those liable to be exposed are properly protected by appropriate and well-fitting breathing apparatus and clothing, and that persons not so protected are excluded from the danger area.

7.6.2 Fugitive emissions

Carson and Mumford[6] also provide indicative data on the rates of fugitive emissions from different sources. A selection of these emission rates for liquids, vapours and gases is given in Table 7.5.

It is clear from this table that by far the largest fugitive emissions come from the glands of conventional seals on pump and compressor shafts and piston rods, followed by the seals of rising stem valves. Actual emission rates from these sources depend on further factors such as shaft speed, diameter, straightness, roundness, and surface condition, bearing

Table 7.5 Emission rates for fluids from various equipment

Equipment	*Emission rate (mg/s)*
Pump shaft seals	
Regular packing without external lube sealant	140
Regular packing with lantern ring oil-injection	14
Grafoil packing	14
Single mechanical seals	1.7
Double mechanical seals	0.006
Bellows seal with auxiliary packing, diaphragm pump (double), canned pump	Nil
Valve stems (excluding pressure relief valves)	
Rising stem valves with regular packing	
Rating $\leqslant$ 300 lb	1.7
Rating > 300 lb	0.03
Non-rising stem valves	0.005
Pressure relief valves, average release for valve vented to closed system without upstream protection by bursting disc	2.8
Compressors	
Reciprocating	
Single rod packing	45
Double rod packing	3.6
Rotary	
Labyrinth seal	45
Mechanical seal	As for pumps
Liquid film seal	0.006
Stirrer seals ≅ Pump seal data × shaft speed (m/s) ÷ 1.91	
Piping and flange connections	
Open ended pipe	0.63
Flange and asbestos gasket	
Rating 150–300 lb	0.056
Rating >300 lb	0.056
Threaded connection	0.056
Welded connection	Nil

alignment, internal pressure, fluid viscosity, out of balance and radial loads, vibration and, above all, on the state of maintenance. On plants handling hazardous fluids shaft seals should be of high quality or eliminated entirely.

The same authors also give typical dust emission rates from various equipment. These are reproduced in Table 7.6 and refer to total dust emissions. Values should be halved for an estimate of the respirable dust release. While again the actual figures depend on many additional factors, those given provide a useful starting point for estimating airborne concentrations of hazardous materials for initial design and for surveys when plant is running.

Under COSHH, occupational hygienists are more involved in the design of installations (especially ventilation) where substances hazardous to

Table 7.6 Dust emission rates from various equipment

Type of equipment/handling	*Emission rate* (mg/s)
Vibratory screens	
Open top	5.5 times top surface area in m^2
Closed top with open access port	
6-inch dia. port	0.11
8-inch dia. port	0.21
12-inch dia. port	0.44
Closed cover — no ports	Nil
Bag dumping	
Manual slitting and dumping	3
Semi-automatic (enclosed dumping but manual bag entry/removal)	
Fully automatic and negative pressure	Nil
Bagging machines (filling)	
No ventilation	1.5
Local ventilation	0.01
Total enclosure and negative pressure	Nil

health are handled. Maximum allowable emission rates from plant and equipment should be included in their design specifications.

7.6.3 Exposure to micro-organisms

In certain industries people may be exposed to bacteria, viruses and spores which can cause diseases. Perhaps the best known of the infections is anthrax, which is caused by a bacillus (a rod-shaped bacterium) which is capable of forming extremely resistant spores. The disease can affect people handling animal products, particularly hair and hides. Other diseases which can result from contact with animals include brucellosis, glanders, Q fever, psittacosis and tuberculosis. Miners can become infected by minute hook-worm larvae, which can penetrate the skin. Sewer, canal, and abattoir workers can be exposed to a number of organisms which cause leptospirosis. However, very few people in the process industries will ever be exposed to these risks, unless animal products are handled.

An infective organism which could be present in almost any working environment is legionella pneumophila, which was first identified in 1976. A violent 'flu-like attack, which culminated in pneumonia, affected mainly people who were staying at a hotel in Philadelphia during a meeting of the American Legion. Of 221 people taken ill 34 died. Not all of these people were staying in the hotel: some had simply been in the street outside, and yet the disease was not transmitted to friends and relatives. It took six months to discover the responsible organism and during that time many wild theories were discussed by the media: a terrorist attack, the leakage of a germ warfare agent, a secret test by the CIA of a biological weapon, and extraterrestrial microbes brought back by space exploration vehicles. The discovery of the organism was delayed partly because it was difficult to culture in the laboratory, and was indeed unlike any other bacterium. It was eventually found in January 1977, and despite the Legionnaires' objections it was called legionella pneumophila, the agent responsible for legionellosis.

Bacteria have the ability to adjust to alien environments and are found in very inhospitable situations. This is partly because a large number of generations appear in a relatively short time. L. pneumophila was living in the hotel's air-conditioning plant and had been passed to people in aerosol droplets. It was later found in cooling towers, condensers, humidifiers, hot-water systems, jacuzzis, and in the air released from steam turbines. Blood samples which had been taken from victims of earlier unexplained pneumonia attacks were re-examined and some were found to contain the antibodies of the organism. This showed that it had been around for at least 30 years. A possible explanation of its ability to survive in a building for many years was found in 1990. There had been 26 cases of legionellosis in a hospital over a three-year period, and ten people had died. The bacteria were eventually found in the spray heads of showers, where they were feeding on a harmless amoeba which is commonly present in such places (and which is very difficult to eliminate). Other sources of nutrients for the organism are algal slime, and organic materials in sediments and sludge. The organism thrives at any temperature between 20° and 45°C but is killed at a temperature of 60°C. Since the original discovery of L. pneumophila, a number of similar bacteria have been identified. The word legionella is now applied to the genus of which L. pneumophila is one species. By 1993 at least 37 different species had been recognised, and 14 serogroups (strains) of L. pneumohila, of which group 1 is most commonly associated with outbreaks of legionnaires' disease.

The risks resulting from exposure to legionella come within the scope of both the Safety at Work Act 1974 and the Control of Substances Hazardous to Health Regulations (COSHH) 1988.[7] The HSE published Guidance Note EH48 in 1987, which was replaced by HS(G)70 in 1990. The 1988 update of this document was published as 'The control of legionellosis including legionnaires' disease', of which a fourth impression was issued in 1996. This guidance contains a brief account of the microbiology and medical aspects, sampling methods, legislation, identification and assessment of risk, and management responsibilities for the application of precautions and for record keeping. There is then a detailed discussion on the design and construction of hot- and cold-water services, with particular emphasis on the avoidance of legionella colonisation. Routine inspection and maintenance are described, and also methods of cleaning and disinfection. Two methods of disinfection are described. In the first, a concentration of 20 ppm of chlorine is maintained in the water for at least two hours, or 50 ppm for at least one hour. Alternatively the temperature of the water in the entire system can be held at 60°C for at least one hour. There is a similar discussion on air-conditioning and industrial cooling systems. The main risk here is the spray from cooling towers, which has been known to infect people 500 m away. Again there is advice on cleaning and disinfection, and also on continuous water treatment. This is intended to prevent scale and corrosion, which can provide protected areas for legionella, and also to eliminate sediments, bacteria, and other organisms. Chlorine can be used at 5 ppm, together with biodispersants, for disinfection, and at 1 to 2 ppm for continuous use.

Advice is also given on spas, whirlpool baths, humidifiers, air washers, and firefighting (sprinkler) systems. Mention is made of the HSE's ACOP on the prevention or control of legionellosis, and also of BS 6700, which is concerned with the disinfection of water systems. There are several other references in the text and also a list of publications which can provide further guidance.

It should perhaps be noted that a number of other new diseases appeared during the last quarter of the twentieth century. These included Ebola fever, which was also found in 1975; Lyme disease, which appeared in the year before; T-cell leukaemia; Rift Valley fever; AIDS; ehrlichiosis; Venezuelan fever; and hantavirus syndrome. A new strain of eschericia coli, designated 0157:47, was identified, and also a new strain of the Creutzfelt-Jacob organism.

7.7 Monitoring the working environment for toxic substances

Advice on this difficult subject which will assist UK employers in meeting their duties under COSHH[7] is given in HSE's Guidance Note EH 42.[13] Similar advice is available in other countries from their occupational health authorities, e.g. OSHA (Occupational Health and Safety Administration) and NIOSH (National Institute for Occupational Safety and Health) in the USA.

Such monitoring involves an initial assessment followed by carefully planned programmes of air-sampling and analysis. The purposes of the sampling programmes discussed in EH 42 are:

1. to estimate personal exposure to hazardous substances, e.g. to investigate compliance with OELs;
2. to investigate the effectiveness of engineering and process control measures.

Sampling is also undertaken for epidemiological investigations and for environmental monitoring outside the working environment.

The measurement of airborne concentrations of any contaminant is complicated by the many variables present. These include the type, number and position of emission sources and their rates, the dispersion of contaminant in the air, and ambient conditions including air velocity, which is specially important for outdoor operations. Variations in personal exposure may occur within shifts, between shifts, between processes and between individuals, and between people doing the same job (such as filling bags) at the same time and place, particularly when the way they work affects the emission. A structured approach to the problem is thus needed, which incorporates the following stages in sequence:

1. an initial assessment
2. a preliminary survey
3. a detailed survey
4. routine monitoring.

If the initial assessment shows that the health hazard is negligible, it is not usually necessary to go further. Preliminary and detailed surveys are exploratory, self-contained and concluded once the prevailing conditions have been established. They may be needed on start-up of a new process, on setting of a new (lower) exposure limit, when the process of control measures have been modified or when unusual operations are to be carried out.

The sampling equipment chosen should match the analytical techniques used and the relevant OELs. For personal exposure monitoring the sampling equipment should, where possible, be worn on the person and the sample taken at a consistent position within his or her breathing zone.

7.7.1 Initial assessment

This can only be made by a trained and experienced person, who needs the following information:

- the substances present at the workplace, raw materials, intermediates, products, contaminants, auxiliary chemicals;
- the airborne form of the substance (dust, fume, aerosol or vapour);
- the health hazards of the substances, and whether combinations of non-hazardous ones could be dangerous;
- whether people could become exposed to these substances through inhalation, ingestion or skin absorption;
- where and when exposure is likely to occur;
- which groups or individuals are most likely to be exposed;
- the likely pattern and duration of exposure;
- what information is available on similar exposures elsewhere.

Simple qualitative tests involving the use of dust lamps and smoke pellets are useful in the initial assessment. In some cases it will be necessary to sample the air and analyse it for possible contaminants.

7.7.2 Preliminary survey

Once people likely to be at risk have been identified, their exposures should be measured by sampling, analytical and control techniques as discussed in 7.7.7. Personal air-sampling techniques should be used initially, complemented where required by static sampling, e.g. to provide further information on the efficiency of engineering controls. The preliminary survey should provide basic quantitative information on the efficiency of process and engineering controls and on the likely extent of workers' exposure to harmful substances. It may reveal shortcomings in preventative measures, work procedures and operator training.

7.7.3 Detailed survey

This is needed where the preliminary survey does not give an adequate assessment of the extent and pattern of exposure, e.g. when:

- the results of the preliminary survey are very variable;
- many people are at risk of excessive exposure;
- personal sampling results are near the appropriate OELs and the cost of additional control measures cannot be justified without further evidence on the extent of the risk. EH 42[13] states that 'detailed surveys may typically take 3 to 10 days to complete in a workplace with between 25 and 200 employees'.

7.7.4 Routine monitoring

After preliminary/detailed surveys and any indicated remedial action has been carried out, further routine monitoring may be needed to check that control measures remain effective, and to reveal trends in exposure patterns. The frequency of monitoring may vary from monthly, where highly toxic or carcinogenic substances are involved, to less than yearly when exposures appear well controlled. The sampling methodology should be well planned to enable results to be analysed statistically.

7.7.5 Sampling strategies

Because of the wide variations possible in exposure patterns, a number of samples are usually needed to reduce errors, and a carefully planned programme is required. EH 42 discusses this in detail and suggests a three-level strategy. The first level employs relatively unsophisticated sampling equipment and experimental techniques and is often suitable for the preliminary survey. The second level should aim to provide accurate measurements of TWA exposures which can be related to the appropriate 10-minute and 8-hour TWA limits, and is more suitable for detailed surveys and routine monitoring. A third-level strategy is occasionally needed if, despite taking all reasonably practicable control measures, personal exposures remain close to the relevant OEL. This usually requires a highly sophisticated sampling programme and rigorous statistical analysis of the results[27].

7.7.6 Interpretation of results

This again is discussed in detail in EH 42. A distinction must be made between substances for which a control limit has been adopted by HSC, those for which a recommended limit is listed and those for which no exposure limits are listed in EH 40. In the last case it is suggested that employers should make their own assessment of the health risk and set tentative limits for their own use, after consulting their employees. EH 42 gives practical guidelines on the actions to be taken when the results of sampling programmes bear particular relations to the relevant OELs. These are summarised in Table 7.7.

Table 7.7 Guidelines on recommended actions following a personal exposure sampling programme

Results	*Action*
A. Preliminary survey; first-/second-level strategy	
< 0.1 limit	Normally none, if exposure is as low as reasonably practicable.
0.1 to 1.5 limit	Investigate process/control measures; make detailed survey.
Some results > 1.5 limit	Investigate; improve control measures, provide respiratory protection to workers with high exposure until improvements made and confirmed by survey.
B. Detailed survey; second-/third-level strategy	
< 0.25 limit	Normally none, if exposure is as low as reasonably practicable.
< 1.25 limit	Investigate and improve control measures; repeat survey using more refined techniques to improve accuracy.
Mean < 0.5 limit, all results < limit	Consider routine monitoring and suitable frequency.
Mean > 0.5 limit some results > limit	Investigate and improve control measures and repeat survey; consider routine monitoring.
Routine monitoring; second-/third-level survey	
All results	Check individual values, mean, 'range', etc. for reliability of compliance with OEL; consider need for corrective action.
Results significantly different from previous survey	Consider need for corrective action, detailed survey or revised basis for survey.

7.7.7 Air sampling and analysis[28]

Being specialised subjects, only broad principles are discussed here. Most precise methods depend on drawing a known volume of contaminated air through an apparatus which may contain a liquid or solid reagent, an adsorbent (such as charcoal or silica gel), or a filter (used mainly for solids). The equipment consists of a sampling head, a pump and a means of measuring the volume of air drawn through the apparatus. For several contaminants the total quantity of one or more present is determined by the extent of a colour change in a specific reagent. With apparatus using filters and adsorbents, the quantity and composition of the trapped contaminant is determined by removing the filter or adsorbent tube and analysing its contents.

Sampling of air for gas and vapour contaminants is fairly straightforward, since all gases and vapours, once mixed with air, are uniformly distributed, and their concentrations are not altered when air is drawn through the sampling tube. Sampling of air for finely divided solids and liquids is more difficult as these tend to be deposited in the sampling tube. Filters for solid contaminants are therefore incorporated into the sampling head itself (Fig. 7.1).

OELs for many airborne solids are quoted on two bases, 'total dust' and 'respirable dust'. The sampling apparatus for respirable dust contains a special elutriator in the sampling head before the filter. This separates the larger particles which would generally be stopped in the nose and upper respiratory tract and allows only the smaller particles (which normally reach the lungs) to pass through.

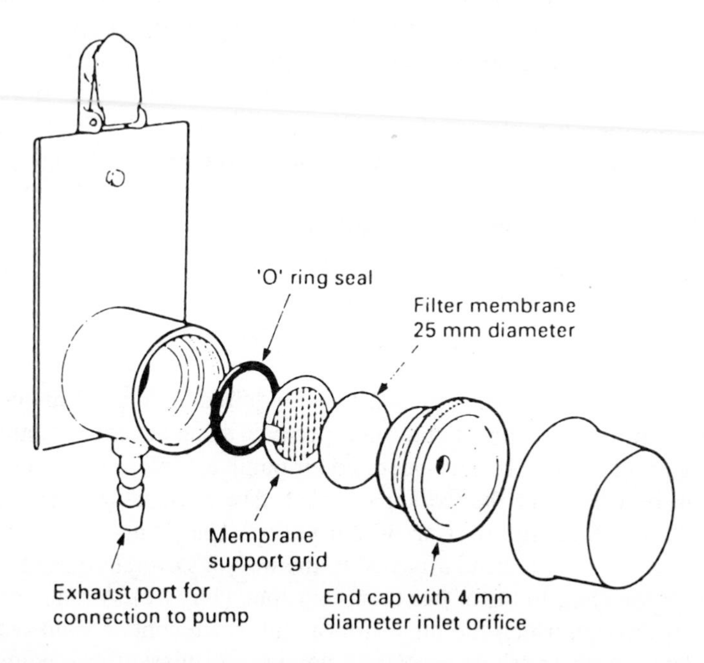

Fig. 7.1 Orifice-type sampling head used for airborne lead. (From HSE Guidance Note EH28).

Some devices designed for use as personal dosemeters depend on the principle of diffusion and adsorption, and require no air pump. An adsorbent, normally charcoal, is contained in the middle of a short glass tube which has plugs of porous material sealed in one or both ends. Vapours present in the air diffuse through the plug at a rate proportional to their concentration in air and are trapped by the adsorbent. The tubes are sealed by caps which are removed immediately before use and replaced immediately afterwards. The adsorbent is later analysed for specific compounds. The main use of these devices, which are not considered as accurate as those using a measured volume of air, is in preliminary surveys of personal exposure [7.7.2]. The tube is fitted to a small holder which is clipped to the lapel or breast pocket of the wearer.

Other monitoring instruments based on physical principles require no measurement of air volume. These have detecting heads which are exposed to the air. Examples are:

- electrochemical measurement of oxygen, carbon monoxide and hydrogen concentrations in air;
- flammable gas monitors which use an electrically heated combustion catalyst (pellistors);
- instruments which depend on absorption of light of a particular wavelength.

Fixed and portable instruments of all types are used. Portable instruments operating on physical principles are employed for checking the atmosphere in confined spaces before anyone is allowed to enter [18.8.1]. The measuring device with batteries is contained in a small case to which a sample probe and a rubber squeeze bulb are connected by flexible tubes. Readings are taken from a dial on the outside of the case. Although easy to operate, these instruments can go wrong, and need to be calibrated and adjusted by trained persons. Typical faults are caused by:

- air leak between instrument and sample probe;
- poisoning of catalyst;
- flat battery.

For accurate measurements such as those required in detailed surveys and routine monitoring, compact lightweight personal sampling equipment which incorporates a low-flow, battery-operated air pump is preferred. Such pumps, which can be carried in the worker's pocket, give reasonably constant flow rates of 20 to 200 ml/min over an 8-hour period. The pump is connected by flexible tube to a sample head attached to the worker's overalls, etc., in a consistent position close to his or her breathing zone (Fig. 7.2). A range of standard tubes through which the air is drawn and which contain solid reagents are available, both as 'short-term' and 'long-term' tubes. Tubes containing reagents which react with one of 30 or more specific contaminants to produce

Fig. 7.2 Personal air sampling equipment. (Courtesy Draeger)

colour changes which are later measured are available. These are generally preferred to liquid reagents in bubblers for personal sampling. For gases and vapours for which no suitable reagent tubes are available, standard adsorbent tubes containing charcoal, porous polymers or silica gel are used. Details of methods recommended by HSE are published in their MHDS series,[29] an up-to-date list of which is given in EH 40.[12]

7.7.8 Detection of toxic/irritant gases and vapours by odour

Odour can often give a useful warning of an emission. Hydrogen sulphide can be detected by smell by most freshly exposed persons at concentrations well below its OEL, but exposure for only a few minutes paralyses the olfactory nerves so that the sense of smell is lost. Common irritant gases such as chlorine, sulphur dioxide and ammonia can be detected by about 50% of people exposed to them below their OELs. Some toxic and asphyxiant gases such as carbon monoxide and nitrogen are odourless.

Individuals vary greatly in their sensitivity to smell, from four times below the mean to four times above it. Lynsky summarises a published study which compared the odour threshold with OELs (TLVs) in the 1982 US list for 214 substances.[30] This placed them in five classes ranging from 'A', for which

90% of distracted persons perceived a warning at the 8-hour OEL concentration, to 'E' for which less than 10% of attentive persons can detect the substance at the 8-hour OEL concentration. It also gives 'odour safety factors', i.e. the ratio of the 8-hour OEL to the odour thresholds for the 214 substances. Substances in category A include acrylates, amyl acetate, hydrogen sulphide, mercaptans and trimethylamine. Those in category E include carbon tetrachloride, chloroform, ethylene dichloride, ethylene oxide, methyl isocyanate, nickel carbonyl, nitropropane, toluene diisocyanate and vinyl chloride. The odour thresholds and safety factors for a number of common toxic organic chemicals listed in Part 2 of Schedule 3 of the CIMAH Regulations are compared with their 8-hour OELs in Table 7.8.

Because of the wide variations in odour, safety factors and people's sense of smell, and the frequent dulling of this sense on long exposure to particular vapours, the apparent absence of odour should never be taken as evidence that an atmosphere is safe to breathe and enter. Smell is no substitute for scientific monitoring. While smell may suggest the need to monitor the air for one chemical, there may be a far greater danger from another present which has no odour.

7.8 Substances hazardous to health, and the law

Numerous regulations about the use and handling of toxic substances in particular industries have been wholly or partly revoked and superseded by the COSHH Regulations. Only some aspects of these and the CIMAH Regulations are considered here.

7.8.1 Scope and practical effects of the COSHH Regulations[7]

The scope of COSHH includes prohibitions, assessment, control of exposure, monitoring exposure at the workplace, health surveillance, as well as information, instruction and training. The substances whose use or import are prohibited under COSHH are those formerly prohibited under other regulations which COSHH will replace.

An assessment by a competent person [7.7.1] must be made before work which may involve exposure of any employee to a substance hazardous to health is started. This should enable the precautions needed to comply with the regulations to be identified. The assessment required is related to the risk.

Exposure should preferably be prevented by elimination or substitution by a less hazardous substance. Where this is not reasonably practicable, adequate engineering controls over materials, plant and processes should be exercised. Personal protective equipment should only be used when adequate engineering controls are not reasonably practicable. Substances to which MELs have been assigned require the strictest safeguards. Those to which OESs are assigned have next priority. While exposure should not exceed OES concentrations, in cases where it does, control may still be con-

Table 7.8 Toxic substances of Group 2, Schedule 3 of CIMAH Regulations

Substance	*Quantity* (tonnes)	*OEL TWA* (ppm)		*Odour threshold* (ppm)	*Odour safety factor*
		8 hour	10 min		
Acetone cyanohydrin	200	–	–	–	–
Acrolein	200	0.1	0.3	0.61	0.16
Acrylonitrile	200	2	–	0.12	17
Allyl alcohol	200	2	4	1.8	1.1
Allylamine	200	–	–	–	–
Ammonia	500	25	35	4.8	5.2
Bromine	500	0.1	0.3	2.0	0.05
Carbon disulphide	200	10	–	92	0.11
Chlorine	50	1	3	3.2	0.31
Ethylene dibromide	50	1	–	–	–
Ethyleneimine	50	0.5	–	1.5	0.32
Formaldehyde (concentration ⩾ 90%)	50	2	2	0.83	2.4
Hydrogen chloride (liquefied gas)	250	5	5	0.77	6.5
Hydrogen cyanide	20	10	10	0.58	17
Hydrogen fluoride	50	3	6	0.042	71
Hydrogen sulphide	50	10	15	0.0081	1200
Methyl bromide	200	15	–	–	–
Nitrogen oxides	50	3	5	0.39	7.8
Phosgene	20	0.1	–	0.90	0.11
Propyleneimine	50	–	–	–	–
Sulphur dioxide	1000	2	5	1.1	1.7
Tetraethyl lead	50	0.10 mg/m^3		–	–
Tetramethyl lead	50	0.15 mg/m^3		–	–

sidered adequate if the employer has identified the cause of non-compliance, and is taking steps to reduce exposure to below the OES as soon as reasonably practicable.

Employers are required to set their own working control standards for harmful substances to which neither MELs nor OESs have been assigned, on the basis of all relevant information.

Situations where the use of personal protective equipment may be necessary include emergencies caused by plant failure, and during routine maintenance.

Engineering control measures include the use of totally enclosed processes and handling systems, those which limit the generation of harmful airborne substances, local exhaust ventilation with or without partial enclosure, adequate general ventilation. and the provision of safe storage and disposal facilities. A thorough discussion of such measures, particularly ventilation as applied to the control of carcinogens in the American workplace, is given by Feiner.[31]

Personnel control measures include reduction of numbers exposed, exclusion of non-essential access, reduction in period of exposure, cleaning walls and surfaces, provision of suitable personal protective equipment, prohibition of eating, drinking and smoking, and the provision of adequate facilities for washing, changing and storing clothing and laundering contaminated clothing.

Procedures for emergencies resulting from loss of containment or control should be established which give maximum employee protection against exposure to substances with health hazards.

Respiratory protective equipment (RPE) must be properly selected for its purpose, of an HSE-approved type or to an HSE-approved standard, and matched to the job and the wearer. Eye protection should comply with the Protection of Eyes Regulations 1974 and BS 2092.

Special protection is particularly needed by maintenance workers when isolating equipment for maintenance. Special respiratory apparatus for escape [22.7] is needed for those who might become trapped in a cloud of toxic gas in works with significant inventories of toxic gases and volatile liquids. The types of protection available for the eyes, respiratory system and skin and criteria for their selection are discussed in Chapter 22.

Both employers and employees should ensure that the control measures provided are properly used or applied, and procedures should be established for regular inspection and maintenance or other remedial action. Records of examinations, tests and repairs should be kept for at least five years.

Monitoring exposure at the workplace, like assessment, should be related to the risk. The methods and strategy recommended have been discussed in 7.7.

Health surveillance of workers, where appropriate, should be given to provide early detection of ill effects resulting from exposure to substances hazardous to health. In such cases health records should be kept for each worker involved. Like other measures, the level of health surveillance will depend on the nature and degree of the risk. Substances and processes considered ‘appropriate’ are listed in Schedule 5. They are those for which health surveillance was called for under former regulations now revoked by COSHH. Health surveillance will generally be supervised by an employment medical adviser (EMA) or an appointed doctor (AD), and may lead to the suspension of employees from work in which their health is adversely affected.

Health surveillance has four objectives: to protect the health of employees, to assist in evaluating control measures, to collect and use data for detecting and evaluating health hazards and to assess the immunological status of employees in work activities involving micro-organisms.

Health surveillance procedures include biological monitoring, biological effect monitoring, medical surveillance including clinical examinations, enquiries about symptoms, inspection by a responsible person, review of records and occupational history.

Employees are entitled to access to their health records.

Information, instruction and training of persons who may be exposed to substances hazardous to health is another important feature of the COSHH Regulations, although most of this is straightforward and self-evident.

7.8.2 Toxic substances and the CIMAH Regulations 1984[22]

Schedule 1 provides 'Indicative Criteria' for toxic and very toxic substances which applies to Regulation 4. This elaborates on the criteria given in Table 7.3.

Two groups of toxic substances to which Regulations 7 to 12 apply are listed in Schedule 3.

Group 1 is a list of the most toxic substances. It contains 32 substances and groups of compounds for which a minimum quantity of 1 kg applies, eight substances with a minimum quantity of 10 kg, 54 substances with a minimum quantity of 100 kg, one substance with a minimum quantity of 500 kg and two substances with a minimum quantity of 1 t.

The 31 substances to which 1 kg applies are all organic and include benzidene, fluoroacetic acid (used as a rodenticide by professional operators) and related compounds, 2-naphthylamine (now banned) and 2,3, 7,8- tetrachlorodibenzo-para-dioxin, the active agent of the Seveso disaster [5.2]

The eight entries to which 10 kg applies include five inorganic ones: arsine, beryllium (powders and compounds), hydrogen selenide, nickel carbonyl, oxygen difluoride and selenium hexafluoride. Nickel carbonyl is an intermediate in the purification of nickel. Beryllium is used in atomic reactors and spacecraft, and in some copper and aluminium alloys.

The 54 entries to which 100 kg applies include five inorganic ones, arsenic trioxide and related compounds, cobalt and nickel powders and compounds, phosphine and tellurium hexafluoride. Many of the organic entries are herbicides or pesticides.[1] They include Aldicarb [5.3], Parathion, a phosphorus-based insecticide used for fruit trees, and Crimidine (LD_{50} 1.25 mg/kg) and Warfarin, both rat poisons, the latter an anticoagulant.

One of the two entries to which 1 t applies is methyl isocyanate [5.3].

Several very toxic pesticides do not appear in Schedule 3 at all, e.g. strychnine sulphate [60–41–3], (LD_{50} 5 mg/kg), used for mole control, and brodifacoum [56073-10-0], (LD_{50} 0.4 mg/kg), another anticoagulant and very potent rat poison. The list of pesticides in group 1 of Schedule 3 will need frequent updating as new ones are developed.

Group 2 is a list of 23 common toxic industrial chemicals to which a minimum quantity of 1 t or more applies. The chemicals and their minimum quantities are listed in Table 7.8. This also gives the short- and long-term TWA OELs for the UK in ppm in cases for which they are available. (The fact that no OELs are published in the UK for four of these 23 common toxic chemicals reinforces the need to minimise exposure to all potentially harmful substances, whether their OELs are published or not.) The lowest actual quantity which applies to any of those listed is 20 t.

Some of these are gases under ambient conditions, while the rest are volatile liquids.

7.9 Treatment of affected persons

This is mainly a medical question which is outside the scope of this book. The relationship between safety management and occupational health professionals, particularly as applying in the USA, is discussed by Kilian[32] and Murphy[14] (an occupational nursing consultant) in Fawcett and Wood's book.

Most works or factories in the process industries, being 'establishments presenting special or unusual hazards', are required to have their own 'first-aid room', with an adequate number of trained 'occupational first-aiders' even though there may be fewer than 400 employees working there.'[33].

The first-aid room is often part of a works medical centre, which is equipped to enable medical examinations to be carried out, or in larger organisations, a department of occupational medicine. This should have a positive role in promoting health and fitness among workers as well as treating injuries, and sickness. It should have close working relations through EMAS[34] with the local health service and hospitals, and an up-to-date knowledge of specialised facilities, such as a burn centre, and skills, such as a retinal surgery service, to which patients can be referred without delay in emergencies. It should also keep up-to-date records both of workers' health and of all materials used on the premises to which they may be exposed.

Small factories which do not present 'special or unusual hazards' are almost all required to have at least one first-aid box or cupboard kept under the charge of a person trained in first-aid. Minimum standards for first-aid boxes and for first-aid training are given in the regulations and clarified by the ACOP[35] and guidance notes[36] prepared for them.

First-aid requirements are specially critical in factories where toxic or corrosive substances are present. A recently published manual provides details of first-aid treatment for exposure to nearly 500 chemicals.[37] Oxygen resuscitation equipment, a stomach pump, emergency showers, special washing facilities, and antidotes to the toxic materials encountered are often required, as well as more common items such as stretchers, warm emergency clothing and blankets.

7.10 How does one decide if a disease is occupational?

While this important question is mainly one for health specialists and lies outside the scope of this book, a short but comprehensive guide (which reflects official American thinking) is published by NIOSH.[38] It includes:

- data for a number of specific disease-producing agents, exposure standards, and a list of occupations with possible exposure to them;
- considerations of medical, personal, family and occupational history;
- clinical evaluation, signs and symptoms of occupational disease and laboratory tests;
- epidemiological data, industrial hygiene sampling and data evaluation;
- aggravation of pre-existing conditions and its legal ramifications;
- sample questionnaires for evaluating respiratory symptoms:
- qualifications of medical and industrial hygiene personnel.

References

1. 'Poisons, economic' in volume 18, *Encyclopaedia of Chemical Technology*, edited by Kirk, R. E. and Othmer, D. F., 3rd edn, Wiley, New York (*ca* 1983)
2. Kletz, T. A., 'The application of hazard analysis to risks to the public at large', *World Cong. Chem. Eng. Amsterdam*, Elsevier, Amsterdam (1976)
3. Hamilton, A., and Hardy, H. L., *Industrial Toxicology*, 3rd edn, Publishing Sciences Group, Acton, Mass. (1974)
4. Hunter, D., *The Diseases of Occupations*, 6th edn, Hodder and Stoughton, London (1978)
5. P. L. 94–69, *The Toxic Substances Control Act*, Environmental Protection Agency, Washington DC (1976)
6. Carson, P. A. and Mumford, C. J., 'Industrial health hazards, Part 1: Sources of exposure to substances hazardous to health', *Loss Prevention Bulletin 067*, The Institution of Chemical Engineers, Rugby (1985)
7. S.I. 1988, No. 1657, *The Comtrol of Substances Hazardous to Health Regulations 1988* and Approved Codes of Practice, *Control of Substances Hazardous to Health and Control of Carcinogenic Substances*, HMSO, London

7a. HSC Consultative Document, *The Control of Substances Hazardous to Health: Draft Regulations and Draft Approved Codes of Practice*, 1st edn, HMSO, London (1984)

8. S.I. 1984, No. 1244, *The Classification, Packaging and Labelling of Dangerous Substances Regulations 1984*, HMSO, London
9. Fawcett, H. H., 'Toxicity versus hazard' in *Safety and Accident Prevention in Chemical Operations*, edited by Fawcett, H. H. and Wood, W. S., 2nd edn, Wiley-Interscience, New York (1982)
10. Tyrer, F. H. and Lee, K., A *Synopsis of Occupational Medicine*, 2nd edn, John Wright, Bristol (1985)
11. *American Industrial Hygiene Association Journal*, **20**, 428–430 (October 1959)
12. HSE, Guidance Note EH 40/89, *Occupational Exposure Limits, 1987*, HMSO, London (1989)
13. HSE, Guidance Note EH 42, *Monitoring Strategies for Toxic Substances*, HMSO, London (1984)
14. Murphy, A. J., 'The role of an occupational health nurse in a chemical surveillance program' in *Safety and Accident Prevention in Chemical Operations*, edited by Fawcett, H. H. and Wood, W. S., 2nd edn, Wiley-Interscience, New York (1982)
15. S.I. 1981, No. 917, *The Health and Safety (First-Aid) Regulations 1981*, HMSO, London
16. S.I. 1985, No. 2023, *The Reporting of Injuries, Diseases and Dangerous Occurrences Regulations (RIDDOR)*, HMSO, London
17. Seville, R., *Dermatological Nursing and Therapy*, Blackwell Scientific, London (1981)

18. Williams, P. L. and Burson, J. L., *Industrial Toxicology*, Van Nostrand, New York (1985)
19. Atherley, G. R. C., *Occupational Health and Safety Concepts*, Applied Science Publishers, London (1978)
20. The Institution of Chemical Engineers, *Nomenclature for Hazard and Risk Assessment in the Process Industries*, Rugby (1985)
21. Hodge, H. C. and Sterner, J. H., 'Tabulation of toxicity classes', *Am. Ind. Hyg. Assoc. Quart.*, **10**(4), 93 (December 1949)
22. S.I. 1984, No. 1902, *The Control of Industrial Major Accident Hazards Regulations 1984 (CIMAH)*, HMSO, London
23. National Fire Protection Association, *Standard 704 M, Identification systems for fire hazards of materials*, NFPA, Boston, Mass. (1975)
24. HSE, Guidance Note EH 44, *Dust in the workplace: general principles of protection*, HMSO, London (1984)
25. MDHS 14, *General method for the gravimetric determination of respirable and total dust*, HMSO, London
26. The International Labour office, *Occupational Exposure Limits for Airborne Toxic Substances – A Tabular Compilation of Values from Selected Countries*, 2nd edn, Geneva (1980)
27. Davies, O. L. (ed.), *The Design and Analysis of Industrial Experiment*, Oliver and Boyd, London (1954)
28. Lee, G. L., 'Sampling: principles, methods, apparatus, surveys', in *Occupational Hygiene*, edited by Waldon, H. A. and Harrington, J. M., Blackwell Scientific, London (1980)
29. *Method for the determination of hazardous substances* (MDH Series, about 40 titles), HMSO, London
30. Lynskey, P. J., 'Odour as an aid to chemical safety: odour thresholds compared with threshold limit values', *Loss Prevention Bulletin 060*, The Institution of Chemical Engineers, Rugby (1984)
31. Feiner, B., 'Control of workplace carcinogens', in Sax, I., *Cancer Causing Chemicals*, Van Nostrand Reinhold, New York (1981)
32. Kilian, D. J., 'The relationship between safety management and occupational health programs', in *Safety and Accident Prevention in Chemical Operations*, edited by Fawcett, H. H. and Wood, W. S., 2nd edn, Wiley-Interscience, New York (1982)
33. S.I. 1981, No. 917, *The Health and Safety (First-Aid) Regulations 1981*, HMSO, London
34. *Employment Medical Service Act 1972 – Guide to the Service*. HMSO, London
35. HSC, *Approved Code of Practice for the Health and Safety (First-Aid) Regulations 1981*, HMSO, London
36. HSE, *Guidance Notes for Health and Safety (First-Aid) Regulations, 1981*, HMSO, London (1981)
37. Leferve, M. J., *First Aid Manual for Chemical Accidents*, English edition translated by Becker, E. I., Van Nostrand Reinhold, New York (1984)
38. Anon. *Occupation and Disease – A Guide for Decision-making*, NIOSH, Center for Disease Control, Rockville, Maryland (1976)

8 Chemical reaction hazards

The reactivity of chemical elements, compounds and free radicals is the very essence of chemistry, which most readers will have studied, and whose memories this chapter may help to jog. Explosivity, flammability and corrosivity are only touched on in this chapter, as the subjects are dealt with in Chapters 9–10. Although self-heating is discussed in this chapter [8.7], autoignition is included in Chapter 10 as an ignition process [10.5.]

8.1 Reactivities of the elements and structural groupings

The reactivity of an element is closely related to its position in the Periodic Table (Table 8.1), which is determined by its atomic number.[1]

The Periodic Table consists of seven periods (each representing a principal quantum number). No. I contains two elements, Nos II and III each contain eight, IV and V each contain 18, VI contains 32, and VII is incomplete and contains only a few radioactive elements which lie outside the scope of this book. The elements are also treated as belonging to eight groups which correspond to the eight elements each in periods II and III. Periods IV, V and VI each contain two groups of eight elements ('A' series) and seven elements ('B' series) and some additional elements. The properties of the first two and the last six elements in periods IV, V and VI have some resemblances to those in similar positions in periods II and III and are treated as belonging to the 'A' series of that group. The others are treated as the 'B' series of groups 1 to 7, apart from three elements in the middle of each period, which are treated as part of group 8B, and the 15 rare earths of period VI which are assigned to group 3B.

The position of an element in the Periodic Table, in particular its group, governs the number of electrons in the outer shell of the atom. An atom

Table 8.1 The Periodic Table (up to Uranium: non-metals shown in italics)

Group	1A	2A	3B	4B	5B	6B	7B	←	8B	→	1B	2B	3A	4A	5A	6A	7A	8A
Period	1																	2
I	H																	*He*
	Light metals														*Non-metals*			
	3	4											5	6	7	8	9	10
II	Li	Be											*B*	*C*	*N*	*O*	*F*	*Ne*
	11	12											13	14	15	16	17	18
III	Na	Mg											Al	*Si*	*P*	*S*	*Cl*	*Ar*
									Transition metals									
	19	20	21	22	23	24	25	26	27	28	29	30	31	32	33	34	35	36
IV	K	Ca	Sc	Ti	V	Cr	Mn	Fe	Co	Ni	Cu	Zn	Ga	Ge	*As*	*Se*	*Br*	*Kr*
	37	38	39	40	41	42	43	44	45	46	47	48	49	50	51	52	53	54
V	Rb	Sr	Y	Zr	Nb	Mo	Tc	Ru	Rh	Pd	Ag	Cd	In	Sn	Sb	*Te*	*I*	*Xe*
	55	56	57	72	73	74	75	76	77	78	79	80	81	82	83	84	85	86
VI	Cs	Ba	[a]	Hf	Ta	W	Re	Os	Ir	Pt	Au	Hg	Tl	Pb	Bi	Po	*At*	*Rn*
	87	88	89	90	91	92												
VII	Fr	Ra	Ac	Th	Pa	U												

[a] The 15 rare earth elements fill this gap

with a complete outer shell of electrons is extremely stable and will be one of the inert gases — helium, neon, argon, krypton or xenon.

8.1.1 Reactivity of metals

The elements of group 1 (1A) are the alkali metals — lithium, sodium, potassium, rubidium and caesium. Their atoms have a single electron in their outer shells, which they part with very readily to form positive ions. Because of this they react violently with water and most non-metals. The affinities of the alkali metals for most non-metals generally increase from lithium to caesium, although the heats evolved per unit weight of these elements in such reactions are generally in the reverse order and greatest for the lightest element. The same applies to the metals of the next two groups.

The atoms of group 2 (2A) elements have two electrons in their outer shells, which they also lose readily. They are the alkali earth metals: beryllium, magnesium, calcium, strontium and barium, which also react vigorously but less violently with water. The atoms of group 3 elements have three electrons in their outer shells and are also active, but besides shedding electrons to form positive ions, they may also achieve stability by sharing their electrons with other atoms. The first member, boron, is a reactive non-metal, followed by aluminium, a metal, and the less common metals scandium, yttrium and the rare earths.

Most of the elements in periods IV, V and VI are metals, and apart from the first three of each period and the rare earths, they are more or less stable in air and are used in engineering. They include the older metals —

zinc, copper, tin, iron, lead, bismuth, mercury, gold, silver and platinum — and the newer ones — titanium, vanadium, chromium, manganese, cobalt, nickel, molybdenum, palladium, cadmium, tantalum, tungsten, osmium and irridium. They show a limited range of reactivity among themselves, from titanium, the most active, through tantalum and tungsten, to the noble metals gold and platinum. Some indication of the activities of metals and non-metals is given by their 'standard electrode potentials'.[2] When placed in order of decreasing potential they form an 'Electromotive Series', shown in Table 8.2.[3]

The orders are somewhat ambiguous, since some elements have more than one position in the series, depending on the valency of the ion formed. Hydrogen is treated here as a metal, as are arsenic and tellurium. Metals higher in the series than hydrogen react with dilute hydrochloric and sulphuric acids, displacing hydrogen. A metal will displace any other metal lower than it in the series from a solution of its salts. When two metals in contact are immersed in an aqueous medium, the one higher in the series is attacked by galvanic corrosion, and the one lower in the series is protected. This is the principle of cathodic protection[13.4].[4]

The series also determines the ease of reduction of metallic oxides, the ease of corrosion in air, the action of water on metals, the solubility and stability of hydroxides and carbonates, the behaviour of nitrates on heating, and the occurrence of the metals in their free state in nature.

As regards corrosion, the alkali and alkaline earth metals rust rapidly in air, sometimes burning. Metals down to copper rust comparatively easily in air, although several metals such as aluminium, beryllium and chromium form rather resistant oxide layers on the surface, which protect the underlying metal from further attack. The alkali and alkaline earth metals displace hydrogen from water, even in the cold, with the evolution of much heat. Magnesium and succeeding metals, except those at the bottom of the list, will displace hydrogen from steam. The properties of individual metals are, however, greatly modified by alloying with others and by heat treatment; thus alloys are available with specific corrosion resistance against various media and conditions.

Many of the metals and their compounds in the middle of periods IV, V and VI are active catalysts for particular reactions. Thus although they do not react themselves, they have the effect of increasing the activities of other elements and compounds.

Table 8.2 Electromotive series of elements

A. *Metals and metalloids*
Li Rb K Sr Ba Ca Na Mg Al Be U Mn Zn Cr Ga Fe Cd In Co Ni Sn Pb
H_2 Sb Bi As Cu Po Te Ag Hg Pb Pd Pt Au
B. *Non-metals*
F_2 Cl_2 Br_2 I_2 O_2 S Te

8.1.2 Reactivity of non-metals

The elements of group 7 (7B) are the halogens, fluorine, chlorine, bromine and iodine, whose atoms have outer shells with one electron short of a full set. They react strongly with atoms with surplus electrons, to form negative ions. This activity decreases from fluorine to iodine. Atoms which need two electrons to complete their outer shells (oxygen, sulphur, selenium, tellurium and the radioactive metal polonium) are rather less active. Besides the inert gases, there are only 15 non-metals, of which some, arsenic and tellurium, are border-line cases and sometimes behave as metals.

Carbon, which forms strong covalent bonds with other carbon atoms, as well as with hydrogen, oxygen, sulphur, nitrogen, phosphorus and the halogens, is the cornerstone of organic chemistry. Single carbon–carbon bonds are less reactive than double bonds which, in turn, are less reactive than triple bonds, although six-membered aromatic rings with alternate single and double bonds have comparable stabilities to chains with single bonds.

While some elements are much more reactive than others, their reactivity usually only manifests itself when one element is brought into contact with other elements and compounds. The greatest activity occurs when elements with opposite and complementary properties are brought together, e.g. caesium and fluorine. In the same way, compounds with opposite and complementary properties show the greatest activity when they mix, e.g. acids with bases and oxidising agents with reducing agents.

8.1.3 Reactivity of some structural groupings

The activity of structural groupings is too large a subject to cover in this book, but three broad and overlapping types of structural groups with high hazard potential are discussed.

- Those which confer instability to the compounds in which they are present, often rendering them explosive. A number of these are listed in Table 9.2.
- Those which render the compound liable to polymerise, with the evolution of heat. These are unsaturated linkages between carbon and carbon, carbon and oxygen, and carbon and nitrogen atoms, as well as three-membered rings containing these atoms. Structural groupings of some of the more common industrial chemicals which exhibit this tendency are given with examples in Table 8.3. Hazardous polymerisations involving some of them are discussed in 8.6.4
- Those which render the compound liable to attack by atmospheric oxygen with the formation of peroxides, hydroperoxides and other compounds containing the –O–O– grouping. A list of these is given in Table 8.4 and their hazards are discussed in 8.5.1.

Table 8.3 Structural groupings liable to cause exothermic polymerisations

Type	*Formula*	*Examples*
Vinyl	–CH=CH_2	Styrene, vinyl chloride, ethyl acrylate
Conjugated double bonds with carbon, nitrogen and oxygen atoms	–CH=CH–CH=CH–	Butadiene, isoprene, chloroprene, cyclopentadiene
	–CH=CH–CH=O	Acrolein, crotonaldehyde
	–CH=CH–C=N	Acrylonitrile
Adjacent double bonds	–CH=C=O	Ketene
	–N=C=O	Methyl isocyanate, toluene di-isocyanate
Three-membered rings	–CH–CH_2 (bridged by O)	Ethylene and propylene oxide, epichlorhydrin
	–CH–CH_2 (bridged by NH)	Ethylene imine
Aldehydes	–CH=O	Acetaldehyde, butyraldehyde

8.2 Reaction rate

The rates of chemical reactions often bear little relation to the heat or energy which they generate. Some very fast ones (e.g. ionic reactions in aqueous solution) produce little or no energy or heat, while others which develop a great deal of heat proceed very slowly in the absence of a catalyst. Reaction rates are studied under chemical kinetics and reaction power under thermodynamics, both branches of physical chemistry.

The simplest chemical reactions for study are uncatalysed, homogeneous ones, which take place entirely in a single gas or liquid phase.

While many ionic reactions in aqueous solutions fulfil these conditions, they are in most cases almost instantaneous, and considerably faster than the speed with which their insoluble products come out of solution. These reactions seldom present serious hazards in themselves. Some which are important in the context of corrosion are discussed in 13.4.

Non-ionic reactions of this simple type usually involve one, two or three molecules, which may be identical or different. The rates at which such reactions proceed depend on the concentrations of the reactants, the temperature and reaction medium, and, in the case of gas reactions, on the pressure. This dependence on reactant concentrations is not a straightforward one. Many reactions proceed in stages, some faster than others. The overall reaction rate is then determined by the rate of the slowest stage, whose equation may be quite different from that of the overall reaction. Many reactions, including those of combustion, are *free radical chain reactions*.

An example of these is the gas phase dealkylation of toluene by hydrogen. The overall forward reaction is:

$$C_6H_5\,CH_3 + H_2 \rightarrow C_6H_6 + CH_4$$

If this were a straightforward bimolecular reaction, the rate would be proportional to the product of the concentrations or partial pressures of toluene and hydrogen (strictly speaking their 'activities'). The rate, however, is actually proportional to the product of the toluene concentration and the square root of the hydrogen concentration. Without going into detail, this rate agrees with the following mechanism:

$H_2 \leftrightarrow 2H^\bullet$	Initiation and termination
$H^\bullet + C_6H_5.CH_3 \leftrightarrow C_6H_6 + CH_3^\bullet$	Propagation
$CH_3^\bullet + H_2 \leftrightarrow CH_4 + H^\bullet$	Propagation

Note: $^\bullet$ signifies a free radical.

The rate r_f for this forward reaction is given by the equation:

$$r_f = k_2[C_6H_5.CH_3][H_2]^{1/2}$$

This shows that the rate of disappearance of toluene equals a constant k_2 *times* the concentration of toluene *times* the square root of the concentration of hydrogen. Since the reaction is reversible, there will be a similar equation for the reverse reaction.

Reactions can have various 'orders', from zero to 3, the order being the sum of the power factors in the rate equation. Thus the simple bimolecular reaction would be a second-order reaction, whereas the actual chain reaction shown above has an order of 1.5. While chemical kinetics is a complicated subject, the designer of a reaction system should know the order of the reaction and the rate constants for the forward (and reverse) reactions over the range of temperatures which might be employed. He or she should also have similar information for any side reactions.

Complications arise when more than one phase is present, and one or more molecular species has to move from one phase to another before or after reacting. The degree of mixing, the interfacial area available and the diffusion rates of the molecules all then have to be taken into account. Where solids are involved, their surface areas are also important.

Catalysis is another very important but complicating factor, since most industrial reactors operate with catalysts which may increase reaction rates one hundredfold or more, and thus promote one desired reaction over other competing ones. Some reactions are autocatalytic, being catalysed by their own products, so that after a sluggish start the reaction rate gradually gathers speed, even though the temperature is maintained constant.

In spite of all these complications, there is a characteristic rate constant for every reaction under every condition.

The rates of most chemical reactions, particularly homogeneous ones, i.e. those taking place in a single liquid or gas phase, increase exponentially with temperature, doubling or trebling with every 10°C temperature increase (Fig. 8.1(a)). There are exceptions to this, such as the third-order gas-phase oxidation of nitric oxide to nitrogen peroxide:

$$2NO + O_2 \rightarrow 2NO_2$$

In this the reaction rate decreases slowly with temperature (Fig. 8.1(b)).

Some catalytic reactions are controlled by the rate of adsorption of one of the reactants on the catalyst surface, which may decrease at elevated temperatures. Their rate may first increase with temperature almost exponentially, then rise more slowly to a peak, giving an S-shaped curve (Fig. 8.1(c)).

The rates of biochemical reactions which involve living organisms (fungi, yeasts, bacteria) also show a maximum at the optimum temperature of the organism, but fall off to zero at slightly higher temperatures, which kill the organism (Fig. 8.1(d)). Moulds and fungi generally die at temperatures of 50°C and higher, but bacteria are somewhat more resistant and survive up to a temperature of 70°C.

The rates of some decomposition reactions initially increase exponentially with temperature until their character changes to an explosion (Fig. 8.1(e)).

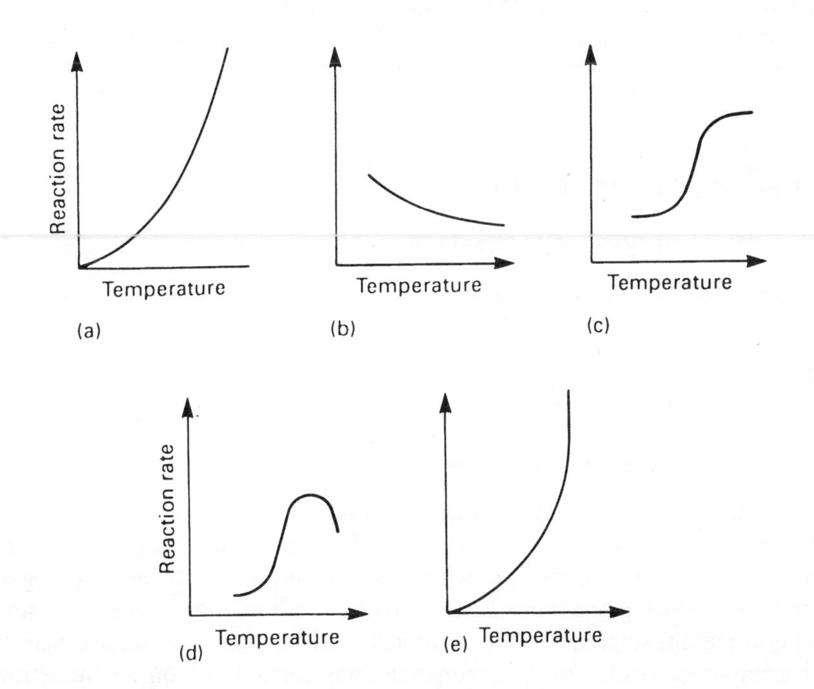

Fig 8.1 Dependence of reaction rate on temperature: (a) most homogeneous reactions in single fluid phase; (b) third-order gas-phase oxidation of nitric oxide; (c) catalytic reaction controlled by adsorption at solid surface; (d) biochemical reactions; (e) decomposition culminating in explosion

8.3 The power of reactions

The main clue to the possible violence of any reaction lies in the heat liberated, the temperature that may be reached and the volume and nature of any gases and vapours formed. Thermodynamics can tell us:

1. how much heat is given out or absorbed in the reaction, and
2. whether it is equilibrium-limited or can proceed virtually to completion.

The first question requires a heat balance to be drawn up between the starting materials and end products. This requires information on heats of reaction, specific heats, latent heats if there is a change of state, as well as work done on or by the reaction system, e.g. through expansion, compression or electrolysis. The heat of reaction is calculated from the heats of formation of the reactants and products. The second question is answered by the equilibrium constant for the reaction at the temperature in question. This is governed by the 'free energy change' of the reaction which can be calculated from the free energies of formation of the reactants and products. The thermodynamic data required can generally be found in the chemical literature[5–7], or else estimated by special methods.[8,9] Broadly speaking, any reaction (of which there are many) that can lead to a rise in temperature of 300°C and/or the production of a significant amount of gas or vapour may pose a significant hazard.[10] This is discussed further in 8.6.6 and the use of computer programs for screening compounds for chemical stability is discussed in 9.5.1.

8.4 Inorganic reactions

Most of the more powerful inorganic reactions come under the headings of:

- oxidation-reduction reactions
- acid-base reactions
- hydrations and hydrolyses

8.4.1 Oxidation-reduction reactions

Reactions between the many oxidising and reducing agents are usually powerful and sometimes explosive. Oxidising agents include oxygen, chlorine, nitric, chloric and hypochloric acids and their salts and hydrogen peroxide. Reducing agents include hydrogen, carbon, most metals and several non-metals, sulphur dioxide, sulphites, nitrites, ammonia and hydrazine.

Examples of oxidation-reduction reactions carried out on an industrial scale include:

- the smelting of many oxide ores of metals such as haematite with coke, $Fe_2O_3 + 2C \rightarrow 2Fe + CO + CO_2$

- the production of sulphur dioxide by burning sulphur in air, $S + O_2 \rightarrow SO_2$
- the production of phosgene by reacting carbon monoxide with chlorine, $CO + Cl_2 \rightarrow COCl_2$
- the Thermite reaction between aluminium and ferric oxide [10.5.1].

8.4.2 Acid-base reactions

These result whenever a 'base' such as a metal oxide or hydroxide or ammonia reacts with an acid to form a salt. Reactions between 'strong' acids such as sulphuric and hydrochloric and 'strong' bases such as sodium hydroxide are very violent when the reactants are concentrated. Industrial examples are the production of ammonium nitrate and sulphate for fertilisers and the pickling of steel wire and sheet to remove oxide layers before applying protective coatings.

8.4.3 Hydrations and hydrolyses

Several acids and bases are formed industrially by the combination of oxides with water, e.g. the production of sulphuric acid,

$$SO_3 + H_2O \rightarrow H_2SO_4$$

and the slaking of lime,

$$CaO + H_2O \rightarrow Ca(OH)_2$$

Many anhydrous salts add water to form crystalline hydrates while titanium chloride is hydrolysed by water to give titanium dioxide pigment,

$$TiCl_4 + 2H_2O \rightarrow TiO_2 + 4HCl$$

8.5 Some hazardous organic reactions and processes

As with inorganic reactions, there are powerful oxidation-reduction reactions (which are mostly thought of as oxidations, reductions and hydrogenations), acid-base reactions involving both organic and inorganic acids, bases and anhydrides, and hydration reactions involving organic acid anhydrides and acid chlorides. There are also important reactions involving the formation and use of organo-metallic compounds, and a great variety of organic syntheses which have little parallel in inorganic chemistry. Two principal differences between organic and inorganic reactions are described below.

- There is virtually no upper limit to the temperature at which inorganic reactions are carried out, and many ore-reduction processes take place at temperatures above 1000°C. Most organic reactions, however, have to be carried out at temperatures ranging from ambient to 250°C, due to the thermal instability of the materials. A few carefully controlled reactions such as ethylene production, with very short contact times (one second or less), are, however, carried out at temperatures of 800°C or more.

- There is nearly always a fire hazard with organic reactions, which are mostly undertaken within totally enclosed plant and pipework, often under pressure. Most inorganic reactions are free from serious fire hazards, and many of them are carried out in plant which is open to the atmosphere.

Certain structural groupings liable to cause instability or polymerisation have been touched on. The polymerisation of some aldehydes (e.g. acetaldehyde, butyraldehyde), when catalysed by small amounts of alkalis, is followed by the elimination of water and further evolution of heat if the reaction mixture is not well cooled. The overall process is one of condensation.

Many compounds containing two or more different reactive atoms or groups such as (active) hydrogen, chlorine, hydroxy-, amino-, nitro-, etc. undergo intermolecular or intramolecular (i.e. between two or more molecules) condensation reactions with the elimination of water, hydrogen chloride, ammonium chloride or nitrogen. Sometimes these condensations are accompanied by the release of considerable amounts of heat.

Among the most hazardous organic reactions[10–12] both on a laboratory and an industrial scale are oxidations, halogenations and nitrations (all of which are very exothermic). Polymerisation and copolymerisation reactions and condensations of the Friedel–Crafts, Claisen and Cannizaro types can be hazardous if temperature control is lost (e.g. through the breaking of a thermocouple). Other potentially very hazardous reactions are those involving acetylene (Reppe chemistry) and organo-metallic compounds of lithium, zinc and aluminium, and other metals high in the electromotive series. These are too specialised to be discussed here. Reactions which generally impose fewer safety problems include esterifications (other than nitration), reduction, amination, hydrolysis, hydrogenation (apart from the hazards of hydrogen under high pressure) and alkylation.

8.5.1 Peroxide formation

Many organic compounds which contain an 'active' hydrogen atom directly attached to carbon can react with oxygen when in contact with air at ambient temperature, to form peroxides and other 'peroxy-' compounds which contain the –C–O–O– grouping. These reactions involve the formation of free radicals which react with oxygen molecules through a chain reaction [8.2.1]. Most peroxy- compounds are unstable and many are explosive [9.3]. Table 8.4 gives a list of organic structural groupings with active hydrogen atoms which form peroxides with oxygen.[13,14]

Peroxy compounds are very 'labile'[15] (liable to spontaneous change). Thus the first one formed (usually a hydroperoxide) may react further to form a

Table 8.4 Structural groupings containing a peroxidisable hydrogen atom[13,14]

Grouping	*Examples*

>C–O–
  |
  H
	Acetals, ethers, oxygen heterocycles

–CH_2
      >C–
–CH_2 |
       H
	Isopropyl compounds, decahydronaphthalenes

>C=C–CH_2–
     |
     H
	Allyl compounds

>C=C–X
     |
     H
	Haloalkenes

>C=C–
     |
     H
	Other vinyl compounds (monomeric esters, ethers, etc.)

>C=C–C=C<
     | |
     H H
	Dienes

>C=C–C≡C–
     |
     H
	Vinyl acetylenes

–C–C–Ar
    |
    H
	Cumenes, tetrahydronaphthalenes, styrenes

–C=O | H	Aldehydes
–C–N–C< ‖ | | O H	N-alkyl-amides or -ureas, lactams

Note: The products of the last two types readily degrade and do not usually accumulate to a dangerous level

peroxide which may then form a dimer or trimer or change in some other way. Since most peroxides are less volatile than the original compound, they are left behind when it evaporates. There have been many small laboratory explosions after (diethyl) ether which readily forms a peroxide with air has been allowed to evaporate in a dish. Sodium, potassium and other alkali metals readily form peroxides in air as do their alkoxides, amides and organo-metallic compounds. Bretherick[13] quotes three lists of compounds which readily form peroxides, according to their hazards.

- List A, giving examples of compounds which form explosive peroxides while in storage, include diisopropyl ether, divinylacetylene, vinylidene chloride, potassium and sodium amide. Review of stocks and testing for peroxide content by given tested procedures at 3-monthly intervals is recommended, together with safe disposal of any peroxidic samples.

- List B, giving examples of liquids where a degree of concentration is necessary before hazardous levels of peroxides will develop, includes several common solvents containing one ether function (diethyl ether, ethyl vinyl ether, tetrahydrofuran), or two ether functions (*p*-dioxane, 1,1-diethoxyethane, the dimethyl ethers of ethylene glycol or 'diethylene glycol'), the secondary alcohols 2-propanol and 3-butanol, as well as the susceptible hydrocarbons propyne, butadiyne, dicyclopentadiene, cyclohexene and tetra- and deca-hydronaphthalenes. Checking stocks at 12-monthly intervals, with peroxidic samples being discarded or repurified, is recommended here.
- List C contains peroxidisable monomers, where the presence of peroxide may initiate exothermic polymerisation of the bulk of material. Precautions and procedures for storage and use of monomers with or without the presence of inhibitors are discussed in detail. Examples cited are acrylic acid, acrylonitrile, butadiene, 2-chlorobutadiene, chlorotrifluoroethylene, methyl methacrylate, styrene, tetrafluoroethylene, vinyl acetate, vinylacetylene, vinyl chloride, vinylidene chloride and vinylpyridine.

Where a peroxide-containing material has to be distilled, sufficient non-volatile mineral oil should be mixed with it before distillation to keep the peroxide concentration at a safe low level at which there is no danger of an explosion or violent reaction.

Serious explosions due to peroxides have been reported in vessels in which diethyl ether, butadiene and dihydrofurane had evaporated.

Small amounts of antioxidants such as phenols and amines often have to be incorporated into organic liquids and solids to inhibit peroxide formation. The free radicals involved in peroxide formation are trapped by these antioxidants, thereby terminating the chain reaction at an early stage. The antioxidant is eventually consumed. Thus it is necessary to check the antioxidant concentration regularly when materials liable to form peroxides are stored, and to 'top-up' with further antioxidant where necessary. In many cases (e.g. with monomers for synthetic rubber and plastics) the antioxidant has to be removed before using the material in the process. The amount of antioxidant-free material should then be kept to a minimum, and any remaining when the plant is shut down should be blended back with inhibited material in the main storage.

Peroxides are involved in many self-heating phenomena of organic solids, such as soya beans, rape seed, oiled fibres and spent grains from brewing, during drying, storage and transport. The problems and prevention of fire in these situations are discussed in 8.7.

8.5.2 Oxidation processes

A variety of oxidising agents are used including oxygen, air, hot dilute nitric acid, hydrogen peroxide, hypochlorous acid, permanganates and chromates.

Several valuable chemical intermediates are extensively made by the selective oxidation of hydrocarbons with air or oxygen. Three types of hazard are encountered with these processes:

- general hazards of handling volatile flammable liquids;
- hazards resulting from the deliberate mixing of hydrocarbons and oxygen inside the plant;
- several of the materials handled (e.g. ethylene oxide, cumene hydroperoxide, acrylonitrile, acrolein) have special hazards.

All these processes are continuous and the main reactions which are highly exothermic are accompanied by even more exothermic side reactions leading to normal combustion products, water and carbon dioxide. Large amounts of heat must be removed from the reactors.

All plants have distillation columns and other equipment for separating the products from the crude oxidation mixture and usually for recycling unconverted hydrocarbons to the reactors. Special instrumentation is used to maintain the gas compositions in the reactors outside the explosive range and to detect fires and hot spots in them.

These potentially very hazardous processes fall into two types, liquid phase and vapour phase, some of the more important being shown in Table 8.5.

Liquid phase processes

In these processes air or oxygen is bubbled into the liquid hydrocarbon (or a solvent) in which the reaction takes place, usually via a peroxide intermediate which is in some cases isolated before reacting further. Temperatures range from 100 to 225°C and pressures of 5–10 bar g. are usually necessary to maintain the liquid state. Cooling is generally achieved by evaporation of part of the hydrocarbon which is condensed and scrubbed from the 'off-gases' and returned to the reactors, which generally contain large inventories of liquid hydrocarbons at temperatures above their atmospheric boiling points. Thus in addition to very high fire loads there is usually the potential for the massive escape of a flash-vaporising flammable liquid followed by an open flammable cloud explosion (OFCE) [10.2.4]. Control of this hazard depends on preventing a major emission. Steps which may be taken include:

- careful plant siting and layout in relation to prevailing wind, ignition sources, concentrations of people, traffic and valuable property (e.g. other plants);
- reducing the inventories of superheated flammable liquids in the reactors as far as possible;
- employing very high standards of design, construction, inspection, maintenance and operation, including hazard and operability studies [14.4];
- reducing the number of flanged joints in the train of reactors, e.g. by replacing several reactors in series with a single vessel;

Table 8.5 Some major hydrocarbon oxidation processes

Feed	*Product(s)*	*Main final products*
Liquid phase processes		
Cyclohexane	Cyclohexanol/cyclohexanone	Caprolactam (nylon) solvents
Cumene	Phenol/acetone	Resins, acrylics
Paraxylene	Terephthalic acid	Polyester fibres
Light naphtha	Acetic acid	Vinyl acetate
Toluene	Benzoic acid	Terephthalic acid
Ethylene	Acetaldehyde	Ethyl hexanol
Ethyl benzene	Ethyl benzene hydroperoxide	Styrene (rubbers and plastics)
Vapour phase processes		
Ethylene	Ethylene oxide	Antifreeze
Butenes or benzene	Maleic anhydride	Unsaturated polyester resins
Butenes	Butadiene	Rubbers
o-Xylene or naphthalene	Phthalic anhydride	Plasticisers for PVC
Propylene	Acrolein	Glycerol, acrylics
Propylene + ammonia	Acrylonitrile	ABS resins
Ethylene + HCl	Ethylene dichloride	PVC

- separating vessels in a reaction train where feasible with automatic quick-closing valves to isolate the contents of each vessel in an emergency;
- special precautions to prevent the unintentional accumulation and concentration of unstable peroxides in any part of the plant.

Not surprisingly, several major fires and explosions have occurred on these plants. Probably the worst was the Flixborough explosion of 1974[16] [4]. This, however, happened while the plant was being started up, and before air was introduced into the reactors and neither peroxides nor oxygen were involved.

There is little danger of combustion occurring in the main body of cyclohexane oxidation reactors during normal operation when air is dispersed as small bubbles in the liquid. Fires have, however, occurred at the point where compressed air is introduced into such reactors,[17] during start-up following a short shut-down. They were accompanied by darkening of the crude liquid product and fortunately did not lead to the escape of flammable materials. This hazard was eliminated by reducing the compressed air temperature to 100°C before it entered the reactors, and by ensuring that the compressed air line sloped continuously downwards from its control valve until the compressed air entered the reaction liquid.

Vapour phase processes

In these a mixture of the hydrocarbon with air or oxygen is reacted on the surface of a solid catalyst. The processes shown in Table 8.5 include two in which other reactants are introduced as well, ammonia in the *ammoxidation* process for acrylonitrile, and hydrogen chloride in the *hydrochlorination* process for ethylene dichloride. The vapour phase processes employ temperatures from 250 to 600°C. The major hazard is explosion of the mixed feed if its composition reaches an unsafe level. All processes rely greatly on instrumentation and automatic devices to prevent this happening. The reactors of most processes are essentially tubular heat exchangers with the tubes packed with catalyst. Special measures are needed to prevent preferential flow in some tubes at the expense of others. Fluidised-bed reactors which avoid this problem and reduce the explosion hazard have been developed, but with limited success, for some of these processes. The heat is removed by special heat transfer liquids or molten salts outside the tubes and used in waste-heat boilers to raise steam. Some plants use a molten mixture of sodium/potassium nitrate/nitrite as a heat transfer medium. This is a highly oxidising material which generally causes a fire or explosion if it contacts hot organic material. There have been several reactor explosions following leakage between the hydrocarbon/air stream inside the tubes and molten salt outside. Their effects can be minimised by good reactor design, and tube corrosion should be monitored by regular analysis of the salt for metals.

Special material hazards

While several intermediates, products or by-products of oxidation processes constitute major hazards, only two with explosion risks, cumene hydroperoxide (CHP) and ethylene oxide (EO), are discussed here. Of the others, acrolein and acrylonitrile pose serious health risks and all create fire, and to some extent, explosion risks.

CHP, which is produced in solution in cumene, is concentrated by evaporation before conversion to phenol and acetone. It is a very unstable material and can in certain conditions decompose at an accelerating rate and then explode. This can happen through contamination with mineral acids, through overheating, or simply through allowing CHP solutions to stand at normal process temperatures without adequate mixing or heat removal. Such an explosion in a process vessel could release its contents and lead to an OFCE [10.2.4].

EO is very volatile, has wide flammability limits, polymerises readily and is both highly reactive and toxic. The vapour can also decompose explosively [9.5.3]. There have been a number of severe accidents on EO plants, some due to explosions with air or oxygen in the oxidation reactors and some to explosions of EO alone in the purification column.

Conclusions

Hydrocarbon oxidation processes which are increasingly used by major transnational companies to make many of the chemical intermediates of synthetic fibres, rubbers, plastics and paints have high potential for major explosions. Their employment makes special demands on managements and on sound engineering, particularly instrumentation.

8.5.3 Halogenation processes

There are two main types of halogenation processes, addition to double bonds and substitution of hydrogen with elimination of hydrogen halide. Reactions with fluorine are the most violent, more so generally than with oxygen, and extreme measures such as dilution with nitrogen are needed to control them.

Chlorinations are also very vigorous, and besides the direct use of chlorine, a variety of chlorinating agents are used including hydrogen chloride (for addition to a double bond), sodium hypochlorite, phosgene, thionyl chloride, sulphuryl chloride and chlorides of phosphorus. Oxy-chlorination processes which use a mixture of hydrogen chloride with air or oxygen with a catalyst in place of chlorine are now also common, particularly for the manufacture of ethylene dichloride from ethylene as a stage in the manufacture of vinyl chloride. Most hydrocarbon gases burn in chlorine, in which they exhibit upper and lower flammability limits. While measures similar to those used for oxidation reactions are needed in chlorination reactions to control reaction temperature and avoid explosive mixtures, the high corrosivity of chlorine and hydrogen chloride in the presence of water and the toxicity of chlorine and many chlorine compounds add to the overall hazard. First-class engineering, instrumentation, materials of construction, operator training and protective and emergency measures are needed for these plants.

When introducing a chlorine-using process on any site where chlorine is not already produced it is worth considering the installation of a special chlorine production plant. This may be less hazardous than importing chlorine in road or rail tank cars and storing it on site.

Bromination reactions pose similar though less acute hazards. They are less common and are usually carried out on a smaller scale than chlorinations. Iodinations are again less vigorous and less common.

Some halogenated organic compounds such as vinyl chloride and methyl iodide are carcinogens, and require special measures for air monitoring and personal protection.

8.5.4 Nitration processes

In nitration processes two basic types of product are produced. Nitrate esters which contain the nitrate group $-O-NO_2$ are formed from compounds containing hydroxyl groups, such as cellulose, glycerine, ethylene glycol and pentaerythritol. Nitro compounds containing the nitro group $-NO_2$ linked directly to a carbon atom are formed from paraffin hydrocarbons such as methane and propane, and from aromatic compounds such as benzene, toluene and phenol. More highly nitrated compounds such as TNT, nitroglycerine and guncotton are manufactured as explosives, whereas less nitrated compounds with lower nitrogen contents have industrial uses — nitrocellulose lacquers, dyestuff and other chemical intermediates, and nitroparaffin solvents.

In the production of nitrate esters and nitro-aromatic compounds a mixture of concentrated nitric and sulphuric acids is used at the lowest possible temperature, to make all except the least nitrated products. The sulphuric acid serves to 'mop up' the water formed as a by-product and thus drive the reaction nearer to completion. These reactions present at least two major hazards.

- The instability of the products, which decompose violently above a critical temperature. Efficient cooling and temperature control of the reactions are hence vital, and emergency arrangements for dumping the contents of a reactor (into a large volume of ice and water) may be needed.
- Exothermic oxidative side-reactions in which the nitric acid is reduced to the very toxic gas, nitric oxide. These are promoted by the presence of water and other impurities in the starting material, e.g. metal fragments and jute in cotton waste. When such side-reactions 'take over' in the batch nitration of cellulose the reactor is said to be 'on fire' because of the copious production of brown fumes with which the exhaust ventilation system is often unable to cope. There is then a serious health as well as explosion hazard, as the writer recalls from wartime work in an explosives factory, when it was sometimes impossible to see from one side to the other of a nitration room because of the fumes. In addition to the damage to the respiratory system, the teeth of workers in such nitration rooms suffer permanent damage.

Added to these hazards are those of human exposure to acid burns, and to the vapours of some nitration products, e.g. nitroglycerine, which causes severe headache for those freshly exposed and a fall in blood pressure. There is also a danger to workers of explosive compounds lodging in their clothing or hair or under their fingernails [9.8.6].

The danger of destructive explosions in batch nitrators for glycerine, in which a tonne or more of nitroglycerine is produced per batch, has been very much reduced by the development of simple short contact-time continuous nitrators.[18]

The nitro-paraffin solvents are mostly made in a continuous vapour phase process,[19] in some cases using nitrogen peroxide and oxygen as the nitrating medium, at temperatures of 150–300°C and pressures of about 10 bar g. These processes have resemblances to vapour phase oxidation processes, and have similar hazards.

8.6 Reactivity as a process hazard

As a potential process hazard, the chemical reactivity of any substance should be considered in the following contexts:

- its reactivity with elements and compounds with which it is required to react in the process;
- its reactivity with atmospheric oxygen;
- its reactivity with water;
- its reactivity with itself, i.e. its propensity to polymerise, condense, decompose or explode;
- its reactivity with other materials with which it may come in contact unintentionally in process, storage or transport;
- its reactivity with materials of construction, i.e. its corrosivity.

The hazards of exothermic reactions occur in several of these contexts, particularly in storage of compounds which tend to polymerise or decompose,

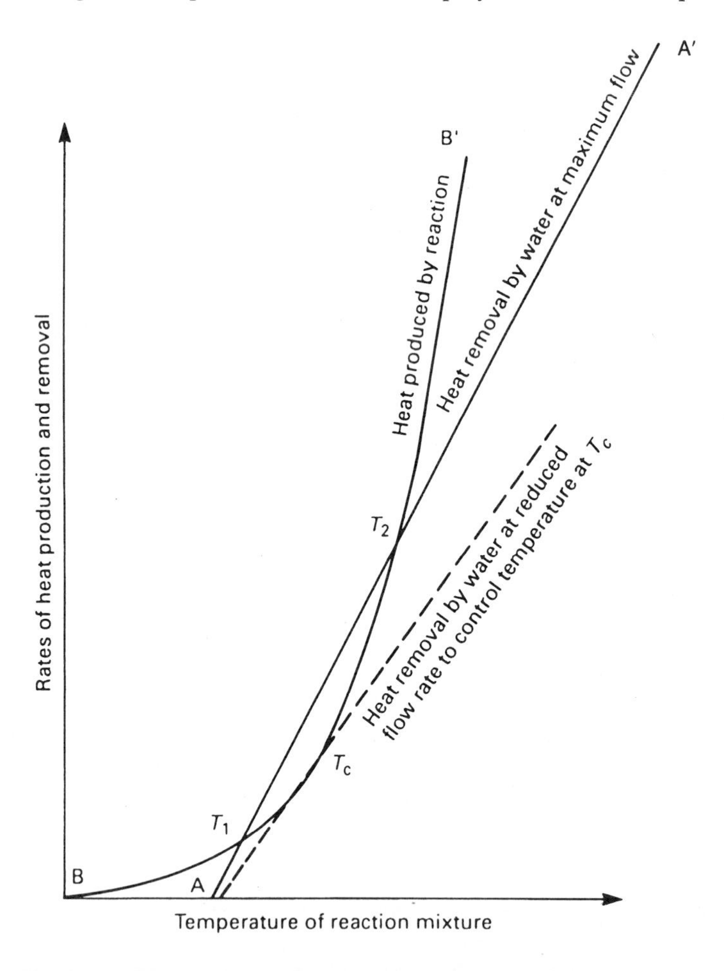

Fig. 8.2 Rates of heat generation and removal from reactor with homogeneous phase exothermic reaction

and in process reactors themselves.[11] The rates of most reactions increase rapidly with temperature [8.2.1], leading to the danger of their getting out of control, with large rises in temperature and pressure and 'loss of containment' of the process materials.

The hazard is shown in its simplest form in Fig. 8.2, where both the rates of heat production and removal from a water-cooled and stirred reactor are plotted against the temperature of the reaction mixture.

If the flow of cooling water is constant (and at its maximum value) the rate of heat removal increases roughly in proportion to the temperature difference between the incoming water and the reaction mixture, giving the line A–A′. The rate of heat generated by the reaction, on the other hand, increases exponentially with the temperature, and is shown by the curve B–B′. This crosses the line A–A′ at a lower temperature T_1 and a higher temperature T_2. At temperatures below T_2, the (maximum) rate of heat removal is greater than the rate of heat evolution, and the temperature will drop to T_1. The temperature can be controlled at some level T_c below T_2 by adjusting the flow of cooling water, but at the temperature T_1 the maximum water flow is required. Operation at this temperature is very stable, since if the temperature is displaced above or below it, it will return to this value without any adjustment in the cooling water rate. But once the temperature rises above T_2, the rate of heat evolution exceeds the (maximum) rate of heat removal. The temperature will then continue to rise, and the reaction goes out of control. The design should therefore provide a substantial difference between T_2 and T_c, to allow an adequate margin of safety.

The curves are not fixed for all time when the plant is designed. The line A–A′ will be lowered if the cooling surfaces become fouled and heat transfer coefficients fall, or if the water pressure falls. The curve B–B′ may be raised by the presence of catalytic impurities or by an increase in the quantity of reacting material in the reactor.

In a very similar way, we arrive at the concept of a 'critical mass' of any material liable to 'self-heating' in storage, above which there will be an uncontrolled runaway reaction.[20] This is based simply on the fact that the rate of heat evolution for a material undergoing a (slow) exothermic reaction will be proportional to the mass or volume of the material, which increases with the cube of a linear dimension, while the rate of heat removal is proportion to its surface area, and thus increases with the square of a linear dimension. In dealing with large piles of solids of low thermal conductivity there is also a greater tendency for hot spots to arise in the interior, where the reaction rate, although still low, may get out of control first. Once this has happened the runaway reaction will spread throughout most of the pile. This critical mass concept, elaborated by Frank-Kamenetskii,[20] has been at the root of some major accidents in the process industries [5.3]. It is discussed further in 8.7 in connection with the self-heating of materials.

8.6.1 Reactivity between reactants in processes

This must be carefully studied when the reaction system is designed, both from the thermodynamic and the kinetic aspects.[21] The information is vital to the design of the process in sizing heat exchangers and determining heating and cooling requirements, not to mention safety.

From the safety viewpoint, the main thing to know is whether the reaction is strongly exothermic, moderately exothermic, mildly exothermic, thermally neutral or endothermic. These expressions are quantified in Table 8.6 in terms of the heat given out or absorbed by the reaction, as J/g of total material fed to the reactor, including reactants, solvents and diluents.

An experienced chemist or chemical engineer can usually make a fair assessment of which of these categories applies by considering the overall chemical equation of the reaction. Reactions involving direct oxidation of hydrocarbons by air or oxygen, chlorination reactions, and ethylene polymerisation without diluents are extremely exothermic. Nitration reactions, and polymerisation of propylene, styrene and butadiene are strongly exothermic. Most condensation and polymerisation reactions of compounds with molecular weights from 60 to 200 are moderately or mildly exothermic. Reactions between aqueous solutions of inorganic salts to form precipitates and esterification reactions between organic acids and alcohols (in the absence of strong dehydrating agents) are usually thermally neutral. The cracking and dehydrogenation of hydrocarbons, and the reduction of most metallic oxides to metals, are very endothermic reactions which require the application of large amounts of heat or energy.

The application of such thermochemical information to evaluating the inherent hazardousness of process units is discussed later in Chapter 11 in connection with the Dow[22] and Mond[23] 'hazard indices'.

Exothermic reactions are usually least difficult to control in continuous processes involving only gases and liquids. In such cases the inventory of at least one of the reactants in the reaction system is usually fairly low, and its supply can be quickly interrupted in case of trouble. Heat exchange surfaces in the reaction system are also usually clean, and it is not difficult to design satisfactory emergency systems for relief of excess pressure or for dumping liquids whose reactions have got out of hand. Even here, however, there is a danger, particularly when starting up, that the reaction fails to start when

Table 8.6 Degrees of reaction heat

Category	*Heat (joules) given out per gram of total reactants, solvents and diluents*
Extremely exothermic	≥3000
Strongly exothermic	≥1200 and <3000
Moderately exothermic	≥600 and <1200
Mildly exothermic	≥200 and <600
Thermally neutral	≥–200 and <200
Endothermic	<–200

expected, but suddenly 'takes off' when the reactor is nearly full of unreacted material. This applies particularly to autocatalytic reactions, many of which depend on the formation of free radicals.

Exothermic reactions are most difficult to control in batch processes where the entire charge of reactants is added at the start of the batch, and where both liquids and solids are present. Here the inventory of reacting material is high at the start of a batch, and the presence of solids may foul heat exchange surfaces and block valves and pipes including pressure relief and blowdown systems. The problem can often be alleviated by carrying the reaction out in a refluxing solvent or inert liquid which boils (usually under pressure) at the desired reaction temperature. Cases have, however, been known when the pipe returning the condensed solvent to the reactor has become blocked with solid, the solvent has all boiled off, and a runaway reaction has developed with disastrous consequences [5.2].

Equilibrium limitations on the reaction are usually first calculated from published free energies of formation, and checked experimentally where necessary. They have an important bearing on the pressure and temperature chosen for the reaction, on the relative amounts of reactants employed, and on measures taken to drive an equilibrium-limited reaction nearer to completion, e.g. by removing one of the products of the reaction such as water as it is formed.

Kinetic aspects which must be studied experimentally affect the reaction temperature and time (hence the size of the reactor), the pressure in the case of gas reactions, the degree of agitation in the case of reactions involving more than one phase, and the catalyst used (if any). They also often affect the relative yields of wanted product and by-products.

8.6.2 Reactivity with atmospheric oxygen

Most hazards caused by reactivity with atmospheric oxygen are dealt with elsewhere in this book. Problems arising from oxidative self-heating are discussed in 8.7. The hazards of flammable process materials are covered in Chapter 10, and problems of rusting and corrosion are described in Chapter 13. There are also many cases where atmospheric oxygen has to be excluded from process plant and storage.

In most continuous organic chemical reactors which operate under pressure, air is automatically excluded, except where it is deliberately introduced as oxidant for a reaction. In some cases more stringent measures are taken, not merely to prevent air entering plant while it is running but also to remove it from the plant before starting up [10.3.2] and to remove oxygen from materials entering the process. This includes the use of oxygen scavengers such as sodium nitrite, sodium sulphite, sulphur dioxide, hydrazine and tertiary butyl catechol. These are generally used in aqueous solution. Such cases include:

- removal of dissolved oxygen from boiler feed water to reduce corrosion of the boiler tubes and hence the risk of boiler explosions;

- prevention of peroxide formation [8.5.1] with many organic compounds;
- protection of oxygen-sensitive materials such as photographic developers, organo-metallic compounds, alkali and alkaline earth metals and their hydrides, yellow phosphorus, titanium trichloride, powdered metals such as zirconium, titanium, uranium and plutonium, and even iron and lead, and special polymerisation catalysts;
- prevention of aerobic fermentations in alcoholic beverages.

The adventitious formation of pyrophoric materials, such as iron sulphide on the inside of crude oil carriers and storage tanks, presents another hazard of reactivity with oxygen [10.5.1]. This needs to be anticipated and prevented as far as possible, e.g. by applying an oil- and hydrogen sulphide-resistant coating to the inside of tanks liable to contain sour crudes. Otherwise it may be necessary to remove all flammable gases and vapours before air is admitted to the tank. Pyrophoric iron sulphide usually reacts harmlessly with atmospheric oxygen to form ferrous sulphate provided it is kept wet. Where problems of this sort arise, the help of an experienced chemist is needed.

8.6.3 Reactivity with water

Many chemicals react violently with water, which is seldom far away, and is widely used in process plants for cooling and cleaning. A short list of these, with the gases formed, is given in Table 8.7.[24]

These materials should only be handled by trained operators, wearing appropriate protection. They must always be stored in sealed watertight containers and kept in a dry place which is not subject to flooding. Plant in which they are used must be carefully designed and operated to prevent any possibility of water entering the process accidentally, e.g. through leaking heat exchangers.

8.6.4 Self-reacting compounds (mainly monomers)

As explosive materials are discussed in Chapter 10, the main compounds considered here are organic monomers, which are liable to polymerise spontaneously with evolution of heat, especially in storage. The same principles may be applied to compounds which are liable to condense, isomerise, transform or decompose without explosion in some other way, such as the α- and β-forms of sulphur trioxide, selenium, and antimony (of which there is an explosive form!). Monomers include styrene and substituted styrenes, vinyl chloride, acrylonitrile, butadiene, isoprene and cyclopentadiene, as well as methyl isocyanate whose runaway reaction caused the Bhopal disaster of 1984 [5.3]. Most of these polymerise spontaneously at low rates at ambient temperature even in the absence of a catalyst. All, however, polymerise much faster in the presence of a catalyst, such as a peroxide formed by the action of atmospheric oxygen on the material itself.

Table 8.7 Materials which react strongly with water

Material and state (s, l or g)	*Action*	*Gas, etc. liberated*
Calcium (s)	Moderate	Hydrogen
Lithium (s)	ditto	ditto
Sodium (s)	Vigorous	ditto
Potassium (s)	Explodes	ditto
Calcium hydride (s)	Vigorous	ditto
Lithium hydride (s)	ditto	ditto
Aluminium alkyls (l)	ditto	Alkanes
Aluminium alkyl halides (l)	ditto	ditto
Zinc alkyls (l)	ditto	ditto
Calcium carbide (s)	Moderate	Acetylene
Calcium phosphide (s)	ditto	Phosphine
Aluminium phosphide (s)	ditto	ditto
Fluorine (g)	Vigorous	Oxygen and ozone
Sodium peroxide (s)	Moderate	Oxygen
Aluminium chloride (s)	Vigorous	Steam and acid fumes
Phosphorus pentoxide (s)	ditto	ditto
Sulphur trioxide (s)	ditto	ditto
Acetyl chloride (l)	ditto	ditto
Phosphorus trichloride (l)	ditto	ditto
Silicon tetrachloride (l)	ditto	ditto
Sulphuric acid (l)	ditto	ditto
Thionyl chloride (l)	ditto	ditto
Titanium tetrachloride (l)	ditto	ditto
Calcium oxide (s)	ditto	Steam
Sodium hydroxide (s)	Moderate	ditto
Potassium hydroxide (s)	ditto	ditto
Activated alumina (s)	ditto	ditto
Activated silica (s)	ditto	ditto
Activated molecular sieves	ditto	ditto

Liquid cyclopentadiene, chloroprene (the monomer of Neoprene rubber), butadiene and isoprene all form dimers at room temperatures, the reaction rates increasing rapidly with temperature. The uncatalysed dimerisation rates for cyclopentadiene, butadiene and isoprene at various temperatures are shown in Fig. 8.3.[25] Although the rates for butadiene and isoprene are low enough for them to be stored for limited periods at ambient temperature, the rate for pure cyclopentadiene is so high that it is normally not stored as such, but produced as it is needed by thermal depolymerisation of its dimer.

These reactions evolve considerable amounts of heat and their rates increase exponentially as the temperature rises. As explained earlier, there is usually a critical temperature and/or a critical mass for such materials in storage above which disaster may lie imminent. In the case of monomers stored as liquids, the temperature of the unpolymerised material will rise until it reaches its boiling point, and then boil vigorously as the rest of it

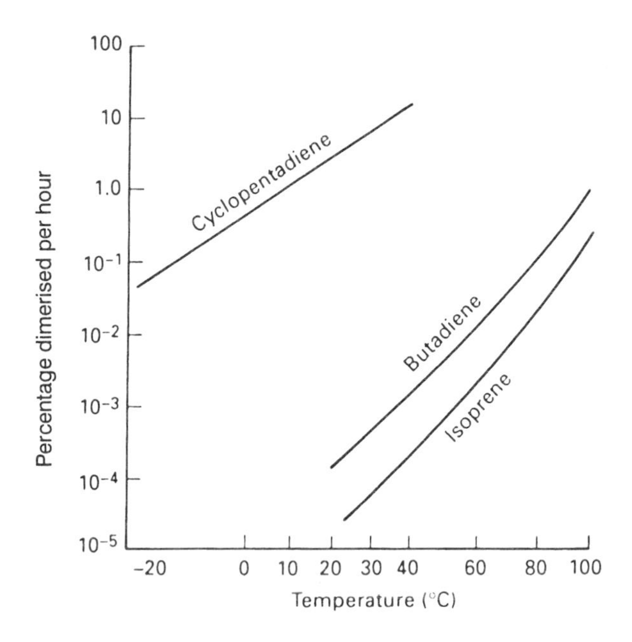

Fig. 8.3 Rates of uncatalysed dimerisation of three-diene hydrocarbons versus temperature

polymerises. With large storage tanks and pressurised spheres, the flow of escaping vapour may be so large that it is impossible to contain or dispose of it safely. Some monomers such as cyclopentadiene polymerise so rapidly at ambient temperature, even in the total absence of a catalyst, that their storage at ambient temperature is impossible.

The writer recalls a 10 m by 2.5 m (dia.) horizontal tank containing C_5 hydrocarbons rich in cyclopentadiene which had been produced 35 years ago as a large-scale experiment. This polymerised exothermally and continued to boil vigorously for three days, causing a serious hazard in the works, despite all efforts to cool it by spraying water over the tank.

Most monomers must be stored out of contact with air, with an inhibitor added which will prevent polymerisation and/or destroy any peroxide catalysts that may have formed [8.5.1]. Inhibitors commonly used are hydroquinone and its monomethyl ether, tertiary butyl catechol and phenothiazine.

Inhibition of many polymerisations (e.g. of acrylonitrile and vinyl acetate) is complicated by the fact that the inhibitor only works effectively in the presence of some oxygen.[26] Some monomers with inhibitors added are stored more safely in a tank which air can enter than under a nitrogen blanket. Oxygen is, however, rigorously excluded from butadiene and chloroprene during the last stages of their manufacture by using an oxygen-scavenging compound such as sodium nitrite. Tertiary butyl catechol is added in storage of butadiene to protect against leakage of oxygen into it with the formation of butadiene peroxide. A dangerous situation arises if inhibitor is lost from a monomer during storage. This can happen through separation of the inhibitor from the product or by reduction in its efficiency.

Since inhibitors are far less volatile than the monomers they protect, the vapour in a tank of stored monomer is free of inhibitor, and liquid condensing from it may form solid polymer. This has caused blockages in vents and pressure relief devices on some monomer storage tanks and vessels.

Loss of inhibitor efficiency can occur either through lack of oxygen as described above or through contact with so much oxygen that all the inhibitor has reacted. Stored monomers thus need to be periodically sampled and analysed to check the effectiveness of inhibition, which must be adjusted when necessary.

Bond[26] reports seven accidents resulting from spontaneous polymerisation of monomers in storage in which the following were involved: styrene, methyl methacrylate, acrylonitrile, acrolein, vinyl chloride and acrylic acid.

Table 8.8 shows the results of calculations made to show the amount of polymer which would have been formed from five monomers stored in bulk at an initial temperature of 25°C, by the time they reached their boiling points. Adiabatic storage with no heat loss to the surroundings was assumed. The table also shows what percentages of the initial material would escape as vapour and be left as polymer, when all liquid had either polymerised or vaporised. Four of the monomers are assumed to be stored at atmospheric pressure in vented tanks. The fifth, butadiene, is assumed to be stored in a pressure sphere with a bursting disc set at 10 bar g. This calculation shows the percentage of polymer formed by the time the bursting disc blows.

8.6.5 Reactivity with adventitious materials

The reactions with which one is mainly concerned here are those which might follow from some incident involving loss of containment, which would increase its consequences. Some types of chemicals which react strongly with each other have been considered already, and care must be taken when planning their storage to eliminate the risk of their coming into contact with

Table 8.8 Adiabatic bulk storage of monomers: calculation of percentage polymerised when liquid boils or bursting disc blows

Monomer	*Acrylo-nitrile*	*Styrene*	*Methyl styrene*	*Vinyl pyridine*	*Butadiene*
Molecular wt	53	104	118	105	54
Boiling point (°C)	77	145	165	158	(81)
Heat of polymerisation (J/g)	1410	680	293	718	1352
Specific heat (av.) (J/g °C)	2.27	1.81	1.89	1.81	2.48
Polymerisation at BP (%)	8.5	30	86	35	(10.1)
Final state:					
Vapour (%)	70	48	7	46	
Polymer (%)	30	52	93	54	

each other as the result of an accident. Thus oxygen cylinders should always be stored well away from acetylene and LPG cylinders and other fuels. Drums of sodium chlorate, sodium nitrate and other oxidising materials should also be stored at some distance from organic materials and powdered metals. Chlorine and acids should be kept far apart from alkalis, alkaline solutions, light metals, carbides, hydrides and other materials with which they might react with dangerous consequences, e.g. fire, explosion or the formation of more toxic materials. This applies to storage in drums and portable containers as well as in tanks, tank farms and silos.

One of the most critical operations where the hazards of incompatible chemicals are at their greatest is in bulk loading of road and rail tank-cars. A useful tabular guide to the compatibility of 24 types of chemicals was published by McConnaughey[27] and is reproduced in Lees's book.[12] This shows unsafe combinations of chemicals for adjacent loading at a glance, and should be considered when planning bulk loading facilities.

Drainage systems need careful study to ensure that spillages of incompatible materials do not mix and react in them to produce toxic or flammable vapours, perhaps some distance from the original spillages. With certain materials one has to consider whether they should ever be allowed to enter a drain at all, particularly a public one. If the answer is 'No', proper means of containing them and disposing of them safely must be provided.[28,29] One must also consider the accidental formation and mixing of substances which have adverse catalytic effects on materials in process or storage. Peroxides have already been mentioned as polymerisation catalysts. The corrosion products of several metals, e.g. iron and manganese, catalyse the decomposition of unstable compounds such as hydrogen peroxide and hypochlorites.

Materials, particularly liquids which react violently with each other, are sometimes used as auxiliary materials in the same process, although they are never intended to come into contact with each other. As an example, a solution of caustic soda is used to remove acid impurities from the products of hydrocarbon alkylation and acid-treating processes. Both plants and storage units must be carefully designed to ensure that mutually antagonistic materials can never mix inadvertently. This requires critical examination by an experienced chemist. The biggest problems often lie in process plant and storage depots which are not recognised as 'chemical', but which contain mutually antagonistic materials. Here there may be no qualified chemists on the staff to appreciate the hazards.

8.6.6 Practical reactivity screening in chemical manufacture

A system of reactivity evaluation set up for a UK company making more than 30 bulk organic chemicals by batch processes in which over 150 stages are involved is described by Coates and Riddell of Sterling Organics.[30] This followed a plant incident involving a reaction (decomposition of a diazonium salt) in which a large volume of gas was evolved. After proceeding smoothly for several years, a batch 'turned nasty', with rapid pressure build-up in a vessel. Investigations showed the reaction only 'misbehaved' when the main reactant was impure.

The system involved the setting up of a Hazard Evaluation Laboratory under a Hazard Evaluation Manager, whose task included the screening of both materials and processes for uncontrollable exothermic behaviour.

Of the many materials and processes used in the works the following received priority for study:

1. all plant processes where any doubt existed regarding the reaction controllability, or which had in the past shown evidence of variable exothermic behaviour:
2. residues from distillations or products residues which were subjected to temperatures in excess of 250°C:
3. raw materials, intermediates and products which contain functional groupings known from experience to have potential for instability;
4. all plant processes involving nitrations, or strongly oxidising conditions, and processes with high gas evolution.

Material evaluation

Most materials tested are typical plant samples. If it is suspected that the material can detonate or deflagrate violently, it is first tested by HSE. Their typical tests include:

- Home Office mortar test, to indicate whether the material exhibits moderate detonation properties;
- detonation/propagation test with thermal ignition in a sealed tube;
- Home Office pressure–time test;
- ballistic-torpedo friction test.

After HSE have cleared that the material is safe for hazard evaluation, the company's testing programme is started. The tests and testing equipment are as simple as possible, and designed for routine use by technicians. The following tests are carried out in sequence, results being assessed after each before deciding whether to proceed to the next.

- *Deflagration test*
- *Simple exotherm test.* This measures the lowest temperature T_e at which the material exhibits exothermic behaviour. After carrying out the test on fresh material, it is repeated on a fresh sample which has been subjected for 24 hours to a temperature 20°C above its normal operating temperature. Compounds which show no exothermic behaviour are not tested further, but where the maximum temperature to which the material is intended to be subjected is within 100°C of T_e, the material is considered to be potentially hazardous, and subjected to the next test.

- *Adiabatic exotherm test.* This is a refinement on the last test and may give values of T_e some 20–30°C lower. Plant material must not be subjected to temperatures closer than 50°C to the value of T_e given by this test unless it satisfies the next test.
- *Long-term hot storage test.* This gives the lowest temperature at which the material will undergo self-heating under isothermal conditions. The storage period is determined by the maximum holding time in the plant. The test is first carried out at a temperature 50°C higher than the maximum operating temperature proposed, and no further tests are done if no self-heating occurs. But if self-heating is found, the test is repeated on fresh material at progressively lower temperatures until self-heating is absent. The plant operating conditions are then adjusted to give a safe margin between the maximum operating temperature and the minimum self-heating temperature.
- *Drying tests.* These are carried out on materials which are dried in contact with air to determine whether the value of T_e determined by the adiabatic exotherm test is affected by exposure of the material to air under drying conditions. The details of the test are adapted to the type of dryer used or proposed. The method and conditions of drying may have to be altered to ensure that the material will not decompose during drying.

Reaction evaluation

Most of the reactions are normally carried out batch-wise on a plant scale, one or more materials being added progressively as the reaction proceeds. With exothermic reactions there is often a danger that the reaction fails to start or proceeds slowly on the first addition when expected, leading to an accumulation of reactants in the mixture. This may then suddenly react too rapidly to be controlled.

The full exothermic potential of a reaction is first assessed by carrying it out in a stirred Dewar vessel without cooling. This allows it to be classed as one of three types:

1. non-exothermic
2. exothermic but not runaway
3. runaway.

Further tests are required when the reaction is of type (3). The possibility of reactions normally of types (1) and (2) degenerating into type (3) as a result of process malfunction or maloperation also have to be considered and sometimes checked experimentally.

The thermal stability not only of the materials used or produced in the reaction but also of the reaction mixture at various stages may require to be tested by the simple exotherm test. If the samples contain materials which are liable to evaporate during the test, they should be contained in sealed tubes fitted with bursting discs.

The heat of reaction and the rate of heat evolution of exothermic reactions are determined by carrying out the reaction in a laboratory calorimeter equipped with an automatic temperature recorder. Expert interpretation of the temperature record shows the point in the reaction where an irreversible runaway reaction starts, and allows temperatures to be set for the operation of alarms, quenching devices, etc. on a plant scale.

The possible effects of unplanned reactions or conditions are determined by carrying out a series of experiments in Dewar flasks.

Where runaway reactions might occur, routines are established for the gradual or stepwise addition of reactants to the reaction mixture, and for checking that the reaction is proceeding as planned before more reactant is added. In some cases this may necessitate special investment, e.g. in microprocessors to control the reaction on the plant scale and in sampling and testing facilities to check the composition of the reaction mixture during the reaction before more material is added.

The information gathered from these laboratory hazard evaluations, valuable though it is in setting safe parameters for plant design and operation, can seldom enable all possible incidents on a full-scale plant to be foreseen. Factors which appear insignificant in a laboratory reaction may become of critical importance on a plant, where conditions of stirring, surface/volume ratios and materials of construction may be totally different. Thus the conclusions from laboratory hazard evaluations often require to be confirmed and extended by further work on a pilot plant, specially built or adapted for the reaction in question. This tends to set the design and operating conditions for the full-scale plant. Once these have been established for a potentially hazardous process they should not be altered without submitting the proposed change to a 'Hazard and Operability Study' where it is examined by a team of staff with experience of all aspects involved, led by a specialist in such studies. Where doubts still exist, the Hazard Evaluation Manager would be consulted. He may then decide on the need for further experiments to study the effects of the proposed change.

8.7 Self-heating hazards of solids

Several organic materials, mainly of natural origin, which are processed in bulk to make foods, textiles, paper, etc. are liable to oxidative self-heating which may lead to combustion when they are dried, stored and transported. Similar problems are encountered in the sometimes violent decomposition of unstable materials such as ammonium nitrate, 'high-strength' calcium hypochlorite, organic peroxides and various nitro-compounds, which is usually preceded by a period of self-heating without atmospheric oxygen.

To put the first problem in perspective, about 8×10^{10} t per year of atmospheric carbon dioxide are continuously being converted into organic matter by natural photosynthesis. Some of this gets burnt, some eaten, some used temporarily by industry, and some buried until it gets dug up again. But probably the bulk of it simply 'decays' at temperatures only a little above ambient, ultimately reverting to carbon dioxide and water. This process 'has to happen', for without decay the cycle of life would long since have ground to a halt. The temperature at which decay takes place is bound to be above ambient, since although decay is much slower than combustion, it uses the same amount of oxygen and produces the same amounts of carbon dioxide, water and heat.

There is fortunately a gap of a few hundred degrees Celsius between the temperature ranges in which decay and combustion take place, so that although decay raises the temperature of the matter involved, it seldom starts to burn. But when this gap is bridged, either by the decay process taking place at abnormally high temperatures or by combustion starting at unusually low ones, decay leads directly to combustion.

We are concerned to prevent our artefacts won from nature from going up in smoke and also usually try to arrest their decay for as long as possible. In spite of this, they do deteriorate in time. In so doing they are sometimes liable to 'self-heat' to the point where they burst into flames. A book by Bowes[31] based mainly on work at the Fire Research Station provides much practical information and underlying theory about this problem.

8.7.1 Mechanisms for self-heating of organic materials

Self-heating of organic materials can take place through various exothermic reactions, most of which depend on atmospheric oxidation. There are also exothermic degradation processes which proceed in the absence of free oxygen. Sufficient has been learnt about most self-heating processes for measures to be taken to prevent their leading to ignition, even when they are not fully understood, as in the case of the farmer's haystack.

It is useful first to gain a general appreciation of the amounts of heat liberated during both aerobic and anaerobic decay processes. As a simple illustration, the theoretical dehydration of softwood (treated as cellulose) to charcoal (carbon) is considered, and compared with the combustion of the wood itself and of the charcoal produced.

Starting with one mol of dry cellulose (taken as 162 g for $C_6H_{10}O_5$), the processes of combustion and of dehydration with their (net) thermal effects are roughly as follows:

- combustion:

$$C_6H_{10}O_5 + 3O_2 \rightarrow 6CO_2 + 5H_2O \qquad +3185\ \text{kJ}$$
$$6C + 3O_2 \rightarrow 6CO_2 \qquad +2410\ \text{kJ}$$

- hence carbonisation (dehydration):

$$C_6H_{10}O_5 \rightarrow 6C + 3H_2O \qquad +780\ \text{kJ}$$

This shows that the simple dehydration of cellulose, if it could be achieved without side reactions, would be a highly exothermic process, yielding 24% of its total net heat of combustion. Another way of looking at this would be to say that even very wet cellulose containing 60% water could in theory be converted isothermally to solid carbon and water vapour without the need for any external heat.

In fact, heat must be supplied during the first stages of charcoal production, when methanol, acetic acid and other combustible liquids are formed by destructive distillation, as well as solids with lower hydrogen and oxygen contents than the original cellulose. The final stages of charcoal production from partly carbonised solids are, however, exothermic and thermally self-sustaining, and raise the charcoal to temperatures at which it readily ignites if air is admitted. Clean timber does not present a self-heating hazard, although at least one fire has been reported where a steam pipe passed snugly through a hole in a thick beam.

There are two types of mechanism for the oxidative self-heating of organic matter.

- Microbiological processes, which occur at usual ambient temperatures in the presence of moisture and nutrients. These can only raise the temperature of the material to the maximum which the organism can withstand, which is still well below usual ignition temperatures.
- Purely chemical processes, of which those involving the initial formation of a peroxide or hydroperoxide are the norm.

Microbiological processes

Two different types of microbiological organism which are present everywhere, aerobic moulds and bacteria, soon get to work. Moulds operate most effectively at a temperature of about 40°C, but are mostly killed at 50°C. In the case of hay they require a moisture content of at least 25% to operate effectively. Bacteria require a somewhat higher moisture content of 40%, and work most effectively at a temperature of about 60°C but die at 75°C.

Parallel with these aerobic processes, anaerobic ones play a part in oxygen-depleted zones of the material. In the case of cellulose, these involve hydrolysis to sugars and fermentation to carbon dioxide, methane, ethanol, glycerol or lactic acid, all exothermic processes.

The generation of impure methane (marsh gas) from decaying vegetation proceeds fast in shallow lakes in tropical countries. If phosphine (PH_3) is also present, then on reaching the surface it may undergo partial combustion as faint flashes of flame known as 'Will-o-the-Wisp'. Marsh gas is not always so harmless. It is recorded that a paddle steamer on an inland lake in Uganda was destroyed by fire resulting from the release of trapped methane by the ship's paddle wheels!

Chemical oxidation

The first step in peroxidation processes is the addition of two oxygen atoms to a C–H bond or an R–C radical. These are 'free-radical chain reactions' which proceed autocatalytically with the accumulation of peroxide bodies. Slow oxidation at moderate temperatures is accompanied by evolution of 5–15% of the final oxidation products, CO, CO_2 and H_2O. Almost all the oxygen consumed at a given temperature can, however, be recovered as these oxidation products by heating to a higher temperature. This suggests that once sufficient oxygen has been absorbed, further self-heating may take place in its absence. The rate constants for peroxidation processes increase rapidly with temperature, but the actual rates in piles of organic matter are restricted by the diffusion of oxygen through the outer layer of the pile and by the availability of reactive sites which oxygen molecules can reach. Nevertheless, these peroxidation processes, and the thermal decomposition of peroxides which succeed them, are sometimes enough to raise the temperature at one or more places in the pile to ignition point.

Besides peroxidation, there are other mechanisms which account for the rapid ignition of fresh surfaces of cool charred hay or activated charcoal with atmospheric oxygen. These may involve the exothermic adsorption of oxygen or water or the presence of chemically active centres on the fresh surfaces. Such self-heating is often prevented by atmospheric weathering of the cooled material spread in thin layers for a few days before bulk storage or shipment.

The effect of moisture on self-heating

Materials which self-heat to the point of ignition are often wet. This is surprising, considering the additional heat which self-heating has to supply to dry the material before it can ignite. Thus wet haystacks are more liable to fire than dry ones, while wood charcoal produced for fuel in tropical countries[32] has to be stored under cover, due to frequent fires which break out in material which has been exposed to rain. It is, however, clear from the thermal data on cellulose quoted earlier that there is no shortage of chemical energy to evaporate a fair quantity of water, provided this energy can be released.

In other cases the risk of self-heating occurring and leading to fire is greatest if the material has been ‘over-dried’. This hazard can be controlled by the design of the dryer, and by the use of instruments which monitor the moisture content of the product. While the influence of moisture on self-heating is rather ambiguous there seems to be an optimum range of moisture contents corresponding to a relative atmospheric humidity of 25–50%, for minimum fire risk through self-heating.

8.7.2 Effects of pile size, initial temperature and time

The critical importance of the mass or size of the pile of material for self-heating has been mentioned earlier [8.6]. For the simplest case of a material undergoing self-heating, the following heat balance equation can be written:

$$\begin{aligned} dq/dt &= kQV\phi T - k'A(T - T_0) \\ &= 0 \text{ for a steady state condition} \end{aligned}$$

where dq/dt is the rate of heat accumulation of the reacting mixture,
Q is the heat of reaction per unit mass of reactant consumed,
V is the volume of material reacting,
A is the surface area of the volume reacting,
T is the internal temperature,
T_0 is the external temperature,
$\phi T = r$ = reaction rate = mass of reactant consumed per unit volume, which is taken as a function of the internal temperature T,
k and k' are constants.

The first term on the right-hand side is the rate of heat production and the second term the rate of heat loss to the surroundings. In most cases the reaction rate r increases with T much faster than $(T - T_0)$.

At sufficiently low values of T_0 or V/A the equation will have a real root in T and a steady state will be set up, with a corresponding internal temperature. But if T_0 or V/A are raised, eventually a condition will be reached where the equation has no real roots, and a steady state condition is no longer possible. The temperature will then rise until the pile ignites.

Although material in storage in real piles, silos, bins, bales and other containers (all subsequently referred to as piles) which is subject to an oxidative self-heating process presents a more complicated picture, the concepts of a mutually dependent critical temperature and mass or size still apply. The dependence of critical size and temperature is discussed in a paper by Boddington, Gray and Walker.[33] As important, however, as the temperature T_0 of the surroundings is the initial temperature at which the pile is made or the bin etc. is filled. This may be quite high if the material comes direct from a dryer. It is then often necessary to cool it to a lower temperature before it is safe to store.

Three factors which arise with self-heating materials complicate the above picture.

- Most self-heating materials have a cellular, granular or fibrous structure with a very low thermal conductivity. Thus when self-heating occurs, the temperature in the interior of the pile is much higher than the surface temperature, which may be closer to the outside temperature than the average temperature of the pile.
- The oxygen concentration in oxidative self-heating processes which depends on the diffusion of atmospheric oxygen into the pile will be highest at the surface and decrease inside with distance from the surface. This factor acts in opposition to the first one.
- The time taken for a self-heating pile to reach a point of rapid temperature rise and ignition may be considerable (i.e. several days). The slow rate of oxygen transfer into the pile, the initial build-up of peroxidic bodies and the need to dry out the pile make the overall process a lengthy one.

Two illustrations of the concepts of critical mass and critical temperature are taken from Bowes's book.[31] Table 8.9 shows the results of ignition tests on mixed hardwood sawdust, samples of which were placed to various depths on a thermostated hot-plate. Several tests were done for each depth of sawdust at different surface temperatures, and the minimum temperature at which ignition occurred was found for each one. The time to reach ignition was also recorded.

Table 8.10 shows the critical ambient temperatures for rectangular and cubical packages of a typical batch of activated charcoal.

Bowes provides mathematical equations with constants derived from laboratory experiments on particular materials to predict their behaviour in real situations. These are too specialised to be given here. He also discusses computer modelling of self-heating processes. The underlying theory is largely based on the work of Frank-Kamenetskii.[20]

8.8 Reactive substances and CIMAH Regulations 7 to 12[34]

Group 3 of Schedule 3 of these regulations [2.5.5], given in Table 8.11, is a list of highly reactive substances with the minimum quantities of each to which Regulations 7 to 12 apply.

Most of these substances are peroxy-compounds [8.5.1]. Their main hazards are explosive decomposition and fire, and of catalysing exothermic runaway polymerisations of monomers. They are also liable to react violently with many other chemicals.

3,3,6,6,9,9-Hexamethyl-1,2,4,5-tetroxacyclononane is one of the least known of several cyclic trimeric peroxides, with powerful and sensitive explosive properties. Of the other substances, acetylene, ammonium nitrate, ethylene oxide, hydrogen and sodium chlorate present hazards of fire and explosion as well as of reactivity, and are discussed in Chapters 9 and 10 as well as in this chapter. The footnote to Table 8.11 suggests that ammonium nitrate and sodium chlorate have safe states which are exempt from the regulations.

Table 8.9 Ignition tests on mixed hardwood sawdust on hot surface

Depth of layer (mm)	*Ignition temperature* (°C)	*Time to ignition* (min)
5	355	9
10	320	24
20	290	50
25	280	105

Table 8.10 Critical ambient temperatures for self-ignition of piles of activated carbon

Critical temperature (°C)		40	60	80	100
Length of shortest side (m)	*Rectangle*	1.3	0.55	0.21	0.09
	Cube	1.8	0.73	0.28	0.14

Propylene oxide is less sensitive and more easily transported than ethylene oxide, but it still needs to be handled with some care.[35] The main hazard of ethyl nitrate is explosion, but its use appears to be limited. The lower nitroparaffins which have similar hazards and are more commonly used are not included in the list.

The list also does not include many common reactive chemicals such as sodium, fluorine, hydrogen peroxide, sodium hypochlorite, hydrazine, sulphur trioxide, phosphorus trichloride and the lower aluminium alkyls. It is,

Table 8.11 Schedule 3, Group 4 of CIMAH Regulations — Explosive substances (for the application of Regulations 7 to 12)

Substance	*Quantity* (t)	*Number* *CAS*	*EEC*
Acetylene	50	74–86–2	601–015–00–0
Ammonium nitrate[a]	5000	6484-52-2	
2,2-Bis (*tert*-butyl-peroxy) butane (concentration ≥ 70%)	50	2167-23-0	
1,1-Bis (*tert*-butyl-peroxy) cyclohexane (concentration ≥ 80%)	50	3006-86-8	
tert-Butyl peroxyacetate (concentration ≥ 70%)	50	107-71-1	
tert-Butyl peroxyisobutyrate (concentration ≥ 80%)	50	109-13-7	
tert-Butyl peroxy isopropyl carbonate (concentration ≥ 80%)	50	2372-21-6	
tert-Butyl peroxymaleate (concentration ≥ 80%)	50	1931-63-0	
tert-Butyl-peroxypivalate (concentration ≥ 77%)	50	927-07-1	
Dibenzyl peroxydicarbonate (concentration ≥ 90%)	50	2144-45-8	
Di-*sec*-butyl peroxydicarbonate (concentration ≥ 80%)	50	19910-65-7	
Diethyl peroxydicarbonate (concentration ≥ 30%)	50	14666-78-5	
2,2-Dihydroperoxypropane (concentration ≥ 30%)	50	2614-76-8	
Di-isobutyryl peroxide (concentration ≥ 50%)	50	3437-84-1	
Di-*n*-propyl peroxydicarbonate (concentration ≥ 80%)	50	16066-38-9	
Ethylene oxide	50	75-21-0	603-023-00-X
Ethyl nitrate	50	625-58-1	007-007-00-9
3,3,6,6,9,9-Hexamethyl-1,2,4,5-tetroxacyclononane (concentration ≥ 75%)	50	22397-33-7	
Hydrogen	50	1333-74-0	001-001-00-9
Methyl ethyl ketone peroxide (concentration ≥ 60%)	50	1338-23-4	
Methyl isobutyl ketone peroxide (concentration ≥ 60%)	50	37206-20-5	
Peracetic acid (concentration ≥ 60%)	50	79-21-0	607-094-00-8
Propylene oxide	50	75-56-9	603-055-00-8
Sodium chlorate[a]	50	7775-09-9	017-005-00-9

[a] Where this substance is in a state which gives it properties capable of creating a major accident hazard

however, expected to be extended when the regulations are amended, when it will probably include sulphur trioxide. The fact that a chemical site may have none of the substances on this list is no proof that it does not contain dangerous quantities of highly reactive chemicals.

References

1. Huheey, J. E., *Inorganic Chemistry*, 3rd edn, Harper & Row, London (1983)
2. Lewis, G. M. and Randall, M., *Thermodynamics*, McGraw-Hill, Maidenhead (1923)
3. Hodgman, C. D. (ed.) *Handbook of Chemistry and Physics*, 32nd edn, Chemical Rubber Publishing Co., Cleveland, Ohio (1950)
4. Uhlig, H. H., *Corrosion and Corrosion Control*, 3rd edn, Wiley-Interscience, New York (1985)
5. National Bureau of Standards, *Selected Values of Properties of Hydrocarbons, Circular 461*, Washington, DC: US Govt Printing Office (1947)
6. National Bureau of Standards, *Selected Values of Chemical Thermodynamic Properties, Circular 500* (1952) and *Circular 270–3* (1969) Washington, DC: US Govt Printing Office
7. Rossini F. D. (ed.), *Selected Values of Physical and Thermodynamic Properties of Hydrocarbons and Related Compounds. API Research Project 44*, Pittsburgh, Pa (1952) and Williams C. C. (ed.) *ibid* Texas (1975)
8. Coffee, R. D., 'Chemical stability' in *Safety and Accident Prevention in Chemical Operations*, 2nd edn, edited by Fawcett, H. H. and Wood, W. S., Wiley-Interscience, New York (1982).
9. Wenner, R. R., *Thermochemical Calculations*, McGraw-Hill, New York (1941)
10. NFPA *Manual of Hazardous Chemical Reactions, 491M–1975*NFPA, Quincy, Mass. (1975)
11. Austin, G. T., 'Hazards of commercial chemical operations' and 'Hazards of commercial chemical reactors', in *Safety and Accident Prevention in Chemical Operations*, 2nd edn, edited by Fawcett, H. H. and Wood, W. S., Wiley-Interscience, New York (1982)
12. Lees, F. P., *Loss Prevention in the Process Industries*, Butterworths, London (1996)
13. Bretherick, L., *Bretherick's Handbook of Reactive Chemical Hazards*, 4th edn, Butterworths, London (1990)
14. *Recognition and Handling of Peroxidisable Compounds*, Data Sheet 655, National Safety Council, Chicago (1976)
15. Swern D. (ed.), *Organic Peroxides*, Wiley-Interscience, New York, vol. 1 (1970), vol. 2 (1971), vol. 3 (1972).
16 *The Flixborough disaster, Report of the Court of Inquiry*, HMSO, London (1975)
17. Alexander, J. M., 'The hazard of gas phase oxidation in liquid phase air oxidation processes', paper in I. Chem. E. Symposium series 39a, *Chemical Process Hazards with special reference to plant design — V*, Rugby (1974)
18. Kletz, T. A., *Cheaper and Safer Plants*, The Institution of Chemical Engineers, Rugby (1984)
19. Waddams, A. L., *Chemicals from Petroleum*, 4th edn, John Murray, London (1978)
20. Frank-Kamenetskii, D. A., *Diffusion and Heat Exchange in Chemical Kinetics*, Translated by J. P. Appleton, Plenum Press, New York (1969)
21. Stull, D. R., 'Identifying chemical reaction hazards' in *Loss Prevention* **4**, 16 (1970) and, 'Linking thermodynamics and kinetics to predict real chemical hazards', in *Loss Prevention* **7**, 67 (1973)
22. Dow Chemical Company, *Fire and Explosion Index — Hazard Classification Guide*, 5th edn, Midland, Mich. (1980)
23. ICI Mond Division, *The Mond Index*, 2nd edn, ICI PLC, Northwich, Cheshire (1985)
24. King, R. W. and Magid, J., *Industrial Hazard and Safety Handbook*, Butterworths, London (1980)
25. Kirk, R. E. and Othmer, D. F., *Encyclopaedia of Chemical Technology*, 3rd edn, Wiley-Interscience, New York (*ca* 1979)

26. Bond, J., 'Violent polymerisations', *Loss Prevention Bulletin 065*, The Institution of Chemical Engineers, Rugby (1985)
27. McConnaughey, W. E. *et al*, 'Hazardous materials transportation', *Chemical Engineering Progress*, **66**(2), 57 (1970)
28. Fawcett, H. H., 'Chemical wastes' in *Safety and Accident Prevention in Chemical Operations*, 2nd edn, edited by Fawcett, H. H. and Wood, W. S., Wiley-Interscience, New York (1982)
29. Ross, R. D., 'Disposal of hazardous materials' *ibid*.
30. Coates, C. F. and Riddell, W., 'A system of hazard evaluation for a medium size manufacture of bulk organic chemicals'. Paper in I. Chem. E. Symposium Series 68, *Runaway Reactions, Unstable Products and Combustible Powders*, Rugby (1981)
31. Bowes, P. C., *Self-heating: evaluating and controlling the hazards*, HMSO, London (1984)
32. Uhart, E., *Potential Charcoal Development in Uganda*, United Nations Development Programme, Vienna (1975)
33. Boddington, T., Gray, P., and Walker, I. K., 'Runaway reactions and thermal explosion theory'. Paper in I. Chem. E. Symposium series 68, *Runaway Reactions, Unstable Products and Combustible Powders*, Rugby (1981)
34. S.I. 1984, No. 1902, *The Control of Industrial Major Accident Hazards Regulations 1984 (CIMAH)*, HMSO, London
35. Gait, A. J., 'Propylene oxide', in *Propylene and its Industrial Derivatives*, edited by E. G. Hancock, Ernest Benn, London (1973)

9 Unstable chemicals

In Chapter 10, which is concerned with combustion mechanisms, we discuss the behaviour of premixed flames. These may be propagating through mixtures of air and flammable gases either as explosions or detonations. In this chapter we are concerned with similar effects involving unstable solid materials. These may be true combustion processes as in the burning of cordite, in which carbon and hydrogen atoms react with oxygen atoms, all of which are in the same molecule. Oxygen may be absent from certain solid explosives, and the energy is released mainly by the breaking of strained valence bonds, often those of nitrogen atoms. Most of these processes release large volumes of hot gases, but this is not always the case. For example silver acetylide decomposes violently to release only silver, carbon and heat:

$$Ag_2C_2 \rightarrow 2Ag + 2C$$

The behaviour of unstable solids is best understood by looking first at materials in which violent decomposition is a desirable property: the explosives.

9.1 The commercial explosives[1,2,3]

Cordites have been used as propellants for guns and rocket motors since the end of the nineteenth century. They comprise a colloidal suspension of nitroglycerine in nitrocellulose. In both of these substances the $-O-NO_2$ (nitrate) group is the oxidant for the carbon and hydrogen atoms in the remainder of the molecule. When a piece of cordite is ignited, a flame is established close to the surface. As the surface layers are consumed, burning progresses at a characteristic rate into the bulk of the cordite. A lump of coal burns in much the same way, except that at the pressures used in guns the burning rate of cordite can be as high as 20 mm/s, but that of coal is perhaps 20 mm/hr. The large volume of hot gases produced by the combustion of cordite can be used to propel a shell out of a gun, in exactly the same way that a pea is ejected from a peashooter. Materials which are used like this are called propellants or gas-generators, and are sometimes also called low explosives. The combustion process is called deflagration, and this term is sometimes, rather confusingly, applied also to gaseous explosions [10]. However, the two processes are indeed analogous: the progression of a flame into a fuel/oxidant mixture.

The explosive decomposition of high explosives progresses at much higher velocities than those which characterise deflagrations. The flame travels through the material accompanied by a shock wave. The shock is maintained by energy from the flame, and in turn transfers energy ahead into the unconsumed material. The propagation of a flame in association with a shock wave is called a detonation. Typical detonation velocities are: TNT, 6700 m/s; PETN 7600 m/s; gelignite, 6000 m/s; picric acid, 6700 m/s; and nitroglycerine 8000 m/s. The latter is a liquid and, as we have seen, a constituent of cordite. However, cordite has never been known to detonate except under extreme conditions of pressure and initiation. A detonation can be established in nitroglycerine by an impact, but the commercial and military explosives are much less sensitive and will detonate only when they receive a shock wave from the detonation of another high explosive. The chain of events leading to the detonation of a high explosive charge usually includes a third class of explosives, the initiators or primary explosives. These can be detonated either by impact or by being heated. In a standard electric detonator, the fusing of a bridge wire ignites a match-head, and the flame from this ignites a primary explosive like mercury fulminate. The burning of this material rapidly becomes a detonation, which is passed on to a small high explosive charge, possibly PETN. This generates a sufficiently energetic shock to set off the main charge. Mercury fulminate is also used in percussion caps, located in the base of cartridge and shell cases. They are initiated by the impact of a firing pin.

The high explosives possess a property called *brisance*: the ability to release energy so quickly that it cannot be absorbed by the movement of an object. If a quantity of cordite is placed on a piece of rock and ignited, the only effect will be a slightly discoloured, hot area. If the same amount of energy is delivered by a hammer blow, the rock will move and much of the energy will be absorbed by the soil. If now the energy is supplied by the detonation of a high explosive, the rock will not have time to move away and may be shattered.

Materials with the sensitivity of primary explosives will very rarely be used in industry, but they may be formed inadvertently. For example, copper acetylide can be formed simply by passing acetylene through a copper pipe. However, many industrial chemicals do have properties similar to those of the propellants and high explosives.

9.2 Industrial chemicals with explosive potential

While explosives are manufactured on a large scale for military purposes, blasting, fireworks and rockets, etc., many compounds which might and sometimes do explode are made in the chemical industry for entirely different purposes. They may possess other desirable properties in themselves, or they may be intermediates in the manufacture of other products.

Some examples are shown in Table 9.1. These include a few compounds which readily release oxygen and form explosive mixtures with organic matter and other fuels, such as the nitrates and chlorates of sodium, potassium and ammonium, hydrogen peroxide and nitric acid. The fuels include porous or finely divided solids such as charcoal, coal dust, sugar, sawdust, sulphur and aluminium powder, and combustible liquids such as fuel oil, ethanol and hydrazine.

The explosion hazards of materials which fall within the scope of the Explosives Act of 1875 are very strictly controlled[1] during manufacture, transport and storage. This has not always been so with compounds made for entirely different purposes. Several serious explosions have occurred in

Table 9.1 Some industrial chemicals with explosive potential

Chemical	*Actual use*	*Intermediate for*
Acetylene	Gas welding and cutting	Vinyl chloride, acrylics, perchloroethylene
Ammonium nitrate	Fertilisers, explosive ingredient	Nitrous oxide
tert-Butyl hydroperoxide	Polymerisation initiator	
Ethylene oxide	Fumigant, rocket fuel	Detergents, paints, antifreeze, etc.
1,3-Dinitrobenzene	Explosive ingredient	Dyestuffs
Nitromethane	Solvent, underwater explosive, corrosion inhibitor	Chloropicrin
Picric acid	Yellow dye	
Trinitrobutyl-toluene	Perfumery ingredient	
Vinyl acetylene		Neoprene rubber
Oxidants		
Hydrogen peroxide	Bleaching agent, antiseptic	Organic peroxides
Manganese dioxide	Batteries	Manganese compounds, ferromanganese
Nitric acid	Descaling concrete	Nitrates, nitro compounds
Nitrogen peroxide	Rocket propellant	Nitromethane
Potassium chlorate	Weedkiller, explosive ingredient	Potassium perchlorate
Potassium nitrate	Explosive ingredient, fertiliser	

the manufacture, transport and use of ammonium nitrate and other industrial compounds and their mixtures (Fig. 9.1). Hence there is always a need to screen industrial chemicals and their mixtures for potential explosivity, and to examine those with this tendency for destructive power and sensitivity to heat, friction, impact and other forms of initiation. Only then can safe conditions for their manufacture, storage, transport and use be established.

9.3 Structural groups which confer instability

Table 9.2, due to Coffee,[4] shows 17 structural groupings, nine of which contain nitrogen, which confer instability to the compounds in which they are present, often rendering them explosive. These groupings are known as 'plosophors'. The list is far from exhaustive, and a more extensive one containing 42 structural groupings which confer instability is given by Bretherick.[5] Some of these are rarely seen outside research laboratories.

Some more complex groupings which create explosive tendencies in organic compounds (including salts of organic bases) are:[6]

Primary nitramine	$-NH-NO_2$
Secondary nitramine	$>N-NO_2$
Nitroso	$-N=O$
Diazosulphide	$-N=N-S-N=N-$
Picrates	$[C_6H_2(NO_2)_3.O]'$
Iodates	$[IO_3]'$

Acetylene and acetylenic compounds are very sensitive to heat, shock and abrasion, and many are explosive.

Fig. 9.1 Crater and destruction caused by ammonium nitrate explosion at Oppau, Germany, in 1921 (Badische Anilin und Soda Fabrik)

Table 9.2 Structural groupings indicative of potential instability[4]

Acetylide	–C≡C–METAL
Amine oxide	$\equiv N^+–O^-$
Azide	$–N\equiv N^+=N^-$
Chlorate	$–ClO_3$
Diazo	–N=N–
Diazonium	$(–N\equiv N)^+X^-$
Fulminate	–(C≡N→O)
N-Haloamine	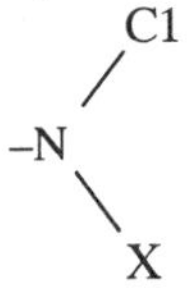 –N(Cl)(X)
Hydroperoxide	–O–O–H
Hypohalite	–O–X
Nitrate	$–O–NO_2$
Nitrite	–O–NO
Nitro	$–NO_2$
Ozonide	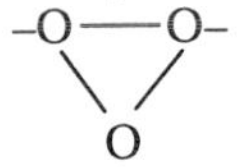 –O—O– (bridged by O)
Peracid	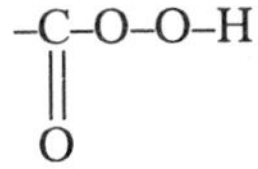 –C(=O)–O–O–H
Perchlorate	$–ClO_4$
Peroxide	–O–O–

Molecular nitrogen, ammonia and amino groups are in themselves very stable, but the compounds formed when ammonia or an amino group reacts with an oxidising agent are among the most unstable; they include lead azide and mercury fulminate[1] [9.1]:

Mercury fulminate O ← N ≡ C–Hg–C ≡ N→O

Lead azide (N=N)>N–Pb–N<(N=N)

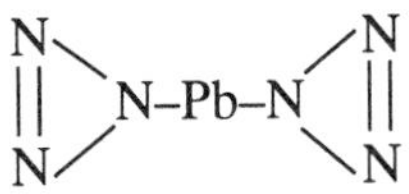

Mercury fulminate is also a health hazard to those working with it.[7] Most azides are explosive and very sensitive.

Hydrazine vapour, $H_2N=NH_2$, is shock sensitive and decomposes violently, and most substituted hydrazines and hydrazones can form explosive compounds.

Nitrate and nitro groups introduce oxygen at a high energy level into organic compounds and provide the main basis for the explosives industry. The explosivity of such compounds depends largely on their oxygen balance and the C:N ratio in the compound. Mono-nuclear aromatic compounds with three nitro groups are common explosives. Most with two nitro groups will explode at mildly elevated temperatures. Some have been used as explosive ingredients, and some have other industrial uses (Table 9.1). All dinitro-aromatic compounds should be treated as potentially explosive and thermally sensitive.

Most aromatic compounds with only one nitro group are fairly stable at ambient temperatures but liable to decompose explosively at higher temperatures. Thus explosions of (mono) nitrotoluenes have occurred during their distillation when heated to high temperatures, or when held at moderate temperatures for long periods. Harris and others of ICI Dyestuffs division have shown that such decomposition on a large scale is preceded by an induction period which increases as the temperature is lowered.[8] At 190°C decomposition occurred after 80 days.

Of the nitroparaffins, nitromethane, which is used both as a propellant and a solvent, has been involved in serious tank-car explosions in the USA. Nitroethane is also explosive, though less sensitive than nitromethane, but becomes more so with the addition of small amounts of amines.

Peroxides and peroxy-compounds, also discussed in 8.5.1, form a very large subject, due to their great number and variety. Bretherick classes peroxides in three main groups, inorganic, organic, and organo-mineral.[5] Most peroxides are hazardous and some are explosive. Rather more than 65 organic peroxides are made and used commercially, the total consumption in 1978 in the USA being about 13 000 t. This does not include production of peroxides as intermediates (sometimes only in dilute solution) in the manufacture of bulk chemicals such as phenol and cyclohexanol by direct oxidation processes. Their main use is as free radical initiators for industrial polymerisations. They are also used as bleaching agents for edible oils and flour, disinfectants, fungicides, cross-linking agents for thermoplastics, curing agents for resin and rubbers, shrink-proofing agents for wool, in the manufacture of epoxy resins, and as the active ingredient in some pharmaceuticals. Only peroxides which can be produced, shipped and utilised with a reasonable degree of safety are made and sold commercially. While the hazards of peroxides should be well known to those making and using them, the risks posed by those made inadvertently by reaction of various substances with atmospheric oxygen [8.5.1 and 8.7] are more insidious.

Although several hazard-classification systems have been developed for organic peroxides, none has been accepted commercially. The Factory Mutual Engineering Corporation classifies organic peroxides according to their fire and explosion hazard into five classes, and gives safeguards for their storage and use.[9] Those at the extremes of the range are:

- *Class I.* These present a high explosion hazard through easily initiated, rapid explosive decomposition. This group may include peroxides that are relatively safe under highly controlled temperatures or in a liquid solution, where loss of temperature control or crystallisation from solution can result in severe explosive decomposition.

- *Class V.* These present a low or negligible fire hazard, and with them combustible packing materials may present a greater hazard than the peroxide itself.

Most of the 'Highly reactive substances' listed in Group 3 of Schedule 3 of the CIMAH Regulations [8.8] are organic peroxy-compounds, and come under the scope of Regulations 7 to 10 for quantities of 50 t or more.[10]

Of the inorganic peroxides, hydrogen peroxide which is made in grades ranging from 3% to 98% concentration (in water), decomposes in all concentrations with the evolution of heat and oxygen in the presence of small amounts of catalytic impurities such as platinum, manganese dioxide and alkalis. In concentrations of 86% and higher, the liquid will detonate with a high-energy source, but many mixtures of lower-grade hydrogen peroxide with organic compounds, which occur when hydrogen peroxide is used as an oxidising agent in a chemical reaction, are explosive.[11]

Another danger of hydrogen peroxide lies in the use of acetone or other ketones as a solvent for reactions in which it is used, since they themselves form peroxides which readily transform into crystalline dimers and trimers, which are explosive and very sensitive.[5,12]

Other inorganic peroxides are made and used on a large scale in household washing powders, but they seldom appear to be involved in explosions. Some little-used organo-mineral peroxides are very dangerous.

The importance of a thorough literature search can hardly be overstressed before using, making or storing unfamiliar chemicals. Thus two serious explosions of propargyl bromide, an acetylenic compound, occurred in the 1960s in the USA. Reports in the Russian literature that it was thermally sensitive and could detonate had apparently been missed.[4]

9.4 Preliminary screening of materials for explosivity

Table 9.2 provides a useful first checklist for materials of known composition. On finding such a grouping in the material being screened, the literature should be consulted, starting perhaps with Bretherick.[5] If any compound contains two such groupings, the probability of it proving explosive is strong.

Two simple but effective preliminary tests are first quoted. These can be applied to materials whose composition is not known, provided that effective precautions are taken.[6]

The first, which should be done before all others, is to take about one-tenth of a gram of the material and drop it onto a hot plate at above 300°C. If it goes *pop, bang, snap* or *crackle*, one is warned of possible trouble. If the sample decomposes or chars the test is continued until no further decomposition is apparent.

A test for instability in liquids is to place about 50 ml in a small beaker located behind suitable barricades. A small 50 watt coil of heating wire, attached by flexible cable to a plug and socket switch in a safe location, is then placed in the liquid. The observer having retired to a safe observation point, the current is switched on and left on until the sample has completely evaporated, decomposed or burnt. The beaker should not be approached again until it is cool and the electricity has been disconnected. This simple test, while not infallible, has detonated compounds whose explosivity more sophisticated tests had failed to reveal.[6]

9.5 Thermochemical screening

When any substance explodes, a considerable amount of heat is usually released, partly in the form of pressure waves and partly as a rise in temperature of the explosion products. The power of the explosion of the substance is related to this quantity, known as its heat of decomposition (ΔH_d). Values of ΔH_d and of ΔH_c, the heat of combustion, are quoted here in MJ/kg 'net' (as recommended by Lees,[3] i.e. assuming the water formed to be in the gaseous state) at 25°C and atmospheric pressure. The reader is warned that values of ΔH_d used by some writers appear to be 'gross' figures (i.e. assuming the water formed to be in the liquid state). Thermodynamicists (unfortunately for me) treat the heat of decomposition as negative when heat is given out and positive when heat has to be supplied when a substance decomposes. Here, to avoid confusion when comparing different heats of decomposition, I reverse this convention and treat ΔH_d when heat is given out during decomposition as positive (except in Fig. 9.2, p.208, which is based on Stull's correlation).

Knowledge of ΔH_d should provide some guide to the explosive potential of any substance. Thus we might (tentatively) expect any substance capable of exploding in the absence of air to have at least a certain minimum value of ΔH_d.

Closely linked with the heat of decomposition is the oxygen deficiency in the molecule compared with that needed to convert all carbon and hydrogen in it to carbon dioxide and water (and to convert any other elements to their combustion products — oxides, carbonates, etc.). For compounds and mixtures containing only carbon, hydrogen, nitrogen and sufficient oxygen to convert all carbon and hydrogen to carbon dioxide and water, the heats of decomposition and combustion should be identical, apart from the effects of small amounts of carbon monoxide, ammonia, oxides of nitrogen and hydrocarbons in the decomposition products. But for compounds and mixtures with a deficiency of oxygen, the heat of decomposition is always less than the heat of combustion. By mixing a material such as nitrocellulose which has an oxygen deficit with nitroglycerine which has an oxygen excess, both the heat of decomposition and the explosive power are increased. The same applies when TNT is mixed with ammonium nitrate.

A theoretical check should also be made on the volume of gases which would be released by the decomposition of unit weight of the compound.

This together with the theoretical heat release provides a rough appreciation of its destructive potential. Thus whilst copper and silver acetylides both decompose explosively with a sharp crack, their decomposition products, carbon and a metal, are both solid, and their main hazard is probably as a source of ignition or initiation for some other flammable or explosive material. Likewise the primary explosives mercuric fulminate and lead azide produce only small volumes of gases per unit weight, compared with propellants such as cordite.

9.5.1 The ASTM 'CHETAH' computer program

The 'CHETAH' (CHEmical Thermodynamic And energy Hazard evaluation) program, developed by the American Society for Testing and Materials, and first published in 1974,[13] is claimed to enable one to predict the maximum value of ΔH_d for any compound or mixture of compounds whose chemical formula and structure are known. The program also estimates three other quantities:[3]

- (the heat of combustion ΔH_c) — (the heat of decomposition ΔH_d) (both taken here as positive when energy is released);
- the oxygen balance (which defines whether the substance contains more or less oxygen than that required for complete combustion);
- the 'y' criterion (proportional to the molecular weight of the substance and the square of its heat of decomposition).

The probability of the substance exploding under the impact of shock was found to depend on all these four quantities, the most important being the heat of decomposition. As a very rough guide, only those substances whose heats of decomposition ΔH_d exceed 0.7 kcal/g (2.9 MJ/kg) are likely to explode when subjected to mild heat or shock. This figure is about 10% more than the amount of heat required to bring cold water to the boil and convert it into steam.

CHETAH can be applied to a mixture of chemicals in a process such as a chemical reactor to characterise the hazard potential of the system as high, medium or low by calculating the heat of reaction ΔH_r or decomposition ΔH_d:

High	$\Delta H_d > 2.9$ MJ/kg
Medium	$2.9 > \Delta H_d > 1.25$ MJ/kg
Low	$\Delta H_d < 1.25$ MJ/kg

Materials with medium or high heats of decomposition should always be treated with great caution.

Stull[14] suggested that a better correlation of the probability of any substance exploding is given by plotting ΔH_d *vs* the difference ($\Delta H_c - \Delta H_d$) for a number of potentially explosive substances. This is discussed in 9.5.3.

9.5.2 On estimating heats of decomposition

The heat given out when a substance explodes is difficult to measure and depends on the composition of the explosion products. If one knows this composition and the heat of formation of the substance, its heat of (explosive) decomposition can be estimated.

The heats of decomposition and compositions of the explosion products of commercial explosives have been well studied and reported. But when a substance which is normally thought of as safe explodes during storage or shipment, there may be considerable uncertainty about its heat of (explosive) decomposition, since nobody has had the opportunity to collect and analyse the explosion products. One can in such cases only estimate a range of values for ΔH_d, the highest value being sometimes many times higher than the lowest.

Most explosive substances contain only the elements carbon, hydrogen, nitrogen and oxygen, while their main decomposition products are solid carbon and the gases carbon monoxide, carbon dioxide, hydrogen, nitrogen, oxygen and water vapour. The heats of formation of the elements carbon, hydrogen, oxygen and nitrogen under standard conditions (usually at 25°C) are by definition zero. The heats of formation at 25°C in megajoules per kg-atom of contained oxygen of the main combustion products are –110.58 for CO, –196.94 for ½(CO_2) and –242.01 for H_2O as vapour. Some other gases such as ammonia, oxides of nitrogen and methane are also usually formed. Of these, methane and ammonia have negative heats of formation while both nitrous and nitric oxides have positive heats of formation. Thus the presence of methane or ammonia in decomposition products rather than elemental carbon, hydrogen and nitrogen raises the heat of decomposition. Methane and ammonia both decompose into their elements at high temperatures. They may be expected to be formed during the initial stages of many explosions and then decompose if a high enough temperature is reached during the explosion. Significant concentrations of methane and ammonia in explosion products thus tend to be characteristic of weak explosions.

Free oxygen is only likely to be present in the decomposition products when the amount of oxygen contained in the substance exceeds that needed to convert all carbon and hydrogen present to carbon dioxide and water vapour. In the usual case, where less oxygen is present, one should consider the proportions in which it is converted to carbon dioxide, carbon monoxide and water vapour. These proportions can have a significant effect on the calculated heats of decomposition. A substance giving a decomposition product containing a given amount of carbon, hydrogen and oxygen in the form of carbon and water vapour will have a considerably higher heat of decomposition than if it is in the form of carbon monoxide and hydrogen. Thus although the heat released when a chemical compound explodes depends (like its heat of combustion in air) on its chemical formula and heat of formation, it also depends on the composition of the combustion products. For compounds which rarely explode this composition is difficult to predict. Many compounds normally thought of as safe and whose heats of decomposition are relatively low are known to explode when sensitised by particular impurities (such as metal oxides) which can affect the mechanism of the explosion and the composition of the explosion products, thus giving a heat of decomposition near to the upper end of its possible range.

Here in estimating the heats of decomposition of several potentially explosive compounds I have used two methods, one which gives low (and in some cases normal) values, the other tending to give high values which may only apply in special circumstances.

Assumptions for estimating lower heats of decomposition

1. Combined oxygen in the substance first forms carbon monoxide until all carbon present is thus accounted for.
2. Any remaining oxygen first converts the carbon monoxide to carbon dioxide.
3. When all carbon present has been thus accounted for, the hydrogen present forms water with the remaining oxygen.

Assumptions for estimating upper heats of decomposition

4. Combined oxygen in the substance first forms water vapour until all hydrogen present is thus accounted for.
5. Any remaining oxygen first converts carbon present to carbon monoxide.
6. Any oxygen still remaining converts carbon monoxide to carbon dioxide.

While assumption (4) may appear rather extreme, it provides some allowance for increases in the heat of decomposition caused by the presence of methane and ammonia in the combustion products.

9.5.3 Heats of decomposition for particular compounds

Table 9.3 shows ten common industrial compounds with explosive potential including three well-known explosives, nitroglycerine, cellulose trinitrate, and trinitrotoluene (TNT). Several appeared in Table 9.1 and were discussed in 9.2 and 9.3. The table shows the formulae of the compounds, the number of atoms of oxygen in the combustion products, and the theoretical deficiency (or excess) of oxygen in the molecule for complete combustion. It also shows the net heats of combustion and decomposition and the differences between these two quantities. For compounds whose heats of decomposition were uncertain, the figures given here were estimated using the assumptions given in 9.5.2. Upper and lower values are given for five compounds but in the cases of acetylene and ammonium nitrate these values coincide and only a single figure is given. Figure 9.2 shows the values of ΔH_d plotted against [ΔH_c–ΔH_d] for all ten compounds, including ranges for the five. An arrow points from high to low probability. By sketching contours of equal probability at right angles to this arrow, acetylene (ΔH_d 8.74 MJ/kg),

Table 9.3 Oxygen deficiency and net heats of decomposition and combustion for known and possible explosives

Compound	*State*	*Empirical formula*				*Atoms in CP**	*Oxygen deficit atoms*	*%*	*Heats (MJ/kg)*			
		C	H	N	O				ΔH_c		ΔH_d	$[\Delta H_c - \Delta H_d]$
Glycerol trinitrate	l	3	5	3	9	8.5	−0.5	−5.9	6.31		6.28	0.03
Cellulose nitrate 13.4% N	s	6	7	3	11	15.5	4.5	29.0	9.13		4.06	5.07
Trinitrotoluene	s	7	5	3	6	16.5	9.5	63.6	14.65		4.69	9.96
2,4-Dinitrotoluene	s	7	6	2	4	17.0	13	76.5	18.89	upper	4.35	14.54
										lower	2.18	16.71
4-Nitrotoluene	s	7	7	1	2	17.5	15.5	88.6	26.03	upper	3.26	22.77
										lower	1.34	24.69
Nitromethane	l	1	3	1	2	3.5	1.5	42.8	10.89	upper	5.07	5.82
										lower	4.47	6.42
2-Nitropropane	l	3	7	1	2	9.5	7.5	78.9	20.75	upper	3.40	17.35
										lower	0.44	20.31
Ammonium nitrate	s	0	4	1	3	2	−1	−50	0.96		0.96	0
Acetylene	g	2	2	0	0	5	5	100	48.33		8.74	39.59
Ethylene oxide	g	2	4	0	1	6	5	83.3	26.75	upper	3.04	23.71
										lower	0.06	26.69

* CP = combustion products

dinitrotoluene (ΔH_d 2.18–4.47 MJ/kg) and ammonium nitrate (ΔH_d 0.96 MJ/kg) would all have about the same probability of exploding. This chart is a fair guide to the explosive probability of the compounds shown and is a far better criterion of this than ΔH_d alone.

2,4-Dinitrotoluene (DNT) with an estimated range for ΔH_d of 2.18–4.35 MJ/kg decomposes at 250°C, the decomposition becoming self-sustaining at 280°C. There is a published report of an explosion of DNT which had mistakenly been held at 210°C in a pipeline. A maximum handling temperature of 150°C has been recommended for it.[5]

4-Nitrotoluene (ΔH_d 1.34–3.26 MJ/kg) is made and used on a large scale commercially. There have been a few reports of explosions in vacuum distillation stills used to separate and purify mononitrotoluenes although most of these were attributed to higher nitrotoluenes or other unstable compounds (aci-nitro salts) derived from nitrotoluenes.[5]

Nitromethane (ΔH_d 4.47–5.07 MJ/kg), which is used both as a propellant and a commercial solvent, is dangerous to handle because it can readily be detonated by shock, high temperatures, the sudden application of gas pressure and forced high-velocity flow through restrictions. Nitropropane (ΔH_d 0.44–3.4 MJ/kg) appears to be generally safe to handle although it is susceptible to thermal decomposition, particularly in the presence of metal oxides.

Ammonium nitrate when heated carefully in a test tube to 170°C decomposes quietly into nitrous oxide and water vapour, giving out some heat at the same time. But if the temperature rises above 250°C, the decomposition

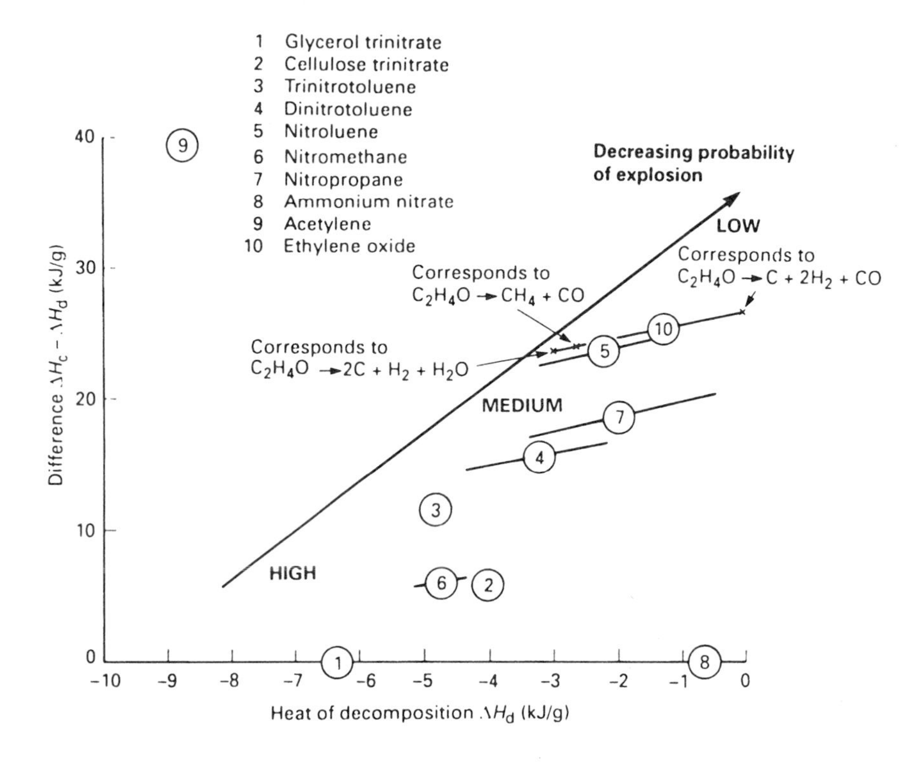

Fig. 9.2 Probability correlation for CHETAH program showing values for substances listed in Table 9.3

may become explosive, and oxygen and nitrogen are formed.[5] The two equations with the heats of decomposition liberated in each case are:

200–260°C	$NH_4NO_3 \rightarrow 2H_2O + N_2O$	ΔH_d 0.46 MJ/kg,
>260°C	$NH_4NO_3 \rightarrow 2H_2O + N_2 + \frac{1}{2}O_2$	ΔH_d 0.96 MJ/kg

There have been a number of ammonium nitrate fires which have burnt freely, liberating oxygen. Although it is practically impossible to detonate ammonium nitrate on a small scale, some of the worst accidental chemical explosions have occurred with ammonium nitrate, with heavy loss of life. Substantial confinement (as at the bottom of a large pile), which allows high temperatures to develop over a period of time as a result of slow decomposition starting at ambient temperature, has been an important factor in some of these explosions. As with other materials which decompose very slowly even at room temperature, the critical mass concept applies to ammonium nitrate[5] [8.6.4 and 8.7].

Another factor contributing to some ammonium nitrate explosions has been the presence of a small amount of hydrocarbon applied to the surface of the 'prills', to reduce their hygroscopicity and caking tendency, and retard their solution when applied to the soil as a fertiliser.[3]

The addition of 7.5% w. of carbon to ammonium nitrate increases the heat of decomposition to 3.64 MJ/kg:

$$NH_4NO_3 + \tfrac{1}{2}C \rightarrow 2H_2O + N_2 + \tfrac{1}{2}CO_2 \quad (\Delta H_d\ 3.64\ \text{MJ/kg})$$

The explosive decomposition of acetylene may be represented by:

$$\underset{g}{C_2H_2} \rightarrow \underset{s}{2C} + \underset{g}{H_2} \qquad (\Delta H_d\ 8.7\ \text{MJ/kg})$$

(It is a surprise to find that acetylene, which contains no oxygen, has a higher heat of decomposition than TNT.)

The explosive decomposition of gaseous acetylene in the absence of air can be initiated by heat or shock at pressures of 2 bar g. and higher, and can escalate into a detonation in pipelines. In the UK acetylene comes under the control of the Explosives Inspectorate when handled at pressures above 24 psig[1] (1.655 bar g.).

The explosive decomposition of ethylene oxide under clean conditions can be represented by:

$$\underset{g}{C_2H_4O} \rightarrow \underset{g}{CH_4} + \underset{g}{CO} \qquad + 181.1\ \text{kJ}$$

rather than by:

$$\underset{g}{C_2H_4O} \rightarrow \underset{s}{C} + \underset{g}{2H_2} + \underset{g}{CO} \qquad \underset{+\ g}{+\ 2.5\ \text{kJ}}$$

From its molecular weight of 44, its heat of decomposition in the vapour state is 2.68 MJ/kg, compared to the upper estimate in Table 9.3 of 3.04 MJ/kg. Had liquid ethylene oxide been able to explode, the heat of decomposition (with methane again formed) would have been 2.01 MJ/kg. The case of ethylene oxide demonstrates the strong effect of methane in the decomposition products in raising ΔH_d to the point where explosion is possible.

While ethylene oxide vapour will explode if ignited, it is stable in the liquid state provided no polymerisation catalysts such as iron oxide are present. But if the liquid polymerises rapidly, the heat evolved will vaporise unpolymerised material, causing an explosion hazard.

Sodium chlorate is another common chemical with a small explosion hazard in itself which is enormously increased by the addition of small amounts of organic matter.

The values of ΔH_d given here are mainly estimated from heats of combustion given in the *CRC Handbook of Chemistry and Physics*.[15] The steps involve:

- calculation of heats of formation from published heats of combustion;
- calculation of net heats of combustion (with water formed in the vapour state);
- estimation of lower and upper values of heats of decomposition using the assumptions stated.

For compounds whose heats of formation or combustion are unknown, it is often possible to estimate them from bond energies between the various atoms.[16,17]

9.6 Stability and sensitivity tests

Tests for thermal and shock stability of potentially explosive substances are needed to establish safe processing conditions. Thermal stability is important for operations such as drying, reaction, evaporation, distillation and other processes involving elevated temperatures. Shock stability is important for size reduction, pumping, blending and transport.

Unfortunately, the stability of many of these materials depends on their concentration, purity and the nature of the impurities present, as well as on their temperature and physical environment. These tests are therefore not infallible. Sometimes an impurity may be present in an industrial compound which considerably reduces its stability. In the manufacture of guncotton, for example, the crude nitration product, after separation and washing to remove spent acids, still contains unstable esters. These are removed by repeated boiling, first with water and then with very dilute alkali. If this is not done thoroughly, or if traces of unstabilised material by-pass this stage, the guncotton is liable to explode during subsequent drying. Similarly, the shock sensitivity of many explosives (e.g. nitroglycerine) depends on whether they are in the liquid or solid state. The solid is less stable than the liquid. Thus stability tests need to be carried out not only on pure materials and commercial products, but also on samples representative of their condition in the processes concerned. This applies to all potentially explosive substances irrespective of their purpose.

Likewise, when considering the process conditions which the tests are intended to validate, a generous allowance should be made for 'upset conditions', with temperatures, etc. outside the normal operating range. Needless to say, processes for hazardous materials should have special safety instrumentation and emergency devices (such as 'dump tanks') to counter hazardous upset conditions.

9.6.1 Thermal stability

Apart from the crude screening tests already mentioned [9.4], a number of special tests are used. These include the use of modified melting point apparatus[6] and thermal analysers which can be employed for differential thermal analysis (DTA) using standard procedures.[18,19] Since these work at atmospheric pressure and may fail to detect exotherms on volatile materials, confinement tests which use thermal stability bombs have also been developed.[4,20] These tests also measure the pressure rise in a confined space.

Other more refined methods include differential scanning calorimetry (DSC) and accelerating rate calorimetry (ARC). These are carried out in the UK by the Royal Armament Research and Development Establishment, EM2 Branch, on behalf of HM Inspector of Explosives.[3,21] The theory and methods developed for thermal analysis, and their limitations, are discussed by Daniels.[22]

9.6.2 Shock sensitivity

Thermally sensitive materials which decompose exothermally are also sensitive to shock or friction, although the stimulus needed may be severe. Tests range from a simple hammer test, through drop weight impact tests[23] (Fig. 9.3), to more severe ones which use explosives for initiation, such as card gap tests.[24,25]

A number of other tests, including the Cartridge Case test for discriminating between detonating and deflagrating explosives and a Bonfire test to investigate the behaviour of the bulk material in a fire, are discussed by Lees[3] and Connor.[21] Lees stresses that the object of carrying out reactivity or explosivity tests should be to obtain a positive result, using extreme conditions where necessary. Less extreme conditions under which it might be handled in practice are then tested.

9.7 Classification of materials with explosive potential

Coffee has proposed classifying such materials by their behaviour in three basic but different stability tests:[4]

- theoretical computation by CHETAH
- thermal analysis by DTA
- impact sensitivity by drop weight.

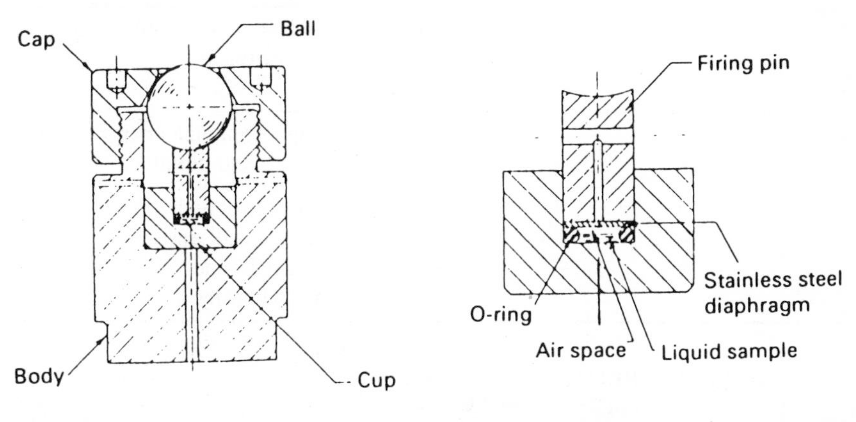

Fig 9.3 Sample holder of Olin drop-weight impact tester. (From Coffee[4], courtesy Wiley-Interscience)

Taking results as either positive or negative, eight combinations are possible. Further examination reduces these to three groups representing low, medium and high potential hazards. These are shown in Table 9.4.

9.8 Explosions of industrial chemicals outside the explosives industry

A few examples of major explosions during the manufacture, storage and transport of chemicals for uses other than as explosives follow under the headings of the chemicals involved. A large number of minor explosions have occurred in the chemical industry, generally during the first years of manufacture of some unstable compound the hazards of which were not fully appreciated. Thus the celebrated Dr Schwartz wrote at the turn of the century about 'Chemico-Technical Factories and Colour Works':[26]

> The progress of chemical reactions on a working scale must not be excluded from inspection, proceed they never so smoothly *in the laboratory*. When working *on the large scale*, quite different and dangerous reactions may arise. Hence the risk of insuring such establishments should not be accepted until a summer and winter working season has passed without an accident from fire and explosion.

Elsewhere the same author in discussing acetylene which was then coming into wide use for illumination states 'between 1897 and 1900, the number of explosions amounted to 32, 17.5, 5.4 and 2.2 per thousand users of the gas'.

Table 9.4 Classification of explosive hazard of materials

Combination number	*Heat of reaction*	*Thermal stability*	*Impact sensivity*	*Overall rating*
1	–	–	–	Stable and incapable of high energy release
2	–	–	+	
3	–	+	–	
4	+	–	–	Capable of high energy release but hard to initiate
5	+	+	–	
6	+	–	+	Capable of high energy release and sensitive thermally and mechanically
7	–	+	+	
8	+	+	+	

	Code
Heat of reaction, $-\Delta H_d$ (CHETAH)	– <0.7 kcal/g + >0.7 kcal/g
Thermal stability (DTA)	– no exotherm + exotherm
Impact sensitivity	– insensitive at 550 in-lb (solids) or 100 kg-cm (liquids) + sensitive at 550 in-lb or 100 kg-cm

9.8.1 Ammonium nitrate

Although Schwartz had warned that 'ammonium nitrate explodes by percussion or at 70°C',[26] several large explosions of ammonium nitrate made primarily for use as a fertiliser occurred between 1918 and 1947. The two most destructive incidents (both described in more detail by Lees[3]) were at Oppau, Germany, in 1921 when 430 people were killed, and at Texas City in 1947 when 552 people were killed and over 3000 injured.

At Oppau two explosions in close sequence occurred at the BASF works where some 4500 t of a mixture of ammonium nitrate and sulphate were stored, when blasting powder was being used to break up piles of the material which had become caked. The explosion created a crater 75 m in diameter and 15 m deep, destroyed the works and 1000 nearby houses (Fig. 9.1), while the air pressure wave caused considerable damage in Frankfurt more than 80 km away.

The Texas City explosion involved two ships in the harbour with cargoes of bagged ammonium nitrate. The first explosion occurred about one hour after a fire was reported (and could not be extinguished) on a ship carrying 2300 t of ammonium nitrate. The explosion set fire to a cargo of sulphur which the second ship moored 200 m away was carrying as well as ammonium nitrate, which exploded 16 hours after the first explosion. The explosions caused fires in many tanks in nearby oil refineries, some of which burned for nearly a week, and caused extensive damage to warehouses, business premises and residential property.

9.8.2 Vinyl acetylene

Vinyl acetylene (C_4H_4) is formed by dimerisation (the combination of two molecules) of acetylene and also in high-temperature cracking and dehydrogenation processes as a by-product of butadiene. Like acetylene, it is a high-energy compound which can decompose explosively into its elements without the involvement of oxygen. It also polymerises readily with the evolution of heat, which can initiate an explosion of the unpolymerised material. For many years it was produced from acetylene as the first step in the manufacture of chloroprene (2-chlorobutadiene), from which the oil- and chemical-resistant rubber neoprene is made. Being an easily liquefiable gas like butane and butadiene, it is generally handled as a liquid, either refrigerated or under pressure. Two accidents are discussed here, one involving vinyl acetylene produced from acetylene, the other the separation of butadiene from vinyl acetylene and other C_4 hydrocarbons.

On 25 August 1965 several explosions occurred at Dupont's Louisville works (known as 'Rubber Town') on the banks of the Ohio river, where about 200 t of vinyl acetylene produced there from acetylene was stored as a liquid. Twelve people were killed (most by the first explosion) and much of the plant was wrecked. A company investigation[27] showed that the first explosion was caused by the mechanical failure and overheating of a compressor which circulated vinyl acetylene gas. The subsequent explosions which occurred over a period of eight hours were initiated by flying metal fragments and fires started by the first explosion, and by transmission through pipelines.

On the evening of 23 October 1969 a butadiene purification column on one of Union Carbide's butadiene plants at Texas City exploded.[28] This column, which operated at pressure of 2–3 bar g., separated butadiene (as overhead product) from 2-butenes and vinyl acetylene left in the butadiene-rich stream from which most other impurities had been removed by extractive distillation with a selective solvent. The boiling points at atmospheric pressure of these hydrocarbons (°C) are:

butadiene	–4.6
trans butene	0.8
cis-butene-2	3.7
vinyl acetylene	5.0

Although vinyl acetylene has the highest boiling point, so that its highest concentration might be expected to be at the bottom of the column, it actually accumulates several trays higher because of the 'non-ideality' of the mixture.

At 09.00 a compressor was running hot and had to be shut down to change a valve. As this was not expected to take long, instead of shutting the whole plant down the feed was stopped but not the steam to the reboiler, and the column was put on 'total reflux' (to save time when the plant was restarted). A laboratory analysis on a sample from the base of the column at 11.00 showed 36.9% vinyl acetylene, which probably remained unchanged until 19.23, when the column exploded, the upper part rising several feet before falling on its side (Fig. 9.4). A cloud of vapour was released and ignited. Fortunately no one was seriously injured. Examination of the column showed the explosion to have centred on the fourteenth tray (from the bottom).

Computer studies showed that a maximum vinyl acetylene concentration of 57–60% would have been expected on the tenth tray. Investigations showed that vinyl acetylene in most concentrations in butadiene would exhibit an exotherm starting at 135–140°C and culminate in an explosion. Such temperatures were just possible in the reboiler tubes. The effects of several possible initiators for the explosion were investigated. Of these, only sodium nitrate lowered the threshold temperature for the exotherm by 20°C. Solids containing 13% of sodium nitrate were found in the reboiler of the column. It had been formed by the oxidation of sodium nitrite a solution of which (following established wisdom) was continuously circulated through butadiene-distillation columns to scavenge traces of oxygen and prevent the formation of butadiene peroxide.

It was concluded that the explosion of vinyl acetylene had probably been triggered by a thermal polymerisation assisted by sodium nitrate.

The following preventative actions were taken for future operation.

Fig. 9.4 Base of butadiene distillation column after internal explosion of vinyl acetylene. (Courtesy Gulf Publishing Company)

- The steam temperature in the reboilers of similar columns was limited to 105°C;
- A high-temperature alarm set at 55°C was installed in the base of each similar column;
- Continuous analysers of vapour from the base and tray 10 of each column were installed and equipped with an automatic alarm and trip to shut the steam valve to the reboiler if the vinyl acetylene concentration exceeded 30%;
- Operation of these columns at total reflux was no longer permitted;
- Sodium nitrite was eliminated as an oxygen scavenger and replaced by an organic inhibitor.

The following practice which the writer initiated on another butadiene plant is worth mentioning. A vinyl acetylene-rich sidestream is removed continuously from a tray in the column where its concentration peaks. This passes to a selective hydrogenation unit where it is converted to butadiene and butenes and the treated stream is returned to the butadiene plant.

9.8.3 Cumene hydroperoxide[29]

One of three phenol plants at Philadelphia, USA, was destroyed on 9 February 1982 by an explosion when 25 000 gallons of cumene hydroperoxide in a tank were being heated by steam. Following this, safe temperature limits were set for heating this unstable peroxide, and arrangements were made for the rapid release of vapour from similar tanks [8.5.1].

9.8.4 Ethylene oxide[30]

On an ethanolamine plant in Kentucky in April 1962 an explosion occurred when ammonia feed to the plant inadvertently entered the ethylene oxide feed vessel and reacted violently. One man was killed and 12 were injured.

9.8.5 Hydrazine derivative[30]

Process equipment in which a hydrazine derivative was being made exploded in February 1961 on Olin Matheson's plant at Lake Charles, Louisiana. Hydrazine, like acetylene, is an unstable compound which can decompose explosively on its own when heated. One man died from injuries.

9.8.6 Two mini-explosions

Schwartz[26] reports the case of a nitroglycerine plant worker who had returned home and was lighting a cigar when an explosion occurred which stripped the flesh from his thumb and forefinger. It was caused by a few milligrams of nitroglycerine trapped under his thumbnail.

Walking past a gas-separation plant which was shut down for maintenance in the late 1950s, the writer was surprised to hear a series of sharp 'pops'. These were explosions of small deposits of copper acetylide which had formed on the outside of copper instrument lines inside the lagging. They resulted from small leaks over a long period of gases containing traces of acetylene. The lines were very cold during plant operation, but the explosions occurred during maintenance when the lagging was removed, the lines warmed up to room temperature, and fitters were working on the plant. They, not surprisingly, downed tools until all the explosive deposits were destroyed. After that the copper lines were replaced by stainless steel.

References

1. Watts, H. E., *The Law Relating to Explosives*, Charles Griffin, London (1954)
2. Wharry, D. M. and Hirst, R., *Fire Technology: Chemistry and Combustion*, Institution of Fire Engineers, Leicester (1992)
3. Lees, F. P., *Loss Prevention in the Process Industries*, Butterworths, London (1980)
4. Coffee, R. D., 'Chemical stability' in *Safety and Accident Prevention in Chemical Operations*, edited by Fawcett, H. H. and Wood, W. S., 2nd edn, Wiley-Interscience, New York (1982)

5. Bretherick, L., *Bretherick's Handbook of Reactive Chemical Hazards*, 4th edn, Butterworths, London (1990)
6. Snyder, J. S., 'Testing reactions and materials for safety', in *Safety and Accident Prevention in Chemical Operations,* edited by Fawcett, H. H. and Wood, W. S., 2nd edn, Wiley-Interscience, New York (1982)
7. Hunter, D., *The Diseases of Occupations*, 6th edn, Hodder and Stoughton, London (1978)
8. Harris, G. F. P., Harrison, N. and MacDermott, P. E., 'Hazards of the distillation of mononitrotoluences', paper in I. Chem. E. Symposium series 68, *Runaway Reactions, Unstable Products and Combustible Powders*, Rugby (1981)
9. *Loss prevention data sheet 7–80, Organic Peroxides*, Factory Mutual Engineering Corporation, 1151 Boston-Providence Turnpike, Norwood, Mass. 06062 (1972)
10. S.I. 1984, No. 1902, *The Control of Industrial Major Accident Hazards Regulations 1984 (CIMAH)*, HMSO, London
11. Kirchner, J. R., 'Hydrogen peroxide', in vol. 13, *Encyclopaedia of Chemical Technology*, edited by Kirk, R. E. and Othmer, D. F., 3rd edn, Wiley-Interscience, New York (*ca* 1979)
12. Swern, D. (ed.), *Organic Peroxides*, Wiley-Interscience, New York, vol. 1 (1970), vol. 2 (1971), vol. 3 (1972)
13. *CHETAH — The ASTM Chemical Thermodynamic and Energy Release Evaluation Program*, DS51, American Society for Testing and Materials, Philadelphia, Pa (1974)
14. Stull, D. R., 'Identification of reaction hazards', *Loss Prevention*, **4**, 16 (1970)
15. Weast, R. C. (editor) *CRC Handbook of Chemistry and Physics*, 1st student edn, CRC Press Inc., Boca Raton, Florida (1988)
16. Wenner, R. R., *Thermochemical calculations*, McGraw-Hill, London (1941)
17. Benson, S. W., *Thermochemical Kinetics*, Wiley, New York (1968)
18. ASTM E537-76, *Assessing the Thermal Stability of Chemicals by Methods of Differential Thermal Analysis*. American Society for Testing and Materials, Philadelphia, Pa (1974)
19. ASTM E487-76, *Test for Constant-Temperature Stability of Chemical Materials*. American Society for Testing and Materials, Philadelphia, Pa (1974)
20. ASTM 476-73, *Test for Thermal Instability of Confined Condensed Phase Systems*. American Society for Testing and Materials, Philadelphia, Pa (1974)
21. Connor, J., 'Explosion risks of unstable substances — Test methods employed by EM2 (Home Office) Branch, Royal Armament Research and Development Establishment', in *Loss Prevention and Safety Promotion in the Process Industries*, edited by Buschmann, C. H., Elsevier, Amsterdam (1974)
22. Daniels, T., *Thermal Analysis*, Kogan Page, London (1973)
23. ASTM D2540-70, *Test for Drop Weight Sensitivity of Liquid Monopropellants*, American Society for Testing and Materials, Philadelphia, Pa (1974)
24. ASTM D2539, *Test for Shock Sensitivity of Liquid Monopropellants by the Card Gap Test*. American Society for Testing and Materials, Philadelphia, Pa (1974)
25. ASTM D2541, *Test for Critical Diameter and Detonation Velocity of Liquid Monopropellants*. American Society for Testing and Materials, Philadelphia, Pa (1974)
26. Von Schwartz, *Fire and Explosion Risks* (first English edition) Charles Griffin, London (1918)
27. Armistead, J. G., 'Polymerisation explosion at Dupont' in Vervalin, C. H., *Fire Protection Manual for Hydrocarbon Processing Plants*, 3rd edn, Gulf Publishing, Houston (1985)
28. Griffith, S. and Keish, R. G., 'Butadiene plant explodes', *ibid*.
29. Garrison, W. G., *100 Large Losses — A thirty-year review of property damage losses in the hydrocarbon-chemical industries*, 11th edn, Marsh and McLennan Protection Consultants, 222 South Riverside Plaza, Chicago, Illinois 60606 (1988)
30. Vervalin, C. H., *Fire Protection Manual for Hydrocarbon Processing Plants*, 3rd edn, Gulf Publishing, Houston (1985)

10 Combustion technology

In this chapter we are concerned mainly with explosions and fires, and the methods by which they can be prevented, controlled, or stopped. Both result from very energetic vapour-phase reactions involving temperatures at which radiation is substantial and usually visible. Combustion is also possible by the direct oxidation of a fuel surface, resulting in glowing or smouldering. These processes are unspectacular but insidious, and may be difficult to control. The reactions in flaming combustion are usually between a gaseous fuel and the oxygen in the air, but other reactants may also be capable of burning in a flame. For example, pure hydrochloric acid can be produced by burning hydrogen in an atmosphere of chlorine, and then dissolving the product in water:

$$H_2 + Cl_2 = 2HCl$$

Vapour-phase reactions can occur in two very different kinds of flame. Explosions are invariably premixed flames, and fires comprise mainly large turbulent diffusion flames. The characteristics and behaviour of premixed and diffusion flames are discussed extensively in the literature,[1–4] but are summarised here in only sufficient detail to provide an understanding of the hazards they present and the ways in which their effects can be mitigated. It will be convenient to look first at premixed flames.

10.1 Premixed flames

Figure 10.1(a) shows an apparatus which is used to measure the concentration of a gaseous extinguishing agent which will suppress premixed combustion.[5] It comprises a spherical steel vessel with a capacity of 6 l. Gaseous mixtures can be injected through connections in the steel ring which separates the two halves. Electrodes pass through insulated bushes in this ring to

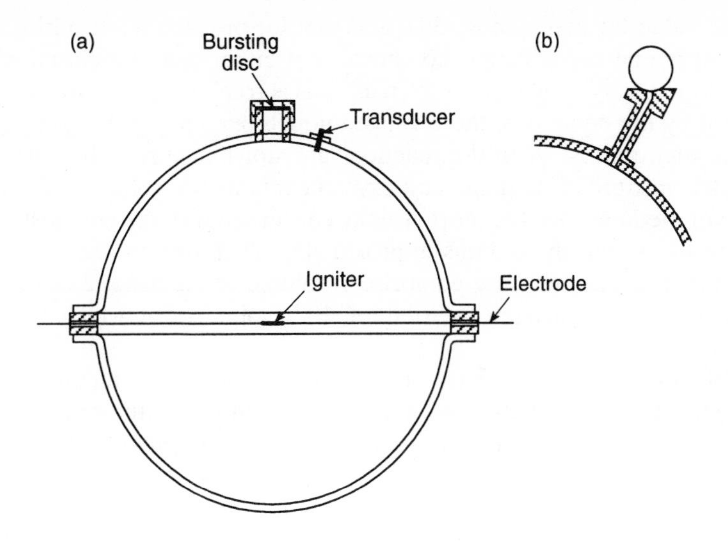

Fig 10.1 (a) 6-litre steel vessel; (b) simple pressure detector

an igniter at the centre. This is a short length of pencil lead over which a high-voltage surface discharge takes place, but the ignition source could equally well be a spark from the same high-voltage generator. The presence of combustion following a discharge can be detected by a pressure transducer. Very low pressure rises can be detected if the transducer is replaced by the simple device shown in Fig. 10.1 (b). This is a steel ball resting on a hole in a block of PTFE. Its weight can be selected so that it is displaced at any pressure above that caused by the ignition source. This equipment can be used to measure minimum ignition energy, limits of flammability, peak concentrations, explosion pressure, and burning velocities, all of which are discussed below.

If we consider now the combustion of methane, the reaction with oxygen is shown by this equation:

$$CH_4 + 2O_2 = CO_2 = 2H_2O - 800 \text{ kJ/mol}$$

Two things should be noted. First, by Avogadro's rule, three volumes of reactants have produced three volumes of end products. Second, a great deal of heat has been released. The heat of combustion is shown as a negative quantity because it is lost from the system (but see also 9.5). This indicates that if the reaction takes place adiabatically in a closed vessel, there must be an increase in pressure. If we look at the combustion of propane in the same way we find that it also burns to produce only carbon dioxide and water vapour:

$$C_3H_8 + 5O_2 = 3CO_2 + 4H_2O - 2044 \text{ kJ/mol}$$

This time the reaction has produced a slight increase, from 6 volumes to 7 volumes. However, if we express the heat of combustion of the two fuels as kJ/g, the value for methane is –50.0 and that for propane is –46.5 kJ/g. The net result is that the adiabatic combustion of either fuel in a closed vessel will result in a very similar pressure rise. This is true of course only when, as indicated by the equations, there is just enough oxygen to react completely with the fuel: that is, when the reactions are stoichiometric. Thus the stoichiometric mixture of methane and oxygen contains two volumes of oxygen for one of methane. We are more usually concerned with the combustion of methane in air, which contains approximately 20% oxygen. The stoichiometric mixture will therefore comprise 1 volume of methane, 2 of oxygen, and about 7.5 of nitrogen, a total of 10.5 volumes, of which 9.5% is methane.

So, let us assume that we have made a mixture of 9.5% of methane in air — the stoichiometric mixture — and have placed it, at atmospheric pressure, in the 6-litre vessel. Let us assume also that the pressure transducer has been attached to a recorder so that we can follow the pressure changes in the vessel. If a spark is now passed which has sufficient energy to ignite the mixture, then a spherical flame will be established at the centre of the vessel and will propagate into the unburned mixture. The sphere will enlarge like a balloon until it reaches the walls of the vessel, and all of the reactants have been consumed. The flame will advance into the unburned mixture at a characteristic rate called the *burning velocity*. Perhaps surprisingly, for most hydrocarbon/air mixtures this is only about 0.5 m/s. The measured maximum velocity for methane/air is 0.4 m/s, but the value will be slightly higher than this in the vessel. This results from the expansion of the burned gases behind the flame, and from the effect of the increasing

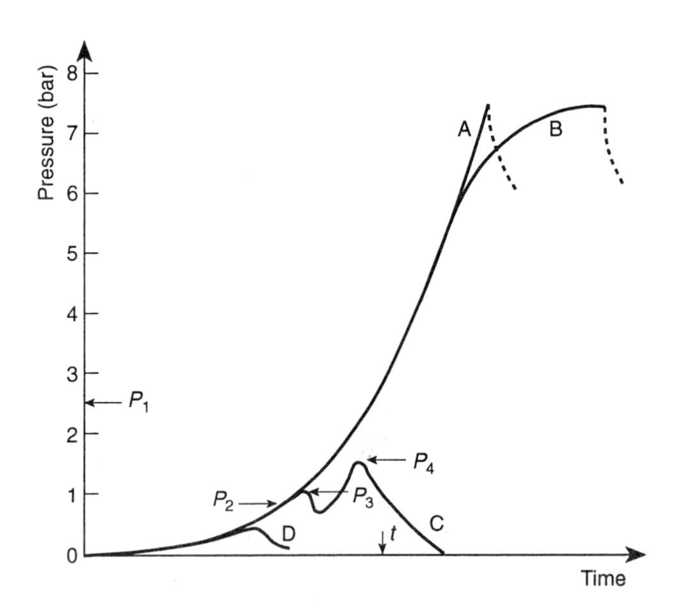

Fig. 10.2 Variation of pressure with time for (A,B) stoichiometric methane/air explosions, (C) a vented explosion, and (D) a suppressed explosion

pressure on the burning velocity. When the fundamental burning velocity is exceeded like this, for any reason, it is usual then to refer to the combined effects as the *flame speed*. However, if we assume that the burning velocity remains constant, then the radius r of the flame sphere will increase linearly with time, but its volume will increase as r^3. The recorded plot of pressure against time will therefore follow a cube law and will look like curve A of Fig. 10.2. It will be seen that the maximum pressure, at the moment when the flame is quenched at the wall, is 7.4 bar. This is a typical value for stoichiometric mixtures of many hydrocarbons. However, the values will be different if the initial pressure is not atmospheric, and in order to avoid possible confusion it is usual to quote an expansion ratio of 7.4. This means that the final pressure will be 7.4 times the initial pressure, or that in unconfined conditions the combustion products will expand to 7.4 times the initial volume.

Curve A of Fig. 10.2 indicates two important characteristics of premixed combustion. First, most process plant would be ruptured by pressures well below the maximum values. Second, a pressure which is capable of causing damage is not achieved until an appreciable time after ignition: perhaps at t_1 on the figure. It is during this time that the presence of the relatively small flame can be detected and its propagation into the bulk of the flammable mixture can be stopped (the explosion suppression technique is discussed later [10.3.5]), or a suitable vent can be opened. In the last two paragraphs we have of course been describing an explosion. In this chapter the word is applied only to combustion processes: a premixed flame is an explosion. However, in Part III, in the discussion of pressure vessel failures, the word carries the same meaning as it does in, for example, the Boiler Explosions Acts.

The minimum quantity of energy, released by an electric spark, which is capable of initiating a methane/air explosion, is 0.29 mJ. This is a very small amount of energy, and static charges are readily generated in some process plant which can release considerably more energy than this. It is indeed possible for a static discharge from a person to release in excess of 10 mJ (discussed in [10.5.1]). When sufficient energy is available, the reaction zone of the spherical flame which propagates from the ignition source is only about 2 or 3 mm thick and is usually bright blue in colour.

The encounter of one methane molecule with two oxygen molecules which is indicated in the equation on p. 219 is a relatively rare event in a mixture of these gases, and cannot possibly account for the explosive reactions in a flame. Even in this comparatively simple system, there are many very complicated branching chain reactions involving hydrogen and oxygen atoms and radicals such as OH, C_2 and CHO. The burning velocity is dependent not only on the rate of heat transfer from the flame to the unburned gases, but also on the active diffusion of these chain carriers into the gases. The effectiveness of a spark as an ignition source also results from its ability to generate ionised particles.

So far we have discussed stoichiometric reactions, but premixed combustion can also occur over a range of fuel/air mixtures. Explosions are possible for example in methane/air mixtures which contain between 5% and 15% methane. These are the limits of flammability, and values for a number of other gases in air are given in Table 10.1, in columns C and D. The existence of these limits results from the presence in the flame's reaction zone of a thermal load. Thus at the lower limit there is an excess of air which is heated to the flame temperature without then contributing to the reactions (this is in addition to the nitrogen which is present even in the stoichiometric mixture). The result is a reduction in the flame temperature to a value (about 1400°C) at which the combustion processes are no longer possible. The values in column F of the table are obtained by dividing the lower limit concentration (C) by the stoichiometric concentration (E). The figure for methane is 0.53, and it will be seen that this is a typical value for a hydrocarbon. The heat of combustion at these two concentrations must be directly related to the amount of fuel which is present, so it is evident that if 0.47 of the heat generated by a stoichiometric flame can be removed, then the flame temperature will be reduced to the lower-limit value, and combustion will cease.

No similar relationship exists between the upper limit and the stoichiometric concentrations, although for many hydrocarbons the D/E value is around 2.5. The explanation is that although excess fuel will also act as a thermal load, much of it will be pyrolised when it is heated to flame temperature. The composition of the mixture will therefore be changed, and if for example hydrogen or acetylene is released, as might well happen, then the values for these gases in columns C and D of the table show that the limit could be widened. The unusually high upper limit for hydrogen is due to the high flame temperature (column I) and to its high diffusivity. The high acetylene upper limit can be explained by its heat of decomposition, which is added to the heat of combustion. This results from the rupture of the strained triple bonds between the carbon atoms: HC ≡ CH. At pressures

Table 10.1 Combustion properties of gases

A	B	C	D	E	F	G	H	I	J	K	L
		Lower limit	Upper limit	Stoichiometric ratio	C/E	Maximum burning velocity	Fuel at maximum velocity	Adiabatic flame temperature	Expansion ratio	Minimum ignition energy	Quenching distance
		%	%	%		m/s	%	°C		mJ	mm
Methane	CH_4	5.0	15.0	9.5	0.53	0.45	10.0	1875	7.4	0.29	2.0
Ethane	C_2H_6	3.0	12.5	5.6	0.54	0.53	6.3	1895	7.5	0.24	1.8
Propane	C_3H_8	2.2	9.5	4.0	0.55	0.52	4.5	1925	7.6	0.25	1.8
Butane	C_4H_{10}	1.9	8.5	3.1	0.61	0.50	3.5	1895	7.5	0.25	1.8
Hexane	C_6H_{14}	1.2	7.5	2.2	0.55	0.52	2.5	1948	7.7	0.25	1.8
Benzene	C_6H_6	1.4	7.1	2.7	0.52	0.62	3.3	2014	7.9	0.22	1.8
Acetylene	C_2H_2	2.6	80.0	7.7	0.34	1.58	9.3	2325	9.0	0.02	
Hydrogen	H_2	4.0	75.0	30.0	0.13	3.50	54.0	2045	8.0	0.02	0.5

above about 2 bar g., an explosive decomposition may start spontaneously, with an expansion ratio of 8.0, and the upper limit can then be regarded as being 100%.

An increase in oxygen concentration above the normal atmospheric value of 21% results in an increase in flame temperature and a widening of the upper limit. In 100% oxygen the upper limit of methane is increased to 60%, but the lower limit is almost unchanged. This is because the thermal load imposed by the excess of oxygen has a similar effect to the load resulting from an excess of air. An increase in pressure produces much the same effect as an increasing oxygen concentration. This is because the partial pressure of oxygen has been increased. Thus, at 200 bar the upper limit of methane/air is again 60%, and the lower limit is reduced to about 4%.

Referring again to Table 10.1, it will be seen that for this representative range of hydrocarbons the maximum burning velocity is around 0.5 m/s (column G). This invariably occurs in mixtures which are slightly richer than stoichiometric (column H). Again this can be explained by the pyrolysis of the slight excess of fuel, and the release of hydrogen. The adiabatic flame temperatures (I) are calculated values for the stoichiometric mixture. The minimum ignition energies (K) are measured values in stoichiometric mixtures, as are the quenching distances (L). If a premixed flame is propagating through a narrow gap, both heat and active species will be lost to the wall. If the gap is equal to, or less than, the values in L, the losses will be too great and the flame will be quenched. It should be noted that in mixtures which are not stoichiometric but are within the limits, the values in G, I and J are reduced, and those in K and L are increased.

The apparatus shown in Fig. 10.1 can be used to measure many of the properties which are shown in the table. However, it was specifically designed to measure the effect of additives to the fuel/air mixtures. As we have seen, the limits are established when the presence of unburned fuel or air imposes a critical thermal load within the reaction zone. This load is in addition to that which already exists through the inevitable presence of nitrogen. Suppose now that we increase the concentration of nitrogen by mixing known amounts with the fuel/air mixture. The result will be to reduce the range of flammable mixtures, the limits coming closer together with increasing concentrations of added nitrogen. This is shown by curve A on Fig. 10.3, where it will be seen that the two limits converge at a nitrogen concentration of 36.5%. This is the minimum concentration which will prevent the ignition of any methane/air mixture. For obvious reasons it is called the *peak concentration* (also known as the inerting concentration and the inhibitory factor). It occurs at a methane concentration of 6.7%, which is on the rich side of stoichiometric (shown by the line marked St on the figure).

Curve B shows the effect of adding carbon dioxide instead of nitrogen: it will be seen that its peak concentration is only 23.5%. One would expect the values for the two gases to be directly related to their specific heats, and this is very nearly so. Carbon dioxide is slightly more effective than would be

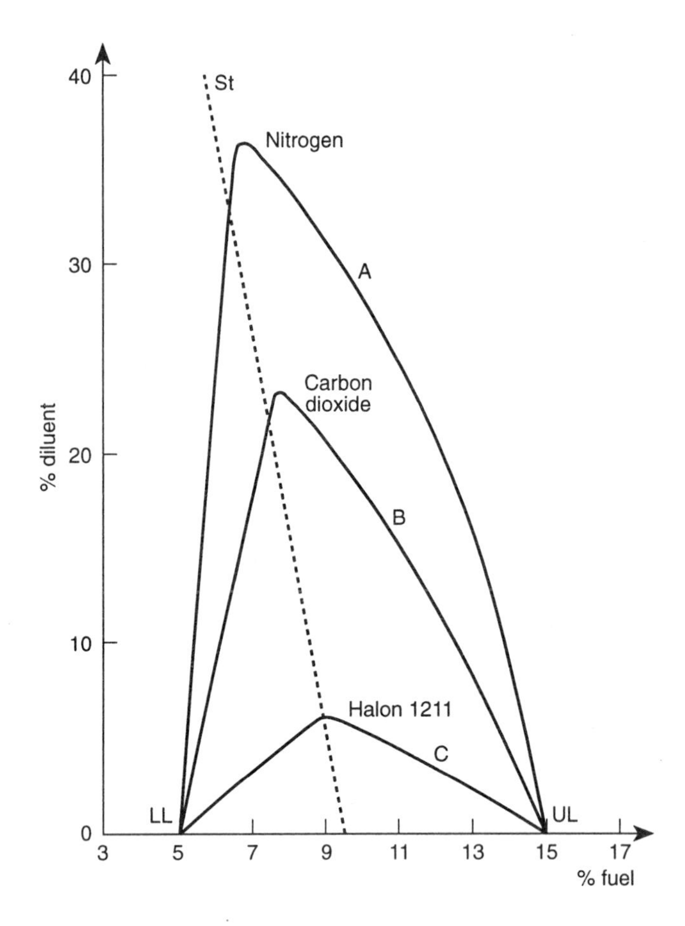

Fig. 10.3 Peak concentrations of nitrogen, carbon dioxide and Halon 1211 for methane/air mixtures

predicted in this way, showing that it must have a chemical as well as a physical effect on the combustion processes. The peak concentration for Halon 1211 ($CBrClF_2$) is 6.5% (curve C), reflecting its extremely active chemical role in the suppression of combustion reactions.[3] (The manufacture, but not the use, of Halon 1211 was banned by the Montreal Protocol.[6] However, ozone-friendly Halons are now available, none of which unfortunately has the same effectiveness as Halon 1211.) A knowledge of peak concentrations is necessary in the design of automatic total-flooding systems for use in areas where explosive atmospheres can exist, and also for purging and inerting procedures (discussed in 10.3.2). Measurements can be made in the apparatus shown in Fig. 10.1, but much of the presently available data was obtained by a technique developed at the US Bureau of Mines between 1925 and 1950.[7,8] Their apparatus, which is still used, comprises a vertical glass tube 50 mm in diameter and 1.5 m long which contains the premixed gases. A spark can be passed between electrodes close to the bottom of the tube and at the same time a sealing disc is removed from the bottom. If a flame travels to the top of the tube, the mixture is within the limits of flammability. However, if the flame moves only part way up the tube, the mixture is regarded as being outside the limits. This is because the test concerns the capability of the mixture to propagate flame, not the capacity of the source of energy to initiate a flame.[7] However, in a practical situation the combustion initiated by an energetic source may be capable of causing a hazardous increase in pressure. Measurements made in the steel vessel, using a range of ignition energies and a suitable pressure detector, may then be more appropriate.[5]

Not all premixed combustion involves flames which are propagating freely throughout a flammable mixture. Nearly all of the natural gas which is consumed in industrial and domestic equipment is burned in premixed flames. Many of the flames are established on self-aspirating burners. These are similar in design to the original bunsen burner, which is illustrated in Fig. 10.4. Gas discharged from a nozzle at the bottom of the tube entrains air through a suitably located orifice, and the two are mixed by turbulence. The velocity of the combined flow through the tube is in excess of the burning velocity of the flame, so the reaction zone is established as a bright blue cone at the end of the tube. It is not possible by this simple design to entrain more than about half of the air needed for complete combustion. Unburned gas passing through the reaction zone then burns by mixing with the surrounding air in a separate diffusion flame, which surrounds the premixed inner cone. (Diffusion flames are discussed in 10.4.) Premixed flames which are attached in this way to a suitable burner may be correctly called stationary explosions.

The main characteristics of premixed flames may be summarised like this:

- the fuel and oxidant gases are already mixed before the flame is established;

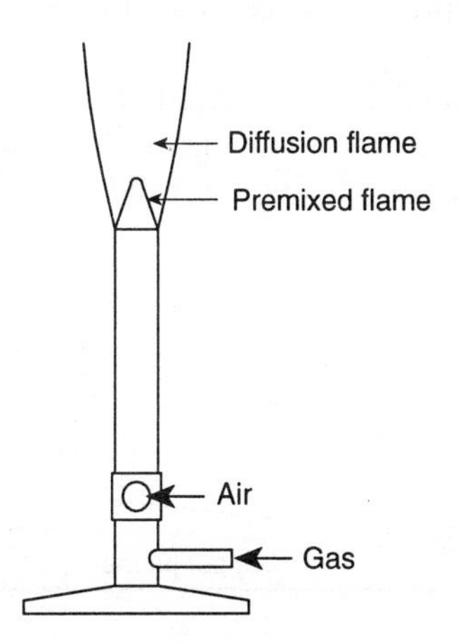

Fig. 10.4 Bunsen burner

- the flame propagates freely throughout the premixed volume (unless it is stabilised on a burner through which the mixture is flowing);
- the flame is usually bright blue in colour;
- combustion is possible in any mixture which is within the limits of flammability;
- the fundamental burning velocity is controlled by the rate at which heat and active species are transferred to the unburned gases;
- the occurrence of premixed combustion in a closed vessel results in an increase in pressure which may rupture the vessel;
- a premixed flame is an explosion (or, in certain conditions, a detonation [10.2.2]).

10.1.1 Premixing

Gases and vapours can form flammable mixtures with air by diffusion. If a gas is ejected violently, or if there is an airflow, turbulent mixing can also be present. As the fuel is progressively diluted with air, a mixture will be formed which is at the upper limit of flammability. With further dilution the mixture will pass through the flammable range and will eventually lie outside the lower limit. This process is illustrated in Fig. 10.5, which is based on an incident which happened at the Rhone–Alps refinery at Feyzin in 1966. Liquid propane (boiling point –45°C) was leaking from a sample point at the bottom of a 120 m^3 sphere. The released vapour, which has a high density than air, was being carried down wind. The drawing shows the possible extent of the flammable volume which surrounded the vapour plume. Unfortunately this extended beyond the boundary fence and over a motorway, where an ignition occurred. A premixed flame then travelled throughout this flammable volume and ignited the rest of the plume, which then burned in a diffusion flame [10.4]. During the very substantial fire which followed, the sphere ruptured and caused a boiling liquid expanding vapour explosion (BLEVE) [10.2.3].

An escaping gas would have behaved in much the same way as the propane. The maximum down-wind ignition distance would depend on the rate of release, the gas density, the wind speed, and on the roughness

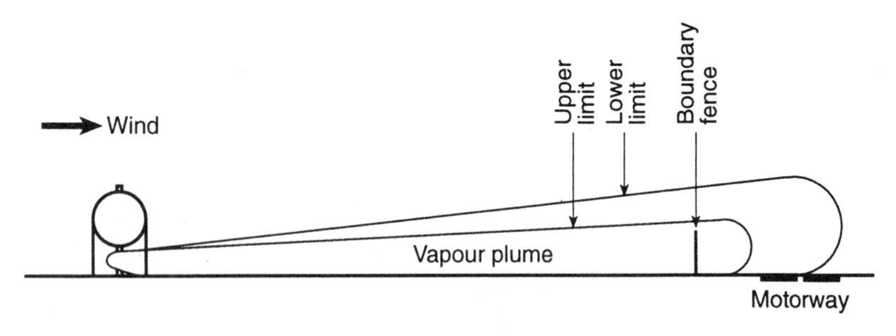

Fig 10.5 Vapour plume from leaking propane at Feyzin

of the ground (which could cause turbulent mixing). Computer programs are available which can be used to predict the behaviour of gaseous releases, and to assist in the emergency planning for such events. A knowledge of the behaviour of releases on a smaller scale is necessary in considering the use of electrical equipment in hazardous areas [10.5.2].

In the Feyzin incident the down-wind ignition distance was about 160 m. Such a distance was possible because of unusual atmospheric conditions, and also because a large volume of vapour was released as the escaping liquid cooled to the boiling point of –45°C. Vapour would also continue to be released as the pool of spilled liquid gained heat from the ground and from the air passing over its surface. However, most flammable liquids are stored at atmospheric pressure and at temperatures below their boiling points. The quantity of vapour which they can release from a spill depends on the vapour pressure at the ambient temperature, and on the heat flow into them. Petrol, with an initial boiling point of around 37°C and a vapour pressure at 20°C of about 0.25 bar, is very unlikely to create a hazardous atmosphere at more than 15 m downwind of a spill. On the other hand, kerosene, for which the corresponding values are about 160°C and 1.3 m bar, is quite incapable of forming a flammable mixture at ambient temperatures. It is necessary to heat a typical kerosene to 43°C before the vapour pressure is sufficiently high to form a lower-limit mixture at the surface. At temperatures above this, the liquid becomes progressively more hazardous. It develops properties similar to those of ambient-temperature petrol when it is close to its boiling point. The temperature of 43°C, at which a lower-limit mixture can form at the surface, is the flash point of kerosene. Flash points are measured by several standard methods (depending on the type of liquid to be tested) but in each case the temperature of a small sample of the liquid is slowly increased until the application of an ignition source causes a premixed flame to flash over the surface. Sustained burning in a diffusion flame is possible at the fire point, which is a temperature a few degrees higher than the flash point.

A knowledge of the flash point of a liquid is clearly of importance in determining the hazards which are created by its use. If it is possible for its temperature to exceed the flash point, then it is capable of causing explosions in addition to being a serious fire hazard. In the legislation covering the storage and use of flammable liquids, the requirements differ according to the flash points. Petroleum products are classified by the Institute of Petroleum into four groups according to the severity of the hazards which they present, and the boundaries between the groups are determined by flash points. Class 0 materials comprise liquefied petroleum gases (LPG) like propane. Class I liquids have flash points below 21°C, like petrol. Class II liquids have flash points between 21° and 55°C (like kerosene) and for Class III liquids the flash points are between 55° and 100°C (diesel fuel for example). Classes II and III are subdivided according to the temperature at which the liquids are used: either (1) below the flashpoint, or (2) above. Liquids with flash points above 100°C are unclassified.[9]

10.1.2 Mists

If a flammable liquid is in suspension as a fine mist, then explosions are possible which have many of the characteristics of vapour/air explosions.[4,10] If the vapour-phase flammability limits are expressed on a weight basis, then they will be almost identical to the mist limits, and the expansion ratios will also be very similar. It should be noted that it is not necessary for the liquid to be at a temperature above its flash point in order to form an explosive mist. At one time mist explosions happened in the crank cases of large marine engines. The mist was formed by the condensation of lubricating oil which had been vaporised on an overheated bearing. Ignition eventually occurred when the bearing reached a sufficiently high temperature. These explosions can be averted by using an optical detector to give warning of the first appearance of a mist [10.6.1]. Explosions are also possible in flammable liquid foams, and have been known to occur as a result of the presence of a film of oil in a compressed air line. In both of these situations, the liquid film is broken down into droplets ahead of the advancing flame.

10.1.3 Solid fuels

It should not be forgotten that solid fuels are also capable of releasing flammable vapour and can therefore create an explosive atmosphere. Suppose for example that the inner surface of some metal ducting has become coated with a combustible solid deposit. If the outside of the duct is heated, perhaps in a fire or by welding operations, then the deposit may reach a temperature at which sufficient vapour is released to form an explosive mixture with air. A suitable ignition source — perhaps part of the heated metalwork — will cause a premixed flame to travel through the ducting. Flash points of solid fuels are rarely measured, but the fire point is sometimes quoted: cooling the surface to the fire point is the normal method of extinguishing fires in these materials [10.4.2]. Explosions due to the vapour from burning solids do sometimes happen during fires in buildings. If a fire is established in a room in which there is little or no ventilation, then the air in the room will be vitiated with combustion products, and its oxygen content will be depleted. Eventually flaming combustion will cease, but flammable vapours will continue to be released from heated surfaces. If air is then admitted, through the failure of a window or the opening of a door, explosive mixtures can be formed. Their ignition will cause a sudden flow of burning gases through the opening: a hazard known to firefighters as flashback.

In the same way that a mist of flammable liquid can create an explosion, so can a suspension of fine particles of a solid fuel. What is probably the earliest report of a dust explosion was given by Count Morozzo.[11] This concerned a violent flour dust explosion in a Turin bakery in 1785. The dust was released when flour was transferred from a silo and the ignition source was a candle in the bakery shop. Similar explosions have happened in coal mines, grain silos, sugar grinding plant, and indeed in almost any process involving the comminution of combustible solids.[12–14] A relatively minor explosion in for example a grinder may trigger a secondary explosion which can wreck a building. This is because a layer of dust which has collected on horizontal surfaces throughout the building can be dispersed into the air by the pressure pulse from the primary explosion. Care is taken in dusty plants to eliminate horizontal surfaces or to remove deposits regularly. In coal mines, deposits of dust are sprayed with a binding material so that they cannot be dispersed, or are covered with a layer of stone dust.

In general, the severity of dust explosions increases with decreasing particle size. The minimum ignition energy and the lower limit concentration (usually called the *minimum explosible concentration*) are reduced, and both the maximum pressure and the rate of rise of pressure are increased. It is unlikely that a suspension of a fuel with a particle size greater than 500 micron can cause an explosion. However, it should not be assumed that processes using particles of a size greater than 500 micron will be safe. This is because attrition during transportation is likely to produce some fines. If it is possible for combustible particles to be suspended as a cloud, then it would be wise to submit a sample for testing to determine its explosibility (however, it should perhaps be pointed out that visibility in a lower-limit concentration is usually considerably less than 1.0 m: a dusty atmosphere in which it is tolerable to work does not constitute an explosion hazard). A series of standard tests can be conducted at a number of laboratories, for example the Fire Research Station in the UK and the Bureau of Mines in the USA (both of which have conducted extensive research on dust explosions). Typical of the results are these obtained for a sample of flour which contained 13% moisture: minimum ignition temperature 390°C; minimum explosible concentration 40 g/m^3; minimum ignition energy 50 mJ; maximum explosion pressure 7.1 bar; maximum rate of rise of pressure 141 bar/s.[15] The latter value is important in the design of pressure relief panels and bursting discs. This value for flour is quite moderate: the value for aluminium flake for example is 1380 bar/s,[16] and for paracetamol powder it was found to be 1042 bar/s.[13] Lists of measurements which have been made for a wide range of materials are published by the test laboratories, a selection of which is reproduced in Reference 13.

10.2 Explosions, detonations, BLEVE, OFCE

10.2.1 Explosions in vessels

If a plant is operating at atmospheric pressure and a stoichiometric explosion occurs in a closed vessel, then a pressure rise to 8 bar is possible [10.1]. However, it is unlikely that the vessel will withstand more than say 0.5 bar above atmospheric pressure, and the vessel will therefore be ruptured. Even if some of the hot gases are vented, perhaps through an open manhole, a dangerous pressure may still be generated. If the explosion happens within a building, it is unlikely that the structure will withstand more than a few m bar. Venting would have little effect even in for example an offshore module unless the equivalent of an entire wall was open. The situation is exacerbated, particularly in the offshore situation, by the presence of clutter. As the advancing flame moves around pipelines, vessels and similar objects, it becomes turbulent. There is then an increase in the area of the reaction zone, resulting in an increase in the rate of consumption of the unburned gases. Thus, although the fundamental burning velocity remains the same, the flame speed accelerates. By the time the flame has travelled the length of a module, the flame speed could be several hundred metres a second. The rate of rise of pressure will be correspondingly high, and safe venting will be increasingly difficult.

A particular problem arises if a flammable atmosphere has penetrated into a number of interconnected vessels in a plant. Figure 10.6 shows two such vessels, which we will assume contain a mixture with about half the stoichiometric concentration of fuel, and having an expansion ratio of about 4. An ignition has occurred in the larger vessel and a spherical premixed flame is propagating away from the source. As the pressure in this vessel increases, unburned gases will flow into the smaller vessel, and the flow will be turbulent as it enters the vessel. By the time the flame has entered the second vessel, the pressure will have increased to say 2 bar. We now have the ignition of a mixture which has an expansion ratio of 4 and an initial pressure of 2 bar. The final pressure could therefore be 8 bar. This will not be achieved because the flow through the connecting pipeline will be reversed and some of the hot gases will escape. However, the increased flame speed caused by the turbulence will ensure that very

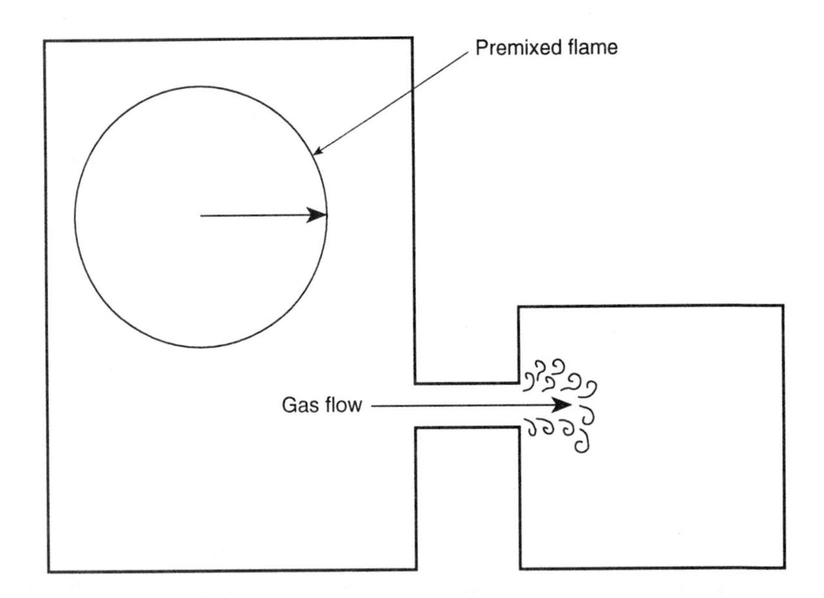

Fig. 10.6 Pressure piling

little of the pressure will be relieved in this way. If a series of other vessels is connected, then the pressure will be further increased as each vessel is involved. This effect is called *pressure piling*.

10.2.2 Detonations

The acceleration due to turbulence described above happened within a relatively large volume. Had it happened in a substantial length of pipeline, and the flammable mixture had been sufficiently energetic, then the propagating explosion could have changed into a detonation. The reaction zone would then have been travelling with a shock wave, and the velocity could have been in excess of 1000 m/s. The very high reaction rate is possible because the unburned gases are raised to a high pressure, and therefore to a high temperature, as the shock wave passes through them. At the same time, the shock wave is sustained by the large amount of energy which is released by the flame. Even if the gases in the pipeline are stationary, or if the flow is laminar, the passage of a premixed flame can produce turbulence. The flame will then accelerate and again it may detonate in a suitable mixture. This happens after the flame has travelled through a certain length of pipe, called the *run-up distance*. The smaller the pipe diameter the shorter the distance, and it is convenient therefore to quote the run-up distance as so many pipe diameters. For methane, butane and many other hydrocarbons the value is about 60.[17] However, this distance can be considerably reduced by the presence of obstructions and bends, particularly where pressure pulses can be reflected back to the flame. Transition to detonation is possible only over a restricted range of mixtures: the limits of detonability are considerably narrower than the limits of flammability. The limits for hydrogen/air are 18.2–59% and for acetylene/air 4.2–50%: these should be compared with the flammability limits in Table 10.1. However, it will be remembered that at pressures above about 5 bar pure acetylene can decompose exothermically. This decomposition can also proceed as a detonation, making the limits 4.2–100%.

The detonation velocity for stoichiometric hydrogen/air is 2200 m/s, falling to 900 m/s at the limits. The velocity in pure acetylene at 5 bar is 1000 m/s, rising to 1600 m/s at 30 bar. The value for stoichiometric hydrogen/oxygen is 3500 m/s (about 7800 mph). As might be expected, the flame temperatures are higher than those in explosions. The expansion ratio for hydrocarbon/ air detonations is about 20, but the pressure effects are directional. Thus, a detonation which has failed to split open a straight run of pipeline may punch a hole at a sharp bend. The destructive capability of detonations is clearly obvious but fortunately they almost invariably occur in pipelines. The so-called spherical detonation, in a large vessel or in the open air, is almost unknown (but see OFCE) [10.2.4]. If a detonation travels down a pipeline into a vessel, it will revert to a normal explosion, which will then consume the gases in the vessel. If the flame then enters another pipeline attached to the vessel, it may pass over to a detonation again after the run-up distance. Some books provide lists of a limited number of gases which are known to be capable of forming detonable mixtures with air. These lists can be misleading. A mixture which has failed to make the transition from explosion to detonation, despite a long run-up, may still detonate if a sufficiently powerful ignition source is used: the shock wave from a detonator for example, or from another detonation.

Detonations are of course possible in liquid and solid explosives: this was discussed in [9.1]. The detonation velocity in nitroglycerine has been measured at between 1300 and 8000 m/s, and for the solid high explosive TNT, the value is 6700 m/s. The distinction between explosions and detonations applies also to the range of substances which are classed as *explosives*. High explosives will detonate if they are triggered by a detonation, usually from a small high explosive charge in a detonator. Primary explosives, or initiators, like mercury fulminate and lead azide, can be detonated by impact or heat and are used in detonators and percussion caps. The propellants (or low explosives) cannot normally detonate but do explode in the sense that they are consumed by a premixed flame at a characteristic burning rate. This process is usually called deflagration, and the word is sometimes applied to gaseous explosions as well. In many propellants (which are used to push a shell out of a gun, or to drive a rocket motor) true combustion takes place, in addition to the exothermic decomposition of the propellant. Cordite, which is a colloidal suspension of nitroglycerine in nitrocellulose, decomposes to release carbon and hydrogen, which are oxidised by the oxygen in the $-NO_2$ groups (all of which are present in the molecules of the two constituents: $C_3H_5(ONO_2)_3$ and $[C_6H_7O_2(ONO_2)_3]_n$). The combustion is on the rich side of stoichiometric, which accounts for the muzzle flash which made the early cordites unattractive, particularly for naval use at night. The burning rates of cordites are between about 4 and 15 mm/s at 70 bar, with flame temperatures between 2000° and 3000°C. The earliest propellant, black powder (gunpowder) comprises two fuels, carbon and sulphur, and potassium nitrate, KNO_3 (saltpetre), as the oxidiser. The very intimate mixing, by prolonged grinding, ensures almost complete premixing. The powder deflagrates readily, but has never been known to detonate.[3,18,19]

10.2.3 BLEVE (Boiling Liquid Expanding Vapour Explosion)

Figure 10.5 shows how a spill of propane from a spherical storage vessel became ignited. The fire which followed involved the pool of liquid beneath the sphere. This pool increased in size and some burning propane entered the drainage ditches. It may be surprising that a large volume of a liquid with a boiling point of –45°C can persist in a fire, but in fact in the initial stages of the fire the burning rate was considerably less than the rate of release of the liquid. (The burning of pool fires is discussed again in [10.4.1].) By the time the fire brigade arrived, the pressure relief valve on top of the sphere had opened: this situation is shown diagramatically in

Fig. 10.7. The fire was spreading towards three similar 1200 m³ spheres containing propane, and four of 2000 m³ containing butane. The brigade's efforts were concentrated on preventing the involvement of these vessels in the fire. They thought that because the first sphere was being vented it was safe, and could simply be left to burn. Sadly, they were wrong. Large flames completely engulfed the vessel but the initial heat transfer to it was reduced by the presence of a layer of thermal insulation. As this gradually disintegrated, the heat transfer increased. The lower part of the vessel was cooled by its contact with the liquid propane and its temperature would be not much higher than the propane boiling point (which would depend on the internal pressure). Unfortunately the upper part of the shell was not cooled and was receiving high levels of radiation. Its temperature steadily increased, and approached 500°C after about 90 minutes. Steel completely loses its load-bearing strength at about this temperature, and the vessel therefore ruptured. There would still be a substantial pressure in the vessel despite the open vent, and the discharge of the contents was extremely violent. The liquid propane was at a temperature well above its ambient-pressure boiling point and it flashed rapidly into a mixture of hot vapour and mist. This ignited rapidly as it continued to mix into the air. The resulting fireball killed 18 of the firemen. Fragments of metal from the ruptured vessel cut the legs off the next sphere, which fell over, releasing liquid propane from its relief valve. Five vessels were eventually destroyed by the fire. (There is a different version of this event in 13.3.)

A BLEVE can happen whenever a sealed or partially vented vessel containing a flammable liquid is heated in a fire. The violent ejection of the contents ensures at least partial premixing and a rapidly expanding fireball: flame speeds as high as 50 m/s have been observed. Many of the reported

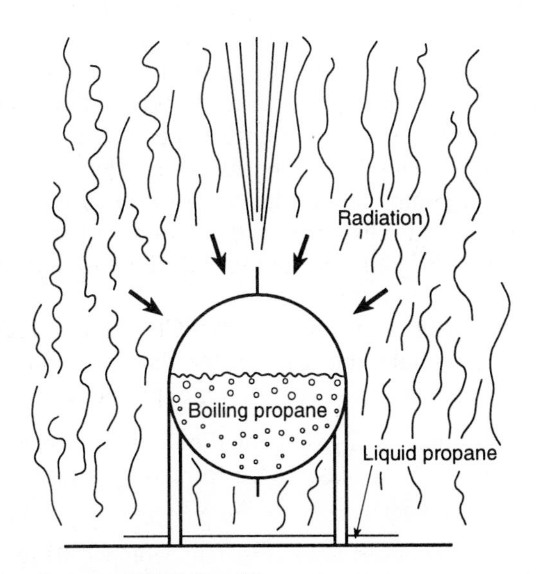

Fig. 10.7 Events leading to a BLEVE (Feyzin)

incidents have involved road and rail tankers. Some of the BLEVEs have happened as little as 10 minutes after the start of the fire, but delays of 30 to 60 minutes are more usual. BLEVEs can also be caused by small butane cylinders and aerosol cans. BLEVEs can usually be averted by the prompt application of cooling water.

10.2.4 OFCE

The events leading to the disastrous fire at the Nypro plant at Flixborough are discussed in Chapter 4. The failure of a temporary pipeline caused the release of about 80 t of cyclohexane at 155°C. Cyclohexane has a boiling point of 80°C, a flash point of –33°C, and was being processed at a pressure of 9 bar. When the liquid was discharged violently from the plant about half of it would flash off into vapour and the remainder would form a mist. Air was entrained into the high velocity jet, and some of the droplets were than vaporised. Premixing occurred in the same way that it happens in a BLEVE, but at Flixborough there was no immediate ignition. A partially premixed cloud was formed above the plant which could have been 200 m in diameter and up to 100 m high. About 45 seconds after the start of the discharge, some part of the cloud encountered an ignition source. Since the cloud was unconfined one might have expected a premixed flame to pass through it, at a speed dependent on the degree of turbulence, resulting at ground level in a pressure pulse and a high level of radiation. What did happen had the appearance of a detonation, with a TNT equivalent estimated to be between 15 and 45 t. The blast caused considerable damage to plant and storage tanks, and led to a 40 000 m^2 fire which burned for several days.

The event which happened at Flixborough was an open flammable cloud explosion: an OFCE. If the cloud had burned as a flash fire with no blast wave, this would have been an OFCF, the final F indicating fire. The cloud was not completely unconfined because the ground was beneath it, the upper part of some of the plant was engulfed by it, and the mass of the air which surrounded the cloud was considerable. The mechanism of these events has not been completely explained but it does seem unlikely that a detonation occurred. At Flixborough the ground level overpressure was between 0.7 and 1.0 bar, and calculations indicated a flame speed in excess of 100 m/s, but not at detonation velocity.

There had been well over 100 large OFCEs before Flixborough, and although there are considerable gaps in the data, it is possible to list the conditions under which such an event becomes probable.

- The released fuel may be a gas, a volatile liquid or a refrigerated liquid, but a large cloud is most likely to be formed by a flashing superheated liquid (as at Flixborough).
- The fuel must be released with sufficient violence to form a premixed cloud.

- There must be a delay before the cloud is ignited, otherwise the result will be a BLEVE. The minimum delay which will result in an OFCE is probably about 10 seconds and the maximum reported delay is 15 minutes. The most likely interval is one or two minutes, during which time the cloud may have drifted a few hundred metres.
- It has been suggested that the minimum size of cloud to produce an OFCE is 10 to 15 t, but one explosion has been reported which involved only 5.5 t (Beek, Netherlands, 1975). Also in some experiments using large balloons, violent explosions were recorded using less than one tonne.[20,21] What is certain is that the larger the amount of fuel in the cloud, the greater is the probability of an OFCE [11.4].

In some references the event is called an unconfined vapour cloud explosion, UVCE, or an aerial explosion.

10.3 Prevention and control of explosions and detonations

Explosion prevention should begin during the initial research on a new process. Someone should be asking questions: Do we have to use a flammable solvent? Why are we processing liquids at temperatures above their flash points? Do we really need to have (as at Flixborough) nearly 200 m^3 of cyclohexane at 155°C and 9 bar? The Research Department should be required to justify the presence of any hazardous materials and processes, and to show that only minimum quantities will be present. Later, after a hazard and operability study (HAZOP) has been completed, and at some time before construction begins (perhaps after a model has been made), the whole project should be stopped until the people who are responsible for the safety of the plant have confirmed in writing that they are satisfied with all of the precautions which have been built into the plant and the process.

It is necessary also to consider the presence of possible ignition sources, and these are discussed in 10.5. However, it should be emphasised here that although the proper control of ignition sources is essential in hazardous areas, it must not be the only precaution. Methods of preventing a loss of containment of flammable materials are of equal importance. The lesson to be learned from past experience is clear: if a flammable leak is allowed to persist, then sooner or later, in one way or another, it will find an ignition source.

10.3.1 Diluting a release

If a flammable gas or a liquid at a temperature above its flash point has been released from a plant, then it is inevitable that mixtures will be formed with air which are within the limits of flammability. This may happen at some distance from the plant, where an ignition source may be present. It is important that the mixture should be diluted to below the lower limit as rapidly as possible (and also that ignition sources should be eliminated.) The simplest way of achieving rapid dilution and safe dispersion is to locate the plant in the open air at a suitable distance from other plant and from the boundary fence. Fortunately the conditions which existed at Feyzin (Fig. 10.5) are unusual, and the plume from a typical let-down on a process plant will be safely diluted in a few metres. Guidance is available on suitable distances between blocks of plant and between storage tanks[9] (and may be required by legislation). It is unusual for recommended distances to be greater than 15 m. In critical situations the dispersion of gases[22] and the maximum ignition distances[23] may be predicted by calculation.

If it is essential to provide some weather protection for the plant and the operators, then a roof supported by pillars should be considered. If this is not acceptable, then perhaps up to three of the sides could be walled. If it is necessary to enclose the plant completely then the possibility of an explosion within the building must be accepted. The maximum possible ventilation should be provided; ignition sources should be eliminated (or properly controlled — see 10.5 below); at least a substantial part of the structure should be capable of providing pressure relief; access should be strictly limited; work on the plant, and particularly hot work, should be strictly controlled by a permit-to-work procedure; and a gas detector system should be installed to provide an early warning of a leak and, if necessary, to shut down the plant automatically. (Most of these precautions are discussed later [10.3.3], [10.6.2], [18].)

It sometimes happens that a plant, from which it is possible to release a substantial volume of a flammable gas, is surrounded by other plant, some of which may contain potential ignition sources. It may be possible then to prevent the spread of a hazardous cloud by means of a steam curtain.[24,25] The plant is surrounded by a wall about 1.5 m high (which is not sufficiently high to impede the natural dispersion of small leaks). A pipe is located on top of the wall from which steam can be released through a number of holes. The individual jets combine to form a curtain. Air is entrained by the steam so that when the flammable gas encounters the curtain it is lifted and diluted, preferably to below the lower limit. It is necessary to have a substantial and reliable supply of steam, and it is necessary also to take precautions against the generation of static charges by the jets. Water curtains can also be effective in diluting and dispersing a gas cloud. Specially designed ground monitors can be used by a fire brigade down-wind of a release to ensure its rapid dilution.

10.3.2 Purging and inerting[4]

If a plant or a pipeline contains for example methane, it will usually be at a positive pressure. The explosion hazard is then the one discussed above: the formation of a flammable atmosphere through leakage from the plant. The situation is different if the plant has to be shut down for maintenance or repair and it becomes necessary to open up some part of it. Before the manway can be safely removed from say a large vessel, it is necessary to ensure that an explosive atmosphere cannot be formed either inside or outside the vessel. One way of doing this would be to replace the whole of the methane with nitrogen or carbon dioxide (the vented flammable gas being discharged to a flare stack) and then to displace the nitrogen with air. The second step could be accomplished simply by leaving manways and other openings uncovered and relying on natural ventilation (remembering, however, that high concentrations of released inert gas can be rapidly fatal). It may also be decided that the rest of the plant should be inerted while the work is in progress. When the work has been completed and the apertures are closed it will be necessary to reintroduce methane to the plant. This can be done safely by first replacing the air with nitrogen, and then displacing the nitrogen with methane. It could be argued that the nitrogen is not needed because there are no ignition sources in the plant. Although the probability of an ignition is admittedly low it still represents an unacceptable risk. If the vessel to be inerted is small and there is a good supply of nitrogen, then this procedure becomes quite feasible: nitrogen can be injected until a nil reading is obtained by a methane detector when the vessel is being freed of methane. Similarly a nil oxygen reading is obtained when the air is being removed. (It should be noted that the familiar flammable gas detector, or explosimeter, cannot be used as the methane detector since no air is present.) However, if a large volume of plant or pipeline must be made safe, and particularly if the plant does not have its own nitrogen supply, then it is necessary to determine the minimum concentration of nitrogen in both methane and air which will prevent the formation of flammable atmospheres.

It will be remembered from Fig. 10.3 that the peak concentration of nitrogen against methane/air mixtures is 36.5%. This is the minimum concentration of nitrogen which will prevent the ignition of any methane/air mixture. Clearly we must ensure that the equivalent of this concentration of nitrogen is achieved whenever hazardous mixing is occurring both inside and outside the vessel. We can determine the necessary concentrations in both methane and air by transferring the information in Fig. 10.3 to a triangular diagram, as has been done in Fig. 10.8. In Fig. 10.3, only the concentrations of methane and nitrogen are shown and the concentration of air can be found by difference. The triangular diagram shows all three constituents.[4]

Each of the three corners of the triangle represents 100% of a constituent. Thus the one at the top is 100% nitrogen, and the bottom line of the large triangle is at 0% nitrogen. The horizontal lines above this represent 10%, 20%, 30% and so on, to the top. Lines drawn parallel to the other two sides show similar concentrations of methane and air, but to avoid confusion they have not been numbered. Any point on one of the sides of the main triangle indicates a mixture of two components, so that A is a

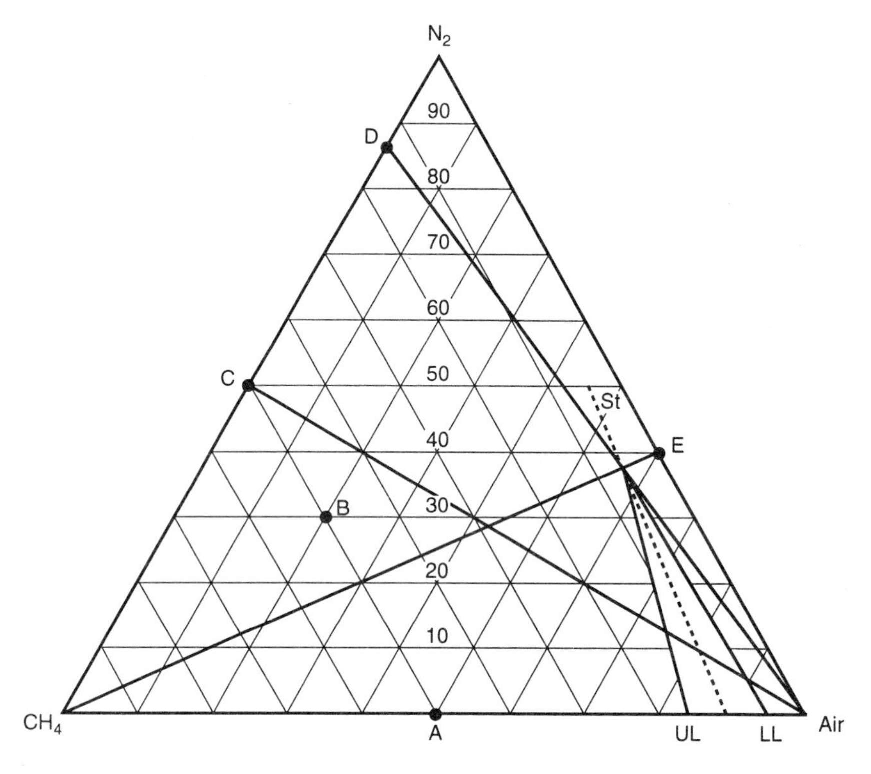

Fig. 10.8 The determination of inerting end points for a vessel containing methane and for the same vessel containing air

50/50 mixture of air and methane. Any point within the triangle represents a mixture of all three components: at B there is 50% methane, 30% nitrogen, and 20% air. The limits of flammability of methane are on the bottom line at 5 and 15% methane (on this diagram the concentration increases from right to left). The flammability envelope with nitrogen has been drawn, with the peak at 36.5%. This is slightly on the rich side of stoichiometric (shown by the St line). Suppose now that we look at the first situation of a plant full of methane, which must be made safe. If we add sufficient nitrogen to make a 50/50 mixture, we will be at point C on the N_2–CH_4 line. If now the plant is opened up, this mixtures will begin to mix with air, both inside and outside the plant. The mixtures formed in this way will all be represented by points along the C–AIR line. This cuts through the flammability envelope, and from the point at which it crosses the St line we can estimate an expansion ratio of about 2. This means an explosion pressure of 2 bar g., which would wreck most plants. Clearly the mixture at point C is not safe. We can find the minimum concentration of nitrogen which will produce an inert mixture with methane by drawing the AIR–D line, which is tangential to the envelope. Mixtures with air along this line cannot be ignited. D represents 86.25% nitrogen, 13.75% methane. If we apply a 20% safety factor, the concentrations we should aim for in the plant are 89% nitrogen, 11% methane.

A tangent to the envelope from the CH_4 corner to the N_2–AIR line gives point E. This represents the minimum concentration of nitrogen in mixtures with air which will inert the air: it will prevent the atmosphere in the plant from forming explosive mixtures when methane is re-admitted. This value is 41% nitrogen, 59% air, and again a 20% safety factor will give 47.1% air. Since air contains 21% oxygen, an oxygen meter can be used to check that no more than 9.9% is present within the plant.

If a large vessel containing methane is to be inerted with nitrogen, less nitrogen will be needed if it is injected at the bottom of the vessel. This is because the density of nitrogen at 1.25 kg/m³ (at 0°C), is higher than that of methane, 0.715 kg/m³. At least some of the displaced methane can therefore be released undiluted (to flare) from the top of the vessel. In the ideal situation, there would be no mixing at all of the two gases, and just sufficient nitrogen would be needed to fill the vessel, having completely displaced all of the methane. If the gas being discharged is monitored for methane content by a suitable detector, a plot showing concentration against time would look like A in Fig. 10.9. In the most unfavourable conditions, which require the maximum volume of nitrogen, the two gases mix completely and continuously as the nitrogen is injected. This would result in a curve similar to B in the figure, and purging would continue until the end point of 11% methane was achieved. If a suitable detector is not available but the rate of injection of nitrogen is known, then the time taken to achieve the end point can be calculated. Clearly it would not be safe to assume that curve A applies, but curve B represents the worst condition and can therefore be safely used. The time to the end point can be calculated from:

$$C = C_0 e^{(-Pt/V)}$$

where

C is required concentration (the end point of 11%),

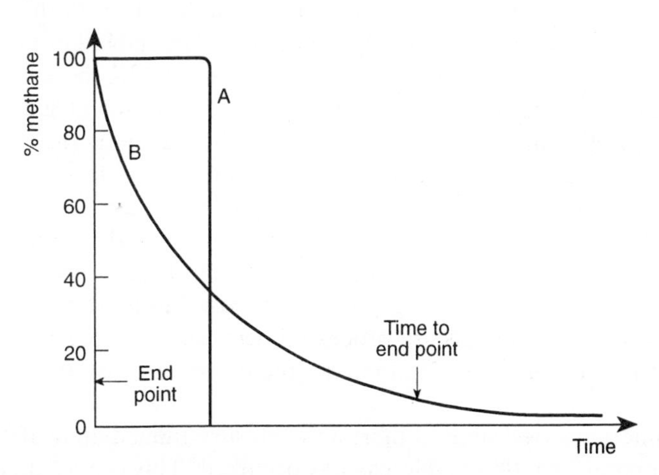

Fig. 10.9 Concentration of methane at the vent of a vessel from which it has been displaced by nitrogen: A, no mixing, B, complete mixing

C_0 is original concentration (100%),
e is 2.718 28,
P is purge rate, m^3/sec (or min),
t is time, sec (or min),
V is volume of vessel, m^3.

Thus, the minimum volume of nitrogen will be used if the concentration of methane in the discharge is monitored continuously, and if the nitrogen is injected at the bottom of the vessel in such a way as to minimise mixing. When air is being inerted with nitrogen prior to admitting methane to the plant, the situation is rather different. The density of air, at 1.29 kg/m^3 (0°C), is slightly higher than that of nitrogen (1.25 kg/m^3), so in theory it would be better for the nitrogen to be injected at the top. However, the advantage would be slight and would hardly justify the reconnection of the pipework.

It is also possible to purge a vessel by repeatedly pressurising with nitrogen. Thus if we have a vessel containing methane at atmospheric pressure, and pump in nitrogen to a pressure of 2 bar g., the mixture will contain 33.3% methane. If this is now vented (safely) down to atmospheric pressure, and again taken up to 2 bar g. by injecting nitrogen, the methane content will be down to 11.1%, acceptably close to our end point. Empty underground petrol tanks are sometimes purged of any remaining vapour by repeated pressurisations with air. This is clearly not entirely safe because the mixture must pass through the flammable range (which is widened by an increase in pressure). It is necessary for an ignition source to be present before an explosion can happen, but this could be provided by pyrophoric iron sulphide, which can be formed in an old tank (ignition sources are discussed in 10.5.1).

Continuous inerting may sometimes be necessary, for example for the protection of fixed-roof methanol storage tanks. The flash point of methanol is 11°C and the unusually high upper flammable limit of 36% is reached when the liquid temperature is 38°C. In the UK there will therefore be an explosive mixture in an unprotected tank for at least half of each year (particularly if the methanol is hot when it enters the tank). The explosion hazard can be removed by replacing the air in the tank with nitrogen. The nitrogen blanketing of a group of tanks can be conveniently achieved by connecting all of them to a gas holder with a floating roof. The holder will adjust to movements into and out of the tanks, and vapour losses will be minimised. Nitrogen cannot be used to inert plant in which hot or finely divided metals are processed because it can react exothermically to form the nitride. Helium, argon or neon must then be used instead.

It is sometimes desirable to inert an enclosure immediately after a loss of containment of a flammable gas has occurred. This can be done by an automatic total flooding system which is actuated by combustible gas detectors. The purpose of the system is to convert the whole of the atmosphere in the enclosure into an inhibiting concentration of a suitable agent, usually one of the volatile chemical extinguishants. The design concentration of the agent is obtained by adding a safety factor to the peak concentration (see 10.1 and Fig. 10.3), together with any additional agent which may be needed to compensate for losses through openings. The required quantity of agent is stored under pressure in suitable containers and then discharged when needed through an array of nozzles. The discharge should be completed in 10 s or less. Detailed guidance on the design of Halon 1211 and Halon 1301 total flooding systems is provided in UK[26] and US[27] standards. When sufficient data are available on the physical properties and behaviour of ozone-friendly alternative extinguishants, suitable amendments can be made to these standards. Total flooding systems are also used to extinguish fires, and the discussion on this usage in 10.4.2 includes a description of the hazards created by all of these systems.

10.3.3 Venting, explosion relief

Curve A in Fig. 10.2 shows the pressure which would be developed by a stoichiometric methane/air explosion in a closed spherical vessel. If the vessel had been a cube instead of a sphere then the flame area would have been reduced progressively once the flame had encountered the walls. The rate of rise of pressure would then fall, following curve B, as separate flames penetrated into the corners. However, the important point is that the maximum pressure would still be the same (just supposing that the vessel could withstand 7.4 bar). Most vessels, particularly those with flat sides, would fail at pressures considerably less than this. If explosions are possible in vessels like this, then their failure can be averted if adequate venting can be provided before the pressure is too high. One or more bursting (rupture) discs may be sufficient to protect a small vessel, but venting panels will be needed for larger plant and buildings. Curve C shows the pressure changes which might be expected in a vented container. P_1 is the pressure at which the plant will be damaged, and P_2 is the pressure at which the vent opens. Because of the inertia of the vent panel there will be a delay before it is fully open, and the pressure will continue to rise to P_3. There will then be a fall in pressure as gas is released, followed by a steady increase until the maximum flame area is established: P_4 is then the maximum vented pressure.

Research into the design of pressure relief panels has been conducted in many countries, and in particular at the British Gas Midlands Research Station.[28] A number of different methods are available for calculating the required vent area[29–32] and some of these are relatively simple. For example, this equation is provided by Cubbage and Simmonds[33,34] for a lightweight panel which is retained by gravity, or the minimum of friction, and is located in the wall of the vessel which has the largest area:

$$A_v = \frac{58 S_0 A_s}{P_m}$$

where A_v is the vent area, m^2,
S_0 is the burning velocity m/s,
A_s is the area of the side containing the vent, m^2,
P_m is the maximum pressure which can be accepted, m bar.

However, this can only be applied when the area of the side A_s is not greater than five times the vent area A_v. Other caveats should be carefully checked[28] and in particular the presence of turbulent combustion will clearly affect S_0. It has been suggested[30] that for a room containing furniture and other objects, S_0 should be increased by a factor of 1.5. Where obstacles are present throughout such a volume, the factor should be 5. If significant turbulence can be present before ignition (due for example to a high-pressure gas leak), then a factor of 8 or 10 should be used. On an offshore platform, where a high turbulence factor may well be appropriate, it is sometimes possible to use lightweight cladding on the whole of the outside wall of a module.

In some situations it is difficult to find a suitable location for a relief panel. For example, security considerations require that a bonded whisky storage is strongly built, with a secure steel door and no openings in the walls. The solution here is to provide a very light roof which is capable of lifting sufficiently far to relieve an ethanol/air explosion. Care must always be taken to ensure that explosion gases are vented safely, away from occupied areas. This may be difficult to achieve in, for example, the offshore module mentioned above. Detailed guidance on the many problems associated with the provision of explosion relief will be found in National Fire Codes No. 68 (USA).[35]

10.3.4 Flame arresters

A propagating premixed flame may fail to penetrate a narrow gap because of the loss of both heat and chain carriers to the walls. Column L of Table 10.1 shows the maximum gap which is capable of preventing the passage of a stoichiometric premixed flame of a number of gases: this is the quenching distance. The value for most hydrocarbon/air flames is around 1.8 mm, but hydrogen, with its high flame temperature and high diffusivity, requires 0.5 mm. A knowledge of quenching distances is used in the design of flame arresters (or flame traps). A large number of passages are provided through which gas can flow, but none of the gaps is greater than about 50% of the quenching distance. In one design a crimped metal ribbon is wound spirally with a flat ribbon to form a robust arrester which can be made to close tolerances. These have about 80% free cross-sectional area, and are to be preferred to earlier devices which relied on a bed of compressed wire gauze. These latter have the disadvantage of a free area of less than 50%, and also the possibility that passages greater than the quenching distance may be present in the bed, or may be opened up by an explosion. Where good mechanical strength and high heat capacity are needed, perforated metal plates can be used, but they have even less free area. Sintered metal discs are also used as arresters and can be made with very small apertures. Unfortunately their flow resistance is high, but they are used for example in catalytic flammable gas detectors [10.6.2]. Simple parallel plate arresters are used to prevent the release of premixed flames from electrical equipment in hazardous areas [10.5.3]. Long runs of pipeline in which explosions are possible are sometimes protected by providing vertical columns filled with pebbles at regular intervals. The distance between the columns is less than the run-up distance for transition to detonation.

Flame arresters are frequently installed in pipelines through which gas is flowing to a burner or is being discharged to a flare. They are also placed in the vents of storage tanks containing flammable liquids. If a methanol tank is not nitrogen blanketed, then vapour/air mixtures which are released through the vent could be ignited at some distance from the tank (perhaps by a fire on a neighbouring tank). A premixed flame could then travel back through the vent and into the vapour space. The resulting explosion within the tank may well blow off the roof and start a fire. The presence of a flame arrester at the vent will prevent the passage of the flame into the tank. However, if vapour continues to be released, a flame may be established at the surface of the arrester. If the arrester then gets too hot it will be unable to remove sufficient heat from the flame to quench it, and the flame will pass through into the tank.

In order that an arrester can remove sufficient heat and chain carriers to cause quenching, the flame must be in contact with it for a certain minimum time. The residence time of the flame must clearly depend on the rate at which the flammable mixture is flowing through the arrester, and on the burning velocity (which must be added or subtracted depending on whether the flame is travelling with or against the flow). There is a critical combined velocity above which the flame will pass through the arrester. This effect has been studied,[36] and it has also been found that it is possible to design arresters which will stop even a detonation,[37] although very little work has been done in this area. The velocity at which a premixed flame approaches an arrester in a pipeline can be considerably reduced by the use of bursting discs. In the arrangement shown in Fig. 10.10, an explosion travelling in either direction will be relieved by the rupture of one of the discs. The drop in pressure will reduce the velocity of the flame and increase the probability that it will be quenched by the arrester.

10.3.5 Explosion suppression

An important characteristic of contained vapour-phase explosions, illustrated in Fig. 10.2, is the relatively slow rate of rise of pressure immediately after ignition. This allows enough time to detect and extinguish the flame

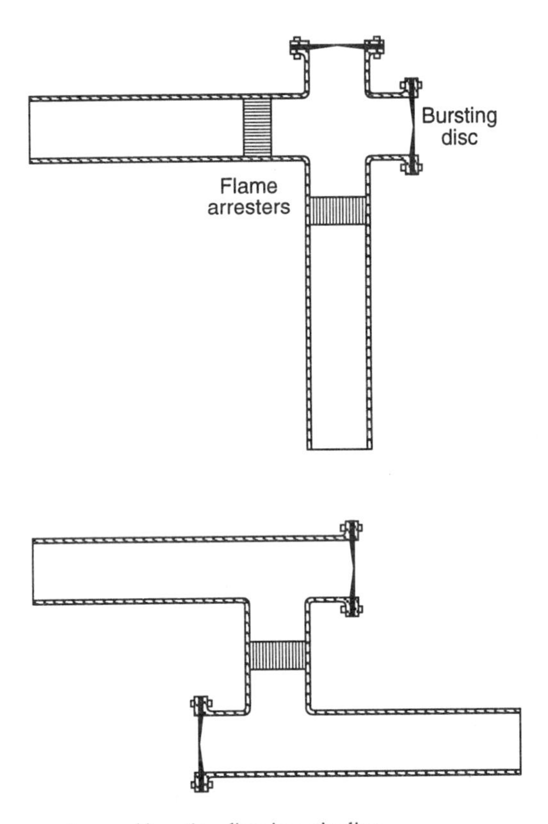

Fig. 10.10 Flame arresters and bursting discs in a pipeline

before the pressure is sufficiently high to damage the container. The explosion suppression technique was originally developed by Glendinning[38] at the Royal Aircraft Establishment, Farnborough, UK, to protect the fuel tanks of military aircraft. No matter what fuel is used, an explosive atmosphere will be present in an aircraft tank at some time during a typical flight (and can be ignited by incendiary ammunition). This is due to the combined effects of changes in both temperature and pressure. When a system was installed in the tank, the light emitted by the premixed flame, or by the incendiary composition, was detected by a photomultiplier. A circuit was then completed through a detonator. The detonator was located within a container of suppressant, usually a Halon, and this was dispersed violently as a mist into the advancing flame. In the industrial equipment which was subsequently developed, the detector is usually a simple pressure-sensitive device. This is mounted in the wall of the vessel to be protected and comprises a metal diaphragm which is perforated with a pinhole. Slow changes in pressure have no effect, but a rapid increase in pressure will cause the diaphragm to snap over and close an electrical contact. Typically a detector will respond to a rate of rise of about 7 bar/s or a static pressure of about 35 m bar, but the settings can be adjusted according to the volume of the vessel. If a detector is likely to be actuated by vibration, then two detectors can be used, mounted at right angles to each other, and connected in series. Again the high explosive charge in a detonator is used to disperse a Halon from a suitable container. The detonator fires about 1 ms after the contacts close, and the Halon mist is discharged at a velocity of about 60 m/s. An extinguishant with a high boiling point is normally used, like Halon 1202 (CBr_2F_2, bp 24.4°C). The premixed flame is extinguished by the combined effects of the chemical extinguishant, the mist acting as an omni-directional flame arrester, and the flame stretch caused by the shock wave [10.4.2]. It is in fact possible to use water, or even the fuel itself, as a suppressant, but the volume will not then remain inerted.

In a complex plant in which an explosion in one vessel could spread down pipelines and into other vessels, additional protection can be provided by high-rate discharge bottles containing agent under pressure. Fast-acting isolation valves can also be used, and 'Armourplate' glass relief panels which can be shattered by a detonator.

The pressure/time curve of a typical suppressed explosion is shown as D in Fig. 10.2. A pressure recording which was made during the suppression of a stoichiometric hexane/air explosion in a 4500-litre vessel gave these results: detection after 60 ms; detonator fired at 61 ms; flame extinguished at 85 ms; maximum pressure 69 m bar.[39]

10.4 Diffusion flames

The apparatus sketched in Fig. 10.1 is used to find the concentration of an extinguishant in a fuel/air/agent mixture which is needed to stop a stoichiometric premixed flame (the peak concentration or inerting concentration). The apparatus in Fig. 10.11 is used to measure the agent concentration in air which will extinguish a diffusion flame.[40] It serves as a useful introduction to a discussion on this kind of combustion.

The apparatus is called a cup burner and is used as a standard method of obtaining design data for automatic systems.[38] The cup is 28 mm in diameter and is located on the axis of a vertical 85 mm glass tube. The cup is kept completely filled with liquid fuel by a levelling device, and there is a laminar flow up the tube of air to which known concentrations of extinguishant vapour can be added. The minimum concentration which will put out the flame is called the *flame extinguishing concentration*. If necessary, measurements can be made in which the fuel is heated in a specially modified burner.

A diffusion flame of this kind supplies radiation to the fuel surface which enhances the rate of evaporation. The vapour then streams up into the reaction zone of the flame. The flame is hollow and the reaction zone is only 1–2 mm thick. In this zone fuel vapour and air are mixed by diffusion, which is a rapid and effective process at flame temperatures. The rate of diffusion controls the reaction rate, hence the name of this kind of flame. The rate of consumption of the fuel is therefore largely dependent on the area of the

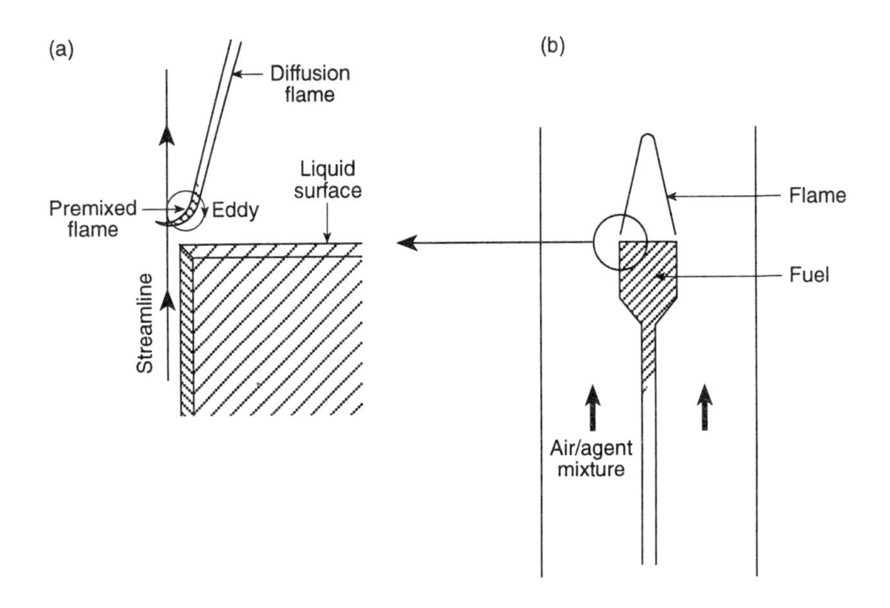

Fig 10.11 (a) Flame and eddy at the edge of a cup burner, (b) cup burner apparatus

reaction zone. As the fuel vapour approaches this zone it passes through a steep temperature gradient. When long hydrocarbon molecules in the vapour are heated, they are broken ('cracked') into smaller ones: for example a decane molecule will produce a pentane and a butane molecule:

$$C_{10}H_{22} \rightarrow C_5H_{12} + C_4H_{10} + C$$

As pyrolysis continues, the carbon atoms which are released at each degradation form graphite rings (C_6) which then clump together to form small solid particles. At the final breakdown into atoms, the hydrogen will stream ahead into the reaction zone but the carbon particles will be heated to flame temperature, perhaps 1600°C, before they begin to react. This causes the yellow/white radiation which is characteristic of most diffusion flames. On pools of liquid which are larger than the cup burner, not all of the carbon is able to enter the reaction zone, and some is released from the top of the flame as soot. The reaction zones of diffusion flames are almost invariably stoichiometric, even in situations where the air supply is vitiated. This can be understood if we assume that after a flame has been established on the cup burner, a slightly more volatile fuel is fed into the cup. An excess of vapour will pass up into the flame and there will be insufficient oxygen in the reaction zone for its complete combustion. Unburned fuel, which has been heated to flame temperature, will then be released from the reaction zone into the surrounding air, where it will react. The result will simply be an enlargement of the flame, and the reactions will remain stoichiometric. A rather similar mechanism ensures a steady burning rate of a pool fire. An increase in the rate of release of volatiles causes an increase in flame area, and the walls of the flame become more nearly vertical. This reduces the radiation flux to the fuel surface, and the evaporation rate is reduced to its original value. In general, although a diffusion flame must always remain close to its source of fuel, it will establish the conditions which are the most favourable to its own existence.

The cup burner results are used to determine the design concentration for automatic systems which may be required to extinguish quite large fires. This extrapolation can be justified for two main reasons:

- the combustion reactions are the same in both small and large flames;
- the critical agent concentrations recorded in a large number of full-scale total flooding tests have always been less than those for the same fuel in the cup burner.

The explanation of the higher agent requirements in the cup burner tests is that the flame is more stable than those in full-scale fires. This is due partly to the fact that the cup burner flame is closed at the top, with no loss of smoke, and hence the whole of the fuel is being burned. More importantly, the flame is being stabilised on the burner by a small premixed region at its base. As the air flows past the top edge of the cup it encounters a stream of vapour at a much lower velocity. The air stream forms a small eddy at the edge, which acts in much the same way as a roller bearing. There is thus a toroidal vortex (a doughnut-shaped eddy) around the periphery which, as it rotates about its circular axis, picks up some of the vapour. A vapour–air mixture is then fed into a small premixed cusp of flame, which is blue in colour, at the base of the diffusion flame. This effect has been maximised in the cup burner by the chamfered edge, which can be seen in Fig. 10.11(a), and which allows the fuel surface to be directly beneath the eddy. There is one practical situation in which extinguishing concentrations are required which are higher than the cup burner measurements. This occurs when a liquid surface diffusion flame is attached to an obstruction in a high air flow, and again recirculating eddies are established behind the obstruction.[41]

10.4.1 Diffusion flames from gases, liquids and solids

A low velocity flow of gas from a small hole will burn in a diffusion flame very similar to the one in Fig. 10.11, and indeed flames like this can be used in the apparatus. The area of the flame is directly dependent on the mass flow rate. Higher flows and larger flames cause turbulence and the degree of premixing at the base of the flame is increased. At very high gas velocities the base of the flame will lift away from the orifice and will be largely premixed. If the flame is about 30 m high or more, then ‘turbulence balls’ will form repeatedly around its envelope, and their consecutive ignitions will cause the characteristic low-frequency rumble.[4]

If the glass tube is removed from the cup burner apparatus and the flame is allowed to burn in the open, it will still be laminar and closed at the top. However, if combustion is established on pools which are larger in diameter, then the flames will flicker and pulsate, and soot will be released. With increasing diameters the flames will become progressively more turbulent, and fully turbulent flames will be present on any pool greater than about 1.0 m in diameter. As is explained above, the burning rate of a pool of liquid remains substantially constant, although some changes might be expected if a fuel like crude oil is burning, because of fractionation. The burning rate is usually expressed as a regression rate: the rate at which the liquid surface falls during the fire. For petrol, kerosene, and many other hydrocarbons the rate is around 3 or 4 mm/min. A 20 m depth of fuel in a tank could therefore burn for three or four days. The rate is higher for more volatile liquids. For butane it is about 8 mm/min, but liquid methane burns at only 7 mm/min, despite the location of the flame very close to the surface. This is because of the low emissivity of the flame and an unusually high latent heat of vaporisation. If the regression rate of petrol, say 4 mm/min, is converted to 4 $1/m^2/min$, the value can be used to estimate the size of a fire resulting from a known leak. Thus, if we know that petrol can be released at 400 l/min, then the area of the burning pool will be about 100 m^2. If methanol is leaking at the same rate, it will feed a pool twice this size: its regression rate is 2 mm/min.

A particularly difficult fire situation results from the release of a flammable liquid at an elevated point on a structure. The liquid streams down over exposed surfaces and burns as a running fire, probably also feeding a pool fire on the ground. The running fire is described as being three-dimensional while the pool is a two-dimensional fire. If a liquid fuel at high pressure is released through an orifice, a spray may be formed. This will burn very readily even if the liquid has a high flash point.

In a pool fire, the heat needed to vaporise the fuel is transferred from the flame mainly by radiation, but with some conduction and convection. When a solid fuel is burning the process is the same, except that sufficient heat must be received at the surface to pyrolyse the fuel, so that it is broken down into flammable vapours. The minimum surface temperature at which a flow of volatiles is maintained which is sufficient to support the flame, is the fire point. For a fuel like wood the heat flux from a single flame may be incapable of maintaining the fire point temperature, and combustion will cease unless heat is supplied from another source. A fire can be maintained in a heap of logs because of the proximity of mutually radiating surfaces, but the flames on a single log which has been removed from the fire will soon be extinguished.

When all of the combustible volatiles have been released from burning wood there remains a porous mass of carbon (charcoal). This may continue to burn by smouldering: direct surface oxidation at a temperature of between about 600° and 750°C. A faint blue flame may sometimes be seen at the surface, but this does not indicate the presence of premixed combustion. It is in fact the colour of a carbon monoxide flame, showing that some of the combustion is proceeding in two steps: $C \rightarrow CO \rightarrow CO_2$. Smouldering may also happen in fuel which is capable of supporting flaming combustion.

If a layer of dust or particulate matter is ignited at one edge, smouldering may be established and will penetrate into the layer at about 5 to 10 mm/min. Again the temperature at the smouldering surface is 600°C or more. Heat transfer from this zone into the unburned fuel causes a temperature gradient. As the material close to the reaction zone is heated to a temperature above about 200° or 300°C, pyrolysis begins. Flammable volatiles are then released, but the flow is insufficient to support a flame. When pyrolysis is complete, a porous charred layer is left into which the smouldering can propagate. Air which is feeding the reactions must diffuse through a layer of ash against the outward flow of combustion products. If there is a forced airflow towards the smouldering surface, there will be an increase in reaction rate and temperature, and the flow of volatiles may then be sufficient to support a flame (a sufficiently high airflow might also disperse the fuel and cause a dust explosion). If a heap of loosely packed materials like shredded or crumpled paper is involved in a fire, smouldering will soon be established within the heap. This is called a deep-seated fire and in some situations it may be particularly difficult to extinguish.

The main characteristics of diffusion flames may be summarised like this:

- the reactants are always gaseous;
- the reactants are mixed in the reaction zone by diffusion;
- diffusion is rate-controlling;
- the reaction is almost invariably stoichiometric;
- the flames are usually luminous due to the presence of carbon particles;
- large flames are usually open at the top and are releasing soot;
- flames must always be located close to the source of fuel;
- most fires comprise large turbulent diffusion flames;
- some premixing may be present in diffusion flames, and may help to stabilise them.

10.4.2 Fire suppression

There are several ways by which diffusion flames can be extinguished. At one time the concept of the 'triangle of fire' was used to illustrate the common methods of firefighting. The three sides of the triangle are fuel, air and heat. If any of the sides can be removed the triangle will collapse and the fire will go out. Fuel is removed by starving, air by smothering and heat by cooling. An examination of the methods described below will show that the triangle does not represent the whole story (unless it can be provided with five sides)

Cooling to the fire point

Most industrial and domestic fires are extinguished by the application of water to the surface of burning solid fuels. When the surface temperature is reduced to below the fire point, flaming combustion will cease. Fires involving liquid fuels with fire points above about 60°C can also be extinguished in this way, by the gentle application of water spray to the surface.

Surface sealing

If foam is applied to the surface of a burning liquid it prevents the penetration of both radiation and air to the surface. The surface of burning solid fuels can also be sealed by the application of specially formulated powders which either melt or sinter. Certain powders are suitable for application to burning metals. A sufficiently thick layer of dry sand, or common salt, will also extinguish a metal fire. Since surface sealing separates the two reactants it can be regarded as both 'starving' and 'smothering'.

Thermal loading

As with premixed flames, the temperature of a diffusion flame can be reduced to a critical value if a sufficiently high concentration of an inert gas is present in the air feeding the flame. The cup burner can be used to measure the extinguishing concentration for flames of both liquids and gases and, with slight modification, of solids also.

Chemical inhibition

The flame extinguishing concentration of chemical extinguishants like the Halons are very much lower than those of inert gases. This is because they act by interfering directly with the chemical reactions in the flame. Chemical inhibition is also achieved by certain finely divided powders, particularly potassium salts. These function by presenting a chemically active surface, within the reaction zone, on which chain reactions are terminated.[3]

Flame stretch

If a flame is distorted so that the reaction zone becomes thinner, the residence time of the reactants in the zone may become too short for the completion of the reactions. If the temperature then falls to a critical value, combustion will cease. This is what happens when the flame on a match is blown out, or when an oil well fire is extinguished by a high explosive charge. The physical removal of the flame from its fuel supply also contributes to the effect.

The flames of burning gases can be extinguished by the chemical extinguishants or, less effectively, by thermal loading using carbon dioxide. However, once the flame has been removed, an explosive atmosphere may be created by the escaping gas. It may be better to use water to cool surrounding plant, and to prevent fire spread, until the leak can be isolated.

The only effective way of dealing with a large pool fire is by the application of a layer of foam, which will also seal the surface against re-ignition.

Small spills can usually be readily extinguished by the chemical extinguishants and possibly also by carbon dioxide, in addition to the use of small foam-making equipment. Foam cannot normally be used against a three-dimensional (running) fire, although a considerable degree of control can be obtained by a water deluge containing AFFF (aqueous film forming foam).[42] Running fires can best be tackled by using the chemical extinguishants, but there is obviously a limit to the size of fire which can be put out by normally available equipment. Again it may be necessary to cool a structure until the leak is isolated, but less damage will be inflicted if AFFF can be added to the water.

We have already noted that burning solids can be readily extinguished by the application of water to the surface. The flames can also be removed by thermal loading and chemical inhibition, but the surface may then remain sufficiently hot to re-ignite. However, liquid agents like the Halons do provide some cooling. They may also penetrate into a particulate heap and inhibit a deep-seated fire. As we have seen, certain of the powder extinguishants are designed to flux or sinter on the hot surface, so that they exclude the air, or they may chemically inhibit smouldering reactions. Also, some powders are specially formulated to provide surface sealing on burning metals: most of the common extinguishants will react violently with the hot metal.

10.4.3 Portable extinguishers

Records kept over many years have shown that on well-run industrial sites no more than 5% of all fires are put out by the fire brigade.[39] The remainder have been successfully tackled by the person on the spot using first-aid firefighting equipment. The vital importance of portable extinguishers, hose reels, and similar appliances, cannot be over-emphasised. Adequate, appropriate and well-maintained extinguishers, and people who know how to use them, are the first line of defence against disaster. Most fires are small when they start, but even quite large ones can be extinguished by a determined person who knows their own capability, and that of the first-aid equipment. The largest pool fire that it is possible for a person to approach without being scorched can be put out by a portable extinguisher, if it is filled with the most efficient powder. Fires larger than this will require foam-making equipment, which could still be used by plant personnel, or may be extinguished by automatic systems.

Regular inspection and maintenance of extinguishers is essential, to ensure that they will operate correctly when they are needed. In particular, powder may become compacted in extinguishers which are exposed to vibration, and these may have to be replaced every few months. This exchange can be combined with training: the returned units are discharged against practice fires, so that their reliability after a certain exposure on the plant is checked, and so that the operators can gain experience in their use. Everyone on the site, without exception, should have regular training with the equipment which is provided in their place of work.

It is important that the most effective extinguishant is used against a particular fire. Thus water, which may be readily and continuously available from a hosereel, cannot be used against a hydrocarbon pool fire: the fuel would float on the water and this could simply increase the area of the fire. To help in the selection of the appropriate appliance, there is a European standard classification of fires.[43]

- *Class A*. Fires involving solid materials, usually of an organic nature, in which combustion normally takes place with the formation of glowing embers.
- *Class B*. Fires involving liquids and liquefiable solids
- *Class C*. Fires involving gases
- *Class D*. Fires involving metals

The American standard[44] differs from this in that Class B includes both liquids and gases, and Class C comprises fires which involve energised electrical equipment (against which the use of water or foam might be hazardous). Classes B and C in the CEN (European) standard are really very similar because all flaming combustion is necessarily in the vapour phase irrespective of whether the fuel is gaseous or liquid. When CEN decided to discontinue the use of the earlier classification (which was the same as the American one) it is unfortunate that the opportunity was not taken to leave liquids and gases in Class B and use Class C for fires which can be extinguished by surface sealing.

The American standard on portable fire extinguishers[44] includes a recommendation, which appears to have been adopted by all manufacturers, for suitable markings. The units which can be used against Class A fires display a large letter A in a green triangle, together with the words 'ordinary combustibles'. Other markings are a B in a red square, a C in a blue circle, and a D in a yellow star. British manufacturers have preferred instead to indicate the contents of an extinguisher by its colour: red for water, cream for foam, black for carbon dioxide, green for Halon, and blue for powder. Once people have received adequate training, either system will help to ensure that they choose the right extinguisher for a particular fire.

Unfortunately not all European manufacturers have adopted the colour coding, and some confusion has resulted. This was resolved in 1996 by the issue of a new British and European standard, BS EN 3, which requires that all new extinguishers are coloured red (although existing colours can remain in use until the units are time-expired). At the same time BSI issued BS 7863 which allows manufacturers to use coloured panels, in the familiar livery, to indicate the contents.

The performance of the different extinguishing agents against the European fire classes, and two others, are indicated in Table 10.2. In the table the performance of each agent is assessed as poor (P), good (G) or very good (VG). No means that the agent should not be used. The word

Halon indicates the possible continued use of the existing Halons, or the use of ozone-friendly replacements with similar properties. It should be noted that although foam is shown as having a very good performance against Class B fires, this applies only to two-dimensional and not to three-dimensional fires. People fires are the particularly distressing incidents in which a flammable liquid has been spilled on someone's clothing and has been ignited. The traditional fire blanket is very difficult to use effectively, but the flames can be extinguished in a few seconds by the prompt application of powder or Halon. It may be unpleasant but not harmful to breathe in a powder, but exposure to Halon vapour should be minimised, and any wetted clothing should be removed.

The performance of an extinguisher can be measured by finding the largest of a series of test fires which can be reliably extinguished by a skilled and fully protected firefighter. The unit is then awarded a rating which depends on the area of the fire. Rating tests have been conducted by the Underwriters' Laboratories (UL) in the USA since 1965.[45] For Class A fires they use a series of nine cubical wooden crib fires, five wood panels and five wood-wool ('excelsior') fires. Their B ratings are obtained from 15 square metal trays of different areas, containing n-heptane. The largest of these is 1600 ft^2, which provides a 640-B rating. It has been established by testing that the skilled operator can put out a fire which is 2.5 times as large as that which can be successfully tackled, using the same extinguisher, by an unprotected ordinary person. The 640-B rating is therefore obtained by dividing 1600 by 2.5. When it was decided to apply ratings to European extinguishers, CEN (predictably) developed a completely different range of test fires.[46] The B fires are burned in a series of round trays and the rating numbers indicate the volume of fuel, in litres, which is used in each tray. The smallest fire uses 8 litres, then the series continues with 13, 21, 34, 55 litres, and so on, with each number being the sum of the previous two numbers (55 = 34 + 21). The fuel area, in square metres, can be found by dividing these numbers by 31.8. It was originally decided that the tests would be conducted by a skilled but completely unprotected operator, so that the highest rating would in fact have applied to the personnel rather than the

Table 10.2 Extinguishant performance against different fires

Fire classification (European)	*Water*	*Foam*	*Carbon dioxide*	*Halon**	*Powder*	*Special powder*
A	VG	G	P	G	P	VG
B	NO	VG	G	VG	VG	
C	NO	NO	G	VG	VG	
D	NO	NO	NO	NO	NO	G
Electrical	NO	NO	G	VG	VG	
People	P	NO	P	VG	VG	

* Original, or similar replacement

extinguisher. Fortunately this idea was dropped. The number of extinguishers, and their ratings, which are required for a particular risk can be found from information in a BS code of practice.[47] For example the minimum total rating for Class B fires, assuming inexperienced operators, is found by multiplying the surface area of the fuel, in square metres, by 80. Since the rating derived from the test fires is only 31.8 times the area, this represents a safety factor of just over 2.5, which is very close to the UL factor. If the required value was found to be say 100, then this could be met by using a single extinguisher with a 113B rating, or two extinguishers each with a 55B rating. However, the code also gives detailed advice on the minimum number of extinguishers and their minimum rating in particular situations.

Ideally an operator should be able to see an appropriate extinguisher from anywhere on the plant, and should not have to travel more than 30 m to get it. The extinguishers should be mounted on stands or brackets with the handles about 1.0 m from the floor. In a place like a warehouse or store it might then be difficult to see the extinguishers, so a conspicuous notice should be provided above them, with suitable wording (FIRE POINT for example). If there are different risks in the same area, then there should be different types of extinguisher at the fire point. There should also be a manual call point [10.6.1] and, where appropriate, a hosereel.

10.4.4 Automatic systems

In 1812, British Patent No. 3606 was granted to William Congreve for a fire extinguishing system which he had installed in the Theatre Royal in London. The system comprised a network of pipes throughout the theatre, connected through a pump to a large tank of water. If a fire started in any part of the theatre, the appropriate valve was opened and the water was then discharged through half-inch holes in the pipes. The advantages of a system like this were quickly recognised. A number of inventors realised that the discharge would be automatic if the network was connected to a gravity tank and if the holes were plugged with a low melting point material. Systems like this contained the essential features of a modern sprinkler system. The earliest practical sprinkler head, which combined a heat detector with a nozzle, was produced in the UK by Stewart Harrison in 1864. Very little interest was taken in what was an excellent design[42] and much of the early development of sprinklers began some time later in the USA. Here, there had been a substantial rise in insurance premiums following the Chicago (1871) and Boston (1872) fires. A piano manufacturer, Henry Parmelee of Connecticut, was determined to get his premium reduced by installing a sprinkler system, and after many attempts, in 1875, he devised a suitable head. This comprised a nozzle over which a brass cap was attached by a solder which had a melting point of 70°C. Its response time must have been longer than that of the Harrison design, but later an improved version was made, and by 1882 some 200 000 sprinkler heads had been sold in the

USA and the UK. Henry Parmelee did get his insurance premium reduced, as did other people who used his systems. Quite substantial reductions can still be obtained, but clearly the insurers need to be certain that the systems are adequate and reliable before they accept a reduced premium. The earliest fire research laboratories were established by groups of insurance companies, in close collaboration with the sprinkler manufacturers, in order to study the performance of sprinkler systems. Methods of testing the heads were developed, and in particular detailed sets of rules were prepared covering the design, installation and maintenance of the systems. Uniform rules for the whole of North America were drafted by representatives of 20 insurers. This collaboration resulted in the formation of the NFPA and the issuing of 'Sprinkler Equipment Regulations', the modern equivalent of which is the NFPA National Fire Code 13: Standard for the installation of sprinkler systems. In the UK, the first rules were prepared by John Wormald of Manchester in 1885. Three years later they were adopted by the Fire Offices' Committee. The first of the FOC's own rules were issued in June 1892 and they have been continuously revised ever since. The 29th edition of the rules was published in 1969 but responsibility for their revision passed to the Loss Prevention Council on its formation in 1985 (from the merger of the FOC with the FPA and FIRTO, the Fire Insurers Research and Testing Organisation). Meanwhile BSI had produced a code of practice for sprinkler installations, based very much on the FOC rules, in 1952 (CP 402.201). This was superseded by Part 2 of BS 5306 in 1979, and complete revision of this standard, embodying all of the FOC requirements, was issued in 1990. The latest LPC rules therefore comprise BS 5306 together with a number of LPC Technical Bulletins. Compliance with all of these documents is necessary, although some departures from them may be accepted by insurers.

The continuing need for frequent amendments to the rules results from very active research and development by both the insurers' laboratories and the manufacturers. This work is aimed partly at improving the reliability and effectiveness of sprinklers, but is driven particularly by the new demands which are placed on the installations. Recent developments have included systems for use against flammable liquid fires; water/foam sprinklers; zoned systems in which a group of heads are opened at the same time; the protection of high-rack storage; the use of plastic pipes; less obtrusive sprinklers for use in, for example, historic buildings; and fast-acting heads which are of particular value in domestic systems. However, it must be remembered that sprinkler systems were designed to last. Many existing systems which are still completely reliable and effective may have been installed in compliance with earlier editions of the rules. Copies of these editions are available should it become necessary to modify or extend these old systems.

Perhaps it should be emphasised that in a conventional sprinkler system each head is a separate detector and actuator. The heads open one by one as they become exposed to the hot gases rising from the fire. They are relatively insensitive detectors and will not normally be actuated unless the flames are between one-third and one-half of the ceiling height. In the most popular design, a glass bulb containing a liquid is clamped tightly against a valve which seals an orifice. An increase in temperature above a predetermined value causes the liquid to expand so as to shatter the bulb. A jet of water is then released from the orifice and breaks into droplets as it strikes a simple impact plate. Droplets are formed over a wide range of sizes. The smaller drops never reach the floor and are carried away in the plume of hot gases. They do, however, perform the useful function of cooling the gases so that they are less likely to cause an ignition. The medium sized drops will probably not penetrate into the fire, but will cool surrounding materials and inhibit the lateral spread of fire. Only the large drops, probably falling from the wetted ceiling, will penetrate the flames and cool the burning surfaces. The fire may not necessarily go out, but the system will have performed three vital functions: to detect the fire, to raise the alarm, and to prevent the spread of fire (or at least to reduce considerably the rate of spread). However, fires differ greatly in size, intensity, rate of burning and rate of spread. Different water application rates are therefore needed for different risks, and the total flow of water, which depends on the number of heads that have opened, will also vary. The required application rate for a particular risk is called the *minimum design density*, expressed in mm/min. Thus the lowest design density which can be used in houses, hotels, schools and similar risks is 2.25 mm/min, or 2.25 l/m^2/min (originally 0.05 gal/ft^2/min). The high-piled storage of hazardous goods requires the highest density of 30 mm/min. There are 13 different grades of sprinkler systems which are intended to deal with the range of fire risks between these two extremes. The extra light hazard class, ELH, is the one delivering 2.25 mm. The ordinary hazard class, OH, includes four groups and covers the handling and storage of ordinary combustible materials which are unlikely to burn intensely: this includes the majority of commercial and industrial risks. Abnormal fire loads in which intense fires can develop rapidly are in the extra high hazard class, EHH. Process hazards in this class are divided into four types, and high-piled storage hazards into four categories.

For each of the total of 13 grades there is an assumed maximum area of operation, AMAO, which is based on the number of sprinklers which are expected to open during a fire. The rules also specify the maximum area to be covered by each sprinkler and the maximum spacings between sprinklers. Within these limits, the design density can be achieved by varying the flow through the heads and the spacing between them. The requirements for the ELH class are: AMAO, 84 m^2; maximum area per sprinkler, 21 m^2 (it is therefore assumed that up to four heads may open); design density 2.25 mm/min; maximum spacing, 4.6 m. At the other end of the range, for EHH process hazards, the requirements are: AMAO, 260 m^2; maximum area, 9 m^2; density, 7.5 to 12.5 mm/min; spacing 3.7 m. The size of the orifice which is opened in a sprinkler is usually 10 to 15 mm in diameter for ELH, 15 mm for OH, and 15 to 20 mm for EHH.

Detailed information on the mechanical design of a system is provided in the rules. The size of the pipes which are used in a system must ensure that an adequate flow of water at the right pressure is available at the most remote group of heads. The connection between the system and the water supply is through an alarm valve. As the name indicates, when the valve opens to supply water to one or more sprinkler heads, it also allows a flow into a small turbine which rotates a hammer against a gong. This alarm is located outside the protected building. In addition to this, it is usual to find flow switches in the various branches of the system so that the alarm, and the location of the fire, can be fed into an electronic system. The capacity, number and reliability of water supplies provide a grading for each installation, which can be either 'single', 'duplicate' or 'superior'. The rules indicate the appropriate grading for the class of installation. For example, a 'single' connection to a town main water will not normally be accepted for an EHH system.

Most systems are full of water right up to the sprinkler heads so that water can be discharged from the moment that the head is actuated. These are called wet systems. In dry systems there is initially no water because of the danger of freezing. Dry systems contain compressed air which is released when a sprinkler is actuated. The drop in pressure allows the alarm valve to open and water is admitted into the system. There is inevitably a delay before the water can reach the open sprinkler but this can be reduced by using a 'pre-action' system. Again the system contains compressed air, but it is released when the fire is detected by a separate, more sensitive, detector system. Water is then already in the pipework behind the sprinkler before it opens. In areas where there is a danger of freezing for only part of the year, an 'alternate wet and dry' system can be used. Each of these systems requires a different design of alarm valve. Details are provided in the rules of the inspections, tests and maintenance work which must be completed at regular intervals: daily, weekly, monthly, quarterly, half-yearly, yearly, and then every 3 and 15 years. The rules, both British and American, may appear to be confusing if they are read piecemeal, particularly if some of the jargon is unfamiliar. Like most standards, these documents are best understood if they are read straight through, and then a return can be made to the most appropriate parts. Some guidance can also be found in books on fire protection.[39,42,48]

Total flooding systems have already been mentioned. An extinguishant, which may be a gas, or a very volatile liquid, is stored under pressure and then discharged into an enclosure through an array of nozzles. The whole of the atmosphere in the enclosure then contains a critical concentration of the agent. If the system is required for a simple fire risk (and therefore only diffusion flames can be present) then the design concentration of the extinguishant can be obtained by adding a safety factor to the flame extinguishing concentration, which is measured by a cup burner.[40] However, it is possible that, after the fire has been extinguished, sufficient flammable vapour will be released from the hot fuel to form an explosive atmosphere.

An ignition (on glowing embers, for example) will then result in the propagation of a premixed flame, which will require a factored inerting concentration [10.1 and 10.3.3] for its suppression rather than the lower value of the diffusion flame extinguishing concentration. It would clearly be safer to design all systems to produce the higher concentration, and this is always done with carbon dioxide systems. Carbon dioxide is a relatively cheap agent but when standards were prepared for systems using the Halons, which are much more expensive, it was accepted that the lower concentration could be used in situations where it could be demonstrated that only diffusion flames could be present.[26,27]

Exposure to high concentrations of carbon dioxide can be rapidly fatal: the gas is toxic and not simply, like nitrogen, an asphyxiant. Doors into buildings protected by these systems are invariably locked whenever the systems are on automatic operation. Coloured lights are sometimes displayed at the doors to indicate the status of the system: live, locked off, or discharged. A permit-to-work procedure should be used to ensure the safety of people who need to work within a protected enclosure. The Halons, and some of their replacements and substitutes, are much less toxic than carbon dioxide, and this has led some manufacturers to claim that they may safely be discharged from a total flooding system into an occupied area. In some cases the claims are backed by the results of human testing, and by demonstrations. However, it is difficult to understand how any discharge into an occupied area can be justified. First, there appear to be no records of any tests involving infants, the elderly, or people suffering from asthma or heart diseases. Second, any lipid-soluble vapour can sensitise the heart to the effects of adrenalin, and this can result in cardiac arrhythmias, which can be fatal. Halons and similar materials have lipid-soluble vapours (as indeed does petrol). Third, we were designed to breathe fresh air and should not be exposed to high concentrations of anything else. Fourth, and most importantly, there is nothing to be gained from discharging a system when people are there. Almost all fires are small when they start, particularly in normally-occupied areas, and people are much better at detecting and extinguishing an incipient fire than is an automatic system. A fire in a waste paper basket can be dealt with quite readily with a glass of water and does not require 100 kg or so of an expensive extinguishant. It is just possible that a large flammable liquid spill could happen in exceptional circumstances in an occupied room. An automatic discharge might then be needed to control a fire, but clearly this situation should be avoided whatever the cost. One of the very few places in which people would have to remain during and after a system discharge is the missile control room of a warship in action. There is no good reason why a normal industrial system should be discharged when people are there: these systems should be set to 'manual control only', when the area is occupied. When people have left the area and the doors have been secured, the system can be switched to automatic operation. Even so, a pre-discharge alarm can be provided, in the unlikely event that someone is there inadvertently when the system goes off. In some places, like a computer room, the system lock-off can be combined with a security lock.

A local application system is designed to discharge the agent directly into a fire from one or more nozzles, so that it acts in much the same way as a hand extinguisher. It is not usually necessary to be concerned about the toxicity of the agent, apart from evacuating the area after a discharge. Foam, water, carbon dioxide, powder and Halon-like materials can all be used in these systems. Very little work has been done on critical application rates for agents other than foam and water, apart from that reported in the appendices of the carbon dioxide and Halon systems standards.[26,27] Reliance must be placed largely on the manufacturers' experience for the proper design of these systems.

Local application systems are frequently used to protect oil quench tanks. They may also be used for pumps handling flammable liquids, and in loading/offloading bays. Small isolated tanks of flammable liquids may be provided with automatic foam systems (and adequate bunds). Halon systems are used for the seals of floating roof tanks. Fixed piping may also be provided through which foam can be pumped into the seal area, and similar systems are used to apply foam into fixed roof tanks and into bunds. Water deluge can also be regarded as a local application mode and is used to cool structures, vessels and tanks which are exposed to radiation from a fire. Only the areas which are receiving radiation need to be cooled, and this can be controlled automatically.

10.5 Ignition

Ignition is the process by which fires and explosions are initiated. Smouldering combustion can be established by heating the surface of the fuel to a critical temperature at which the rate of reaction is able to accelerate. All other ignitions are in the vapour phase and the reactants are premixed. In an explosive atmosphere, combustion then continues to be premixed, but in the more usual ignition processes there is a transition from premixed flames to diffusion flames.

If a flammable gas is escaping from a plant, or if a liquid at a temperature above its fire point has been spilled, then mixtures within the limits of flammability can be formed at some distance (possibly downwind) from the leak. An ignition within this zone will establish a premixed flame. This will become the ignition source for the diffusion flame which will be formed around the rest of the plume. This is what happened at the Feyzin incident, already described in 10.1.1 (and illustrated in Fig. 10.5). A pool of flammable liquid at a temperature below its fire point cannot be ignited in this way. It is necessary that at least some part of its surface is heated to this temperature so that a local ignition can be achieved. This can be done by using a wick: perhaps a piece of cloth which is partially immersed in the liquid, the exposed part of which has become wetted by capillary action. If the wick is then heated by a small flame, sufficient vapour can be released to permit a premixed ignition. If at the same time a small volume of the liquid has been heated to a temperature at or above the fire point, a diffusion flame will be established. Heat transfer to adjacent fuel surfaces will cause the flame to spread, and it will eventually cover the whole of the pool. (As it happens, the advancing diffusion flame is preceded by a small premixed flame. This flickers over the surface as the temperature reaches first the flash point and then the fire point.)

In order to ignite most solid fuels it is necessary to apply sufficient heat to cause pyrolysis, and the release of a flow of flammable volatiles which is sufficient to form at least a lower limit mixture. Sustained burning is possible only when the flow is sufficient to support a diffusion flame, which itself is capable of maintaining an adequate heat flux back to the surface. The ignition of a solid fuel is easier if at least some part of it has been reduced to small pieces (which are analogous to a wick). A match will readily ignite wood wool (excelsior) but has no effect on a wood block.

The control of ignition sources is an important part of fire prevention, but it should never be assumed that a hazardous process is safe because there are no apparent sources. It must always be assumed that if a release of flammable gas is possible, then sooner or later, in one way or another, it will be ignited. An adequate level of fire safety can be achieved only when equal effort is applied to the prevention of a loss of containment, and to the elimination of ignition sources. It is unfortunate that the information media tend to regard the ignition source as the only significant cause of a fire tragedy. If people have died in a fire because of overcrowding, too many flammables and too few exits, then the media are very likely to report that the cause was a match (or whatever). When 31 people died in the London Underground fire, a headline in the *Daily Telegraph* said 'King's Cross fire caused by cigarette'. This does give the impression that the blame for the tragedy rests with a careless smoker. However, an examination of methods of ignition is essential to the proper planning of fire prevention.

10.5.1 Sources of ignition

Electric sparks

A premixed flame can be established in a flammable gaseous mixture if a certain minimum volume is heated to a critical temperature. For a near-stoichiometric methane/air mixture the minimum volume is about 4 mm^3, so that if the volume is spherical it has a diameter of about 2 mm. Table 10.1 shows that this is in fact also the quenching distance, but this is to some extent coincidental.[4] The table also gives the minimum ignition energy as 0.29 mJ, which is sufficient to heat this volume to about 55°C. Since the autoignition temperature is 595°C, this slight increase will have little effect on the ignition processes. The effectiveness of a spark is due mainly to the fact that in its high temperature plasma there is almost complete ionisation.

This produces excited species, many of which are chain carriers of the combustion reactions. These are released into the flammable mixture in just sufficient numbers to form the smallest viable flame: 2 mm in diameter.

The minimum ignition energy measurements were made using capacitive sparks. More energy would have been needed in inductive sparks, which are usually oscillatory and of longer duration. The gap between the spark electrodes was greater than the quenching distance, and voltages around 5 kV were needed to cause a discharge across it. It might therefore be thought that electrical equipment operating at lower voltages would not be capable of producing an incendive spark. Unfortunately this is not so. Because of the high currents which are usually involved, an arc can be formed when a circuit is broken. Metal is vaporised from hot spots on the contacts, providing a low resistance path through which the current can flow as the contacts are separated. The quenching distance is readily exceeded, and a heavy current will cause the arc to spread beyond the edges of the contacts.

Plant which is handling flammable liquids and gases must inevitably depend on electrical equipment: pumps, heaters, valves, lights, instruments and controls. Methods of reducing the probability of an ignition are discussed in 10.5.2 and 10.5.3.

Static charges

The electrons in the outer orbits of atoms and molecules are less strongly attached than those closer to the nucleus, and may be capable of being detached. If two different materials are touching each other, some of these electrons which are at the surface of one may drift over to the surface of the other. If the materials are then separated, the one with extra electrons will have a negative charge and the other will have a positive charge. Almost everything we do, or that is happening around us, is generating charges in this way. The voltage of the charge depends on the number of electrons and the area of the surface which they occupy, and can be between about 1.0 V and 1.0 kV. The word static is used because the electrons remain on the surface and are not flowing through a normal electric circuit. However, a circuit is present because there are no perfect insulators: electrons can flow slowly from the negative charge to the earth, and then back to the positive surface, the earth acting as an enormous reservoir for all such exchanges. Only a minute amount of energy is involved and the whole process is normally quite innocuous. The trouble starts when the separation of two surfaces is continuous. This happens for example when a petroleum product is flowing through a metal pipeline. The conductivity of these liquids is low, so a charge can be carried with the flow. If the liquid is filled into a metal container which is insulated from the ground, the charge will then accumulate on the metal. In these conditions a current flow into the metal of around 1.0 microA has been measured. If the container is a typical 150-litre drum, the voltage on its surface will increase at about 1.0 kV/s. The energy which is stored in a static charge is given by:

$$E = \frac{1}{2}CV^2$$

where E is in Joules and C is the capacitance (the ability to store electrons) in Farads. The capacitance of a drum is about 100 picoF, and if we assume that after 10 s the voltage is 10 kV, then the calculation becomes

$$E = \frac{1}{2}(100 \times 10^{-12})\,(10\,000)^2$$

This gives a value for E of 5 mJ, which is about 20 times the minimum ignition energy for most hydrocarbon/air mixtures (Table 10.1). Also, the spark resulting from a discharge to an earthed object would be 5 or 6 mm long, so the resulting flame would not be quenched. In a similar situation where powder is being filled into a drum, the current flow can be as high as 0.1 mA. If the drum is made of a plastic material then not all of the charge will be released in a spark, but in the case of the liquid fill only 1/20th of the charge would be sufficient to cause an ignition. The human body can readily acquire a charge (even by walking on a thick carpet). Some industrial activities can result in a 10 kV charge. Since the body's capacitance is about 200 picoF, a 10 mJ discharge is possible, which is potentially hazardous.

Static charges can be generated by the surface contact of solids, liquids or gases in any combination, with the possible exception of gas/gas. However, there have been reports of the ignition of high-pressure hydrogen leaks (from early ammonia plants) which appear to have resulted from a static discharge to earth. These are some examples of the other combinations:

- *Solid/solid*: powder flowing in a pipeline; flow of paper or plastic sheets; movement of conveyor belts; grinding and milling; movement of people.
- *Solid/liquid*: liquid flow through pipelines, filters and pumps; settling of particles (rust for example) in a liquid; splash filling of tanks.
- *Solid/gas*: fluidised beds; pneumatic conveying; movement of aircraft and vehicles; discharge of a carbon dioxide extinguisher.
- *Liquid/liquid*: the stirring of two liquids which are not miscible; drops of one liquid falling through another.
- *Liquid/gas*: liquid sprays (including water); discharge of wet steam; gas bubbles rising through a liquid.

The ancient Greeks found that if a piece of amber is rubbed with a cloth, it will attract small particles. The amber acquires a negative static charge and induces a positive charge on the particles, which are then attracted. The Greek word for amber is elektron.

The most common way in which static charges are generated in industry is by the transfer of liquids. While the liquids remain within an earthed pipeline there is no problem, but their discharge into a vessel can be hazardous. In particular splash filling should be avoided. Even when a hydrocarbon is correctly filled into a tank through a line at the bottom, the presence of accumulated water may cause a liquid/liquid charge generation.

The charge will then collect at the surface of the hydrocarbon and, if the conductivity is low, it will leak only slowly to the walls of the tank. If the rising surface approaches an earthed metal object, a spark may pass. Also, if the tank is dipped using a metal tape, then a spark may pass either from the surface to the tape, or from the tape to the edge of the dip hole. A delay of at least an hour should be observed before a tank like this is dipped. Another common cause of static discharges is the emptying of powders from plastic sacks into batch process vessels. A number of explosions have happened in vessels which already contained a flammable solvent.

There are no perfect insulators so all static charges are being continuously dissipated. If the rate at which a charge is generated is less than the rate of dissipation, the process will be safe. In the case of flowing liquids this can be achieved by limiting the velocity. Thus, when petrol is being filled into road tankers the velocity in the filling arm is restricted to 1.0 m/s until the outlet is covered. Higher velocities become possible by the use of so-called antistatic additives. These do not in fact prevent the generation of a charge but they do increase the conductivity of the liquid. This results in a reduction of the relaxation time: the time needed to dissipate the charge.

The proper bonding and earthing of all parts of a hazardous plant are essential. If an earth line had been clipped to the 150-litre metal drum mentioned above, a charge could not have accumulated. Guidance on the avoidance of hazardous static charges is given in British Standard BS 5958 and in a number of other publications.[4,9,22,48,49]

Mechanical sparks

Mechanical sparks are small fragments of hard materials at high temperatures. They can be formed by the grinding of metal, by shot blasting and when one hard substance impacts against another. High temperatures are needed to cause ignition because of their small size and the short contact time as they pass through the flammable atmosphere. Some metal fragments will react with the oxygen in the air and attain very high temperatures, but they may not then be capable of causing an ignition because they are surrounded by oxygen-depleted air. At one time 'non-sparking' tools were used extensively in the petroleum industry. They were made of materials like copper, beryllium–copper, aluminium bronze and brass. They were never completely safe because a piece of grit embedded in a hammer head, or in a tool which was dropped, could still cause a spark. Also the spanners were dangerous to use because they could slip off a nut which was being tightened hard. These tools are rarely used now, except when it is necessary to work near to a hydrogen leak: it will be rembered that the minimum ignition energy for hydrogen/air is 0.02 mJ.

Large and very incendive sparks are released by the collapse of a steel structure, or by the impact of a crashing car or aircraft. A great deal of energy is released by these events and the ignition of released flammable materials is very probable.

Autoignition

The molecules of a gas are in continual movement, each one hitting another molecule or the walls of a container several billion times a second. Their average velocity is directly related to the absolute temperature, but there is a velocity distribution within the population of molecules. A small fraction will be travelling so fast that, in an explosive atmosphere at room temperature, there will be a number of encounters between fuel and oxygen molecules that are sufficiently energetic to cause them to react. (Indeed, this can happen at any temperature above absolute zero.) The heat which is released by these reactions is rapidly dissipated and has no measurable effect on the ambient temperature. However, if the temperature of the gas mixture is gradually increased, the number of reacting encounters will also increase, as will the released heat. At a certain critical temperature, heat will be released faster than it can be dissipated to the outside. The rate of reaction will then accelerate up to the point where a premixed flame is established. The critical value is the autoignition temperature, AIT. This is measured under standard conditions by introducing a gaseous mixture into a small spherical flask which has been heated to different temperatures, and noting whether, after a brief induction period, a flame can be seen. Liquid fuels are assessed similarly, by placing a measured volume in the flask. The published AIT values apply only to the particular test conditions and ignitions could possibly occur at lower temperatures in unusual situations. For example, the AIT of petrol is 280°C, but the ignition of petrol vapour has been observed inside a rusty exhaust pipe, after a delay of several minutes, at 200°C. However, in general the use of the published values will provide a reasonable margin of safety. The methods used in obtaining AIT values are described in British and US standards: BS 4056 and ASTM D2155. (It should be noted that in the BS the term 'ignition temperature' is used.)

Hot surfaces

In the AIT measurements described above, ignitions occur because the gases are in contact with the hot inner surface of a small flask. There must be some convection during the induction period, but all of the gas will remain in the flask, and so its temperature will increase. The situation is different when a hot surface, perhaps a piece of electrical equipment, is present in a large volume of premixed gases. An ignition at the AIT value will not be possible, unless a small volume of gas is trapped in the equipment. This is due mainly to the flow of gas past the hot surface in convection currents, which reduces the contact time. To this must be added the effects of quenching. Suppose for example that a methane/air mixture is in contact with a hot surface at 595°C, which is the AIT of the mixture. If a premixed flame is established at the surface, its temperature will be 1875°C. Heat losses to the surface will therefore be substantial and the flame will be quenched. It is necessary for an ignition to occur at some distance from the surface (probably in excess of the quenching distance of 2 mm) before sustained combustion is possible. A temperature in excess of the AIT is therefore necessary. Some experiments were made using a 6% methane/air mixture in a cubical container. An ignition was obtained when a metal strip at the bottom of the box was heated to 1066°C. When the strip was moved to the side of the box, where the convective flow would be greater, the minimum ignition temperature was increased to 1150°C.[50] However, methane does have an unusually high AIT. The values are considerably lower for other alkanes, and tend to reduce with increasing molecular weight. Thus the value for butane is 365°C, pentane is 285°C and hexane 233°C. Kerosene has an AIT of 210°C, and carbon disulphide has an exceptionally low value of 100°C. (These values were measured by the method described in BS 4056 and are quoted (with many others) in BS 5345: Part 1.)[51] The mechanisms by which the self-heating of organic materials can happen have already been discussed in Chapter 8. We are concerned here with the conditions under which self-heating can result in smouldering or flaming combustion. In the reports of fires in which self-heating is believed to have preceded ignition, it is usually recorded as spontaneous heating. The cause of the fire will then be listed as spontaneous combustion. Examples of such fires are discussed briefly here.

When the solvents have evaporated from a coat of paint, and the surface becomes touch-dry, the final hardening results from polymerisation reactions of the oils and resins which remain. These are oxidation reactions and are exothermic, but the heat is readily lost to the painted surface and the air. If some spilled paint has been wiped up with rags, which are then thrown in a heap, sufficient oxygen will be present for the same reactions to take place. In this situation the heat will not be readily dissipated, and if it is generated faster than it is lost, there will be an increase in temperature. This will increase the reaction rate, and the process will accelerate. A sufficiently high temperature can be reached for the establishment of smouldering. When this has penetrated to the surface of the heap, flaming combustion can then take over and a substantial fire may ensue. Spontaneous heating due to oxidation reactions is possible when fibrous materials become wetted by a wide range of animal and vegetable oils, particularly the first three of this list: cod liver, fish, linseed, corn, cottonseed, olive, peanut, pine, soya, tung and whale. A number of materials used in particular industries are subject to spontaneous heating, particularly if they are stored in large heaps. This behaviour is usually well known and adequate precautions are taken. The materials include charcoal, animal feeds based on cornmeal, fertilisers, fish meal, manure, the red skin removed from peanuts, rags, roofing felt, scrap rubber and waste paper. The initial heating of substances like fish meal, manure, rags and waste paper may be due to the action of thermophilic bacteria, which can survive in temperatures up to 75°C (and were at one time the cause of many haystack fires).

It is perhaps surprising that the surface of coal can be oxidised at room temperature. This is particularly so if the coal has recently been broken, perhaps by being dropped onto a heap. The heat generated at the surface of a single lump will soon be lost, but if the lump is buried in a heap then we may again have the situation of heat being produced faster than it can be dissipated. There will then be a local increase in temperature leading to smouldering. Large heaps of coal are checked regularly for hot spots, usually by buried thermocouples, or by thermal imagers. However, an experienced person can usually detect the presence of smouldering by the smell. Some coals have a greater affinity for oxygen than others and are therefore more susceptible to spontaneous heating. The probability of an ignition depends on this property and on the size of the lumps, the moisture content, the time since the coal was broken up, the ambient temperature and the size of the heap. In general, lumps larger than about 25 to 75 mm give little trouble, and stacks of less than 200 t and less than 2.5 m high have not been known to ignite.

The great majority of combustible liquids and solids are capable of hazardous spontaneous heating only when the temperature has been increased to a point nearer to their AIT. A common example of this effect is the occurrence of fires in 'oil soaked lagging'. A pipe carrying a hot liquid is usually insulated to prevent heat losses, often by use of a fibrous material like mineral wool. A leak at a flanged joint or from a valve may cause the liquid to soak into the lagging. As the liquid spreads, individual fibres become coated with a thin film. A very large area of liquid is then in contact with the trapped air, at the temperature of the pipe. If the liquid is combustible, the rate of oxidation may eventually be sufficiently high to ensure that heat is generated faster than it is lost. An accelerating temperature increase can then lead to smouldering (an interesting situation in which a liquid fuel is involved). As the smouldering penetrates through the lagging, the outer protective layer may begin to burn. The delay between the leak and the appearance of a flame can be a week or more. Ignitions have not been known to happen when the pipeline temperature is less than 100°C. Liquids with high molecular weights, like heavy fuel oil, ignite more readily than for example the lighter petroleum fractions.

Radiation

The radiation received at ground level from a flare stack can be surprisingly high. People working in petrol refineries or an offshore platforms are sometimes worried that an escape of flammable gas could be ignited by the radiation. This is not possible because gases are completely transparent to all but a small part of this radiation. The amount that they do absorb is at characteristic wavelengths by which the gases can be identified, but which has a negligible effect on their temperature. Most liquids are also almost completely transparent, but solid fuels can accept the whole spectrum of thermal radiation (unless the surface is a reflector), with an appreciable increase in temperature.

A typical wooden surface can be heated to its fire point, of 350°C, by a radiation flux of about 12 kW/m^2. If a small flame is applied to the wood it will start to burn, so these are the conditions necessary for the 'pilot ignition' of wood. The AIT, at which combustion will start in the absence of an ignition source, is about 600°C, which can be produced by a flux of 28 kW/m^2. These values can be compared with the maximum value for UK sunshine, about 0.7 kW/m^2, and the flux which will cause pain after a 3 s exposure, which is about 10 kW/m^2. It is therefore reasonable to assume that if the radiation is painful it is also hazardous. It should also be noted that if the sun is focused through a 100 mm diameter lens to form an image 5 mm in diameter, then the radiation intensity will be increased about 400 times (ignoring any losses in the glass). The flux then becomes 280 kW/m^2, which is well above the critical value for the autoignition of most fuels.

It is sometimes necessary to calculate the flux which will be received by a combustible surface from a flame. This could be done if for example a wooden structure is to be placed close to a flare, or if we need to know the minimum safe separation distance between buildings in order to prevent the spread of fire. It is sometimes sufficient to assume that the radiation is coming from a single point within a flame, but in more accurate calculations the flame is assumed to be conical, cylindrical, or spherical. Unless the radiating and receiving surfaces are parallel, a configuration factor (or view factor) must be used. It is usually sufficient to calculate the rate of release of heat from the fire and then to assume that one-third of this appears as radiation. Detailed advice on making these calculations, and tables of configuration factors, will be found in various publications.[2,22,52,53]

Radio transmissions

The use of portable radio transmitters is banned whenever explosives are being used. The reason for this is that the wire leads connected to an electric detonator could act as a dipole (or a fraction or a multiple of one) which would resonate at the transmission frequency. The wire which is within the fusehead could then be at the centre of the dipole, where the current flow may be sufficient to fire the detonator. A similar current, oscillating at the transmission frequency, could be induced in any other piece of metal, even if some part of it is earthed. The current flow would then be incapable of causing any significant temperature increase, but sparking from the end of a fortuitous dipole to earth would just be possible. It is very unlikely that such a spark could be incendive. However, if the plant is close to a broadcasting transmitter, or if mobile radar equipment has been brought near to it, it is possible that some part of the plant could act as a tuned circuit (again despite earthing) and some sparking could then occur.[4]

Compression

In a diesel engine, the rapid compression of a fuel/air mixture in the cylinder causes the temperature to increase to a value above the AIT. Ignition by compression is also possible in for example a pipeline which has not been adequately purged before maintenance work is done. If a volume of explosive mixture is present at a closed end of the line, it may be ignited by a rapid increase in pressure when the plant is brought on line.

Lightning

In the North of Scotland, the average frequency of lightning strikes is 1.0/km^2/decade and this increases to 7.0/km^2/decade in south-east England. (The numbers have to be averaged over a long period because of the variation from year to year due to the 11-year sunspot cycle.) Both rates are sufficiently high to require the proper protection of plants which are handling hazardous materials. There is a similar distribution in the southern half of Canada and most of the USA, but in the South, and particularly in Florida, the frequency can be five times greater than the maximum elsewhere.

The ignition hazard arises not simply from a direct strike, but also from the currents which can be induced in metal structures at some distance from the strike. If plant is not properly bonded and earthed, sparking can occur at any gap in a route to earth. The current in the lightning stroke can be taken to earth through a suitable conductor. Although both the voltage and current are extremely high, the duration is short and the temperature increase of the conductor is negligible. Very detailed guidance on estimating the probability that a structure will be hit, and on the provision of adequate protection, is given in BS 6651.[54] Detailed practical guidance can also be found in the *NFPA Handbook*.[48]

Pyrophoric materials

If a cloud of extremely fine metal powder is released into the air it may ignite and burn in a spectacular flash. Materials which behave like this are said to be pyrophoric. The only such material likely to be encountered in industry is iron sulphide. If petroleum products are being stored and processed, and particularly during the refining of crude oil, hydrogen sulphide and other sulphur compounds which are present can react with rusty steel, or even with the steel itself. The iron sulphide which is produced is a very fine powder which can be pyrophoric. If a storage vessel has been drained, the sludge which is exposed at the bottom may contain a high proportion of the sulphide. When this begins to dry, smouldering can be established which can ignite any remaining petroleum deposits. When natural gas is used to feed gas turbines, it is usually passed through a filter to remove any grit which could abrade the turbine blades. Care should be taken when the filters are opened up for cleaning in case there is a pyrophoric ignition.

Chemical reactions

Many chemical reactions release sufficient heat to be capable of causing ignitions when the reactants are in contact with a fuel. The heat of decomposition of some unstable materials can also be sufficient for these materials to be sources of ignition. Chemical reaction hazards are discussed in some detail in Chapter 8, but mention should be made here of the thermite reaction, which has been the ignition source in a number of incidents:

$$2Al + Fe_2O_3 = Al_2O_3 + Fe$$

Temperatures up to 2500°C can be achieved by this reaction, which is sufficiently high to melt the iron which is one of the reaction products. Thermite mixtures, usually with some additional iron or steel powder, are used to weld steel. A pyrotechnic device is needed to ignite the mixture. However, the significance of this reaction is that when aluminium paint has been applied to rusty steel, both of the reactants are present and are in close contact. After the paint has dried, a glancing blow with a hammer can release a shower of incendive sparks.

10.5.2 Classification of hazardous areas

When flammable liquids and gases are being processed it is almost inevitable that electrical equipment will be present. On a small plant it may be possible to arrange that a flammable liquid pump is driven by an electric motor to which it is connected by a shaft passing through a sealed opening in a wall. This is not practicable on most sites and it must be accepted that an item of plant which is very likely to develop leaks is located close to electrical equipment in which there is a high probability of incendive sparks. Clearly it is necessary to make the plant as secure as possible against a loss of containment and also to render the electrical equipment acceptably innocuous. The concept behind the technique of area classification is to determine the probability that a flammable atmosphere will be present, to decide the extent of the hazardous area and then to select electrical equipment for use in that area which has an acceptable probability of causing an ignition. Thus, in an area which contains only well constructed and maintained all-welded pipelines, electrical equipment which has a relatively high probability of sparking can be used. On the other hand, in the vapour space of a methanol storage tank the only acceptable equipment would be quite incapable of producing an incendive spark.

Standards concerned with area classification and 'safe' electrical equipment have been published by a large number of organisations.[55] Examples are BSI, IP, HSE and ICI/ROSPA in the UK; NFPA in USA; CPP in France; DNV in Norway; SS in Sweden; BCI in Germany; DGL in the Netherlands; and ASA in Australia. Perhaps the most widely used standard in the UK (and in many other countries) is that produced by the Institute of Petroleum.[9] Part 1 of their Model Code was published in 1946 and contained the earliest definition of safe and hazardous areas. It was revised several times up to the 1965 edition, which was reprinted, with several amendments, up to 1982. A completely revised edition based on new concepts was issued in 1990 as part 15 of the code. In any company where the IP Code is used, it should be assumed that any plant which has been designed since March 1990 will be classified in accordance with the new code. Some of the requirements of the code are less stringent than those in previous versions but, particularly where LPG is handled, it may be advisable to check the zone dimensions of earlier plant against the new provisions.

The code is much more detailed than any other standard and does take into account the more general guidance provided by the European Committee for Standardisation (CEN), the International Electrotechnical Committee (IEC), and in particular the British Standards Institution (BSI). BS 5345[51] (which is in nine parts) contains guidance on the classification of hazardous areas and on the selection, installation, and maintenance of suitable electrical equipment (called 'electrical apparatus' in the standard). The equipment (apparatus) is specified, again in nine parts, in BS 5501.[56] Similar guidance on equipment and on area classification is contained in NFPA Codes 70[57] and 497.[58] Here we will be discussing mainly the 1990 IP Code, and the two British standards, since these between them provide the most recent and most comprehensive approach to the subject. A superficial examination of, in particular, the IP Code will reveal what appears to be a complicated and confusing document. However, if the code is read carefully from the beginning it will be seen to provide a logical and clearly written exposition in which the authors have taken great care to explain in advance exactly what is being discussed in each section. In both the code (Section 1.4) and BS 5345 (7.1) it is advised that area classification should be undertaken by an experienced team. The team should comprise, or should be able to consult with, people who have full knowledge of the process and equipment, of safety and loss prevention, and of electrical engineering. Experienced operational and engineering judgement is particularly necessary in assessing the 'grade of release', which is concerned with the frequency and duration of leaks. The team may also be able to decide that some areas are less hazardous than would be indicated by a strict application of the code. For example, the area mentioned above containing only all-welded pipelines may well be accepted as being non-hazardous. The results of a catastrophic failure of a tank, which would release a large volume of flammable liquid, can usually be ignored in determining the dimensions of a hazardous area, because this is a very unlikely event. Flammable gases and liquids are released continuously into furnaces but it would be unreasonable to classify the furnace in strict accordance with the code. Also, no classification is needed if a flammable liquid cannot be heated to a temperature above its flash point, and cannot be released as a spray; nor for any area where the total release of any liquid is less than 50 litres.

It is important to emphasise that the discussion on the IP Code which follows provides only a brief outline, and describes only its more significant aspects. The wording of the code should be studied carefully in order to determine the interpretation, relevance and precise application of each requirement before an area classification is attempted.

A loss of containment (a leak) is called a release in the code and it has already been noted that the grade of release depends on the frequency and duration of the leak. Three grades are recognised:

- continuous grade, in which the release is continuous or nearly so;
- primary grade, in which a release is likely to happen either regularly or at random times during the normal operation of the plant;
- secondary grade, in which a release is unlikely to happen and if it does it will be of limited duration.

As a rough guide, a continuous release is likely to be present for more than 1000 hours a year, and a primary release for between 10 and 1000 hours a year. A secondary release should not be present for more than a total of 10 hours a year, and then only for short periods. It should be noted that in determining the grade no account is taken of the rate of release. This is considered later when the dimensions of the hazardous area are assessed. In the hazardous area, three types of zone are recognised, depending on the probability of the presence of a flammable atmosphere:

- Zone 0, in which an explosive atmosphere is always present, or is present for long periods;
- Zone 1, in which an explosive atmosphere is likely to occur in normal operation;
- Zone 2, in which an explosive atmosphere is not likely to occur in normal operations, and if it does it will exist for only a short time. (It has been suggested that 'a short time' could be no more than a total of 10 hours a year.)

Clearly the three zones can be directly related to the three grades, but this is not always the case. For example, if a pit exists in a Zone 2 area into which a heavy vapour can be released, then the pit must be Zone 1 because the vapour can collect in the pit and remain there for more than a short time. Figure 10.12, which is taken from the IP Code, shows how zones should be established around a fixed roof tank containing a liquid at a temperature above its flash point (petrol for example). The vapour space is Zone 0 (even though with petrol the mixture would normally be over-rich). In normal operations vapour will be released from vents or other openings in the roof, so a Zone 1 is shown here to a height of 3 m. Heavy vapour falling from the edge of the roof will soon be diluted, so the zone extends to only 2 m below the edge. Beneath this, and in the bunded area, an explosive atmosphere is unlikely to occur in normal operations, so this is Zone 2 up to the height of the bund wall. However, heavy vapours could collect in the pit on the left of the tank and, as explained above, this must be a Zone 1 area.

Figure 10.13 shows the Zone 1 area around a petrol tanker when the hatches are open and it is being filled from the top. The dimensions of the zone may be surprisingly small but they are based on experimental data.[9] It

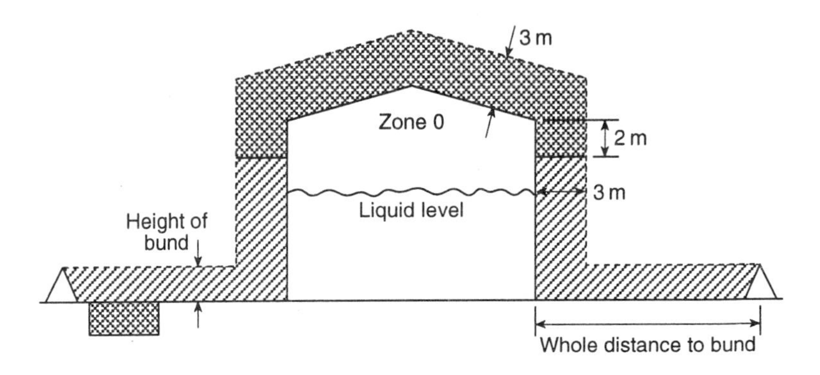

Note Because of the possibility of mist, spray or foam formation the ullage space of Class II(1) and III(1) tanks should also be regarded as *Zone 0*. It is recommended that the area surrounding any vents or openings on the roof of such a tank be regarded as *Zone 1* to a diameter of 1 m.

Fig. 10.12 Area classification (Figure 3.1 from the *IP Model Code* Part 15, courtesy Institute of Petroleum)

was found that when a tanker was being loaded at 2.5 m^3/min, the hazardous area in all directions was less than 1.0 m. The dimensions in Fig. 10.12 were based on similar data, but the corresponding values in other standards may not necessarily be the same.

The IP Code contains a total of 23 diagrams showing zone dimensions around a wide range of hazardous equipment and operations, including vent pipes, road and rail tanker loading bays, tanker discharge areas, jetties for loading and discharging, drum filling, petrol metering pumps/dispensers, offshore and land drilling and wirelining operations, mud tanks, and shale shakers. However, for facilities which fall outside this range, perhaps because of operating pressures and temperatures, or fluid volatility, an alternative procedure must be used: consideration of 'the individual point source'. It is this part of the code which contains many new concepts. Some of the earlier standards, and in particular API 540,[59] did make allowances for the pressure within the equipment and for the effects of ventilation. The IP Code gives greater attention to these and other effects, and provides methods of quantifying them. Consideration is given to the physical properties of the liquid fuels and particularly the relationship between the vapour pressure and the flash point. The concept is used of the 'vapourization potential of the petroleum material'. The effects of ventilation are also taken into account.

A new term is used in relation to 'point source releases' (leaks, for example, from flanges, seals, glands, vents and sample points). If there are a number of potential leakage sites, as on a pump, the item is called a 'composite leak source'. It has already been explained that the extent of a hazardous area is determined by the distance from the leak to the farthest occurrence of a lower limit mixture. In the IP Code this is called the 'hazard radius' and values are given for various situations.

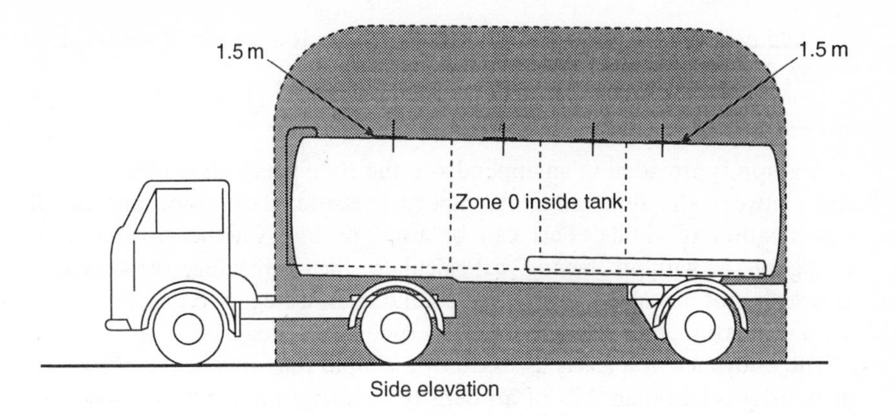

Notes 1 This classification applies also for bottom loading with open venting.
2 Internal *Zone 0* for compartments applies for all classes of petroleum.

Fig. 10.13 Area classification (Figure 3.7 from the *IP Model Code* Part 15, courtesy Institute of Petroleum

The code lists the six standard IP classes of petroleum products, which have already been described above [10.1.1]. However, the code recognises that the way in which a liquid vaporises can be of greater significance than its flash point. Flammable fluids are therefore divided into four categories.

- *Fluid category A.* A liquid which on release would vaporise rapidly and substantially. This includes LPG, and any liquid at a temperature sufficient to produce more than about 40% vaporisation on release.

- *Fluid category B.* A liquid not in category A but at a temperature sufficient to cause boiling on release.

- *Fluid category C.* A liquid not in A or B, but is above its flash point, or can form a flammable mist.
- *Fluid category D.* A flammable gas or vapour.

A graph is provided in an appendix to the code which shows the relationship between the fluid category, vapour pressure, flash point, and the IP classification of fluids. This can be used to find whether a fluid is in Category A without having to calculate the percentage vaporisation. However, hot unstabilised crude oil should be in Category B despite its high vapour pressure, and stabilised (dead) crude is in Category C.

The buoyancy of a gas is assessed by a simple rule: it is lighter than air if its density is less than 0.75 of air density. Hydrogen and methane are both lighter than air but hydrocarbon vapours from ethane (C_2) upwards are all heavier than air.

The procedure for deciding the extent of a hazardous zone for an item of equipment which is not included in one of the 23 'direct examples' therefore involves a knowledge of the vaporising potential of the fluid, the rate and volume of the release, its buoyancy and the degree of ventilation. The grade of release is not considered: the distance to the lower limit is not affected by the frequency of the leak. However, an assessment of the grade of release will have been used in deciding the type of zone. The extent of the zone is determined in three steps.

- *Step 1.* The fluid category of a potential leak is found. For most plants, including much of the equipment at refineries, this will be Category C.
- *Step 2.* The hazard radius is found for each item of equipment, using a number of tables (and guidance) in the code. The tables include pumps; drains and sample points; compressors; instrument and process vents; piping systems; liquid pools due to spillage; sumps; interceptors and separators; pig launchers and receivers; and surface water draining systems. As an example, the hazard radius for a standard pump handling a Category A fluid is 30 m, reducing to 15 m for Category B, and 7.5 m for Category C. If the pump is used for finished Category C products in atmospheric storage, and at no more than 100 m^3/hr, the radius is 3 m. If the hazard radius is inconveniently large it may be possible to reduce the size of a potential leak, or to provide a wall to reduce the spread of vapour. The code gives guidance on such walls.
- *Step 3.* The hazard radius dimension is used to construct a diagram similar to the ones in Fig. 10.14, using the table in the figure. A plan view of the hazardous area would show two circles, the radius of the larger one being the hazard radius (dimension D_1 on the figure). The two drawings apply to a leak of a gas which is heavier than air, either from a point above the ground (on the left) or

above an elevated platform (right). They apply only when the leak is situated in an unrestricted open area. When drawings like this have been completed it will be found that for Category C products the dimensions are about half those shown in the earlier IP Code. For Category B, the dimensions are about the same, but for Category A products the dimensions are about twice those in the earlier standard, although still less than the requirements of some other standards (that produced by ICI/ROSPA[60] for example).

A rather simpler drawing than those in Fig. 10.14 applies when the escaping gas is lighter than air. The hazard radii are smaller than some in Fig.10.14 — up to only 5 m — and again form the main (D_1) dimension.

The code then considers releases into areas which are not in the open air, and which are either 'sheltered' or 'enclosed'. In a sheltered area the ventilation, which can be natural or artificial, must be 'adequate'. It has been

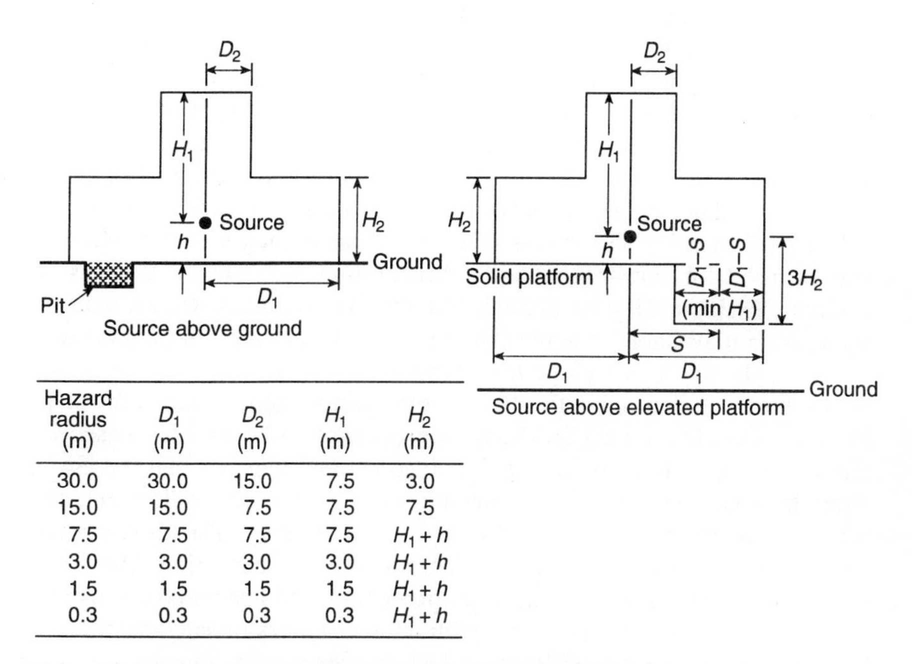

Hazard radius (m)	D_1 (m)	D_2 (m)	H_1 (m)	H_2 (m)
30.0	30.0	15.0	7.5	3.0
15.0	15.0	7.5	7.5	7.5
7.5	7.5	7.5	7.5	H_1+h
3.0	3.0	3.0	3.0	H_1+h
1.5	1.5	1.5	1.5	H_1+h
0.3	0.3	0.3	0.3	H_1+h

Notes 1 h is height of source above ground level or solid platform floor.
2 D_1 for *open area* by definition is equal to the *hazard radius.*
3 *Hazard radius* is determined from tables in 5.6 to 5.14.
4 *Hazardous areas* shown should be classified as *Zone 1* or *2* depending on the *grade of release.*
5 All pits and depressions in *hazardous areas* should be considered in accordance with the definition in 6.5 second paragraph and 6.5.1.3.
6 S is distance from source to edge of solid platform.

Fig. 10.14 Area classification (Figure 6.2 from the *IP Model Code* Part 15, courtesy Institute of Petroleum)

agreed internationally that this means 'the achievement of a uniform ventilation rate of at least twelve air changes an hour with no stagnant areas'. This flow may not be sufficient to prevent the formation of a flammable mixture but it will ensure that once the leak has been isolated the mixture will soon be dispersed. Expert advice may be needed to confirm that this flow has been achieved and that stagnant areas are not present (additional artificial ventilation may be needed for this). Guidance is also available in the literature.[55,61] If the ventilation is 'inadequate' then the area is classified as enclosed. However, not all areas inside a building need to be classed as enclosed, because of the presence of general or local ventilation systems. It is possible for the ventilation rate to be sufficiently high to ensure the continuous dilution of a leak, so that a large hazardous volume cannot exist. The volume may then be so small as to be negligible and the whole enclosed area can be given a higher zone number. This is called dilution ventilation. At lower ventilation rates it may be possible to achieve 'adequate ventilation' and to classify the area as sheltered. If the ventilation rate is just sufficient to ensure the comfort of people working there, about six changes an hour, then the area will have to be classified as enclosed. Tables are provided in the code which show how the three grades of release — continuous, primary and secondary — are combined with the three types of ventilation — inadequate, adequate and dilution — to determine the zone number in an enclosed area. It is possible also to have a pressurised volume, like a control room, within a classified enclosure, and this can be separately zoned.

In Part 2 of BS 5345[51] it says (7.31) that 'established industry codes . . . may be used for carrying out area classification . . . '. The IP Code is an excellent example of such a code, but the BS continues 'where industry codes are not available or applicable, the procedures shown in Figure 1 may be used'. The figure provides a multiple-choice tree showing the steps which are necessary to determine the extent of the various zones. One of the steps (which occurs five times) is 'Determine distance to below LEL based on release rate and other characteristics'. It sometimes happens that the use of a standard method results in a difficult or even impossible requirement, and the method can then be regarded as 'not applicable'. The distance can instead be obtained by an actual measurement or by calculation. The possibility that the whole of a classification procedure could be based on calculation has been studied by an inter-institutional group comprising members of the Institions of Chemical, Electrical, Gas and Mechanical Engineers.[62]

One interesting suggestion made by the group is that the extent of a hazardous area could be determined by calculating the fatal accident frequency rate (FAFR). It might for example be agreed that the hazardous area boundary is drawn through points where the predicted FAFR is no more than 10% above the background rate. Four methods are reviewed of calculating the rate of release, and the dispersion, of liquids, vapours and gases, and references are given to other methods. Methods are provided for calculating: the flow of gas and vapour at sonic and subsonic velocity; liquid flow; flashing flow; discharge coefficient; the fraction of liquid which will flash off; the fraction which will be released as a spray; the pool size when a liquid is released; the vaporisation of volatile and cryogenic liquids; and the dispersion of gas and vapour jets, liquid jets, and passive gas clouds. Using standard hole sizes, calculations were made of the distance to the lower limit for leaks from flanges, compressors, pumps, drains, and sampling points. The results are presented in a number of tables. Some of these are well below those shown in the various codes, but several do exceed the published values. An outstanding example is that of a jet of liquid acetone escaping through a 25 mm^2 hole in a flange at 20 bar. The calculation showed that it would travel 94 m. Hydrogen released in the same way would create a lower limit concentration at 13 m, but with a hole area of 250 mm^2 the distance is 40 m. The group also examined the available data on ignition. They found that the probability that a gas leak of no more than 1.0 kg/s will be ignited is 0.01. This increases to 0.07 for a major leak of 1 to 5 kg/s and to 0.3 for a massive leak of more than 50 kg/s.

The group used the various mathematical models to calculate the estimated frequency of fires and explosions in process plants. The value for explosions was found to be 0.0009 per plant year. An examination of historical records showed that the frequency of explosions in refineries was 0.000 42 per plant year. However, many plant explosions are quite small and may not be reported. It was therefore considered that there was good agreement between estimated and actual values, and that the calculation method had been validated. The work of the inter-institutional group still continues.

10.5.3 Electrical equipment in hazardous areas[63]

Heat is generated in most electrical equipment. The outer surface therefore becomes hot and the temperature may become sufficiently high to ignite flammable mixtures. It will be remembered that an extensive list of AIT values is given in BS 5345[51] Part 1 (but called simply 'ignition temperatures' in the standard) [10.5.1]. It is clearly important that if a leak of a particular gas could come into contact with some electrical equipment, then the maximum temperature of the equipment should be less than the published AIT. Equipment is certified by a number of testing laboratories not to exceed one of six internationally agreed temperatures and is placed in a *temperature class* accordingly. The maximum temperatures permitted in the six classes are shown in Table 10.3.

It will be remembered that area classification provides a measure of the probability of the presence of a flammable atmosphere, and this must be related to the probability that any electrical equipment within the zone will become an ignition source. When the early 'flameproof' equipment was developed it was accepted that the flammable mixture would be able to penetrate the outer casing and become ignited inside. However, the casing was made sufficiently strong to withstand the pressure, and the gaps in it were less than the quenching distance, so the probability of an external ignition would be low. Table 10.1 shows that quenching distances differ quite widely,

Table 10.3 Temperature classes of electrical equipment

Temperature class	*Maximum surface temperature (°C)*
T1	450
T2	300
T3	200
T4	135
T5	100
T6	85

and also that it might be difficult to make equipment with gaps less than the smallest of these. For this reason, equipment is divided into two groups and three subgroups. Group I is equipment for use underground and does not concern us here. Group IIA is suitable for gases of which propane is typical, with a quenching distance of 1.8 mm and a minimum ignition energy of 0.25 mJ. Ethylene typifies Group IIB (0.65 mm, 0.12 mJ). Hydrogen is in Group IIC (0.5 mm, 0.02 mJ). The table of gases in BS 5345 Part 1 which shows AIT values also lists the Apparatus Group required for each gas. No subgroups are indicated for acetylene and carbon disulphide because flameproof equipment cannot be used, but intrinsically safe equipment is permitted (and is described below).

All equipment for use in hazardous areas must be approved by a recognised testing laboratory and must carry an internationally agreed identification. The symbol used for flameproof equipment is the letter d, which is called the 'type of protection'. In the UK the national certifying authority is the British Approvals Service for Electrical Equipment in Flammable Atmospheres (BASEEFA) and their identification mark is Ex. The full identification on for example an electric motor which has been tested in the UK might be: Ex d IIB T4, followed by the BASEEFA certificate number. This indicates flameproof equipment suitable for Group IIB gases with AIT values not less than 135°C.

Nine different types of protection, including flameproofing, have been agreed internationally and are briefly described below, together with their symbol letters and the zones in which they can be used. Full details will be found in BS 5501,[56] and in IEC, CENELEC, and other standards.

Intrinsically safe

Type Ex ia for Zones 0, 1, 2 and type Ex ib for Zones 1 and 2. The amount of electrical energy within this equipment is restricted to a level at which it cannot cause an ignition either by sparking or heating. It is important that the energy in both the equipment and the interconnecting wiring is restricted, so that the entire system within the hazardous area is intrinsically safe. If the system is connected to other equipment outside the area, for example when a detector network is connected to its control panel, it is usual to have a diode safety barrier between them, to limit the electrical energy which can pass into the system. When intrinsically safe equipment is tested and certified, consideration is given to the fault conditions which could lead to a potentially hazardous release of energy. The number of faults which are taken into account determines whether the equipment will be certified Ex ia or Ex ib. Only Ex ia equipment can be used in Zone 0. Some very simple items of equipment, like a thermocouple, or a photocall, do not need to be certified provided that they cannot generate or store electrical energy in excess of 1.2 V, 0.1 A, 20 microJ, and 25 mW.

Flameproof

Type Ex d, Zones 1 and 2. From the explanation given above it is clear that the two main features of this equipment are: a) the capability to withstand an explosion (where the maximum pressure could be 8 bar); and b) the limitation of the size of the gaps in the casing to no more than the quenching distance. In fact the standard[56] specifies both the maximum width and minimum length of the gaps for each of the three groups of gases, IIA, B and C. It should be remembered that the gases emerging from the gap after an ignition will still be quite hot, and a small solid object close to the gap could be heated to the AIT of the surrounding mixture. The standard therefore limits the distance between the flanges of the equipment and any solid obstacle. It may sometimes happen that a part, or all, of the gap becomes blocked by paint or even a sealing material. This could be hazardous because all of the gas may then be released through a small area for a relatively long time, and the flame may not then be quenched (because the gap will be hot).

Increased safety

Type Ex e, Zones 1 and 2. Equipment can be certified in this category if it does not normally get hot or produce arcs and sparks, and has been given increased security against failures which could cause these faults. Examples of the measures which can be taken include: high-integrity insulation; temperature de-rating of insulation; terminals which are unlikely to become loose; thermal cut-outs; increased clearances around moving parts.

Pressurisation and continuous dilution

Type Ex p, Zones 1 and 2. Pressurisation can be used to prevent a flammable atmosphere from entering an enclosure which contains potential ignition sources. A control room which is located within a classified area is a common example, and is described in BS 5345.[51] Continuous dilution will ensure that when a flammable gas enters a hazardous enclosure it is diluted to below the lower limit concentration. For example, the air flow which is needed to cool a large electric motor can also provide continuous dilution. The method can also be used when equipment with an internal leakage source, like a gas analyser, is located in a hazardous area. In this situation it is possible to limit the leakage rate by means of a flow restrictor. The standard then gives a simple way of calculating the flow of air which is needed for a known leakage rate:

$$Q = R \times \frac{C}{100} \times \frac{100 \times F}{L}$$

where

Q is required air flow, m^3/hr,
R is maximum leakage rate of flammable gas, m^3/hr,
C is percent concentration of gas in the leak (which could be 100%),
L is lower limit concentration,
F is safety factor (if it is decided to dilute to 20% of the lower limit, then $F = 5$).

Suppose for example that the gas analyser is being fed with a mixture of 30% hydrogen, lower limit 4%; 20% methane, lower limit 5%; and 50% nitrogen. An orifice restricts the maximum flow to 0.55 m^3/hr. The calculation for the hydrogen, assuming a safety factor of 5, is:

$$Q = 0.05 \times \frac{30}{100} \times \frac{100 \times 5}{4} = 1.875\ m^3/hr$$

A similar calculation for the methane gives $Q = 1.0\ m^3/hr$ and the two values must be added together, giving a total of 2.875 m^3/hr.

The level of protection provided by pressurising and diluting systems is very dependent on the reliability of the systems. The standard provides guidance on this and also requires that the protected equipment should be monitored for a loss of pressure or flow. The action to be taken when there is a failure depends on the type of leak and the zone in which the equipment is located. Actions could vary from the simple sounding of an alarm to the shut-down of all equipment which is capable of causing an ignition.

Special protection

Type Ex s, Zones 1 and 2. When the earliest standards for 'safe' electrical equipment were written, they were little more than descriptions of the best available designs. It was recognised that the development of new methods of protection could be inhibited because the equipment would not comply with the standard. The Type s category was therefore introduced, which permits the acceptance of new designs provided that it can be demonstrated that they meet the requirements of Zone 1 or 2 equipment. The standard also accepts that under unusual circumstances Type s equipment could be suitable for Zone O use.

Encapsulation

Type Ex m, Zones 1 and 2. A piece of equipment is encapsulated when it has been embedded ('potted') in a fire-resistant insulating material. The material should not be capable of being fractured by internal fault conditions.

Normal operation

Type Ex N, Zone 2. Here the British standard differs from international practice by using a capital N instead of a lower case letter. Equipment of this type cannot cause an ignition in normal operation at its rated duty, and a fault which could cause an ignition is unlikely to occur.

Oil immersion

Type Ex o, Zone 2. This method is sometimes used to protect switch gear. Contacts which can spark are immersed in a non-volatile oil.

Powder filling

Type Ex q, Zone 2. This method is rarely used now. The equipment is filled with a finely ground powder, usually quartz, which acts as a flame arrester.

10.6 Detection

10.6.1 Fire detectors

Fires can be detected by sensing one (or more) of a number of emissions from the flames: heat, radiation, smoke, and other products of combustion. The heat is usually that present in the rising plume of hot gases above the fire, and is normally detected at ceiling height. Radiation from flames is mainly in the range of wavelengths from 0.1 to 30 microns. Visible radiation is between about 0.35 and 0.7 microns, so most of the flame radiation is in the UV and IR bands. Radiation detectors must be capable of discriminating between these flame emissions and the continuous background radiation from the sun. The solid and liquid particles in smoke range in size from about 0.5 millimicron to 10 microns in diameter. Particles larger than 0.3 micron can be detected by their ability to obscure or to scatter light. Unfortunately the particles which are released in the pre-ignition heating of a fuel (at temperatures below the fire point) are generally in the range 0.5 to 1.0 millimicron, and those from smouldering combustion are about 0.1 to 1.0 micron. Optical detectors will therefore fail to sense the pre-ignition processes and will be insensitive to smouldering combustion. The so-called ionisation detectors respond very readily to these small particles, and indeed their sensitivity increases with decreasing particle size (whereas the optical detectors show the reverse effect). However, the smallest particles do tend to coalesce into larger clumps with time, so that 'old' smoke which has followed a tortuous route from a smouldering fire may contain particles in the visible range. Conversely, large droplets will evaporate and become smaller with time.

As a rough guide, a fire of 100 kW or more may be needed to actuate a ceiling-mounted heat detector. The response time can be as low as 20 to 40 s, or about 2 min for a sprinkler. A smoke detector in the same situation might respond in a few seconds to a fire of 10 kW or less. A flame (radiation) detector could be triggered instantly by a single small flame, but a practical design could be expected to pick up a 3 kW fire at 20 m.

Most detectors respond to what is happening where they are located, and are therefore point detectors. A line detector will sense a fire at any position along its length. Heat-sensitive line detectors are often integrating: they will respond to a relatively small temperature increase over a long length, or to a higher temperature over a short length. The radiation detectors will respond to a flame anywhere within a certain volume (which is often conical in shape) and are sometimes called space, or surveillance, detectors.

It should not be forgotten that people are excellent fire detectors. They will respond rapidly to incipient fires, assess the possible consequences, and take appropriate action. The existence of a fire, and its location, can be signalled back to a control centre by actuating a manual call point. These are traditionally of the break-glass type, but more robust units can be provided offshore and in similar locations (detector systems are described later). [10.6.3]

Some examples from the wide range of available detectors are described briefly below.

Heat detectors

Some simple detectors depend on the melting of a small volume of solder to make or break an electrical circuit. These are usually inexpensive, but cannot be re-set. Some sprinklers are actuated by the failure of a soldered linkage. Other sprinklers depend on the expansion of a liquid to shatter a glass bulb. The expansion of a liquid or a gas in a length of tubing can be used to close (or open) a switch. Some detectors do depend on the expansion of a piece of metal, but the change in length is very small, and some mechanical amplification is needed. Much greater movement can be obtained from a bi-metallic strip. Thin sheets of two metals which have different coefficients of expansion are fixed together (usually by compression welding) and then cut into strips. When a strip is heated it will curl, with the metal having the greatest coefficient of expansion on the outside of the curve, and the movement can be used to open or close electrical contacts. If a disc is cut out of the sheet and then dished, it will snap over at a certain temperature, and again can move a switch.

Changes in electrical properties can be used to detect an increase in temperature. One line detector comprises a thin stainless steel tube containing a concentric wire. The gap between the wire and the tube is filled with a glass which exhibits a sudden drop in resistance at a high temperature. There is a corresponding change in the capacitance of the line with temperature, and this can also be monitored by a suitable circuit. Line detectors are also made from plastic materials. In one design a cable 2.7 mm in diameter contains two spirally wound conductors separated by a temperature-sensitive material, and is available in lengths up to 200 m.

Thermistors are semiconductors in which the resistance decreases rapidly with increasing temperature, for example from 10^5 ohms at 20°C to 10 ohms at 100°C.

Rate-of-rise detectors

If a point detector has been designed to operate at say 58°C (one of the standard sprinkler ratings), then if there is a very slow increase in temperature the device will actuate as soon as this temperature is reached. On the other hand, if there is a rapidly developing fire, the temperature of the hot gases surrounding the detector may be several hundred degrees by the time it operates. Clearly, a device which is sensitive to the rate of rise rather than to the actual temperature will respond more rapidly. This can be achieved by the use of say two thermistors (or two bimetallic strips) one of which is thermally insulated and the other is exposed directly to the atmosphere. The device is actuated at a predetermined temperature difference between the two sensors. However, a simple rate-of-rise detector might fail to sense the presence of a smouldering fire. Because of this, UK and US standards require that they respond also to a fixed temperature, but this is not a requirement in France.

The sensitivity of a heat detector depends on the rate at which heat is being transferred to it, and the quantity of heat needed to cause its actuation. The most sensitive devices will have a low heat capacity, a high heat transfer coefficient and a large surface area.

Optical detectors

Obscuration refers to the ability of solid objects in a beam of light to reduce the amount of light which arrives at a target (a photocell for example). For large objects the amount of obscuration simply depends on the cross-sectional area of the object. The ability of particles to obscure light when they are less than about 10 microns in diameter decreases as the size is reduced (and as their diameter approaches the wavelength of the light). When the diameter is less than about 0.3 micron, the obscuration is negligible. However, the wide distribution of particle sizes in ordinary smoke ensures that its presence can be detected in quite low concentrations. An obscuration detector comprises a light-tight enclosure into which smoke can pass through baffles, and containing a light source and a photocell. A reduction in the amount of light reaching the cell indicates the presence of smoke. The smoke particles can also be detected by their ability to scatter light. When a beam of light is directed into a cloud of particles, some of the light is scattered sideways. This can be detected by a photocell placed where the maximum amount of scattered light is received. This method has the advantage that the photocell is operating only when smoke is present, and is likely to last longer than a continuously illuminated cell in an obscuration detector. The life of the light source can be prolonged by using a light-emitting diode instead of a filament bulb, and also by pulsing the light. This kind of detector is more sensitive to the grey smoke which is produced by smouldering, than to black smoke. It should be capable of detecting smoke with an obscuration of 3%/m (there are a number of ways of expressing obscuration, but this is self-explanatory).

Optical beam detectors

If a beam of light is projected from one end of a room to a photoelectric device at the other end, the presence of smoke anywhere in the path of the light can be detected. The arrangement is also very sensitive to differences between the refraction caused by the plume of hot gases from a fire and that of the colder air surrounding it. This results in flickering, or even complete displacement, of the light arriving at the photocell, which is readily detected. Visible light is not used because the detector would be swamped by the background, and IR radiation is usually employed instead. This can conveniently be a laser beam, which also overcomes the difficulty of focusing a light beam. Problems can be caused by vibrations within a building, and by the presence of intruders (even a bird).

Ionisation detectors

These devices do not detect ionisation and are therefore sometimes called combustion gas detectors, although even this is not strictly accurate. The name derives from the use of a radioactive source (an alpha emitter) to ionise some of the air in a small chamber. The application of a potential between two electrodes causes a flow of positive and negative ions to the opposite electrodes. The ions which constitute this small electric current can be captured on the surface of the particles which are released by combustion processes. This results in a reduction of the flow, which can be measured. The actual current flows are extremely small, between 1 and 100 picoamps, and their detection became possible only after the development of a cold cathode tube by Meilie, a Swiss physicist.[64] Modern detectors use solid state circuitry, but the currents being detected are still very small and false alarms can be caused by radio transmissions (from mobile telephones for example). The detectors may also false-alarm if the radioactive source becomes coated with moisture, or if the ion flow is displaced by a current of air. Compensation is provided for changes in atmospheric conditions by having two similar ionisation chambers in each detector. One of these is open to the atmosphere (and the combustion products) but the other has only a pin hole in its casing. Both will respond equally to changes in pressure, temperature and humidity, but only the open one will respond rapidly to the combustion products. Because of the small current flows, these detectors can run for a year or more on a small battery, and are ideal for domestic use.

The radioactive source is usually a foil disc about 10 mm in diameter containing two thin layers of americium 241, each protected by a layer of gold. The two chambers are arranged back-to-back with the disc between them. The alpha radiation cannot penetrate the detector case (and indeed is absorbed by a thin layer of moisture, as is noted above). Some low energy gamma rays do pass through the case, but at a distance of 100 mm this radiation is only a small fraction of the background radiation. The detectors do not constitute a health hazard in normal use and are exempted from most of the legislation covering radioactive substances. However, it would be inadvisable to store a large number of them close to an occupied area.

Flame detectors

The radiation from a flame will travel to a detector at the speed of light. Flame detectors therefore have the fastest possible response time. They are surveillance devices: a single unit will cover a large volume. They are also very sensitive and will respond to quite a small flame. However, it is usually necessary for the detector to 'see' the flame, although some reflection at IR wavelengths may be picked up by an IR detector. Smouldering combustion is unlikely to be detected, although again an IR unit may respond to an exposed glowing area.

Sunlight contains the complete spectrum of radiation from the far UV to very long IR, and it is necessary for a flame detector to ignore this radiation while still being capable of responding to a fire. Fortunately some solar radiation is absorbed as it passes through the atmosphere making it sufficiently different from flame radiation to enable detectors to be selective. Some manufacturers prefer instead to rely on detecting the flickering which is present in most flames. The flicker frequency varies from about 15 Hz for a 10 cm^2 pool fire to 2 Hz for pools greater than 1000 cm^2 (although with large flames there are also small fluctuations due to areas of turbulence).[65] Detectors are normally set to respond to frequencies between 5 and 15 Hz, and will ignore any continuous radiation. False alarms have been caused by reflections from rotating machinery, and by car headlights being reflected from railings.

Ultraviolet detectors can be made solar-blind by being operated at wavelengths below 0.29 micron, where there is almost complete atmospheric absorption. Unfortunately only a small fraction of the energy radiated from flames is in this region and it is necessary to use a very sensitive photocell. A Geiger-Muller tube is frequently used, in which the current amplification can be as high as 10^{12}. The tube is mounted behind a small UV window in a suitable housing, and can usually cover a conical volume with a solid angle of 90° or more. Typically it will detect a 3 kW fire at 20 m. Radiation from lightning strokes and other transients can be ignored by incorporating a time delay in the circuit. Unfortunately, welding operations will usually trigger the device, and indeed welding on one offshore platform has been known to cause false alarms on a neighbouring platform. X-rays and the radioactive sources used for inspecting welds will also be detected. It is important that the window in the detector is kept clean because UV radiation is readily absorbed by a film of oil (hence its use in preventing sunburn).

Infrared detectors can also be designed to respond to the wavelengths at which IR solar radiation is absorbed in the atmosphere. Water vapour and carbon dioxide molecules absorb strongly at between 1.5 and 1.8 microns and also between 4.2 and 4.5 microns. These are also the wavelengths at which these molecules will radiate when they are heated in a flame. Photocells sensitive to these wavelengths are the obvious choice for use as IR flame detectors. However, these detectors would then respond to radiation from car exhausts, chimneys and hot surfaces. As we have seen, this can be avoided by designing them to respond only to radiation at the correct flicker frequency. It is usual also to set the response time to about 3 s so that transient effects will be ignored. Again, detectors will usually cover a conical volume with a 90° solid angle, but dirt or oil on the window will have little effect on the sensitivity. In one design the photocell is mounted above a rotating mirror, which enables the device to scan a large volume.

Heat, smoke and flame detectors are specified in BS 5445, which is the English language version of the European standard EN 54 (except that, as noted above, the latter permits the use of rate-of-rise detectors which do not have a fixed temperature response). Brief specifications are also contained in the American standards NFPA 72A and 72E.

10.6.2 Flammable gas detection

Most of the available gas detectors depend on measuring the heat released by the oxidation of low concentrations of gas on a catalytic surface. In early designs the catalyst was a small spiral of platinum wire which was heated to 900°C by an electric current. The heat generated by the reactions at the surface of the wire increased its temperature, and also increased its electrical resistance. The resulting reduction in the current flowing through the wire was detected by a Wheatstone bridge: a simple but very sensitive device developed by Sir Charles Wheatstone in the early years of the nineteenth century for finding faults in telegraph systems (see Fig. 10.15).

The platinum catalyst had a short life, mainly because at the high operating temperature metal was lost by evaporation. Catalysts operating at lower temperatures were developed by encapsulating the heater wire in a ceramic bead (perhaps only 1 mm in diameter) coated with a catalytic material. This often now comprises a layer of thorium oxide on to which palladium has been deposited. This is effective at temperatures between 400 and 500°C, and is usually called a pellistor. In the head of a detector, a pellistor is

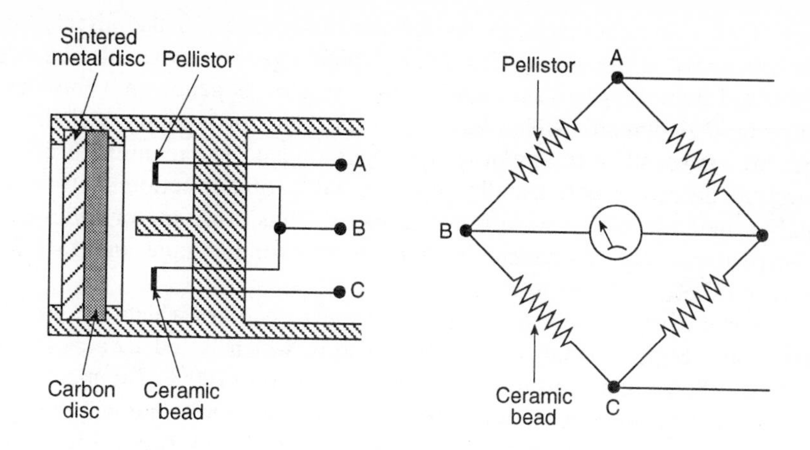

Fig 10.15 Pellistor flammable gas detector and Wheatstone bridge

connected in series with a ceramic bead assembly which is identical except that it has no catalytic surface. These form two arms of a Wheatstone bridge (which is still used in many detectors) and this arrangement ensures that changes in ambient conditions do not affect the response. With a typical pellistor, a 17% change in the concentration of the gas results in a change in the voltage drop across the pellistor of about 25 mV, which is readily detected. Figure 10.15 is a diagram of a pellistor detector, showing the arrangement within the unit, and the associated Wheatstone bridge. The meter which is connected across the bridge is usually calibrated from 0 to 100% of the lower limit of the gas against which the device was calibrated. When detectors are used in a process area it is often arranged that they alarm at two levels. For example, with an alarm at say 10% of the lower limit a search is made for the leak, but at 20% the plant is shut down.

If the gas/air mixture which enters the detector is within the limits of flammability, then clearly the pellistor will act as an ignition source. The passage of a premixed flame outwards into the flammable atmosphere is prevented by the sintered metal disc, shown in the diagram, which acts as a flame arrester. Like most catalysts, the pellister can be poisoned, particularly by gases which are adsorbed strongly or which react to leave a solid oxide. It is the purpose of the carbon disc, also shown in the figure, to filter out some of these gases. Silicon compounds are particularly troublesome and are contained in many spray lubricants. These should never be discharged near to a gas detector. Other materials which will poison the catalyst include phosphorous compounds, hydrogen sulphide and the fumes from welding.

The heat released by the combustion of a unit volume of many gases at the lower limit is much the same. This is not surprising when one remembers that there is a critical lower-limit temperature, and that the combustion products of many fuels are very similar. It might therefore be expected that the readings given by a gas detector would be accurate for a wide range of fuels. Unfortunately this is not so because the response is also affected by the diffusivity of the gases. The rate at which a gas can penetrate the two discs, and then the gases surrounding the pellistor, depends on its molecular weight. A detector which has been calibrated for methane will therefore be less sensitive to any other hydrocarbon gas. The manufacturers of portable detectors will usually supply a table of correction factors, to enable their use with a wide range of gases. Thus, for a meter calibrated with pentane (C_5H_{12}), typical factors are 0.51 for methane and 1.47 for nonane (C_9H_{20}).

In certain situations a gas detector can give readings which are grossly inaccurate. Suppose that in a gas/air mixture which is fed to a portable detector, the gas concentration is increased from 0 to 100%. The meter will show an increasing reading up to 100% of the lower limit, and will then go off-scale. The current passing through it will continue to increase to about the stoichiometric mixture, and will then fall gradually to zero as the gas concentration increases to 100%. The meter will therefore fall back again to the 100% lower-limit reading at a gas concentration which might well be about 80%. It will continue to show a reading above 0% on the scale until there is no air at all in the mixture. This potentially hazardous situation can be avoided by several methods. If a sample is being drawn from a container into a portable meter by a long probe, it can be arranged that air is entrained into the probe to dilute the sample. If the addition of air causes an increase in the meter reading, then the sample contains a fuel-rich mixture. The detector itself can also be modified by including another sensing device, on one arm of the Wheatstone bridge. The heat losses from this device increase with increasing thermal conductivity of the sample. An excess of gas with a conductivity different from that of air would then unbalance the bridge. It is clearly important to ensure that an appropriate detector is used whenever it is possible that high concentrations of gas will be present. The sensitivity of a detector may also be affected by temperature, pressure, humidity, air movements and vibration,[66] and it may be necessary to take account of these conditions.

Recent developments in methods of gas detection have included long-life pellistors, semiconductor sensors and the measurement of infrared absorption.[66] Conventional detectors are specified in BS 6020.

10.6.3 Detector systems

The simplest fire detector system comprises a detector with normally-open contacts in series with an alarm and a source of electrical power. A short circuit in the leads to the detector would give a false alarm, and a break anywhere in the circuit would make the system inoperative. The existence of these faults can be indicated at a control panel by using the end-of-line device shown in Fig. 10.16.

The exact values of the two resistors are not important, provided that one is substantially greater than the other. Suitable circuitry in the control panel

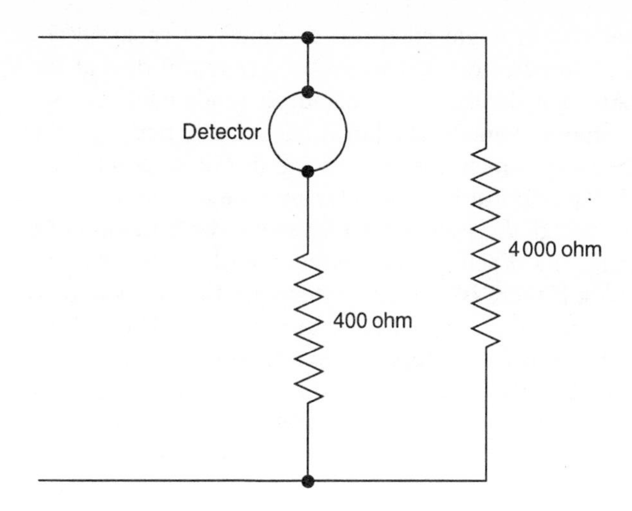

Figure 10.16 End-of-line device

can monitor the resistance in the circuit and give an appropriate indication, probably using coloured lights (the actuation of the detector will also sound an alarm);

4000 ohms	Normal condition	Green light: System healthy
400 ohms	Detector activated	Red light: FIRE
4000+ ohms	Open circuit	Amber light: Fault
0–400 ohms	Short circuit	Amber light: Fault

A separate distinctive alarm may also be sounded to call attention to a fault condition. A loss of sensitivity of a detector cannot be discovered by this simple system and reliance must be placed on regular inspection. Unfortunately the opposite fault, that of a false alarm, occurred sufficiently frequently in early equipment to cause serious problems. This was particularly true when the alarm was sounded at a Fire Station, or when the receipt of an alarm signal actuated the shut-down of a process. The frequency of such false alarms is considerably reduced by the use of a voting system. The actuation of a single detector simply provides a warning, and no action is taken until a second detector has responded. However, a single actuation of a manual call point should be accepted at once since a human detector is not expected to false-alarm. Voting systems are frequently used offshore because the partial or complete shut-down of the process may be necessary when a fire or a gas release is detected.

The great advances which have been made in electronic equipment, and particularly in the use of microprocessors, have led to the development of detector systems which are both versatile and reliable. These are sometimes called intelligent systems but they do not think for themselves and are better called programmable systems. The most important advance is in the use of addressable detectors. These can be individually interrogated from the control panel. A typical system might comprise a number of separate circuits each of which has 120 detectors which can be scanned in turn at the rate of 30 a second. As each detector is contacted it sends back its own address, its type (heat, smoke, flame), whether it has a fault condition, the smoke density or temperature which it is sensing and its output conditions. (Is it reporting a fire? Is its indicator light on?) Despite the amount of information which is being handled, the system will respond to the actuation of a manual call point within 0.1 s. Each detector circuit is a loop, so that the detectors will still function if there is a break. If there are two breaks, some of the detectors will not be able to signal back to the panel. This will be indicated as a fault and the lost detectors will be identified. Isolators are installed at regular intervals in the loop, and if a short develops, the isolator on either side of it will open. Again the lost detectors will be identified. Since analogue detectors are used (giving a continuous signal), the system can be arranged to give a pre-alarm warning. Thus, if a number of heat detectors in an area have sensed a temperature above the normal operating conditions, it may be that some corrective action is needed. A warning can be sounded and the information can be conveniently displayed as a bar chart on a VDU. The system's memory can also be used to check the recent history of temperature rises in the area.

Standards for detector systems are available in many countries. A particularly comprehensive one was developed by the BSI to take account of the advances in the use of microprocessors, stored programmes, addressable analogue detectors, radio linkages and other developments. This is BS 5839. An interesting innovation in the Standard is a classification of systems depending on their type and usage, and in particular whether they are protecting life or property. Other factors which are taken into account in selecting a system are: the probability of ignition; the rate of spread of fire in the room of origin; the value of the contents; the probability of fire spread to other areas; and the time taken for this to happen. All of these are weighed against the value of the time which is gained by the installation of a system. The Standard also defines the zones into which a building is divided, the siting of detectors and manual call points and the use of sounders (alarms). If a system is to respond reliably to a fire it is essential that it is properly maintained for what may be many years before this happens. The standard therefore requires that a responsible person is appointed, and provides a comprehensive list of duties for them.

References

1. Gaydon, A. G. and Wolfhard H. G. *Flames, Their Structure, Radiation and Temperature*, 3rd edn. Chapman & Hall, London (1970)
2. Drysdale, D., *An Introduction to Fire Dynamics*, 2nd edn, John Wiley, Chichester (in preparation)
3. Wharry, D. M. and Hirst, R., *Fire Technology: Chemistry and Combustion*, 3rd edn, Institution of Fire Engineers, Leicester (1992)

4. Hirst, R., *Fire Technology: Fire Dynamics*, Institution of Fire Engineers, Leicester (1996)
5. Hirst, R. and Savage, N., 'Measurement of inerting concentrations', *Fire Safety Journal* **4**, 147 (1981/82)
6. *Protocol on Substances that Deplete the Ozone Layer*, signed in Montreal, 16 September 1987, came into force 1 January 1989. Published in the UK as *Command Paper No 1* (1988) (CM 283), HMSO
7. Coward, H. F. and Jones, G. W., 'Limits of flammability of gases and vapours', Bureau of Mines Bulletin 503, 1952
8. Zabetakis, M. G., 'Flammability characteristics of combustible gases and vapours', *Bureau of Mines Bulletin 627*, 1965
9. Institute of Petroleum, *Model Code of Safe Practice in the Petroleum Industry*, Parts 1 to 19 (see for example p.2 of Part 19: Fire precautions at petroleum refineries and bulk storage installations), John Wiley, Chichester (various dates)
10. Griffiths, J. F., and Barnard, J. A. *Flame and Combustion*, 3rd edn, Chapman & Hall, London (1995)
11. Count Morozzo, 'Account of a violent explosion in a flour warehouse in Turin, December 14, 1785,' *The Repertory of Arts and Manufacturers 2.416* (1795)
12. Palmer, K., N., *Dust Explosions and Fires*, Chapman & Hall, London (1993)
13. Field, P., 'Dust Explosions', *Handbook of Powder Technology*, Volume 4, Elsevier, Oxford (1982)
14. Bartknecht W., *Explosions*, Springer-Verlag, Berlin (1989,
15. Raftery, M. M., 'Explosibility tests for industrial dusts', *Fire Res. Tech. Paper 21*, HMSO, London (1975)
16. Jacobson, M., Cooper, A. R. and Nagy, J., *Explosibility of metal powders*, US Bureau of Mines, RI 6516, Washington (1964)
17. HSE, HS(G) 11, *Flame arresters and explosion reliefs*, HMSO, London (1980)
18. Fordham, S., *High Explosives and Propellants*, 2nd edn, Pergamon, Oxford, (1980)
19. Davis, T. L., *The Chemistry of Powder and Explosives*, John Wiley, New York (1972)
20. Kletz, T. A., 'Unconfined vapour cloud explosions. An attempt to quantify some of the factors involved', *11th Loss Prevention Symposium*, New York, 1977 (American Institution of Chemical Engineers)
21. Marshall, V. C., 'Safety of control rooms — a strategic view', *Meeting of Belgium Pet. Inst.*, Antwerp (1976)
22. Lees, F. P., *Loss Prevention in the Process Industries*, 2nd edn, Butterworths, London (1995)
23. Cox, A. W., Lees, F. P., and Ang, M. L., *Classification of Hazardous Locations*, Institution of Chemical Engineers, Rugby (1990)
24. Cairney, E. M. and Cude, A. L., 'The safe dispersal of large clouds of flammable heavy vapours', *Major Loss Prevention in the Process Industries*, Institution of Chemical Engineers, Rugby (1971)
25. Simpson, H. G., 'The ICI vapour barrier', *Power and Works Engineering* (May 1974)
26. BSI, BS 5306, *Code of practice for fire extinguishing installations and equipment on premises*, Part 5, Halon systems. Section 5.1 Halon 1301 total flooding systems, Section 5.2 Halon 1211 total flooding systems. British Standards Institute, London
27. NFPA, National Fire Codes: No. 12A, *Standard on halogenated fire extinguishing agent systems — Halon 1301*; No. 12B, *Standard on halogenated fire extinguishing agent systems — Halon 1211*. National Fire Protection Association, Boston, USA (regularly updated)
28. Harris, R. J., *Gas Explosions in Buildings and Heating Plant*, Spon, London (1983)
29. Rasbash, D. J., 'The relief of gas and vapour explosions in domestic structures', *Fire Research Note 759*, HMSO (1969)
30. Rasbash, D. J., Drysdale, D. D. and Kemp, N., 'Design of an explosion relief system for a building handling LPG', *I Chem E Symposium Series No. 47*, Institution of Chemical Engineers, Rugby (1976)
31. Cubbage, P. A. and Marshall, M. R., 'Pressures generated by explosions of gas-air mixtures in vented enclosures', *Institution Gas Engineers Communication No. 926* (1973)

32. Fairweather, M. and Vasey, M. W., 'A mathematical model for the prediction of overpressures generated in totally confined and vented explosions', *19th Symposium (International) on Combustion*, Butterworths, London (1982)
33. Cubbage, P. A., and Simmonds, W. A., 'An investigation of explosion reliefs for industrial drying ovens — 1, top reliefs in box ovens', *Trans Institution of Gas Engineers 105*, 470 (1955)
34. Cubbage, P. A. and Simmonds, W. A., 'An investigation of explosion reliefs for industrial drying ovens — 2, back relief in box ovens, reliefs in conveyor ovens', *Trans Institution of Gas Engineers 107* (1975)
35. NFPA, National Fire Codes No. 68, *Guide for Explosion Venting*, National Fire Protection Association, Boston, USA (reprinted annually)
36. Palmer, K. N., 'The quenching of flames by wire gauzes', *7th Symposium (International) on Combustion,* Butterworths, London (1959)
37. Cubbage, P. A., 'Flame traps for use with town gas/air mixtures', *Gas Coun Res Comm GC 63*, London (1959)
38. Glendinning, W., British Patent 643188 (1948)
39. Hirst, R., *Underdown's Practical Fire Precautions*, 3rd edn, Gower Technical (1989)
40. Hirst, R. and Booth, K., 'Measurement of flame-extinguishing concentrations', *Fire Technology*, **13** (4) 296 (1977)
41. Hirst, R., Farenden, P. J. and Simmons, R. F., 'The extinction of fires in aircraft engines — Part 1, small scale simulation of fires', *Fire Technology*, **12** (4) 266, Boston (1976)
42. Nash, P., and Young, R. A., *Automatic Sprinkler Systems for Fire Protection*, 2nd edn, Paramount Publishing, Borehamwood (1991)
43. BSI/CEN, BS 4547 and EN 2, *Classification of fires*, European Committee for Standardisation, Paris: English version, British Standards Institute, London
44. NFPA, National Fire Codes No. 10, *Standard for portable fire extinguishers*, National Fire Protection Association, Boston, USA
45. UL, Standards for Safety 711, *Classification, rating, and fire testing of Class A, B, and C fire extinguishers and for Class D extinguishers or agents*, Underwriters' Laboratories, Chicago, USA.
46. BSI, BS 5423, *Specification for portable fire extinguishers*, British Standards Institution, London
47. BSI, *Code of practice BS 5306 for fire extinguishing installations and equipment on premises*, Part 3: portable fire extinguishers, British Standards Institution, London
48. NFPA, *Fire Protection Handbook*, current edition, National Fire Protection Association, Boston, USA
49. Hughes, J. R., *Storage and Handling of Petroleum Liquids*, 3rd edition, Charles Griffen, London (1988)
50. Laurendean, N. M., 'Thermal ignition of methane-air mixtures on a hot surface: a critical examination', *Combustion and Flame*, **46** (29) (1982)
51. BSI, BS 5345, *Code of practice for the selection, installation and maintenance of electrical apparatus for use in potentially explosive atmospheres*, Part 1: basic requirements for all parts of the code, British Standards Institution, London
52. Shields, T. J. and Silcock, G. W. H., *Buildings and Fire*. Longman Scientific, Harlow (1987)
53. I.Chem.E., *Thermal Radiation Monographs*, Institution of Chemical Engineers, Leicester (1990)
54. BSI, BS 6651, *Code of practice for the protection of structures against lightning*, British Standards Institution, London
55. Gale, W. E., 'Module ventilation rates quantified', *Oil and Gas Journal*, Dec 1985, p. 39
56. BSI, BS 5501, *Electrical apparatus for potentially explosive atmospheres*, British Standards Institution, London
57. NFPA, National Fire Code 70, *National Electrical Code*, National Fire Protection Association, Boston, USA
58. NFPA, National Fire Code 497, *Electrical Installations in Chemical Plants*, National Fire Protection Association, Boston, USA

59. API, RP 540, *Recommended practice for electrical equipment in petroleum plants*, American Petroleum Institute, New York, USA
60. ICI/ROSPA, *Electrical Installations in Flammable Atmospheres*, Royal Society for the Prevention of Accidents, London
61. Marshall, A. R., 'The effect of ventilation on the accumulation and disposal of flammable gases', *I. Chem. E 4th International Symposium*, Harrogate (1983)
62. Cox, A. W., Lees, F. P., and Ang, M. L., *Classification of Hazardous Locations*, Institution of Chemical Engineers, Rugby, 1990
63. Palles-Clark, P. C., 'History of area classification and the basic content of BS 5345 (Part 2)', *IEE Colloquium Digest 1982/26*, Institution of Electrical Engineers, London, 1982
64. Meilie, C. H., 'The ionization chamber smoke detector', *SEV Bulletin 23*, **43**, 1952
65. Portscht, R., 'Uber das flacken von flammen', *6th International Seminar on Fire Detection*, Aachen (1971)
66. Cullis, C. F. and Firth, J. G., *Detection and Measurement of Hazardous Gases*, Heinemann, London (1981)

11 Fire and explosion hazard rating of process plant

The hazards of substances (which range from those of salt to nitroglycerine) and the plants which process them differ widely in degree and type. Plants with high hazard potential need greater spacing, greater care in their design and operation and more safety and protective features than less hazardous ones. Several methods are used to rate the fire and explosion hazards of process plants. The results have important financial implications. Faced with choosing one of several processes with different capital and operating costs, and different degrees of hazard, managements need to know how these will affect the project costs.

Dow Chemical Company's 'Fire and Explosion Index' (F&EI) *Hazard Classification Guide*, now in its sixth edition, is probably the best-known method. The degree of hazard of each unit of a process plant is rated as a single number. The third edition[1], published as a manual by the American Institute of Chemical Engineers,[2] forms the basis for the more complex later ones[3,4] and for the Mond Index[5–7] which was developed by ICI plc. Both the third edition of Dow's guide (chosen for its comparative simplicity) and the second edition of the Mond Index[6] are described and discussed here. Later editions of the Dow guide use different numerical bases for the index (Table 11.1) and include credit factors for various loss-control measures, estimates of maximum probable property damage and other damage parameters. Those using the Dow guide occasionally should keep to the same edition to prevent confusion. The fourth edition is described by Lees.[4] The methods can be used as an aid to feasibility studies, process design, layout and mechanical design and can be applied at any subsequent stage in the plant's life.

To use the Dow guide one needs a process flow diagram of the plant, a plot plan, a copy of the guide and a good understanding of the process. To apply the Mond Index, one also needs detailed cost data for the installed equipment, pipework, buildings and structures, a drawing compasss and a calculator (or computer and program).

Table 11.1 Dow Fire and Explosion Index Range and Degree of Hazard

Degree of hazard	*Index range*			
	3rd edn (1973)	*4th edn (1976)*	*5th edn (1980)*	*6th edn (1987)*
(Mild)		0–20		Not
Light	20–40	1–50	1–60	applicable
(Moderately heavy)	60–75			
Moderate	40–60	51–81	61–96	
Intermediate		82–107	97–126	
Heavy	75–90	108–133	128–156	
(Extreme)	90 and up			
Severe		134-up	159-up	

This chapter concludes with a section [11.4] on estimating the potential loss from vapour cloud explosions from projected and existing plants.

11.1 The Dow Fire and Explosion Hazard Index, 3rd edn[1,2]

The description given here is necessarily abbreviated and several finer points have had to be omitted. It nevertheless gives all the essential steps with a worked example, including the safety features called for in the design.

The first step is to divide the plant into units. Here a unit is defined as a part of a plant which can be readily and logically characterised as a separate entity. One should start with the units as defined by the process design and, if possible, split them into smaller ones which do not overlap significantly. Figure 11.1 is a very simplified flow diagram of a hypothetical plant making alcohols from propane oxidation which is divided into six units (Table 11.6). Recycled propane vapour plus make-up propane from pressure storage is mixed with a small proportion of compressed air to give a mixture well above the upper explosive limit before entering a vapour phase reactor at 315°C and about 22 bar g. The oxygen and part of the propane react exothermally to give a mixture of alcohols, aldehydes and ketones, with nitrogen and unreacted propane. The hot gases leave the reactor at 538°C and, after cooling, pass through an absorber where water is used to recover the alcohols, aldehydes and ketones. These are separated from the water in a stripping column and passed to three columns in series where acetaldehyde, acetone and methanol are separated as overhead products. The gas leaving the absorber and consisting principally of nitrogen and propane passes to a second absorber where most of the propane is removed by circulating absorption oil. Propane is recovered from the absorption oil in a pressurised stripping column and recycled as vapour to the preheater. Vapour phase hydrocarbon oxidation processes were discussed in 8.5.2.

The F&EI for each unit is evaluated from the following factors:

1. the material factor for the unit, MF
2. special material hazards factor
3. general process hazards factor
4. special process hazards factor.

(In later editions of the guide (1) and (2) are combined and the MF is a function of three NFPA hazard ratings.[8]

11.1.1 The Material Factor (MF)

This is a number (generally from 1 to about 60) which denotes the intensity of energy release from the most hazardous material or mixture of materials present in significant quantity in the unit. MF is calculated as:

$$MF = \frac{\Delta H_c}{2326}$$

where ΔH_c = net heat of combustion (kJ/kg).

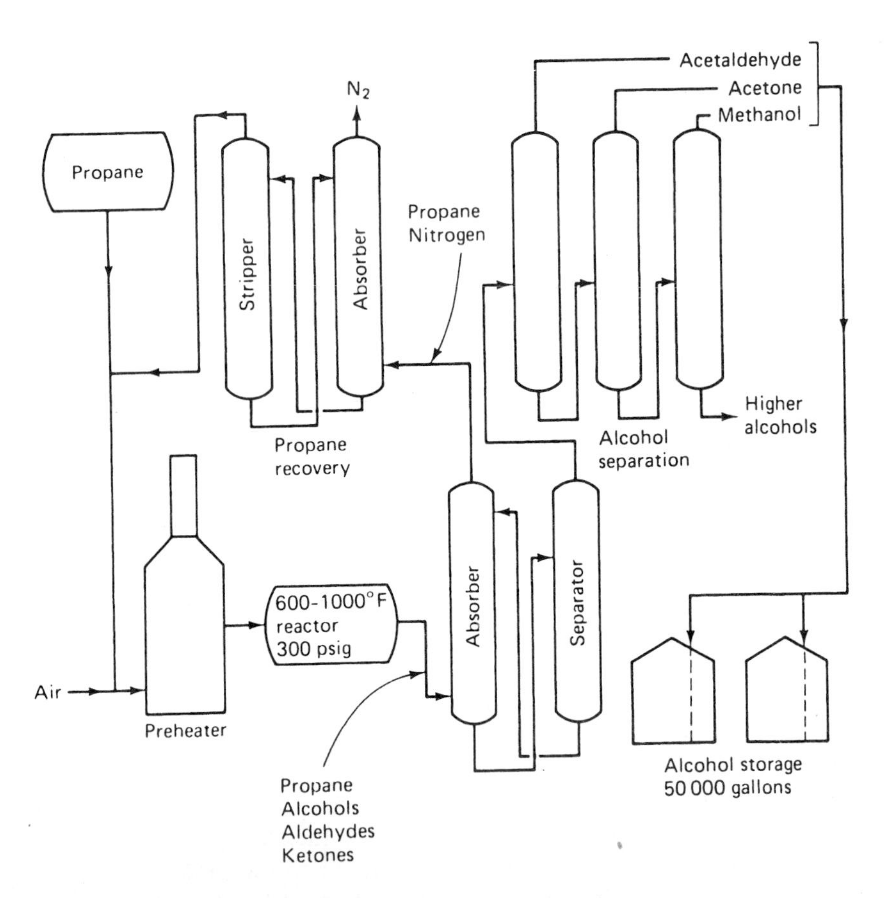

Fig. 11.1 Flow diagram of hypothetical alcohols from propane plant

For combinations of reactive materials such as oxidising and reducing ones, the heat of reaction is used instead of the heat of combustion. MFs for several materials are given in Table 11.2.

11.1.2 Special material hazards

An additional percentage as given in Table 11.3 must be added to the MF of materials which have certain hazardous properties, provided these have not already been taken into account in determining the MF. Where ranges are given, judgement is needed to decide the percentage to be added.

11.1.3 General process hazards

A further percentage must be added to the MF for each of the applicable process characteristics given in Table 11.4.

Table 11.2 Material Factors for common chemicals

Compound	*MF*	*Compound*	*MF*
Acetone	12.3	Isopropyl alcohol	13.1
Acetylene	20.7	Methyl chloride	5.5
1,3-Butadiene	19.2	2-Nitrotoluene	11.2
Carbon disulphide	6.1	Potassium perchlorate	0.0
Chlorine dioxide	0.7	Stearic acid	15.7
Diethyl ether	14.7	Styrene	17.4
n-Dinitrobenzene	7.2	Sulphur	4.0
Dowtherm A	14.0	Triethyl aluminium	18.9
Ethyl acetate	10.1	Urea	3.9
Ethylene oxide	11.7	Vinyl chloride	8.0
Hydrogen	51.6	Xylene	17.6

Table 11.3 Percentages to be added to MF for special material hazards

Special property of material	*Percentage to be added to MF*
A. Oxidising	0 to 20
B. Reactive with water to produce combustible gas	0 to 30
C. Subject to spontaneous heating	30
D. Subject to spontaneous polymerisation	50 to 75
E. Subject to explosive decomposition	125
F. Subject to detonation	150
G. Other unusual hazardous properties	0 to 150

Table 11.4 Percentages to be added to MF for general process hazards

Process characteristic	*Percentage to be added to MF*
A. Handling and physical change only	0 to 50
B. Continuous reactions	0 to 30
C. Batch reactions	25 to 50
D. Multiplicity of reactions in same equipment	0 to 50

11.1.4 Special process hazards

A further percentage must be added to the MF for each of the applicable special process characteristics given in Table 11.5.

11.1.5 Evaluation of the F&E indices

The indices for the six units of the hypothetical alcohols plant are evaluated on standard calculation sheets which are combined in Fig. 11.2. The results are summarised in Table 11.6.

The reaction unit has the highest F&EI followed by the propane storage unit which is assumed to consist of pressure spheres. The only units whose F&EIs are affected by the plant scale are propane storage, propane recovery and alcohol storage. The F&EIs of all other units would be the same for a pilot plant as for a large commercial one (a possible weakness in the method).

Table 11.5 Percentages to be added to MF for special process hazards

	Special process characteristic	*Percentage to be added to MF*
A.	Low pressure	0 to 100
B.	Operation in or near explosive range	0 to 150
C.	Low temperature	15 to 25
D.	High temperature. Use one only	
	1. Above the flash point	20
	2. Above the boiling point	25
	3. Above the auto-ignition temperature	35
E.	High pressure	
	1. 17 to 200 bar g.	30
	2. Above 200 bar g.	60
F.	Processes or reactions difficult to control	50 to 100
G.	Dust or mist explosion hazard	30 to 60
H.	Greater than average explosion hazard	60 to 100
I.	Large quantities of combustible or flammable liquids in unit (use one only)	
	1. 7.5 to 22.5 m^3	40 to 55
	2. 22.5 to 75 m^3	55 to 100
	3. 75 to 190 m^3	75 to 100
	4. More than 190 m^3	100 or more
J.	Other unusual process hazards	0 to 20

Table 11.6 Units of alcohols from propane plant with estimated F&EIs

	Unit	*F&EI*	*Degree of hazard*
1.	Propane storage	56.3	Moderate
2.	Reaction area	76.5	Heavy
3.	Alcohol absorber	28.0	Light
4.	Alcohol separation	15.5	Mild
5.	Propane recovery	39.0	Light
6.	Alcohol storage	27.8	Light

11.1.6 Selection of preventative and protective features

Protective systems may be considered as passive, i.e. systems without moving parts such as blast-walls and knockout vessels, or active, i.e. systems with moving parts such as valves and switches. Systems without moving parts give fewest problems, but all need regular maintenance and testing, which must be budgeted for. The simplest ones are usually the best. When many rarely used protective devices are installed, some readily fall into disrepair and are not available when a real need arises. The checklists which follow are intended as guides for selection and not imperatives to be followed blindly.

Dow's preventative and protective features are grouped in three categories:

1. basic features needed in any plant with an F&E hazard;
2. features whose application depends on the F&EI of the unit (Table 11.7);
3. special preventative features (Table 11.8) which relate to specific hazards listed in Tables 11.3 to 11.5.

Table 11.7 Preventative and protective features related to F&E hazard

Legend	*Priority*
Feature optional	1
Feature suggested	2
Feature recommended	3
Feature required	4

Feature	*Priority for calculated F&EL*					
	0 to 20	20 to 40	40 to 60	60 to 75	75 to 90	over 90
A. Fireproofing	1	2	2	3	4	4
B. Water spray						
a. Directional	1	2	3	3	4	4
b. Area	1	2	3	3	4	4
c. Curtain	1	1	2	2	2	4
C. Special instrumentation						
a. Temperature	1	2	3	3	4	4
b. Pressure	1	2	3	3	3	4
c. Flow control	1	2	3	4	4	4
D. Dump/blowdown/spill control	1	1	2	3	3	4
E. Internal explosion relief and/or suppression	1	2	3	3	4	4
F. Combustible gas monitors						
a. Signal alarm	1	1	2	3	3	4
b. Actuate equipment	1	1	2	2	3	4
G. Remote operation	1	1	2	3	3	4
H. Building ventilation	see Dow code					
I. Building explosion relief	see Dow code					
J. Diking	1	4	4	4	4	4
K. Dust explosion control	see Dow Guide					
L. Blast and barrier walls/separation	1	1	2	3	4	4

FIRE AND EXPLOSION INDEX	NAME				
LOCATION				JOB NUMBER	
PLANT Alcohol				CHARGE	
UNIT		Propane storage		Reaction area	
MATERIALS		Propane		Propane	
CATALYSTS		None		None	
REACTIONS		None		$C_3H_8 + O_2 \rightarrow CH_3OH + CH_3COCH_3 + CH_3$(	
SOLVENTS		None		None	
1. MATERIAL FACTOR FOR:		Propane	19.9	Propane	1
2. SPECIAL MATERIAL HAZARDS	% factor suggested	% factor used	19.9	% factor used	
A. Oxidizing materials	0-20				
B. Reacts with water to produce combustible gas	0-30				
C. Subject to spontaneous heating	30				
D. Subject to rapid spontaneous polymerization	50-75				
E. Subject to explosive decomposition	125				
F. Subject to detonation	150				
G. Other					
Add percentages A-G for special material hazard (SMH) total		0		0	
((100 + SMH total)/100) × (material factor) = subtotal no. 2 →			20	0.2 →	1
3. GENERAL PROCESS HAZARDS					
A. Handling and physical changes only	0-50	25			
B. Continuous reactions	25-50			50	
C. Batch reactions	25-60				
D. Multiplicity of reactions in same equipment	0-50				
Add percentages A-D for general process (GP) total		25		50	
((100 + GP total)/100 × (subtotal no. 2) = subtotal no. 3 →			25	→	
4. SPECIAL PROCESS HAZARDS					
A. Low pressure (below 15 psia)	0-100				
B. Operation in or near explosion range	0-150			100	
C. Low temperature: 1. Carbon steel 50° to −20°F	15				
2. Below −20°F	25				
D. High temperature (use one only)					
1. Above flash point	10-20				
2. Above boiling point	25	25		25	
3. Above autoignition point	35				
E. High pressure: 1. 250-3000 psig	30			30	
2. Above 3000 psig	60				
F. Processes or reactions difficult to control	50-100				
G. Dust or mist hazard	30-60				
H. Greater than average explosion hazard	60-100				
I. Large quantities of combustible liquids (use one only)					
1. 2000-6000 gallons	40-55				
2. 6000-20,000 gallons	55-75				
3. 20,000-50,000 gallons	75-100				
4. Above 50,000 gallons	100+	100			
J. Other					
Add percentages A-J for special process (SP) total		125		155	
((100 + SP total)/100) × (subtotal no. 3) = F & EI →			56.3	→	76

Figure 12.2 Combined Dow F&EI calculation sheets for units of alcohols plant (courtesy Dow Chemical Co.)

DATE							
Alcohol absorber		Alcohol separation		Alcohol storage		Propane recovery	
MATERIALS AND PROCESS							
Reactor products Propane		Alcohols, ketones, aldehydes		Alcohol		Propane	
None		None		None		None	
None		None		None		None	
Water		None		None		Absorption oil	
Propane	19.9	Methanol	8.6	Methanol	8.6	Propane	19.9
% factor used	19.9	% factor used		% factor used		% factor used	
0		0		0		0	
0.2 →	20	0.2 →	8.6	0.2 →	8.6	0.2 →	20
0		0		25		0	
0		0		25		0	
→	20	→	8.6	→	10.7	→	20
				50			
10				10		10	
		25					
30						30	
		55				55	
				100			
40		80		160		95	
→	28	→	15.5	→	27.8	→	39

Many of the same features appear in both Tables 11.7 and 11.8. The method used here in selecting them is explained in 11.7.

1. **Basic preventative and protective features**

A. Adequate water supply for fire protection;
B. Structural design of vessels, piping, structural steel;
C. Overpressure relief devices;
D. Corrosion resistance and/or allowances;
E. Segregation of reactive materials in process lines and equipment [8.6.1];
F. Electrical equipment grounding;
G. Safe location of auxiliary electrical gear;
H. Normal protection against utility loss (alternate electrical feeder, spare instrument air compressor, etc.);
I. Compliance with various applicable codes
J. Fail-safe instrumentation;
K. Access to area for emergency vehicles and exits for personnel evacuation;
L. Drainage to handle probable spills plus fire-fighting water;
M. Insulation of hot surfaces that reach 80% of autoignition temperature of any flammable in the area [10.5.1];
N. Adherence to the appropriate electrical codes;
O. Limitation of glass devices in flammable or hazardous service;
P. Building and equipment layout appropriate to hazard;
Q. Protection of pipe racks, instrument cable trays and supports from fire exposure;
R. Provision of accessible block valves at battery limits;
S. Protection of cooling tower(s)
T. Protection of fired equipment against accidental explosion and fire [10.3];
U. Special precautions for critical rotating equipment;
V. Appropriate construction of control rooms and their separation/isolation from control laboratories, electrical switchgear/transformers and potential sources of hydrocarbon release.

2. **Features dependent on F&EI of unit**

Table 11.7 gives minimum preventative and protective features which must be considered, but whose application depends on the probability and expected intensity of a fire or explosion as indicated by the F&EI calculated for the process unit. The features should be interpreted rather broadly, as explained in the guide and summarised as follows.

A. *Fire protection of structural supports*
This applies only when flammable liquid might be retained in the area, the fireproof rating depending on the expected depth of liquid retained. Drainage is, however, generally preferred to fireproofing of structures as a means of protection against pool fires.

B. *Water spray protection of equipment and area*
Water spray is mainly needed for protection against liquid and solid fires. Directional spray may be used to protect pumps, vessels and cable trays. Area spray is used in addition for F&EIs of 60 and over.

Table 11.8 Safety features for particular hazards (based on Table VII of Dow guide[1,2])

Hazard type	*Protective features*
2. **Special material hazards**	
A. Oxidising materials [8.5.2]	Keep separate from combustible material and store in fireproof area
B. Reacts with water to produce combustible gas [8.6.3]	Protect from water and ignition sources, ventilate gas formed and comply with appropriate electrical code
C. Subject to spontaneous heating [8.7]	Provide cooling
D. Subject to spontaneous polymerisation [8.6.4]	Provide polymerisation inhibitor system, cooling and over-pressure relief
E. Subject to explosions	Design equipment to contain or safely relieve explosion; provide temperature and/or pressure control if effective and consult recognised authority
3. General process hazards	
A. Handling and physical change only	For loading and unloading, provide excess flow and remotely operated valves, alarms for inadequate electrical grounding, purge procedures for vessels and lines, special fire-extinguishing systems and ventilation
B. Continuous reactions	Prevent reactant unbalance by appropriate instruments or other method; provide safe over-pressure relief, over/under-temperature alarm or automatic shutdown; consider measures to keep out hazardous impurities
C. Batch reactions	Same as (3B) with special procedures/instruments to avoid hazards during turnaround
D. Several reactions in same equipment	Same as (3C) with positive separation of reactants when not required
4. Special process hazards	
A. Low pressure	Provide instrument interlocks to avoid hazardous pressure range and alarms to indicate approach to hazardous condition, e.g. pressure, oxygen concentration
B. Operation in or near explosive range	Design equipment to contain or safely relieve explosion; consider explosion suppression, dilution or inerting to avoid explosive range and back-up instrumentation for process control
C. Low temperature	Provide special vent or dump systems
D. High temperature	Arrange to minimise flow of flammables; consider flammable gas monitors for alarm, shutdown or deluge actuation; provide special vent or dump system
E. High pressure (>17 bar g)	Provide quick-acting and safe vent/dump systems and remote or instrument operated valves to minimise consequences of line or equipment failure; consider flammable gas monitors for alarm, shutdown or deluge actuation
F. Processes or reactions difficult to control	Design equipment to contain or safely vent worst situation; consider dump, vent and quench systems
G. Dust hazard	Same as 4B

Water curtains serve to reduce the movement of vapour clouds and may be installed between furnaces and potential points of vapour release. Sprinkler piping, supports and nozzles are vulnerable to explosions and should be robustly constructed where this danger exists.

C. *Special instrumentation*

This covers instruments installed for safety rather than process control, e.g.

- interlocks between flow controllers which cause one to 'fail safe' if another flow fails,
- analysers which initiate alarms or shut down equipment when a hazardous condition is approached,
- remotely operated valves to stop the flow of flammable fluids in case of fire.

The need for such special instrumentation starts with F&EIs of 40, when it may be added to normal control loops. At F&EIs of 60 or more, additional safety instrumentation which is quite separate from control instrumentation is needed, and at F&EIs of 75 or more, the installation of redundant safety instrumentation should be considered. Safety instrumentation needs careful selection to have most effect.

D. *Dump, blowdown or spill control*

This covers systems designed to remove hazardous materials quickly from danger spots in an emergency. It includes fixed deluge equipment to wash away spilled materials, pump-out equipment, sloped drainage, headers to vent and flare stacks and the injection of inert and 'short stop' materials to control 'runaway' reactions.

E. *Internal explosion*

This category covers techniques which prevent explosive mixtures from forming inside process equipment or — where this is impossible — those which relieve, contain or suppress the explosion and/or eliminate sources of ignition. Many of these techniques depend on the use of inert gases [10.3].

F. *Combustible gas monitors* [10.6.3]

These may sound an alarm, actuate protective equipment such as sprinklers or ventilation fans or close valves on process lines. They should be located between points of possible release of flammable fluids and sources of ignition (e.g. furnaces) and in areas where natural ventilation is poor.

G. *Remote operation*

This is specially needed for operations which are too hazardous for personnel to be allowed in the same area. All process plants with control rooms depend to a large extent on remote operation.

H. *Building ventilation*

Three types are covered in the Dow guide, which refers to the appropriate NFPA standards:

- removal of smoke and hot gases during a fire;
- air change in areas where flammable liquids are handled;
- removal of airborne dust and harmful vapours in areas where their formation cannot be entirely prevented.

I. *Building explosion relief*
Operations where there is an explosion risk should preferably be located in separate detached buildings. The area of explosion vents needed depends in the expected explosion intensity [10.3.3].

J. *Diking*
This includes the sloping of ground at an angle of at least 2° away from tanks and other possible areas of spillage towards impounding basins capable of holding the largest spill which could occur. Where dike (bund) walls are provided round tanks their average height should not exceed 1.9m. Bunded areas should be provided with means of draining water in such a way that flammable spills do not escape.

K. *Dust explosion control*
This generally means providing explosion relief panels, the ratio of whose area to the volume relieved depends on the maximum expected rate of pressure rise. This in turn is largely dependent on the particle size of the dust [10.3].

L. *Blast and barrier walls and separation*
Blast walls are used to confine damage from mishaps with operations involving high pressures, potentially explosive materials and mixtures, and the use or processing of explosives. They may also be used to protect sprinkler valves. Where space is limited barrier walls may be used to separate flammable storage areas from process units.

3. **Preventative features for specific hazards**
These are given in Table 11.8.

11.1.7 Safety features for the example given

Table 11.9 shows the special features stipulated in Table 11.8 for each unit together with the corresponding feature and degree of priority indicated in Table 11.7. The features appropriate to the specific hazards of each unit were first selected from Table 11.8. Their priorities were then assessed from Table 11.7 and the F&EI of the unit. For units with F&EIs of 40 or over, any other features in Table 11.7 which could apply are also listed with their priorities.

11.1.8 Some comments on the Dow guide, 3rd edn

The different protective and preventative measures suggested by the guide seem reasonable although their priorities are more questionable and their details still have to be decided. As examples, the fireproofing of supports for propane pressure-storage vessels and the use of water spray to prevent their tops from overheating in a fire would now be generally regarded as essential rather than merely 'suggested' (A2) or 'recommended' (B3). This is hardly surprising, since the guide's recommendations are based on combinations of rather general hazards and not on specific cases. Final

Table 11.9 Safety features for each unit with priorities

Unit	*F&EI*	*Feature and priority*	
		From Table 11.8 with 11.7	*From Table 11.7 only, units (1) and (2)*
1.	56.3	3A not applicable	A(2), B(3), D(2), E(3) L(2)
		4D C(3), F(2), J(4)	
		4I same + G(2)	
2.	76.5	3B C(4), D(3)	possible A(4), B(4) and L(4)
		4B C(4), E(4)	
		4D C(4), F(3), J(4)	
		4E C(4), D(3), F(3), G(3)	
3.	28.0	4D C(2), F(1), J(4)	
		4E C(2), D(1), F(1), G(1)	
4.	15.5	4D C(1), F(1), J(1)	
		4I same + G(1)	
5.	39.0	4D C(2), F(1), J(4)	
		4E C(2), D(1), F(1), G(1)	
6.	27.8	4I C(2), D(1), F(1), G(1)	

decisions and details of the safety features needed require far more information than that given in the very sketchy flow diagram shown in Fig. 11.1. Safety features must, moreover, be considered realistically. Most can themselves fail or go wrong and compete with other items for management attention.

The Dow guide is essentially a condensation of much experience which would fill several volumes. One risk in using it against which it warns lies in taking an oversimplified view and applying it without considering the details of the case. Its main virtue is that it enables a reasonable profile both of the hazards and of the safety features needed for a plant to be quickly drawn from preliminary process information. This is particularly important during preliminary studies when alternative investment opportunities are being considered. Here the method enables process hazards and the cost of safety features needed to combat them to be taken into account at the earliest stage. The guide is also a useful adjunct to other hazard studies, particularly in the checklists which it provides, as a project takes shape and its design proceeds.

11.2 The Mond Index[5,6]

The 'Mond Fire, Explosion and Toxicity Index' was first presented by D. J. Lewis in 1979[5]. The second edition of the Mond Index which is discussed here was published as an 80-page booklet by the Explosion Hazards Section of the Mond Division of ICI.[6] It is developed from the third edition of Dow's F&EI guide and deals primarily with F&E hazards. Toxicity is considered only as a complicating factor. Like the Dow Index, it is a rapid hazard-assessment method for use on existing chemical plant, during process and plant development, and in plant design and layout.[7,9] When used during development design and layout, it highlights features requiring further study, thus enabling problems to be recognised and often eliminated, with financial saving. Its use generally should lead to a rational expenditure on safety items.

The Mond Index extends the calculation procedures of the Dow Index to highlight particular hazards. Thus it provides separate indices for fire, internal explosion and aerial explosion (OFCE) potential, as well as an overall hazard rating. Only its general features are given here. Readers intending to use the Mond Index are advised to attend one of the short courses which are offered on it. A computerised version is also available for use on an IBM PC.

The first step, as with the Dow Index, is to divide the plant into units, and it is better to start with too many than too few. The next step is to determine the material factor B, which provides a numerical base for the indices. The base is then modified by many other considerations contained in the following sections:

1. special material hazards, M
2. general process hazards, P
3. special process hazards, S
4. quantity hazards, Q
5. layout hazards, L
6. acute health hazards, H

Standard calculation sheets (Fig. 11.3) are used. These list the conditions with hazard potential with suggested penalty ranges for each of the six hazard factors.

Most penalties are positive, but 'negative penalties' can also be applied, e.g. to gases which rise rapidly and to marginally flammable materials such as trichloroethylene. From the seven hazard factors for each unit, the overall or equivalent Dow Index and the three special hazard ratings are calculated.

The equivalent Dow Index, D, is given by the formula:

$$D = B(1 + M/100)(1 + P/100)(1 + [S + Q + L + T]/100)$$

The three special indices for fire, F, internal explosion, E, and aerial explosion, A, are first calculated for basic standard protection only. Offsetting factors for various forms of special protection are then applied and final adjusted values of these three indices are then calculated. Finally an overall risk rating is calculated, before and after offsetting this for the special protective measures applied or considered.

11.2.1 Dominant material and material factor *B*

The dominant (key) material on which the material factor is based is next selected. It is defined as:

> That compound or mixture in the unit which, due to its inherent properties and the quantity present, provides the greatest potential for an energy release by combustion, explosion or exothermic reaction.

The material factor, *B*, is in most cases the net heat of combustion of the material in air, expressed as thousands of Btu per pound (2326 kJ/kg). For reactive combinations of materials, the heat of reaction is used if it exceeds the heat of combustion. This material factor is practically the same as that given in the third edition of Dow's guide.

11.2.2 The six hazard factors

The first three, *M*, *P* and *S*, of these factors are elaborations of those given in the third edition of Dow's guide.

The **Special Material Hazards Factor (*M*)** is applied to take into account any special properties of the key material which may affect either the nature of the incident or the likelihood of its occurrence. Ten properties are listed, with appropriate penalties. They include any tendencies of the key material to act as an oxidant, to polymerise spontaneously, to decompose violently, to detonate, etc. One property designated (*m*) represents the mixing and dispersion characteristics of the material and also features in the aerial explosion index. The highest penalties recommended are for unstable materials which can deflagrate or detonate.

The **General Process Hazards Factor (*P*)** relates to the basic type of process or other operation being carried out in the unit. Six main types are listed, including material transfer, physical change only and various types of reaction with different characteristics.

The **Special Process Hazards Factor (*S*)** reflects 14 listed features of the process operation which increase the overall hazard beyond the basic levels already considered. These take account of operating temperature and pressure, corrosion, erosion, vibration, control problems, electrostatic hazards, etc.

S is evaluated on the assumption that the plant has an adequate control system for normal operation. Credits for more sophisticated safety features such as explosion suppression and combustible gas monitors are applied later.

The **Quantity Hazards Factor (Q)** represents the quantity of combustible, flammable, explosive or decomposable material in the unit which is treated as a separate factor in the Mond Index. It is related to the total quantity *K* in tonnes of such material in the unit. *K* also features in the fire index.

The **Layout Hazards Factor (L)** is another separate factor in the Mond Index. The normal working area *N* of the unit in square metres also features in the fire index, and is defined 'as the plan area of the structure associated with the unit, enlarged where necessary to include any pumps and associated equipment not within the plan area of the structure'.

The height H in metres above ground at which flammable materials are present in the unit also features in the aerial explosion index. L also includes factors for the relation of ventilation rates to flammable vapours which could escape, and 'domino effects' involving the spread of incidents from one unit to another.

The **Acute Health Hazards Factor** (T) is not intended to reflect health hazards as such, but rather the delay caused by the toxicity of escaping materials when tackling a developing or potential fire or explosion. The factor is the sum of penalties for skin effects and inhalation.

11.2.3 Calculation of indices

The equivalent Dow Index (third edition) whose formula was given earlier is not used for interpretive purposes but features in later calculations.

The **Fire Index** (F) relates to the amount of flammable material in the unit, its energy release potential and the area of the unit and is given by:

$$F = B \times K/N$$

F values in the range 0–2 class as 'Light', 5–10 as 'Moderate' and 100–250 as 'Extreme'. The Fire Index is related to the Fire Load which would equal $2442 \times F$ in kJ/m^2 ($215 \times F$ in BTU/ft^2) if all the available combustible material were consumed. In practice often only 5–10% is consumed before the incident is controlled.

Typical fire durations are related to the fire load for four different materials, solid combustibles, heavy crude oil, flammable light liquids, and LPG/LNG, and decrease in that order. For a fire load of 100 000 Btu/ft^2 the fire duration ranges from 10 minutes to 2½ hours, and for a fire load of 1 million Btu/ft^2 from 40 minutes to one day.

The **Internal Explosion Index** (E) is a measure of the potential for explosion within the unit and is given by:

$$E = 1 + (M + P + S)/100$$

An internal explosion index of 0–1.5 is categorised as light, one of 2.5–4 as moderate and one above 6 as very high.

The **Aerial Explosion Index** (A) relates both to the risk and magnitude of a vapour cloud explosion [10.2.4] originating from a release of flammable material, usually present within the unit as a liquid at a temperature above its atmospheric boiling point. This index includes quantitative and qualitative factors, and is given by:

$$A = B(1 + m/100)(QHE/1000)(t + 273/300)(1 + p)$$

The **Overall Hazard Rating (R)** is used to compare units with different types of hazards, and is given by:

$$R = D\,(1 + [0.2 \times E \times (AF)^{1/2}])$$

MOND INDEX 1985
* * * * * * * * * * * * * * * * * *

PAGE NO. 1
FILE NO.
NAME
DATE

LOCATION

PLANT

UNIT

MATERIALS

ADDITIONAL INFORMATION

COMMENT NUMBER

PRESSURE = psig TEMPERATURE t= DEG.C

MATERIAL FACTOR (Section 5)

KEY MATERIAL OR MIXTURE :
FACTOR DETERMINED BY :
MATERIAL FACTOR : B=

SPECIAL MATERIAL HAZARDS (Section 6)

	RANGE		FACTOR INITIAL	FACTOR REVIEW
1.OXIDISING MATERIALS	0 TO 20			
2.GIVES COMBUSTIBLE GAS WITH WATER	0 TO 30			
3.MIXING & DISPERSION CHARACTERISTICS	-60 TO 100	m		
4.SUBJECT TO SPONTANEOUS HEATING	30 TO 250			
5.MAY RAPIDLY SPONTANEOUSLY POLYMERISE	25 TO 75			
6.IGNITION SENSITIVITY	-75 TO 150			
7.SUBJECT TO EXPLOSIVE DECOMPOSITION	75 TO 125			
8.SUBJECT TO GASEOUS DETONATION	0 TO 150			
9.CONDENSED PHASE PROPERTIES	200 TO 1500			
10.OTHER	0 TO 150			
SPECIAL MATERIAL HAZARDS TOTAL		M		

GENERAL PROCESS HAZARDS (Section 7)

1.HANDLING & PHYSICAL CHANGES ONLY	10 TO 50	
2.REACTION CHARACTERISTICS	25 TO 50	
3.BATCH REACTIONS	10 TO 60	
4.MULTIPLICITY OF REACTIONS	25 TO 75	
5.MATERIAL TRANSFER	0 TO 150	
6.TRANSPORTABLE CONTAINERS	10 TO 100	
GENERAL PROCESS HAZARDS TOTAL		P

MOND INDEX 1985
* * * * * * * * * * * * * * * * * *

PAGE NO. 2
FILE NO.

SPECIAL PROCESS HAZARDS (Section 8)

1.LOW PRESSURE(BELOW 15 PSIA)	50 TO 150	
2.HIGH PRESSURE	0 TO 160	p
3.LOW TEMP.:1.CARBON STEEL +10C TO -25C	0 TO 30	
2.CARBON STEEL BELOW -25C	30 TO 100	
3.OTHER MATERIALS	0 TO 100	
4.HIGH TEMP.1.FLAMMABLE MATERIALS	0 TO 35	
2.MATERIAL STRENGTH	0 TO 25	
5.CORROSION & EROSION	0 TO 400	
6.JOINT & PACKING LEAKAGES	0 TO 60	
7.VIBRATION,LOAD CYCLING,ETC.	0 TO 100	
8.PROCESSES/REACTIONS DIFFICULT TO CONTROL	20 TO 300	
9.OPERATION IN OR NEAR FLAMMABLE RANGE	25 TO 450	
10.GREATER THAN AVERAGE EXPLOSION HAZARD	40 TO 100	
11.DUST OR MIST EXPLOSION HAZARD	30 TO 70	
12.HIGH STRENGTH OXIDANTS	0 TO 400	
13.PROCESS IGNITION SENSITIVITY	0 TO 100	
14.ELECTROSTATIC HAZARDS	10 TO 200	
SPECIAL PROCESS HAZARDS TOTAL		S

QUANTITY HAZARDS (Section 9)

MATERIAL TOTAL TONNES	K
QUANTITY FACTOR	Q

LAYOUT HAZARDS (Section 10)

HEIGHT IN METRES		H
WORKING AREA IN SQUARE METRES		N
1.STRUCTURE DESIGN	0 TO 200	
2.DOMINO EFFECT	0 TO 250	
3.BELOW GROUND	50 TO 150	
4.SURFACE DRAINAGE	0 TO 100	
5.OTHER	50 TO 250	
LAYOUT HAZARDS TOTAL		L

ACUTE HEALTH HAZARDS (Section 11)

1.SKIN EFFECTS	0 TO 50	
2.INHALATION EFFECTS	0 TO 50	
ACUTE HEALTH HAZARDS TOTAL		T

MOND INDEX 1985
****
PAGE NO. 3
FILE NO.

OFFSETTING INDEX VALUES FOR SAFETY & PREVENTATIVE MEASURES
****

A.CONTAINMENT HAZARDS (Section 16.1)
1-PRESSURE VESSELS
2-NON-PRESSURE VERTICAL STORAGE TANKS
3-TRANSFER PIPELINES A)DESIGN STRESSES
B)JOINTS & PACKINGS
4-ADDITIONAL CONTAINMENT & BUNDS
5-LEAKAGE DETECTION & RESPONSE
6-EMERGENCY VENTING OR DUMPING

PRODUCT TOTAL OF CONTAINMENT FACTORS K1=

B.PROCESS CONTROL (Section 16.2)
1-ALARM SYSTEMS
2-EMERGENCY POWER SUPPLIES
3-PROCESS COOLING SYSTEMS
4-INERT GAS SYSTEMS
5-HAZARD STUDIES ACTIVITIES
6-SAFETY SHUTDOWN SYSTEMS
7-COMPUTER CONTROL
8-EXPLOSION/INCORRECT REACTOR PROTECTION
9-OPERATING INSTRUCTIONS
10-PLANT SUPERVISION

PRODUCT TOTAL OF PROCESS CONTROL FACTORS K2=

C.SAFETY ATTITUDE (Section 16.3)
1-MANAGMENT INVOLVEMENT
2-SAFETY TRAINING
3-MAINTENANCE & SAFETY PROCEDURES

PRODUCT TOTAL OF SAFETY ATTITUDE FACTORS K3=

D.FIRE PROTECTION (Section 17.1)
1-STRUCTURAL FIRE PROTECTION
2-FIRE WALLS,BARRIERS
3-EQUIPMENT FIRE PROTECTION

PRODUCT TOTAL OF FIRE PROTECTION FACTORS K4=

E.MATERIAL ISOLATION (Section 17.2)
1-VALVE SYSTEMS
2-VENTILATION

PRODUCT TOTAL OF MATERIAL ISOLATION FACTORS K5=

F.FIRE FIGHTING (Section 17.3)
1-FIRE ALARMS
2-HAND FIRE EXTINGUISHERS
3-WATER SUPPLY
4-WATER SPRAY OR MONITOR SYSTEMS
5-FOAM & INERTING INSTALLATIONS
6-FIRE BRIGADE ATTENDANCE
7-SITE CO-OPERATION IN FIRE FIGHTING
8-SMOKE VENTILATORS

PRODUCT TOTAL OF FIRE FIGHTING FACTORS K6=

MOND INDEX 1985
****
PAGE NO. 4
FILE NO.

EQUATIONS
=========

EQUIVALENT DOW INDEX (for initial assessment and review)

D = B(1+M/100)(1+P/100)(1+(S+Q+L+T)/100))

FIRE INDEX

INITIAL ASSESSMENT AND REVIEW F= BK/N

OFFSET F*K1*K3*K5*K6

INTERNAL EXPLOSION INDEX

INITIAL ASSESSMENT AND REVIEW E=1+(M+P+S)/100

OFFSET E*K2*K3

AERIAL EXPLOSION INDEX

INITIAL ASSESSMENT AND REVIEW A=B(1+m/100)(1+p)(QHE/1000)(t+273)/300

OFFSET A*K1*K2*K3*K5

OVERALL RISK RATING

INITIAL ASSESSMENT AND REVIEW R=D(1+(.2E*SQUARE ROOT(AF)))

OFFSET R*K1*K2*K3*K4*K5*K6

INDICES COMPUTATION
===================

INDEX	INITIAL		REVIEW		OFFSET	
	VALUE	CATEGORY	VALUE	CATEGORY	VALUE	CATEGORY
D						
F						
E						
A						
R						

Figure 11.3 Calculation sheets for Mond Index (courtesy Explosion Hazards Section, ICI Mond Division, Northwich)

An Overall Hazard Rating of 0–20 is categorised as light, 100–500 as moderate, 1100–2500 as high and over 12 500 as extreme.

11.2.4 Criteria and review

Ranges of the four indices for different degrees of hazard are given in Table 11.10.

The most important index is the overall hazard rating *R*. Experience from applying the full method to operating plants has shown that it is uncommon for a unit, after a complete assessment, to have an overall hazard (*R*) with a category rating greater than 'high'. It is therefore reasonable to assume that a unit assessed at this level can be operated in a satisfactory manner given full regard to the hazards indicated by the assessment. Offsetting usually reduces the overall hazard category by one or two levels and gives a clearer picture of the relative importance of the different protective measures which could be taken.

When the initial assessment is unfavourable, the estimates should be refined by the use of better data. The effects of possible changes in materials of construction, sizes and types of equipment and process conditions, and reduction in inventory should also be considered. When changes thus indicated have been made and all factors have been reviewed, their new values are entered in the 'Reduced Value' column of the form with a note on the reason for the change. The final stage of the index calculations in which the hazards are reduced by applying special safety features and protective measures is done on the basis of these reduced values.

The scope for reductions in the indices by design changes is greatest before the design is finalised. On existing plants most improvements result from the incorporation of the safety features and preventative measures contained in the offsetting section. However, reducing inventories has a significant effect on fire potential and can usually be achieved on new and existing plants.

11.2.5 Index reduction by offsetting measures

Safety features and preventative measures may reduce the probability or magnitude of an incident (sometimes both). The booklet classifies them in

Table 11.10 Mond Index ranges for various degrees of hazard

Potential hazard category	*Fire* *F*	*Internal explosion* *E*	*Aerial explosion* *A*	*Overall hazard* *R*
Mild or light	0–2	0–1.5	0–10	0–20
Low	2–5	1.5–2.5	10–30	20–100
Moderate	5–10	2.5–4	30–100	100–500
High	10–30	4–6	100–400	500–2500
Very high	20–50	>6	400–1700	2500–12 500
Extreme	100–250		>1700	12 500–65 000
Very extreme	>250			>65 000

these two groups and devotes 20 pages to discussing them. Factors are suggested by which the values of the appropriate index should be multiplied when a safety feature or preventative measure (which is additional to the basic standard) is introduced. Before such measures can be evaluated the basic standards which would apply to the design, construction, operation and personnel training have to be defined. As examples, the basic standard for pressure vessel design is taken as Pressure Vessel Construction Category 3 of BS 5500, and the basic standard for process control instrumentation is the minimum compatible with operation under normal design conditions (i.e. without alarms or trip systems).

Three broad categories of safety features and preventative measures which reduce the *probability* of an incident, and the symbols used for the product totals of their sub-factors, are:

A. features which improve containment of process materials K_1
B. features which improve the safety of process control K_2
C. features which improve safety awareness of personnel K_3

There are several possibilities in each category. The factor for each category is the product of the suggested values for the features and measures which apply.

Three more broad categories of safety features and preventative measures are considered to reduce the *magnitude* of any incident. These are:

D. fire protection K_4
E. isolation of process materials K_5
F. firefighting K_6

The factor for each category is obtained in the same way as for the first group. Brief descriptions of the features and measures considered in each category are given in the calculation sheet. Where only the basic standards apply, a factor of 1 is used.

The factors K_1 to K_6 are calculated for the actual or proposed protective features. The offset indices are then obtained by multiplying the original (reduced) indices by the appropriate offsetting factors.

Offset Fire Index $= F \times K_1 \times K_3 \times K_5 \times K_6$
Offset Internal Explosion Index $= E \times K_2 \times K_3$
Offset Aerial Explosion Index $= A \times K_1 \times K_2 \times K_3 \times K_5$
Offset Risk Rating $= R \times K_1 \times K_2 \times K_3 \times K_4 \times K_5 \times K_6$

The benefits given by the protective features are assessed by comparing the degrees of hazard for the original and the offset indices. These benefits apply only when the protective hardware is maintained and is in proper working order and when the management procedures on which the benefits depend are followed. Neglect of either will cause the indices to revert to their original values. (The importance of maintaining special protective features was clearly demonstrated by the Bhopal disaster [5.3].)

11.3 Plant layout and unit hazard rating[7]

The layout of process installations is a complex task, even where there are no serious hazards. The site and layout chosen must satisfy commercial criteria as well as being socially and legally acceptable (in the eyes of the local planning authority assisted in the UK by advice from HSE and its Major Hazard Assessment Unit). While industrial planners and plant contractors deal with layout problems frequently, most personnel in the industry, including safety specialists, usually have to live with existing layouts. Risk analysis [14] is often required. The final layout is usually a compromise. Mecklenburgh in the I. Chem. E. guide *Plant Layout*[9] put it thus:

> In general it is true to say that the most economical plant layout is that in which the spacing of the main equipment items is such that it minimises interconnecting pipework and structural steelwork. . . . As a general rule, as compact a layout as possible with all equipment at ground level is the first objective consistent with access and safety requirements.

'Consistency with safety requirements' is open to a wide range of interpretations. Absolute safety being impossible, one has to decide what degree of risk is acceptable.[10] If this were judged on purely economic grounds and insurance was 100% efficient, we would have reached a state where the layouts adopted minimised the overall costs, including capital, operation, maintenance and insurance. In practice only a few large enterprises (which often carry much of their own insurance) have the knowledge needed to achieve such optimum layouts.

Up to quite recently, safety distances were decided largely on the basis of personal judgement and company policy. As Yelland put it in 1996, 'The existing literature in this field is wide, diffused and often contradictory.'[11] Minimum safety distances have, however, been specified for the storage of highly flammable liquids,[12] liquid petroleum gases[13] and explosives[14][9]. The Oil Insurance Association of Chicago has for many years published general guidelines and recommended spacings for plants and plant units in the oil, gas and petrochemical industries.[15]

Since Flixborough and other more recent disasters in the process industries, far more attention has been given to plant layout and siting.[16] Lees[4] deals with the subject in depth and gives a comprehensive bibliography on layout. Liston's chapter in Fawcett and Wood's book[17] is useful reading for those faced with layout problems. Figure 11.4(a), taken from this reference, shows the cross-section of a recommended 'in-line' arrangement of process equipment, with the relative positions of roads, pipebridges, pumps and plant structures. This has roadways flanking the equipment on either side, giving excellent access for firefighting, etc. Next to the road on the left is a gantry-way with a crane on rails to handle heavy equipment for maintenance, with buried cooling water headers running below it. This provides useful space for tube bundles and other equipment to be stored or worked on during maintenance. Even where a gantry cannot be justified, an open strip between the road and the plant is recommended. Next to the right lies an area depressed by about 200 mm, where fractionating towers, heat-exchangers, vessels and other stationary process equipment are located. This area is divided by low walls (about 200 mm) running across it to isolate spills of process liquids which drain into special sewers. Process pipes run on raised racks to the right of the depressed area, thus allowing unhindered passage to persons under them. Air-cooled heat-exchangers are located above the pipe racks. Pumps and other rotating equipment are located in a line to the right of the pipe racks and adjacent to the road on the right flank of the in-line arrangement. Other components of the unit such as control rooms, furnaces and reactors may be located at either end of the in-line arrangement. Figure 11.4(b) from the same reference shows an integrated plan of several 'in-line' arrangements of process units, with a main pipe-rack at one end separating them from furnaces, reactors, compressors and boiler plant, etc.

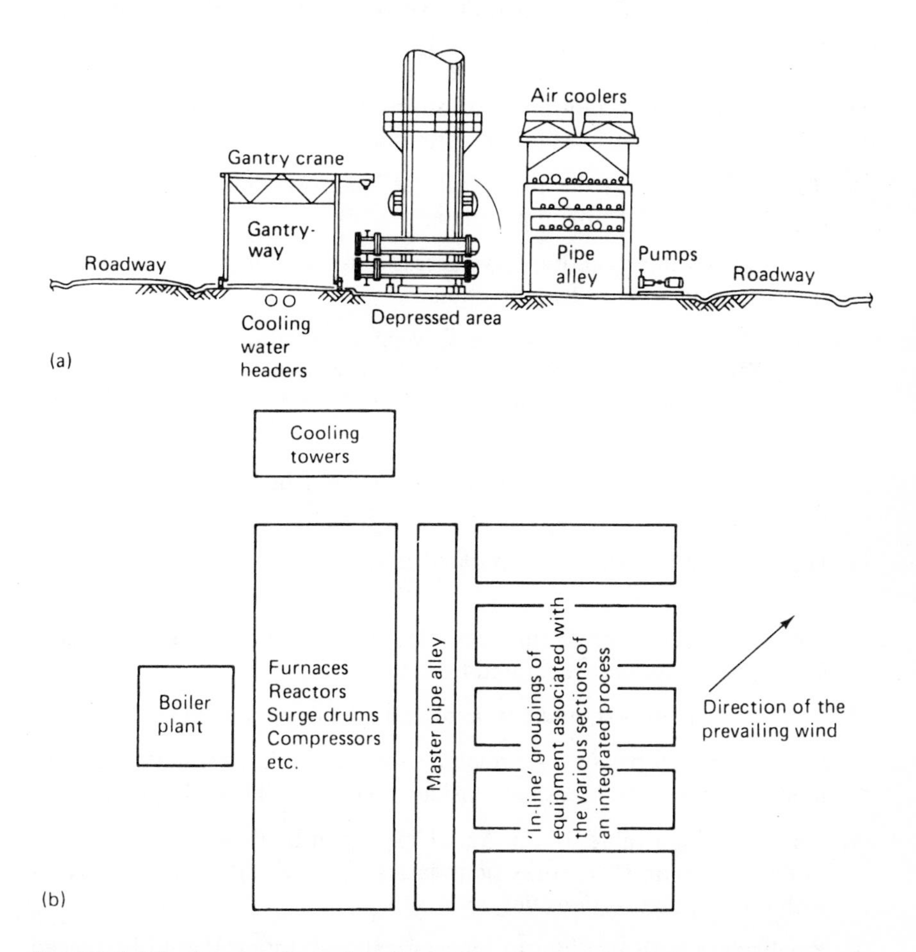

Fig. 11.4 (a) Cross-section of in-line arrangement of process equipment. (b) Example of applying the in-line arrangement to integrated processing units (from Liston[17], courtesy Wiley-Interscience)

Few of these references give recommendations on the separation of plant and storage units of various degrees of hazard from one another and from boundaries, office and other buildings, etc. This can only be done if one has some numerical scale such as the Dow or Mond Index for the degree of hazard of units themselves. Dow's guides do not give explicit spacing recommendations, although some latter ones[3] allow the maximum probable property damage for any proposed layout to be estimated, and the layout to be revised to reduce excessive damage estimates.

After publication of the Mond Index its author, D. J. Lewis, gave a further paper on its application to plant layout and spacing problems.[7] The spacings proposed in this method are intended to ensure that incidents apart from disasters will produce only moderate damage to adjacent plant sites, with minimal effects outside the works boundary. The spacings needed to prevent knock-on effects arising from disasters such as major explosions, major tank froth-overs and catastrophic vessel failures are not considered to be reasonably practicable.

Lewis refers to the Overall Risk Rating before allowance for offsetting factors as R_1, and after allowance for offsetting as R_2. He first recommends the following basic principles for an optimum layout.

1. Roads should enter the plant from at least two points, preferably on opposite sides of the plant perimeter.
2. Emergency vehicles should have access to units with moderate and high fire risks from at least two directions.
3. Control rooms, amenity buildings, workshops, laboratories and offices should be close to the site perimeter, and next to units of mild or low risk to shield them from higher-risk units.
4. Pipebridges should not allow incidents to be transferred easily from one unit to another. Key pipebridges should be assessed for hazard potential.
5. High-risk units should be separated from each other by units of lower risk.
6. Ignition sources such as furnaces, electrical switchgear and flare stacks should be adequately separated from units.
7. Medium-risk units should not be placed next to populated buildings.
8. Units with high risk of toxic release should be well separated from populous buildings and areas both inside and outside the works perimeter.
9. Units with high values of the Aerial Explosion Index A_2 should be separated from plants or works boundaries by areas of low-risk activities with low population densities.
10. Pipebridges with medium to high overall risk rating should be placed where they are at least risk from transport accidents and incidents on tall process units.

11. Units previously treated separately for hazard ratings may be combined into larger ones, providing (a) the risks are similar, (b) potential losses are not excessive and (c) the reassessed rating R_2 is not excessive.
12. The process flow should be as logical as possible to minimise pipework.
13. Pipebridge routes should be chosen to minimise the spread of incidents and should allow as much control of flows as possible if there is an incident.
14. Storage units should be well separated from operational areas and as far as possible from roads and railways within the site.

These principles allow an initial layout to be prepared with a nominal inter-unit spacing of 10 m. This layout is related to the flow of materials and control and maintenance requirements. The basic pipebridge and access routes are marked on it and the nominal inter-unit distance is then replaced by distances chosen logically. After explaining how spacing distances are to be measured (e.g. whether from the bund wall of a storage tank or from the tank itself), Lewis gives five tables of spacing distances involving process and storage units.

In cases where process and other considerations do not allow the recommended spacing to be used, special means such as fixed water sprays should be considered to reduce the effects of heat radiation or the spread of fire or to control the escape of toxic or flammable materials.

After considering the effects of serious releases of toxic and flammable materials (over 5 t for the latter), and units of high Aerial Explosion Index A_2, Lewis recommends that these be located near the centre of a site.

A problem which can arise with Lewis's method is that the process units selected for the Mond Index tend to be small. Two or more may occupy the same structure and interpenetrate, with no clear boundary between them. Lewis therefore suggests that small units may be amalgamated into larger ones. His proposals on distances are consistent although his figures are open to debate. His paper contains no actual examples of the application of his method to plant layout.

11.4 Maximum probable property damage from vapour cloud explosions (OFCEs)

At present about two or more OFCEs occur world-wide every year. Because of their enormous destructive potential, insurers are concerned both to identify installations where one might occur and to estimate the maximum resulting loss. International Oil Insurers (IOI), London, have published EML (estimated maximum loss) 'Rules',[18] recently updated as 'guidelines'[19] for use by insurance surveyors. International Risk Insurers (IRI), Chicago, have also published their own guide[20] which differs in several respects from the IOI rules, for estimating the 'maximum probable property damage'

(MPPD). (The terms EML and MPPD appear to have identical meanings. Both refer to property losses including plant inventories, but exclude personal injuries and business interruption losses.)

CWA Information and Research Ltd of London offer computer programs for EMLs resulting from OFCEs based on the IOI rules, and also for 'instantaneous fractional annual losses'[21] caused by fires and explosions on plants handling flammable materials.

The method outlined here is a fairly simple one (which the writer developed in the 1970s for an international insurance broker). It aims to combine the best features of the IOI and IRI methods and serves as an introduction to more elaborate procedures.

11.4.1 Flammable materials with OFCE potential

Broadly speaking, all flammable gases and flammable liquids (flash point <40°C) are assumed capable of supporting an OFCE, when present in appropriate quantity and condition. C_1 to C_4 hydrocarbons are classed as gases and most C_5 to C_9 hydrocarbons are flammable liquids.

Gases

Gases vary considerably in their fire and explosion hazards [10]. Some like ammonia, methyl chloride and carbon monoxide which have low heats of combustion are unlikely to support OFCEs on their own, but may contribute when mixed with other gases. Some vapours such as acetaldehyde are very readily ignited, e.g. by a hot steam pipe. Butadiene, vinyl acetylene, formaldehyde, acetaldehyde, acrolein and ethylene oxide are capable of spontaneous exothermic polymerisation under appropriate conditions [8.6.4], while acetylene, vinyl acetylene and ethylene oxide can explode in the absence of air [9.2].

To qualify as an OFCE hazard, a gas needs to be flammable and present in one of the following forms.

1. As gas under a pressure of 35 bar g. or more. This form is generally restricted to gases with critical temperatures below 20°C. Examples are hydrogen, carbon monoxide, methane and ethylene.

2. As refrigerated liquid. This form can cover the entire range of flammable gases from hydrogen to butane and ethylene oxide.

3. As liquid under pressure. This form is generally restricted to gases with critical temperatures above and boiling points below 20°C. Examples are C_3 and C_4 hydrocarbons, ammonia, ethylene oxide, ethyl chloride and vinyl chloride.

The gases involved most frequently in vapour cloud explosions are C_3 and C_4 hydrocarbons, present initially in form (3).

Liquids

To qualify as an OFCE hazard, a flammable liquid must generate a large amount of vapour very rapidly when it is released. This usually means that it is initially under pressure and at a temperature at least 10°C higher than its atmospheric boiling point. Most hydrocarbon liquids at their boiling points under a pressure of 20 bar g. vaporise completely when the pressure is suddenly reduced to atmospheric. Their heats of combustion (gross) lie between 42 000 and 50 000 kJ/kg. For non-hydrocarbon organic liquids, the percentage vaporised under similar conditions lies between 25 and 85%, with heats of combustion between 10 000 and 40 000 kJ/kg. These differences partly explain why hydrocarbon liquids have been far more frequently involved in OFCEs than non-hydrocarbons.

Liquid hydrocarbons which have been involved in OFCEs include gasolines, intermediate light fractions and naphthas in the gasoline boiling range, the light ends of crude oils, aromatics and cyclohexane.

Organic liquids which could in theory support an OFCE include the lower alcohols, aldehydes, ketones, esters, ethers, oxides, acid anhydrides, as well as some compounds containing chlorine, sulphur and nitrogen. Only a few of these have actually caused an OFCE.

11.4.2 Sources of OFCEs

While the usual point of escape has been a pipe or pipe fitting, the sources of the material escaping have generally been vessels, due to their much larger inventories compared to pipes (except for pipelines). Sources are therefore grouped as:

1. storage vessels and tanks
2. process vessels (including columns and reactors)
3. pipelines (generally large or long).

Storage vessels and tanks

Pressurised storage vessels containing more than 10–20 t of liquefied flammable gases are nearly all potential sources of OFCEs. In theory the entire inventory of a full vessel could escape and vaporise in the event of a rupture. Most liquid storage tanks under ambient conditions do not qualify except for:

- storage tanks containing flammable liquids immiscible in water which have been fitted with steam coils or other means of heating;
- storage tanks containing a material capable of spontaneous exothermic polymerisation causing part of the contents to boil [8.6.4];
- cryogenic storage tanks containing liquefied flammable gases including LNG and LPGs — as with pressurised storage, the entire contents could escape and vaporise in the event of a rupture.

Provided the storage vessels and tanks are well isolated by automatic or remotely operated valves, the maximum amount of an escape is usually limited to the contents of the largest single tank or vessel whose contents qualify for OFCE potential.

Process vessels and equipment

Two types must be considered, high-pressure gas-filled equipment and moderate to low-pressure equipment in which most of the flammable material is in the liquid state.

High-pressure gas-filled equipment is typical of hydrocrackers and HP hydrogenation processes, OXO plant reactors, and HP polyethylene, ammonia, synthetic methanol and ethanol plants. The inventory of flammable gas in an HP train is calculated on the assumption that the entire contents of the high-pressure system downstream of the compressor will escape in the event of rupture, unless proven means of automatically isolating parts of the system rapidly in case of rupture have been installed.

Process vessels containing liquid include feed and buffer vessels, reactors, settlers and separators, liquid–liquid extraction columns, gas-absorption columns, absorbers and driers, distillation columns with associated reboilers, condensers, reflux drums and heat exchangers. In estimating the maximum amount of flashing liquid which could escape from any vessel in the event of rupture, it is assumed to be filled with liquid to its maximum normal operating level at the time of the escape. The corresponding volume of liquid, V_{max} is calculated from this level and the vessel dimensions, after deducting the volumes of vessel internals, packing, catalyst, etc.

The maximum weight of liquid is then given by:

$$W_L = d_L V_{max}$$

where W_L is maximum liquid inventory, t
d_L is flammable liquid density under operating conditions

To judge the extent to which the contents of connected vessels contribute to the escape, the main process lines containing flammable liquids, both upstream and downstream of the hypothetical point of escape, need to be examined. Here it is assumed that (properly maintained) non-return valves downstream of the point of escape close when an escape occurs, as do automatic and remotely operated isolation valves designed to close in the event of an escape.

The position of branches on vessels connected to the pipe or vessel from which the escape occurs is important. Only the liquid contents of these vessels which lie above these branches need normally be considered, except where there are internal baffles or there is reason to expect considerable flashing and entrainment in the vessel. This is allowed for in 11.4.3.

The probability of escape from other parts of the process unit cannot be ignored. Pumps, stirrer glands, heat exchanger tubes and pipework are more likely to fail than the shells of pressure vessels.

11.4.3 Maximum plausible size of explosion

The maximum plausible size of explosion is described as tonnes of TNT equivalent, which is related to the quantity of flammable gas or vapour escaping by the equation:

$$W_e = W_v Hf/4652$$

where W_e is equivalent weight of TNT, t
W_v is weight of flammable vapour released, t
H is heat of combustion (gross) of flammable vapour, kJ/kg
f is explosive yield factor
4652 is heat of combustion of TNT, kJ/kg

The main problem lies in assessing a realistic value for f, which allows for the facts that part of the vapour will escape without ignition, and that part will burn before the rest explodes. For estimating MPPD, a value for f of 0.05 is assumed. This is based on historical studies of many OFCEs. For escapes of flammable liquids at temperatures above their atmospheric boiling points, W_v is calculated as:

$$W_v = \frac{2W_L C_p(T_1 - T_2)}{H_v}$$

with a maximum value of $W_v = W_L$
where W_L is weight of liquid escaping, t
C_p is specific heat at $(T_1 + T_2)/2$ of liquid escaping, kJ/kg
T_1 is temperature of liquid in plant
T_2 is atmospheric boiling point of liquid
H_v is heat of vaporisation of liquid at T_2, kJ/kg

Here the weight of vapour flashing at the point of release is arbitrarily doubled to allow for the further evaporation of liquid spray in air. For liquefied flammable gases with boiling points below 20°C:

$$W_v = W_L$$

Where it is necessary to take account of vapour released from a series of connected vessels by flashing of liquid inside them, the following expression is used:

$$W'_v = \frac{W'_L \cdot C_p(T_1 - T_2)}{H_v}$$

with a maximum value of $W'_v = W'_L$.

W'_v and W'_L apply to the flammable liquid contents of other vessels connected to the vessel considered as the source of the vapour cloud, and which lie below the connecting nozzles of the vessels to which W'_L refers. The total vapour released is then given by:

$$W_{VT} = W_v + W'_v$$

11.4.4 Maximum probable property damage (MPPD)

Assessment of maximum probable property damage is based on the assumption that the vapour cloud explodes with its centre either at the point of release or at some other centre within the works boundary [11.4.5] where it might cause even greater damage. Here one needs an accurate plot plan of the works, detailed cost data for the installed equipment, pipework, buildings and structures, and for the inventories of process materials in process and storage, as well as a calculator and drawing compass.

Either the IOI[18, 19] or the IRI[20] method may be used for estimating property damage. The latter is more detailed although the IOI method appears adequate for most purposes considering the approximate nature of these estimates. It assumes an average property damage of 80% within a circle of radius R centred on the probable centre of the explosion, and an average property damage of 40% due to blast alone within a concentric annular area between radii R and $2R$. To the latter figure a further 40% is added for refinery and petrochemical plant and storage areas which contain significant inventories of flammable materials, to allow for damage caused by secondary fires. The radius R is related to the power of the explosion. Table 11.11, which has been adapted from the IOI method to match the present treatment, gives values of R and $2R$ for various TNT equivalence.

Having selected one or more potential sources of OFCE emissions and estimated their TNT equivalence, circles with the radii indicated in Table 11.11 are drawn on the works plan with the sources as centres. The damage occurring within the inner circle and in the annulus for each source is then estimated and the two are added to give the total MPPD for each source. The MPPD is usually calculated on the basis that it includes the inventory of process materials. Whether it does so or not should be clearly stated. Where there are several potential sources within the same works, one is mainly interested in the source giving the highest MPPD.

The question of cloud drift before ingition is mainly of interest when there are areas with higher densities of property value within or adjacent to the works perimeter than those near the source of escape. This should be apparent from the works plan with the property values marked on it.

Table 11.11 Assumed OFCE damage in target areas

Explosion equivalence (t TNT)	*Circle radii* (m)	
	R, inner circle 80% damage	*2R, outer circle 80% blast damage in annulus*
5	80	160
10	100	200
20	125	250
30	145	290
40	160	320
50	170	340

11.4.5 Ignition sources, drift and other factors

While most large escapes of flammable vapour with OFCE potential have reached an ignition source and burnt and/or exploded, several such escapes have occurred without igniting, and at least one cloud drifted 0.5 km before exploding. The further a cloud drifts, the more of its flammable vapour will have been diluted by air to below the lower flammable limit, thus reducing the explosion potential. (On the other hand, some limited drift in which more concentrated flammable vapour is diluted to below the upper explosive limit may increase the force of an explosion.) It is important to appreciate that there are some very large potential OFCE sources each containing over 1000 t of liquefied flammable gases under pressure (such as C_3 and C_4 storage spheres). The contents of one of these if released suddenly could travel several kilometres in a light breeze (e.g. over an estuary) before exploding.

There are thus several reasons for considering possible points of ignition for any potential OFCE source which is being studied. The ignition point is unlikely to become the centre of the explosion because only the edge of the flammable vapour cloud, which is usually very large and may cover an area the size of a football pitch, has to reach it for ignition to occur. The generally accepted ignition source of the Flixborough explosion was thought to have been the furnace of another plant about 100 m from the point of the vapour escape. Yet the probable centre of the explosion was much nearer the point of escape than to the furnace. If there are clearly identifiable permanent ignition sources within 150 m of a large potential OFCE source, the cloud is unlikely to travel far before it ignites.

In making the study the name of the material being considered as an OFCE source is marked on the plan, together with its ignition temperature in air [10.5.1]. Potential points of ignition for the source material are then considered and marked on the plan. Methane, ethane and propane with ignition temperatures over 450°C require a strong ignition source such as a furnace, an open flame (flare, burning pit, welding torch), a road vehicle or a high-voltage cable. C_5 to C_9 hydrocarbons except aromatics have ignition temperatures in air between 230 and 300°C, and could be ignited by high-pressure steam pipes, hot oil pipes and other hot objects.

MPPD studies should be made for all 'major hazard' installations with OFCE potential. It is important that these be done at the design stage and that their implications are fully appreciated then.

References

1. Dow Chemical Company, *Fire and Explosion — Safety and Loss Prevention Guide — Hazard Classification and Protection*, 3rd edn, Dow Chemical Company, Midland, Mich. (1972)
2. *Dow's Process Safety Guide*, American Institute of Chemical Engineers, New York (1973)
3. Dow Chemical Company, *Fire and Explosion Index — Hazard Classification Guide*, 4th edn (1976), 5th edn (1980), 6th edn (1987), Dow Chemical Company, Midland, Mich.

4. Lees, F. P., *Loss Prevention in the Process Industries*, Butterworths, London (1980)
5. Lewis, D. J., 'The Mond Fire, Explosion & Toxicity Index — A development of the Dow Index', paper presented at the AIChE Loss Prevention Symposium, Houston, 1–5 April 1979
6. ICI Mond Division, *The Mond Index*, 2nd edn, Imperial Chemical Industries plc, Explosion Hazards Section, Technical Department, Winnington, Northwich, Cheshire CW8 4DJ (1985)
7. Lewis, D. J., 'Application of the Mond Fire, Explosion & Toxicity Index to Plant Layout and Spacing Distances', *Loss Prevention*, **13**, 20 (1980)
8. NFPA Standard No. 325M, *Fire Hazard Properties of Flammable Liquids, Gases and Volatile Solids*, National Fire Prevention Association, Quincy, Mass. (1969)
9. Mecklenburgh, J. C., *Plant Layout*, Leonard Hill Books, Aylesbury (1973)
10. Dransfield, P. B., Lowe, D. R. T. and Tyler, B. J., 'The problems involved in designing reliability into a process in its early research or development stages', *I. Chem. E. Symp. Ser. No. 66, 67, 1981*, Rugby (1981)
11. Yelland, A. E. J., 'Design Considerations when Assessing Safety in Chemical Plant', *The Chemical Engineer*, September (1966)
12. HSE, *Guidance Note CS/2, The storage of highly flammable liquids*, HMSO, London (1977)
13. HSE, *Guidance notes for the storage of liquefied petroleum gas at fixed installations*, HMSO, London
14. S.I. 1951, No. 1163, *Stores for Explosives Order*, HMSO, London
15. Oil Insurance Association, *Publication 631, General Recommendations for Spacing in Refineries, Petrochemical Plants, Gasoline Plants, Terminals, Oil Pump Stations and Offshore Properties*, Chicago, Illinois (1972)
16. The Institute of Petroleum, *Model Code of Safe Practice*, Part 15, Area Classification, I. P., London (1990)
17. Liston, D. M., 'Safety aspects of site selection, plant layout and unit plot planning' in *Safety and Accident Prevention in Chemical Operations*, edited by Fawcett, H. H. and Wood, W. S., 2nd edn, Wiley-Interscience, New York (1982)
18. International Oil Insurers, *EML Rules*, London (1976)
19. International Oil Insurers, *The evaluation of estimated maximum loss from fire or explosion in oil, gas and petrochemical industries with reference to percussive unconfined vapour cloud explosion*, London (1985)
20. International Risk Insurers, *Vapor cloud loss potential estimation guide, Chicago (1975)*
21. Whitehouse, H. B., 'IFAL — a new risk analysis tool', *I. Chem. E. Symposium No. 93*, Rugby (1985)

Part III
Preventing mechanical failures

12 Hardware hazards

12.1 Introduction

The process industry's prime safety objective is safe containment. No matter how toxic, explosive, flammable or asphyxiant, if the material being processed is adequately contained then the potential hazard cannot arise and the plant remains in a safe state.

Not only does loss of containment lead directly to a major hazard with possible fatalities and injuries, but adequate containment is most important to the economical operation of process industries. Even if the released substance is not hazardous, the effect on the environment, the cost in time, trouble and money of clearing up and consequent production delays are a considerable spur to the adequate containment of materials. All major process industry disasters have had their root cause in loss of containment, be it from static storage tanks, pressurised systems, pipework, heat exchangers, boilers, distillation columns, containment cylinders etc.

The process industry is well aware of and well able to cope with small spillages, leaks and minor emissions; it is the large uncontrolled and usually unexpected (though not unpredictable) loss of containment that has led to the most serious disasters. It is true to say that had the containment loss been predicted, responsible organisations would have taken appropriate steps to eliminate it. But this is not to say that the total content release could not have been predicted.

All total losses of content have had their basic causes within the field of mechanical engineering. Some very specialist mechanical engineers have spent a lifetime analysing why mechanical failures have led to losses of containment, used that knowledge to predict how future process plant may fail and then taken the further step of using their expertise to eliminate or at least reduce the possibilities of loss of containment. Therefore this chapter analyses how plants can lose their safe containment, and goes on to consider reasonably practicable means of maintaining it.

12.2 The prediction of hardware failure leading to loss of containment

Reliability and risk analysis are described in Chapter 14. The very heart of safety in process plant where a major accident can occur lies in the CIMAH Regulation 7,[1] in which Schedule 6 requires that a report is written. The most significant paragraph (5a) specifies that this report should describe the potential sources of a major accident and the conditions or events which could be significant in bringing one about. It is a legal requirement for major hazard plants, but it is also essential for the more numerous less hazardous process plants because the term 'reasonably practicable' demands that an assessment be undertaken. The same Schedule 6 at paragraph 5c asks for a description of the measures taken to prevent, control or minimise the consequences of a major accident. What is needed is an account of how the plant could fail, followed by assurances that it will not fail, backed up by reasons and technical arguments.

The root cause of a failure often takes considerable time to come into the public domain. In the meantime there seems to be a clamour for popular journalism to attribute it to some cause such as a closed valve, metal overheated in a fire, an internal pressure rise, or a failure due to corrosion. These bland speculations are not in themselves incorrect, but they do little to get to the base cause, and it is essential to find the base cause to prevent future failures. The real question for the safety practitioners is 'why did the containment material catastrophically fail?' Valves are often left closed or open, or corrosion occurs in process plant, or pressure rises, but these factors very rarely cause a disaster because containment is not breached and only when containment is breached will the disaster occur.

12.3 Reasons for failure

Loss of containment, failures of cranes and lifting equipment and general structural failures — be they in cooling towers or motor vehicles — will only occur because of one of six main causes:

- over-stress
- over-strain
- metal fatigue
- brittle fracture
- creep
- stress-corrosion cracking (SCC)

In addition some plant or structures will fail because they become unstable, i.e. they topple over.

The forensic examination of failed process plant over the last 40 years has not revealed any new or startling facts. The six base causes of loss of containment occur due to poor design, poor fabrication, poor erection and construction, and inadequate maintenance and operation.

The best forensic investigations are done with an open-minded attitude, peeling the layers away as if peeling an onion. Conundrums and puzzles will naturally arise, but if each theory, no matter how fanciful, is considered and the impossible is eliminated, then by logic that which remains is probable. Design, fabrication, erection, maintenance and operation can only be based upon the natural strengths and weaknesses of the containment material.

12.3.1 Failure due to over-stress

Prevention may be better than cure but the first essential is that we know why things fail. Most structural components fail due to over-stress. Stress is defined by:

$$\frac{\text{load}}{\text{cross-sectional area (CSA)}} \quad (\text{N/m}^2)$$

$$\text{Strain is equal to: } \frac{\text{change in length}}{\text{original length}}$$

There is a relationship between stress and strain which is called Young's Modulus:

$$\text{Young's Modulus (N/m}^2\text{)} = \frac{\text{stress}}{\text{strain}}$$

Most materials behave in an elastic manner, whereby if the load is increased then the deflection caused by that load is also increased in direct proportion. This is well illustrated by placing a load on to an elastic band or spring balance, but is perfectly demonstrated by a beam that deflects under a variable load. Common sense shows that this relationship between load and deflection extension cannot last for ever because we all know that at some time items break or snap. Metal, wood, plastic, composite materials and fibres that are stretched elastically eventually reach a point where they are no longer fully elastic, and the grains or molecules forming the substance begin to yield or take up a new structural shape. For most metals and plastics, should a sufficient load be applied, there will be a permanent set and the material will remain in the new or distorted position. It is considered that the material flows plastically when the load is applied, and when that load is withdrawn the material stays in the new shape. This region is called the plastic region, and if there is deformation within this plastic region, the cross-sectional area will have reduced but without the load being reduced, and so failure is inevitable. Some materials such as mild steel exhibit a sharp change between fully plastic and fully elastic and this point is usually referred to as the yield point. The stress–strain curve of mild steel is shown in Fig. 12.1.

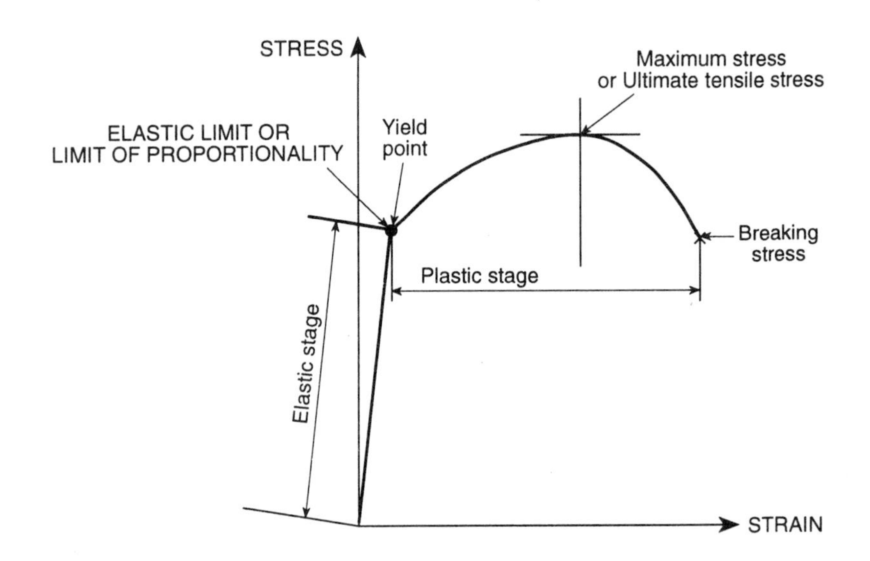

Fig 12.1 Stress–strain curve of mild steel

Metals other than mild steel exhibit a gradual transition between fully plastic and fully elastic states. This is well illustrated by the metals copper, stainless steel and aluminium, where the transition from elastic to plastic is normally designated as a 0.1% proof stress, or the load which brings about a change in cross-section of 0.1%. Some materials such as cast iron and glass go immediately from a fully elastic state to a final breaking state without going through a plastic range.

Once a material is tested and the elastic and plastic ranges are clearly defined, then the stress engineer is able to extrapolate a safe working regime. In the infancy of stress analysis, because the material most commonly used was cast iron, it was common to consider the ultimate tensile strength (see Fig. 12.1), and merely err on the safe side by only allowing known stresses to be one quarter of it. With the production of steels which are much more ductile and the development by sophisticated quality assurances that can produce steels to adequate consistent safety standards, it is more important to keep the steel within its safe elastic limit. It is now customary to use stresses up to two-thirds of the yield stress when dealing with low carbon or mild steel, and two-thirds of the proof stress when dealing with stainless steel, aluminium or high-tensile steels. The design stresses for most metals and plastics are contained within UK and CEN standards.

After a structure had been fabricated it was a good reassurance to engineers to proof test it to prove that all was well. In the old days when they used the ultimate tensile strength divided by four as the working stress it would be totally acceptable to use a test load of twice the design load. But at present design stress is based upon two-thirds of the yield; and if it was tested to twice the design stress, then the test load would cause the stress to exceed the yield point and would deform the structure during the testing process. So it is acceptable in modern-day engineering terms to apply a test load of one and a quarter of the design load.

It has already been stated that over-stress is the main cause of structural failure, and it can be understood by considering the basic equation of load divided by cross-sectional area. If the load goes up, then failure will occur whether the load is applied by weight, pressure or bending stress. In like manner if an original thick piece of material should become thinner either by corrosion or erosion this will lead to increased stress. As corrosion/erosion increases the point of failure will be reached. In layman's terms it may be quite acceptable to consider failure as having been caused by corrosion, but the base root cause is thinning which has led to over-stress.

As material becomes hotter it becomes less strong. Up to a certain temperature metal and plastic will still experience a perfectly elastic safety regime as shown in Fig. 12.1, but for a given stress or load the strain, or change in length, will be enhanced as the temperature rises. Increased temperature also has the ability to reduce the yield point or the proof stress of material whereby it is more easily deformed the hotter it becomes.

In March 1987 at BP in Grangemouth, Scotland, a hydrogen reformer unit was being started up when hydrogen leaked from the high-pressure separator at 1500 psi to the low-pressure separator designed for 150 psi. The rapid rise in pressure caused substantial over-stress and the main reformer vessel failed catastrophically with sufficient energy to throw one tonne pieces over a distance of half a mile. This incident caused the process company to be fined in the Scottish Courts a total of £750,000. The basic reason for failure was loss of containment due to over-stress.

Over-stress can be caused by pulling or tensile loads, compressive loads or sheer loads. Bending stresses are produced by a given load applied over a moment arm. High bending stresses can be useful when using a crowbar to move a large load, but can be catastrophic if large bending forces are applied to process plant.

When the wind blows against a tall distillation column, or a static storage tank then on the down-wind side bending stresses will cause buckling or compressive forces, whereas on the up-wind side there will be tensile forces attempting to stretch the material. Pipework on process plant is often subjected to bending stresses, caused by the self-weight of the fluid within the pipework, or by thermal expansion and compression stresses. In its simplest form bending stress is equal to the moment arm divided by the modulus of the section.

The modulus of the section is a function of rigidness or inflexibility of the structure. For commonly used structural steel members, beams, channels, angles etc. the modulus can be readily found by reference to steel stockist textbooks, or can be calculated from first principles. It was high bending stresses caused by the thermal expansion due to the fire at Flixborough which caused the catastrophic failure of the eight-inch pipe referred to in the official Flixborough report.[2]

12.3.2 Failure due to over-strain

Strain is a function of the change in length of a component. For tensile (pulling) applied loads the strain, if it is within the elastic limit of strength, rarely if ever leads to failure. But if the strain is of a compressive nature then buckling and critical failure can occur. When a thin, slender pillar such as a scaffold pole is subjected to large loading it will buckle or deform until instability occurs. When a containment vessel is subjected to a negative pressure or a vacuum then the skin of the vessel will deform inwards, at first behaving in an elastic measure, until the point of instability occurs. Once a point of no return is reached catastrophic failure will ensue. Much research work has been undertaken on deformation due to compressive loads, and within good stress analysis reference books there will be solutions to buckling problems.

The No. 5 reactor at the Flixborough site exhibited six-foot long cracks in the area near to the 20-inch inlet nozzle. Those cracks were formed by a buckling compressive force being driven into the skin of the vessel by the expansion bellows.

This produced a change in length which was too great for the geometry of the shell to sustain and it cracked catastrophically. Likewise the mitred pipe on the replacement at Flixborough was subjected to compressive stresses at start-up which caused considerable change in length. These were relaxed at shut-down but increased progressively over a seven-cycle period until catastrophic failure finally occurred when the mitred bend jack-knifed and took up a hairpin shape [4].

Cranes are often used in process plant either for regular movement of goods and equipment, or when structural alterations or modifications are undertaken. Compressive stresses within cranes cause them to be vulnerable to excess strains, and therefore it is essential that cranes (and other auxiliary equipment brought onto process plant) should be thoroughly checked for design concepts and adequate maintenance, and used only within their safe envelopes.

High temperatures can affect the resistance to strain in most metals and plastics. A structure which can safely sustain large changes in shape at ambient temperatures may be unsafe when the temperature rises. The investigating team that tried to recreate the failure that was the cause of the Flixborough disaster carried out some of their experiments at room temperature rather than at the elevated temperature of 160°C that was appropriate during the running of the Flixborough plant. The anticipated buckling in the test rigs did not occur because due consideration had not been paid to the temperature effects. Likewise plant that has safely sustained pressure adjusted at ambient temperatures has failed catastrophically at high temperatures because the heated metal could not sustain the compressive strains.

Large non-pressurised storage tanks have failed, dumping their contents due to over-strain buckling induced by high wind conditions.

12.3.3 Metal fatigue

A component may be safely able to sustain a constant level of stress. But should that stress be applied and taken off and then reapplied, failure may occur without any further increase in the stress level. This fluctuating stress causes failure by fatigue. The word fatigue suggests a sleepiness or a quiescent time rather than the dynamic situation when forces are being applied and then relaxed. Maybe the name came about because the failure did not present itself immediately and people were lulled into a false sense of security.

Low-cycle fatigue

Low-cycle fatigue can occur in a small number of cycles when very large loads are applied. It is a party trick to bend a paper clip and count the number of reversals to the point of failure. The stress engineer will realise that the load being applied to the paper clip is taking the metal past its yield point and will anticipate failure occurring within ten stress reversals. There have been cases of large 25 m diameter spheres failing catastrophically due to a very small number of stress or load reversals. Some structures, when they are slightly overloaded, will take up a better shape (e.g. a cylinder will deform towards the shape of a rugby football which is better equipped to resist stress); a sphere, on the other hand, is a naturally perfect shape, and therefore, should any local deformations occur, there will be stress intensifications caused by local bending moments. The next time the sphere is loaded or pressurised, the local deformation will have made it a less perfect shape, and so the stress will be intensified. If the same load is applied again, increased deformation will again take place. Hence, large spheres have been known to fail catastrophically within a very small number of stress reversals or applications of pressure.

High-cycle fatigue

High-cycle fatigue is the more usual phenomenon whereby moderate stresses are applied and relieved over many cycles, sometimes measuring tens of thousands. This will happen in vibrating machinery, such as motors, pumps, glands or furnace tubes.

The solution to metal fatigue should always be to reduce the main tensile stresses to the minimum. Even for low-cycle stresses if the stress is reduced to two-thirds of yield strength, then there will be little opportunity for low-cycle fatigue. On the other hand, if a component is expected to suffer a large number of stress reversals then the stress levels must be reduced still further. Design standards and reference books will advise on the stress levels and the cycle numbers that will be appropriate for various steels, composite materials or plastics.[3]

12.3.4 Brittle fracture

Brittle fracture occurs very suddenly and, to the layman, seemingly without warning, allowing a rapid release of energy. In simple terms brittle fracture occurs because the molecular structure of the material does not slip, either because of the material structure or because it does not have sufficient time to do so. The term 'fast fracture' is also a form of brittle fracture. The stress–strain curve of Fig. 12.1 shows that materials such as mild steel have a perfectly elastic state, and within this fast fracture will not occur. But materials such as cast iron will go from an elastic state directly to a brittle fracture state with no warning when stresses become sufficiently high.

Cast iron is capable of failure in a fast or brittle manner even at room temperatures. It is now known that even mild steel can fail in a similar manner to cast iron should the temperature be sufficiently low. For example a level of stress that could be safely applied to ordinary low carbon or mild steel at normal or ambient room temperatures, would lead to brittle fracture if applied at subzero temperatures. Figure 12.2 shows a typical brittle fracture.

The grain or molecular structure of materials such as stainless steel, copper and aluminium will stay ductile and not exhibit fast or brittle fracture no matter how low the temperature. Certain steels that are held at high temperatures can undergo a molecular change, often associated with nitrogen adhesions, which will change the structure from a ductile to an embrittled state and the crystallised structure of the metal may be altered so that fast or brittle fracture occurs.

Materials that are susceptible to fast fracture will fail more quickly or at a lower level of stress if the loads are applied as an impact or blow. For example it is known that cast iron will break up readily under a percussion blow whereas it will sustain a steady-state load quite happily.

When structures are welded or fabricated there are locked-in welding or residual stresses. Such stresses will not lead to tensile failures but, unless relieved, they can cause premature failing due to fatigue and brittle fracture. It is therefore a sensible precaution to relieve residual tensile stresses either by heat treatment, stress relieving or work stress relieving to avoid brittle fracture and failure due to fatigue.

12.3.5 Creep

In 12.3.1 it is shown that material which is heated becomes less strong. But it has come to be realised that, even allowing for the reduction in strength, if a material is held for a sufficient time at an elevated temperature under a given stress or load regime, molecular flow will take place within the grains of the structure so that it takes up a new shape.

It has been known for centuries that lead flows very slowly even at ambient temperatures. The lead roof of an ancient church may be measured and the difference in thickness from top to bottom will indicate that the lead has gently flowed over the centuries, becoming thicker at the bottom and thinner at the top. The remedy is to remove the roof every 200 years and to rehang it so that the thicker lead is at the top. Over the next 200 years the thickness will be equalised and all will be well. In certain types of process plant the solution is not so easy. For example high-temperature furnace tubes are susceptible to high-temperature creep phenomena, whereby the metal in the tubing of the furnace gently and persistently creeps to change its shape or diameter. This of course induces a thinning of the tube, an increase in stress and an increase in the creep rate. Creep-induced thinning has led to catastrophic bursting of furnace tubes in steam boilers and process-cracking furnaces. The remedy naturally will lie in using materials

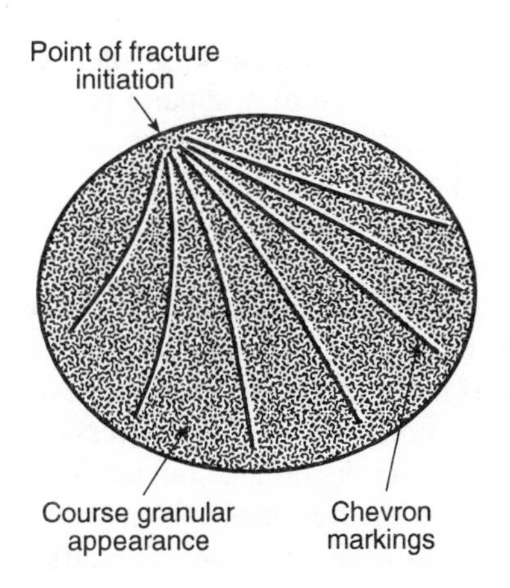

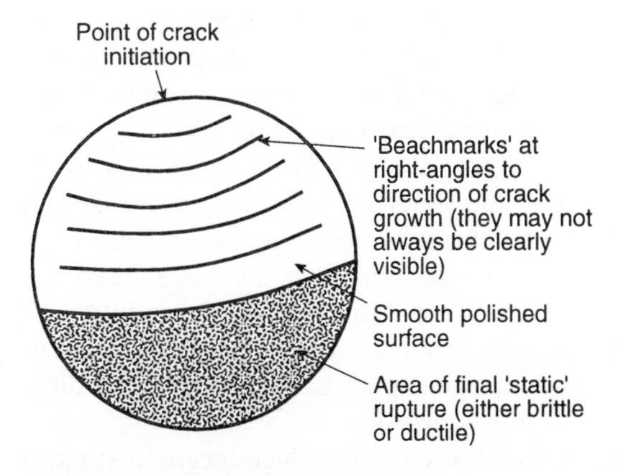

Fig. 12.2 Typical brittle fracture

that are less prone to creep (e.g. high-tensile, high-alloy steels), reducing the stress level and fixing a finite period of safe operation. Creep failures are dependent on time, temperature and stress level.

12.3.6 Stress-corrosion cracking (SCC)

A level of stress which would not normally be sufficient to cause cracking or failure of process plant material, could under certain environmental circumstances lead to cracking. This is best described by illustration. For example, should a stainless steel fork be bent or stressed then it is unlikely to fail by cracking. If the same fork (unstressed) was placed in vinegar (acetic acid) then again it is unlikely to corrode or form any degree of cracking. On the other hand if the fork were stressed *and* immersed in acetic acid then there is a great likelihood that surface-breaking defects or cracks would occur. The stress level which will cause stainless steel in acetic acid to crack is very low. Chlorides, iodides or bromides in conjunction with stainless steel would have the same effect.

It is well known that brass exhibits stress-corrosion cracking when subjected to stresses that would not be harmful in normal atmospheric conditions, if there are small traces of ammonia present. Aluminium behaves similarly if it is under stress and at the same time is in contact with caustic soda.

The phenomenon known as nitrate stress-corrosion cracking gained much publicity after the Flixborough inquiry revealed that some of the cracks in the No. 5 Reactor were due to stress corrosion. Research has shown that when mild steel is in contact with nitrates, SCC will only occur when stress levels are well above the yield point. If mild steel is stressed to a level above its yield strength then it is susceptible not only to nitrate stress-corrosion cracking but also to ductile failure, low-cycle fatigue failure and, if the stresses are compressive, to failure due to over-strain. Numerous cracks have been found inside large spheres that have been used to contain anhydrous ammonia. It is known that in normal working conditions ammonia does not cause SCC in mild steels. Nevertheless cracks up to 19 mm deep in spheres with casings 22 mm thick have been found. It became fashionable to attribute these cracks to SCC induced by contamination with small amounts of oxygen. An old wives' tale suggested that if the ammonia was dosed with water then this so called 'oxygen-induced stress-corrosion cracking' would be inhibited. To the scientific mind it was always difficult to see how water could stick to the upper surfaces of the sphere in order to prevent oxygen attacking it. This theory has become defunct as laboratory experiments have failed to produce genuine stress-corrosion cracks in the metal used for ammonia-containing spheres until it is stressed to a level far beyond its yield point. Similar cracking to that found in ammonia spheres has been found in LPG spheres, oxygen spheres and nitrogen spheres. And it has been difficult to attribute all these cracks to stress corrosion.

It is now widely acknowledged that there are very high localised stresses in spheres that cause, or have been caused by, local stress intensification.

These cracks have progressively grown during loading or pressurising or depressurising the systems. (HSE research has found local stress raisers in spheres giving rise to local stress levels three times the normal level.)

Nevertheless SCC in some process plant, particularly stainless steel clad vessels that have been cleaned with chloride solutions, led to failure and consequent high replacement costs. Stress-corrosion cracking should never be lightly dismissed, but neither should it be used as a camouflage dissuading or distracting the discerning engineer from prime root causes.

12.3.7 Instability

Some process plant has failed catastrophically merely due to its becoming unstable and toppling over. This has occurred when the centre of gravity of the structure has moved outside the stabilising base. The remedy naturally is to be aware of where the centre of gravity of process plant lies, and, if it cannot be contained within its natural base, then the structure must be bolted down or supported by adequate means.

12.4 Safety of machinery

The process industries use many types of machines including centrifuges, mixers, pumps, autoclaves, conveyors, packing machines etc. Many of these machines can cause trapping hazards to workers who can be entangled or struck by ejected parts, crushed and cut, stamped, or abraded by pressing against moving components. These hazards can occur during construction, installation, commissioning, general operating processes or during cleaning, adjustment, maintenance or decommissioning. Machinery can also cause harmful effects by emitting toxic or flammable substances, or by emitting dangerous levels of noise. It can also produce hot or cold burns, and cause electrical hazards.

It is only natural that there should be some attempt to safeguard machinery or at least reduce the level of risk to an acceptable minimum. There is a hierarchy of safety measures with a demand that the highest safety measure is achieved and only going to the next recommendation if it is not reasonably practicable to apply the former. Naturally the first necessity is to identify the hazard and the risk. And from there we can go on to attempt to make the environment safer by:

1. elimination of the hazard;
2. substitution of a less harmful substance for a harmful one, or a less harmful way of undertaking the work;
3. safeguarding;
4. provision of personal protective equipment;
5. management training and supervision.

This section considers the reduction of risk by means of safeguarding. As with all safety matters safeguarding encompasses a hierarchy of measures, the highest being reserved for the most dangerous operations, with lesser safeguards acceptable when the hazard is low.

The first and easiest form of safeguarding is total periphery fencing, to stop workers gaining access to dangerous parts. Usually access will be required through this periphery guarding, and for any access to dangerous parts safety interlocks should be supplied. A guard which is movable or has movable parts, can be interlocked to the machine's power supply or to its control system. That power system could be electric, pneumatic or hydraulic. For the least hazardous situations it will be sufficient to provide for a single-control system of interlocking. For more hazardous or risky work situations dual-control systems of interlocking but without cross-monitoring would be acceptable. The most dangerous or risky operations should be safeguarded either by using dual-control systems with cross-monitoring or guard-inhibited interlocking systems.[4]

Many modern machines are operated by a computer system. Because computer systems can work in such a variety of ways, which cannot all be monitored, and because there is a notorious history of computer malfunctions or failures, it is not acceptable to operate an interlock guarding system by computer. All interlock systems should be hard-wired via circuits outside the computer's control. The safeguarding of machinery is not of itself complex, but it is diverse, and it would be appropriate to consult a textbook such as the *British Standard Code of Practice for Machinery Safeguarding BS 5304*.[4]

Centrifuges are used to extract liquids from mass material by centrifugal force. If a centrifuge was allowed to operate too fast the stresses on the bearings, the centrifuge cage and the outer containment vessel could well be excessive. In the past there has been total bursting or explosions of such centrifuges. It is therefore essential that an over-speed governor is fitted to a centrifuge and that the outer casing and lid are designed so as to take the maximum loads even if the over-speed governor should fail. Over the years there have been a large number of cases where operators have opened the lids of centrifuges to remove the contents before the centrifuge has finally come to rest; as they placed their arms within the centrifuge case they have been entangled, resulting in horrific injuries and fatalities. It is therefore essential that centrifuges have interlocks on the door so that the centrifuge lid cannot be opened until the internal workings have come to a complete rest. Two British standards give valuable advice on the industrial and commercial application.[5,6]

Autoclaves are used to apply pressure and heat. They are used in ceramic manufacturing, brick curing, rubber moulding, dyeing, cleaning and general processing. Naturally the outside skin of the vessel must be adequately designed, fabricated, tested and maintained, but the one vulnerable part is the access door. Within the UK some autoclaves are four metres in diameter and 30 metres or more long. When filled with a compressive fluid such as steam the energy levels are considerable. There have been extensive explosions that have totally wrecked factories and thrown parts of the autoclave weighing more than one tonne distances of three-quarters of a mile due to the door becoming unlatched, when under full pressure. To safeguard the autoclave from violent explosions, it is essential that pressure shall not be applied until the autoclave door is shut and fully locked. The door should be interlocked so that it cannot be opened until the pressure has been turned off, and the contents have been fully vented to the atmosphere. Accidents have happened at autoclaves when the pressure has been reduced to a very low level. For example if a one-metre diameter door was pressurised only to 0.05 bar, then there would be a residual load of approximately 150 kg. If the door should be unlatched even at these low pressures, it would fly open violently striking the operators or other nearby persons. It is therefore essential they can be partially opened but still latched (primary latching). This prevents the door from swinging fully open should there be any small residual pressure still present within the autoclave.

Even if no residual pressure remains, severe scalding and/or chemical burns can occur when autoclave doors are opened. This is due to the contents which may be either extremely hot or corrosively burning, splashing onto operators. The natural remedy is to safeguard the autoclave by allowing the door to become unlocked only when the contents have been fully removed or when the temperature of the contents has dropped to a level where scalding would not take place. The HSE has produced valuable advice on safeguards at autoclaves and dyeing vessels.[7]

References

1. S.I. 1984, No. 1902, *Control of Industrial Major Accident Hazards Regulations 1984*, HMSO, London
2. Report of the Flixborough Inquiry, HMSO 1975
3. BS 5500, *Unfired Pressure Vessels.*
4. BS 5304, *Machinery Safeguarding.*
5. BS 767, *Baskets and Bowls of Centrifuges.*
6. BS 4402, *Laboratory Centrifuges Safety Requirements.*
7. HSC, *Guidance Note PM 73, Safety at Autoclaves*, HMSO, London

13 Corrosion hazards and controls

13.1 Introduction

Corrosion and erosion of metals are a time-dependent wasting process whereby the original material is attacked on its surface; it either becomes universally thinner, or more often, the surface becomes pitted and so thinning occurs in a random manner. Almost all process plant can be susceptible to corrosion or erosion and if this were allowed to progress with no intervention or remedial action, the point would come where the plant failed.

Plant failures of whatever sort may be costly in terms of loss of production and loss of profits; they could merely be embarrassing, or they could lead to major accidents and considerable loss of life. Failures due to corrosion leading to loss of production and profit can often be compensated for by adequate insurance. The claims for compensation will often be negotiated in private, and the world at large never gets to hear about it. This is not always so if the corrosion or erosion problem causes embarrassment. Someone's ego is hurt, and they will go to elaborate lengths both to cover up the initial cause of the erosion or corrosion and/or hide the potential outcome. It is often time consuming and difficult to get at the base causes when individuals are deliberately covering up their own misdemeanours, ignorance or laziness. Occasionally corrosion can be part of a regime where a major accident happens, and in all probability in this case there will be enough independent expertise brought to bear to understand the root causes and prevent future occurrences.

This chapter deals mainly with metal corrosion, followed by a few notes on the corrosion and deterioration of non-metallic construction materials.

13.2 Safeguards

The natural way to reduce opportunities for corrosion or erosion will be to protect the surface where the corrosion takes place. For this reason much process plant is painted or covered in various epoxy resins, or even glass. The type of material which readily corrodes, e.g. ordinary low carbon steel, is cheap and easy to replace should corrosion become too severe. On the other hand, the types of material that resist corrosion such as copper, aluminium or stainless steel are relatively more expensive, and this increased cost has to be justified by a reduction in production costs and/or an increase in safety. As has been said, almost all material erodes, corrodes or deteriorates given time. What matters when considering health and safety is 'does the deterioration cause a hazard and is the risk of that hazard occurring unacceptable?' Chapter 14 deals with reliability, hazard and risk assessment. What we are now considering is 'does it matter if the "stuff" comes out of the process plant?' If it only has nuisance value then there is no health and safety consideration, but if the stuff coming out is toxic, flammable or explosive then there may be a risk to health and safety.

The process industry is well able to cope with small leaks. It does that almost every day of the year. What it cannot cope with is catastrophic failures where the complete content of a system is released. Should that release be toxic, flammable or explosive then remedial action must be taken to reduce the risk of the failure occurring.

Corrosion of itself is not dangerous; it is the consequences of the corrosion which may be dangerous. There is a field of mechanical engineering entitled fracture mechanics which considers whether the material will merely leak or whether it will break before it leaks. Leaks often give the operators of process plant an opportune warning of impending danger and they can react to it. What they cannot react to is total catastrophic failure. The clever bit for the safety expert is to discern whether a plant is likely to break before it leaks or whether small leaks are possible.

Most corrosion results in leakage which can be mopped up and may only be harmful for a very short time. The system of discerning whether a plant is likely to fail catastrophically or stay in a leaking state is dealt with more fully under design and fabrication considerations in Chapter 15.

Corrosion leading to either an overall thinning or local pitting will reduce the thickness of the structure, and this may lead to one of the failure modes described in 12.3; e.g. it may cause an increase in the stress in the remaining material, cause elastic or plastic deformations that might lead to buckling, reduce the thickness and increase the opportunities for fatigue or brittle fracture, and it can play an important part due to failure by creep. Corrosion of itself never causes failure; it only sets the scene for one of the aforementioned failure modes to occur. Often failures have been attributed to such things as human error, or lack of maintenance, but these are only flags to point the lay reader in a certain direction. It may be right that all plant fails because of human error, but that does not progress the art or science of making the next plant safe so far as is reasonably practicable. In like manner, the narrow margin between leak and total failure breakage does not depend on whether or not the plant corroded, but on one of the six basic failure modes [12.3]. Modern engineering design can ascertain and predict the type of failure mode that might occur and suggest that it may be acceptable to allow a process item to fail in a ductile manner, producing a small leak, whereas it would not be acceptable to allow a piece of process equipment containing hazardous substances to fail catastrophically in a brittle manner.[1]

13.3 Anecdotal examples of corrosion-related failures

A whisky distillery in Scotland was coming under pressure from the local authority to treat its effluent rather than dump it into the nearby river. An elaborate effluent-treating plant was obtained and because the effluent was slightly acidic it was decided by the design engineers to use stainless steel to reduce the potential harm due to corrosion. The distillery, wanting to save money, asked for the storage vessel which was some 5 m in diameter and 5 m high to be constructed of steel and for the internal lining to be epoxy resin-coated so as to preclude internal corrosion. Again finding this solution was expensive they ordered a mild steel vessel and undertook the lining themselves. In the end, still trying to save money they omitted the epoxy lining and so the storage vessel went into service in raw mild steel containing effluent which was slightly acidic. Galvanic corrosion began to occur between the parent plate and the main vertical welds. The distillery carried out no monitoring of this plant because in their opinion it was not hazardous, and so the galvanic corrosion continued to the point where there was locally a very thin ligament of steel left and at that point the stresses were so very high that an open crack began from the very base of the vessel, and ran to the very top. This caused the vessel to fail in a ductile manner due to over-stress and when the vessel opened up or petalled open the contained effluent flooded into the courtyard striking and drowning a passer-by. The cause of the failure was over-stress but the root cause was inadequate design; either a material not subject to corrosion should have been supplied so that corrosion would not occur, or the surface of the material should have been coated.

Chlorine is a very valuable commodity used to sterilise organic organisms in water. Chlorine of itself does not corrode mild steel and so it is common practice to store chlorine in mild steel vessels with no protection against corrosion. In one of British Nuclear's power stations chlorine was being used to sterilise incoming cooling water. Unfortunately water was inadvertently sucked into the chlorine storage vessel when a partial vacuum was created, partly due to the chlorine being pumped out and partly due to low temperatures. The mixture of chlorine and water caused corrosion that created pinholes right the way through the main containment body.

Although the holes were very small, the chlorine inside the vessel seeped slowly out until a total of two tonnes had been dissipated into the outside atmosphere. The vessel did not fail catastrophically because the stresses within the structure were low; it did not fail from over-strain because the corrosion had not thinned the material generally to a point where it had deformed; it did not fail through brittle fracture because the material had not become sufficiently cold; neither did it fail because of fatigue because there were no stress reversals. The corrosion in this instance merely resulted in a leak of chlorine which entered and was dissipated into the outside atmosphere with no harm to personnel. There was no major disaster because the chlorine merely leaked before the vessel broke, but there was considerable embarrassment and costs incurred for repair and reinstatement. If the chlorine inside this vessel had been very cold then there would have been the opportunity for total catastrophic brittle fracture, but that could have been predicted by engineering models being undertaken by experienced stress engineers.

One of the UK's major petrochemical companies found catalytic reformer tubes bulged and split. They rightly diagnosed the type of corrosion as hydrogen embrittlement. But having established what type of corrosion it was they did not go the 'extra mile' and see why this had happened. Just because there had not been a catastrophic failure the company had merely reinstated the same type of tubes and the same fabrication mode. The HSE encouraged them to carry out a thorough examination of the cause and the consequences of this hydrogen embrittlement and it was found that the original quality assurance during fabrication had been lacking. The vessels should have been stress-relieved after they had been welded. A detailed analysis showed that they had been 'merely lucky' that a major disaster had not occurred because the split of a reformer tube could have led on to a break prior to leaking if, for example, the hydrogen embrittlement had occurred in another part of the structure. The company had not heeded any warnings of the mechanical reliability of the plant; having found a failure they had put it down to corrosion rather than analysing what failure modes might be prevalent in that structure.

The Feyzin Explosion that happened in France in 1966 caused the deaths of 18 people with 81 injuries. Through a system of 'Chinese whispers', there has grown up within the process industries a concept that this vessel failed because heat had been applied to the top part of this very large sphere where it was not cooled by its LPG liquid. On one side there was flame impingement and on the other side gaseous LPG and so the theory went the metal warmed up to the point at which failure occurred, and the vessel BLEVED. Mechanical engineers began to query this concept believing that hot mild steel, of which the sphere was constructed, was unlikely to fail catastrophically in a brittle manner; rather it was likely to petal open, maybe even experiencing a large hole up to four feet long and two feet wide, but not producing an instantaneous brittle fracture shattering the vessel into more than six pieces. Detailed metallurgical and engineering investigations indicated that the failure had occurred at the equator. It was known that the sphere was 95% full at the time of the fire and the time from the first fire impingement until the final failure was less than 12 minutes, far less time than would have allowed the LPG in the sphere to have burned off. In any case the safety valves had not opened when the fire reached the sphere, but opened at the same time as the catastrophic disintegration of the sphere. The point of failure was closely examined. It was known to have originally been 22 mm thick, but it had corroded to barely 14 mm. The safety valve was set at ten bar, and this was the approximate failing stress should the thickness of the sphere be reduced to 14 mm. The Feyzin disaster can be attributed to corrosion. The question may be asked: 'What was corrosion doing inside an LPG sphere which is not usually susceptible to such corrosion?' It was known that just before the major release of contents from the bottom valve a workman had opened this valve to take a sample, and it is well reported that he was burnt when he opened the valve. It is commonly believed that it was a cold burn due to ice within the bottom valve, but detailed investigations show that he was burnt by a caustic solution in the bottom of the vessel. It is therefore surmised that the corrosion reducing the shell locally in the equator area from 22 mm to 14 mm was either caused by caustic soda, or iodides entering into the sphere. Although corrosion set the vessel up for total catastrophic failure, the vessel actually failed in a fast fracture or brittle fracture mode rather than leaking before it broke. Over-stress caused a small leak; the leaking LPG locally supercooled its structure thus allowing fast or brittle fracture to occur.

The second largest peace-time explosion in Britain occurred in South Wales at a refinery in 1994. Due to a severe electric storm computers and electric systems had cut out, and the plant had to be restarted on numerous occasions. The plant was adequately protected by safety valves and dump lines leading to off-site flare stacks. Sensibly within the system there were knockout drums which separated out water and other liquids from in-trained gases so that only flammable and toxic gases would be sent to the flare stack. Naturally such flare lines and safety devices are very rarely used. It is estimated that some 15 tonnes of flammable gases were released from a 12-inch diameter line and these gases accumulated within the plant, found a source of ignition and violently exploded. The 12-inch pipe leading from the knockout drum to the main flare stack had failed due to severe corrosion. The company were aware that the pipe was experiencing general corrosion, but they had assessed the pipe using only very basic principles of stress analysis for a very simple pressure vessel. In fact the pipe had corroded from 12 mm to something like 0.5 mm thick, but in some places holes had perforated the complete surface. Although the discharge pressure from the knockout drum was very small, probably only in the region of four bar, this pressure caused catastrophic failure of the pipe. When considering such a very thin metal the normal procedures for brittle fracture analysis, ductile failure analysis and plastic buckling are very vague. The company knew the pipe was corroding, but they had merely taken a chance that the corrosion would not lead to catastrophic failure and they were wrong in their premise. Fortunately there was no loss of life and only minor injuries, but the loss of production and rebuilding of the plant has been a large burden on the operating company and has affected the UK balance of payments. The HSE successfully prosecuted the refinery owners for breaches of the Health and Safety at Work Act.

13.4 Galvanic corrosion

Most forms of metal corrosion in aqueous media can be explained by the theory of galvanic corrosion (named after Luigi Galvani), which also suggests ways of controlling it. When two dissimilar metals which are in electrical contact are immersed in an aqueous solution, atoms of the more active or negative one (the anode, see Table 8.2) tend to lose electrons and dissolve, forming positive ions. The electrons flow as an electrical current to the less active or more positive metal (the cathode) which attracts dissolved ions of any more positive metals, as well as hydrogen ions. These ions are discharged on reaching the cathode if their metal (hydrogen included) is more electropositive than the anode metal. Metal ions on being discharged plate out as metal on the surface of the cathode, whereas hydrogen ions may form bubbles of hydrogen gas after reaching it. In the presence of dissolved oxygen, hydrogen ions combine with it to form water as they are discharged.

If the cathode and anode are separated and connected through an external circuit, positive current (a fictional concept) is said to flow from the cathode through the circuit to the anode (whereas in fact electrons flow in the reverse direction).

The two metals, the electrical connection between them, and the solution form a galvanic cell. A cell with a zinc anode and iron cathode at which hydrogen ions are being converted to water is shown in Fig. 13.1.

The rate at which the anode dissolves is proportional to the current flowing. The voltage of the cell is at a maximum when no current flows, and decreases as the current increases. The (maximum) 'open circuit' voltage of a cell is greater, the wider the separation between the two metals in the electromotive series. The behaviour of galvanic cells also depends on many other factors, including the acidity or alkalinity of the solution (its pH), the concentrations of various ions in the solution, the concentrations of dissolved oxygen and other gases in the solution, the formation of insoluble corrosion products, and the presence of 'complexing agents' which combine with various metal ions.

While corrosion of the above type which is dependent on the presence of hydrogen ions proceeds fastest in acid solutions, i.e. where their concentration is high, some metals, such as aluminium, zinc and chromium, whose oxides can behave as acids as well as bases, react with hydroxyl ions and dissolve to form negative ions (anions) in alkaline solutions.

Examples of cells with two dissimilar metals, where one dissolves at the anode to form positive ions, include copper and iron pipes connected

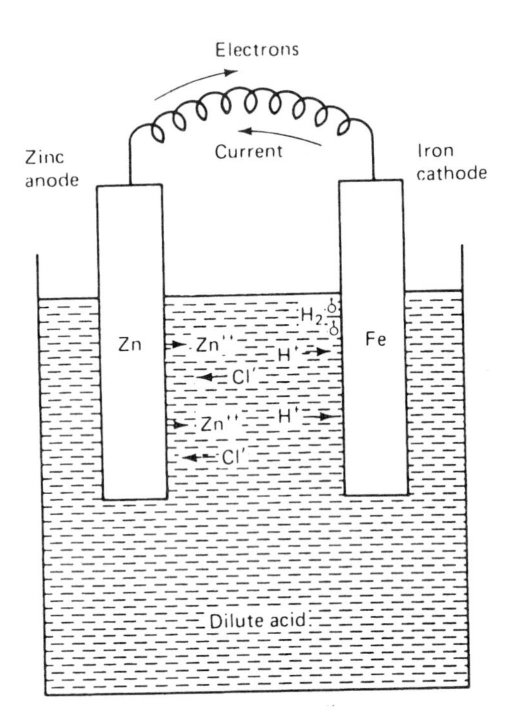

Fig. 13.1 A simple galvanic cell

together in domestic water systems, and bronze tubes in a steel shell in a heat exchanger. In such cases it is always the more negative metal which corrodes. This may be reduced by inserting some electrically insulating material between the two metals at the point of contact.

With galvanised steel, the zinc corrodes preferentially. Similarly, 'sacrificial' anodes of magnesium or zinc may be attached to steel items exposed to an aqueous environment to protect them. Taking this a stage further, a separate electrode of high corrosion resistance (e.g. a 'non-consumable' platinised-titanium anode) may be placed in the same aqueous medium as the metal (usually steel) item which it is intended to protect, and an external voltage applied between the steel and the anode to oppose the natural voltage of the cell, thus preventing a current from flowing. This is the principle of 'cathodic protection' which is widely applied, e.g. to ships' hulls, tanks and pipelines.

Dissimilar metal cells are also found on a micro-scale as the result of electrically conducting impurities on a metal surface. Galvanic cells are also formed between cold-worked metal in contact with annealed areas of the same metal. Another class of galvanic cell in which the anode and cathode are of identical materials is the concentration cell, of which there are two types. The first is the salt concentration cell, which is found where there is a difference in the concentration of dissolved salt containing ions of the metal exposed, at different parts of the exposed surface. Here the metal in contact with the lower concentration of ions is the anode and dissolves preferentially.

The second and more important type of concentration cell is the differential aeration cell. An example is two iron electrodes in dilute sodium chloride solution, the electrolyte round one being well aerated, that round the other being unaerated. The difference in oxygen concentration causes a current to flow from the cathode, which is well aerated, to the less aerated anode where corrosion occurs. The discharge of hydrogen ions at the cathode leaves the solution there short of them and increases the concentration of hydroxyl ions, making the solution slightly alkaline. The hydroxyl ions on meeting ferrous ions formed at the anode undergo a secondary reaction, precipitating ferrous hydroxide whose saturated solution has a pH of 9.5. This settles over the anodic area and reacts with any free oxygen present forming rust (hydrated ferric oxide) and creating a stagnant oxygen-depleted zone below it. Iron thus continues to dissolve in the oxygen-depleted zone below the deposit, while hydrogen ions continue to be discharged and form water in the more aerated areas.

Differential aeration cells cause pitting corrosion of steel, stainless steels, aluminium, nickel and other so-called passive metals, especially when they are exposed to salt water (Fig. 13.2). Other typical forms of corrosion caused by differential aeration are water-line corrosion which occurs on partly immersed metal surfaces, just below the water-line (Fig. 13.3), and crevice corrosion which causes damage in threaded connections and other crevices where the oxygen concentration is lower than elsewhere.

Temperature differences between different parts of the same aqueous medium which is contained within a tube or shell of the same metal can cause e.m.f. differences between hot and cold parts, thus setting up a thermal type of galvanic corrosion cell (found in the corrosion of heat exchangers and boilers).

13.5 Corrosion of iron and steel in aqueous media

The corrosion of iron and steel in aqueous media takes place as a network of short-circuited galvanic cells on the metal surface. The solution of iron to form ferrous ions and electrons at the anode is rapid in most media

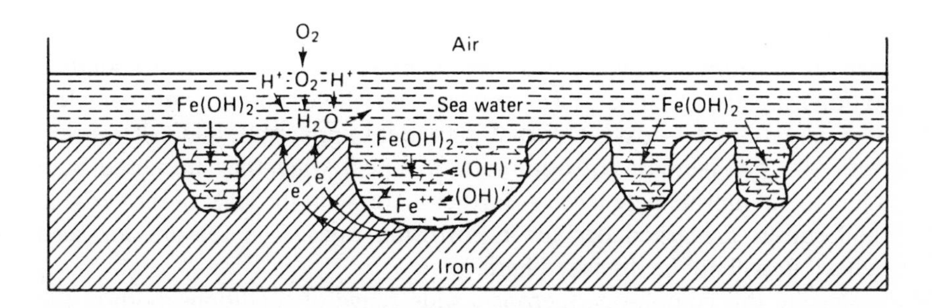

Fig. 13.2 Pitting corrosion caused by differential aeration

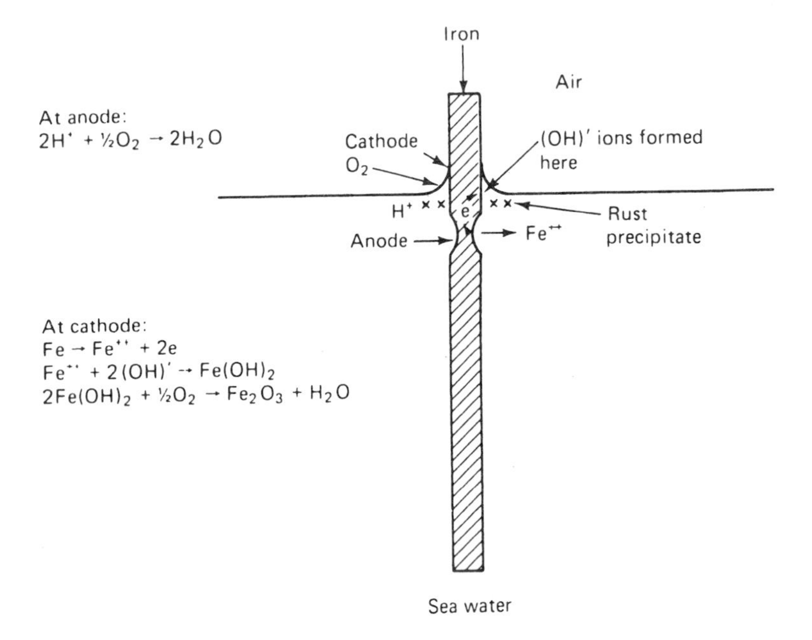

Fig. 13.3 Water-line corrosion

provided the electrons can escape. In deaerated solutions the cathodic reaction is:

$$e + H^+ \rightarrow \tfrac{1}{2}H_2$$

This is fast in acids, but very slow in neutral or alkaline solutions. In the presence of air dissolved oxygen reacts with hydrogen atoms adsorbed on the iron surface:

$$2H + \tfrac{1}{2}O_2 \rightarrow H_2O$$

This proceeds as fast as oxygen reaches the metal surface (often hindered by a barrier of ferrous hydroxide).

Two kinds of variables affect the corrosion of iron and steel in aqueous media. First, there are the environmental variables such as pH, oxygen partial pressure, temperature, liquid velocity relative to the metal, and the concentrations of dissolved substances in the water. Second, there are factors inherent in the object exposed to the medium, its composition and structure, its previous history and heat treatment, surface condition, any bonding to other metals, and mechanical stresses. The whole subject is very complex. Several variables — oxygen partial pressure, temperature and salt concentration — first increase corrosion rates to a peak when they are raised from a low to an intermediate level. A further increase in the level of the variable then causes a marked decline in corrosion rates. The reasons for such behaviour will not be discussed here.

Despite this complexity, there is a wide range of conditions, both of the aqueous medium and of the exposed metal, over which the steady corrosion rates of unprotected iron and steel settle after a few days' exposure within the fairly narrow range of 0.07 to 0.11 mm/yr. This applies to relatively still water which is non-scaling with regard to calcium and magnesium carbonates, with a pH range of 4 to 10 and a temperature range of 10–25°C, in contact with air. It applies to most types of iron and low-alloy steels. Superimposed on this background are the trends caused by the common variables which are discussed next. The effects of bacteria, hydrogen and other variables are discussed in 13.6.11.

13.5.1 Air and oxygen

In completely deaerated water the corrosion rate is negligible (less than 0.005 mm/yr). This increases as the partial pressure of dissolved oxygen in the water increases up to about half an atmosphere. It then decreases sharply as the oxygen partial pressure rises further. This does not occur in salt water where corrosion rate increases steadily with oxygen partial pressure.

13.5.2 Salt concentration (NaCl)

The corrosion rate roughly doubles on passing from fresh water to seawater (*ca.* 3%NaCl), but falls back to its original value as the salt content is increased to 10%. Dissolved salt increases the pitting tendency.

13.5.3 Temperature

In a system open to the air, the corrosion rate roughly doubles as the temperature increases from 15°C to 80°C. It then falls to well below its original value as the water reaches its boiling point and dissolved oxygen is expelled.

13.5.4 Water velocity

On passing from still water to about 0.3 m/s in an open system, the corrosion rate may increase by a factor of two or three, but a further increase in velocity in the absence of salt reduces the corrosion rate to below its original value, until quite high velocities are reached. Erosion and cavitation then begin to play a part, and penetration increases, sometimes spectacularly.

13.5.5 Calcium and magnesium carbonates

Many natural waters contain calcium and magnesium carbonates kept in solution by dissolved carbon dioxide. When this is lost, e.g. by exposure to air, heat, or the addition of alkalis, a protective scale layer forms on exposed metal surfaces which greatly reduces the corrosion rate. Waters are often deliberately treated so that they have a slight scaling tendency, but unless this is carefully controlled, there is a risk that a thick layer will form leading to restrictions in flow, etc.

13.5.6 pH and acids

In the absence of weak acids (e.g. acetic, carbonic), the corrosion rate is practically unaffected by pH over the range 4–10, but rises rapidly at lower pH values, when surface deposits of corrosion products dissolve giving greater oxygen access to the surface. Hydrogen is also formed at low pH values. Corrosion rates fall at pH values above 10 but may rise again at pH values above 13. Weak acids in which solid corrosion products are soluble increase the lower end of the pH range from 4 to 5 or 6. The preferred pH for corrosion control is usually about 8.

13.5.7 Metal coupling

Coupling of iron or steel to a more noble metal increases its rate of corrosion within a critical distance from the metal junction. This distance may be as little as 5 mm in soft drinking water and as much as 500 mm in sea-water. Coupling to a less noble metal similarly reduces corrosion.

13.5.8 Varieties of iron and steel

As the corrosion rate of iron or steel in natural waters, including sea-water over the pH range 4–10, is controlled by the rate of oxygen diffusion to the surface it is practically unaffected by the type of iron or steel. This should therefore be chosen on other grounds, e.g. price and mechanical properties. In the acid (pH <4) and extreme alkaline (pH >13.5) ranges, however, where corrosion is faster and proceeds with evolution of hydrogen, a relatively pure iron corrodes at a much lower rate than iron or steel with higher contents of carbon nitrogen, sulphur, phosphorus and other impurities.

Cast iron water pipe may last much longer than steel for two reasons:

- it is made with a greater wall thickness;
- in the case of grey or 'ductile cast' iron containing spheroidal graphite, the corrosion products tend to cement the residual graphite flakes together, thus allowing the completely corroded pipe to continue functioning for many years so long as it is not disturbed.

13.5.9 Metal composition

Although this has little effect on the corrosion rates of iron and steel in natural waters and soils, it becomes important in other conditions. The addition of small amounts (0.1–1.9%) of chromium, copper or nickel markedly reduces the rate of atmospheric corrosion. An increase in the carbon content of a steel may cause a slight increase in its corrosion rate in sea-water, although not in fresh water.

Stainless steels (containing >12% Cr), high-silicon irons and high-nickel alloys give a very marked reduction in corrosion rates over low-alloy iron and steel for most ordinary applications, but are often only economical in special applications (such as chemical duties).

Sulphur and phosphorus increase corrosion rates in acid solutions, but manganese counteracts the effect of sulphur.

Heat treatment has little effect on corrosion in the usual environments in which oxygen diffusion is controlling, but in handling acid oil-well brines, marked local corrosion often occurs near welds.

13.6 Other types of metal corrosion

The types of corrosion just discussed were dominated by galvanic action. Other types, some of which are shown schematically in Fig. 13.4, are now considered.

13.6.1 Corrosion–erosion

This combination of corrosion and erosion results from a liquid or gas flowing at high velocity, with or without suspended solids or liquid droplets, and can lead to the rapid failure of valves and other equipment. A valve stem subject to erosion by steam or gas often appears to have been 'wire-drawn'. The use of harder materials, corrosion inhibitors, and cyclones and settlers to exclude hard gritty particles from the flowing liquid all reduce erosion. Water droplets can be excluded from steam turbines by raising the inlet temperature. Sometimes the system has to be redesigned to reduce the energy loss through the affected item. The pressure drop is spread by placing other resistances to flow (e.g. a length of smaller diameter pipe) in the fluid path. Liquid erosion is often caused by cavitation.

13.6.2 Cavitation–erosion

Cavitation results from the collapse of vapour bubbles formed within a liquid behind a fast-moving metal surface, or in liquid chokes and valves where sonic velocities are reached, often accompanied by a noise like grinding shingle. An example was the failure of an automatic by-pass valve on a minimum flow return line from the discharge to the suction of a firewater pump serving a large works. It is sometimes found with pump rotors and impellers as a series of small but deep pits (Fig. 13.5), which can lead to the break-up of fast-moving parts. Here it can be avoided by limiting the speed of the impeller and/or increasing the liquid head at the pump suction. Materials most resistant to cavitation are hard, have good fatigue and corrosion resistance, small grain size, and are able to work-harden under repeated stressing.[2]

13.6.3 Fretting–corrosion

This occurs between two surfaces in contact, one or both of which are metal, and subject to slight slip. It is often caused by vibration and leads to

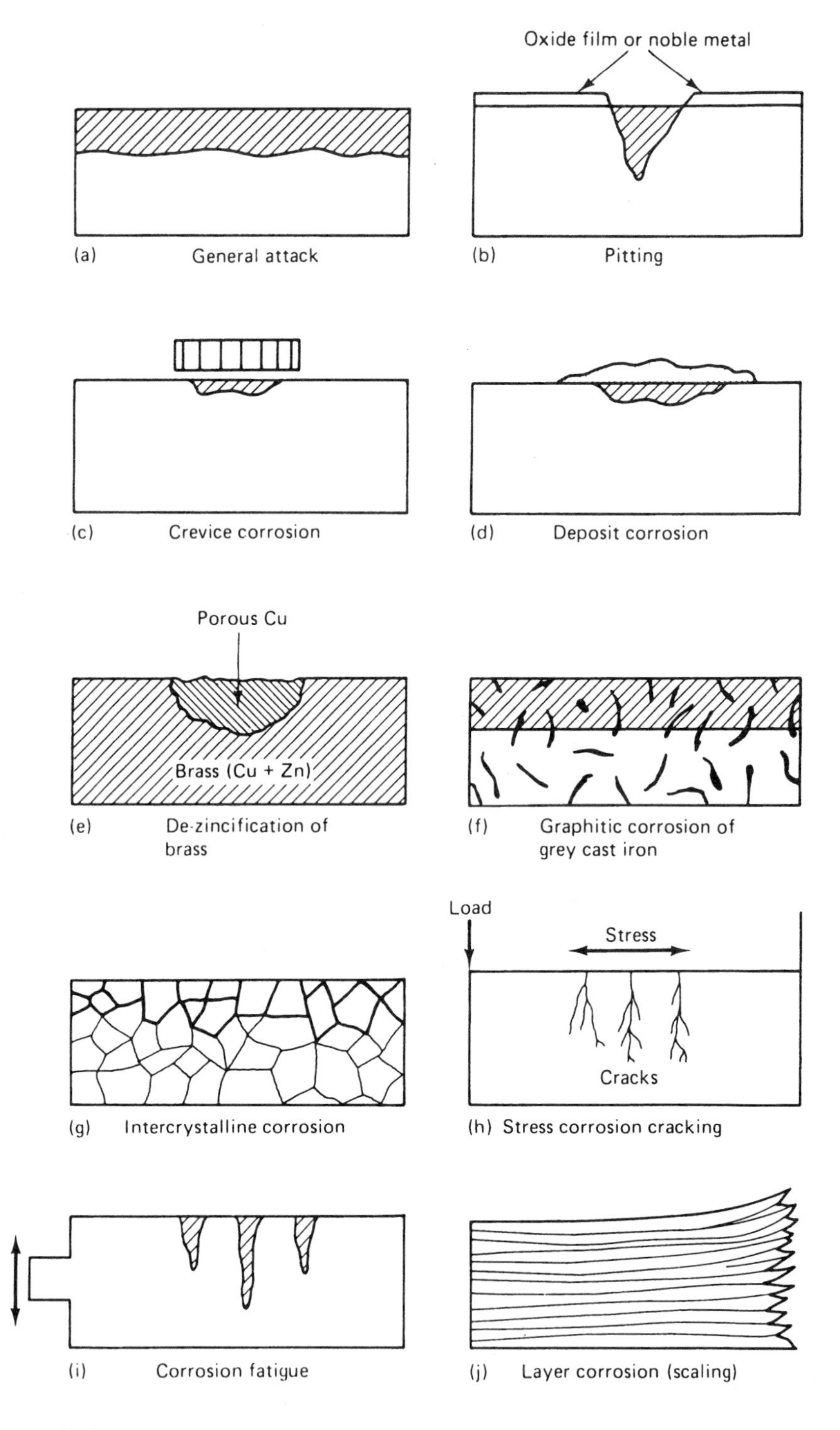

Fig. 13.4 Schematic illustration of different types of corrosion (courtesy Gösta Wranglin and Institut för Metallsbrydd, Stockholm)

Fig. 13.5 Cavitation damage to pump rotor (courtesy the Director, National Engineering Laboratory)

pitting which is evident when the corrosion debris has been removed. It has caused failures of suspension springs, bolt and rivet heads, and parts of vibrating machinery. Stationary ball bearings under static load have been destroyed by fretting caused by vibration. Fretting–corrosion of steel requires oxygen and is practically eliminated in a nitrogen atmosphere.

13.6.4 Dezincification of zinc-containing alloys

Dezincification is a disguised hazard which can occur when zinc-containing alloys are exposed to aqueous media at high temperatures under stagnant conditions, especially if the medium is acid and the surface covered with a layer of porous scale. Apart from tarnishing, the item may appear undamaged. The area affected is porous, and has lost its ductility and much of its strength. Alloys containing less than 15% of zinc, or with small additions of tin, arsenic, antimony or phosphorus, are less subject to dezincification and corrosion inhibitors in the water also help.[3] Copper alloys containing aluminium are subject to a similar form of corrosion, the aluminium dissolving preferentially.

13.6.5 Intergranular corrosion

This occurs at the grain boundaries of a metal, with rapid loss of strength and ductility. The grain boundaries are anodic and the grains cathodic. It occurs with 18-8 stainless steels which have been heated to temperatures of

500–800°C, and then exposed to a corrosive environment. Annealing such items at 1100°C for one hour for every inch of metal thickness, followed by rapid water quenching, usually prevents it. For items which have to be welded and cannot be properly heated afterwards, alloys stabilised with titanium or columbium and of low carbon content should be used. Copper – aluminium alloys which have not been properly heat treated are also subject to this form of attack.

Non-electrochemical grain boundary attack can occur when nickel is heated in an atmosphere containing sulphur.

13.6.6 Stress-corrosion cracking

This occurs when a metal surface is simultaneously exposed to a high static stress and a specific corrosive environment. The stress may be residual (arising from cold working or heat treatment), or externally applied. Stresses developed during fabrication can largely be relieved by heat treatment.

Most structural metals, particularly heavily loaded high-strength steels, are subject to stress-corrosion cracking in some environments such as solutions of inorganic chlorides, alkalis and nitrates. The observed cracks may be intergranular or transgranular, depending on the metal and the environment and they concentrate the stresses, thus starting fatigue or brittle failures. Stress-corrosion cracking was common in riveted steam boilers where it was known as 'caustic embrittlement'.

13.6.7 Corrosion fatigue

This is analogous to stress corrosion and occurs when a dynamically loaded metal component is subjected to repeated tensile stresses in a corrosive environment. These may be well below the critical stress needed to cause failure. A number of branching cracks appear over the affected surface. Complete separation eventually occurs either by brittle fracture at ambient or low temperatures or by yielding or shear at higher temperatures.

13.6.8 Hydrogen attack

This takes various forms, *embrittlement, blistering* and *cracking*, the last resembling stress-corrosion cracking. It results from the diffusion of atomic hydrogen through steel and other metals, forming molecular hydrogen on reaching voids or imperfections in the metal. If this happens at a large number of small centres, it causes embrittlement, but if it occurs at a small number of voids, it may generate sufficient pressure to raise blisters or cause cracks.

Atomic hydrogen is formed at the cathodic areas of galvanic corrosion cells before it combines to form molecules. It can also be formed when cathodic protection is applied to metals, and at high temperatures by the dissociation of hydrogen molecules. Sulphur and arsenic compounds promote and accentuate hydrogen attack, which is described as sulphide cracking when hydrogen sulphide is responsible.

Hydrogen damage can occur at high temperatures and high partial pressures, e.g. in catalytic reformers, and in plant and equipment handling natural gas and oil-brines containing hydrogen sulphide. Chromium and molybdenum increase the resistance of steels to such attack.

13.6.9 Scaling and high-temperature oxidation

These become pronounced with steel at temperatures above 500°C, but its resistance at higher temperatures can be improved by the addition of chromium as shown in Table 13.2[3].

Silicon, nickel, aluminium and yttrium also improve the high-temperature resistance of steel alloys and extend their use up to 1350°C. Nickel alloys containing 20% chromium have good oxidation resistance and mechanical properties up to 1150°C and are used for supporting furnace tubes and for gas turbine blades.

13.6.10 Carburisation of ethylene-cracking tubes

The austenitic 18/8 and 25/20 tubes of ethylene-cracking furnaces which operate at temperatures of 700–900°C suffer rapid internal attack (carburisation) if the hydrocarbon feedstock is sulphur-free. This is prevented by the addition of small amounts of sulphur compounds to the feedstock.

13.6.11 Sulphate-reducing bacteria

These bacteria which are present in many soils and air-free natural waters cause rapid corrosion of steel, particularly buried pipelines, oil-well casings and pipe from deep water wells, at pHs of 5.5 to 8.5. They reduce inorganic sulphates to sulphides in the presence of hydrogen or organic matter, and are assisted by an iron surface, cathodic areas of which supply atomic hydrogen. The bacteria use this to reduce sulphate ions, producing ferrous hydroxide and sulphide. Chlorination, aeration and specific bactericides are used for control. These bacteria must be eliminated from water injected into oil-wells to assist oil recovery, to prevent the production of hydrogen sulphide which converts 'sweet' reservoirs into sour ones.

Table 13.2 Chromium contents of steels for various maximum temperatures (in air)

Cr in Cr-Fe alloy (%)	*Max. temperature (°C) in air*
Nil	500
4–6	650
9	750
13	750–800
17	850–900
27	1050–1100

13.6.12 Atmospheric corrosion or rusting

The rate of rusting depends on the air humidity and temperature and its content of sulphur dioxide, sea-salt and other local contaminants. In an industrial area the rate may be more than 100 times the rate in a desert or arctic region. Iron and steel generally need protection by surface coatings, and aluminium alloys by anodising. Wrought iron (which contains a small quantity of silicate slag) has better resistance to atmospheric corrosion than steel. An inscribed iron pillar erected by a Hindu king near Delhi has remained 'rust-free' for 1500 years (Fig. 13.6). Its surface is covered with a thin film of magnetite (Fe_3O_4). Factors which have contributed to its preservation are the dry and until recently unpolluted climate, its low sulphur and high phosphorus content, and its large mass and heat capacity which reduce diurnal temperature fluctuations.

Handrails which are exposed to rain sometimes fail from corrosion from the unprotected inner surface which become anodic as a result of oxygen depletion [13.5.1]. Insulated pipes for steam and process fluids suffer corrosion where they are exposed to leaks and dripping water, particularly if the insulation contains soluble salts which can lead to stress corrosion.

13.6.13 Zinc embrittlement of stainless steels[4]

Highly stressed stainless steel at temperatures of 750°C and above are liable to fail rapidly in contact with small amounts of molten zinc. This could happen in a plant fire if molten zinc from hot galvanised sheet dropped onto a red-hot stainless steel pipe.

Fig. 13.6 A 1500 year old iron pillar at Delhi

13.7 Passivation

Some metals high in the electromotive series, e.g. chromium, nickel, molybdenum, titanium and zirconium, normally corrode at very low rates and are said to be passive. Although not fully understood, passivity is generally consistent with the formation of a very thin, tough, adherent and almost invisible surface oxide film. Passivity can break down in certain chemical environments, e.g. chloride solutions and organic acids.

According to a wider definition, lead in sulphuric acid, magnesium in water and iron in inhibited hydrochloric acid are also passive. Here the passivity is due to a surface layer of an insoluble salt.

Iron and steel are rendered passive by exposure to oxygen above a certain partial pressure and to aqueous solutions of some oxidising compounds. Of these, sodium chromate is added in small amounts to circulating cooling water for process plants and power stations, and nitrites are used to protect petroleum product pipelines and petrol storage tanks in which small amounts of water are present. Other passivators which are only effective in the presence of atmospheric oxygen include alkalis and alkaline salts such as caustic soda, sodium phosphate, sodium silicate and sodium borate and neutral and slightly acid solutions of sodium benzoate, sodium cinnamate and sodium polyphosphate.

Passivation may also be effected by applying a small voltage between the metal and the liquid in contact with it and making the metal the anode. This is anodic protection, the reverse of cathodic protection [13.4], and is sometimes used to protect steel tanks containing sulphuric and nitric acids

13.8 Corrosion-resistant metals and alloys

The passive properties of chromium, nickel, etc. are also conferred on their alloys, giving a wide range of stainless steels, high nickel and other alloys which have excellent corrosion resistance in specific environments. Their cost increases with their corrosion resistance. This encourages the use of non-metals such as chemical glassware, graphite, plastics, and steel with special linings such as glass, rubber and PTFE. Corrosion data for metals and non-metals in various environments are given in 'Perry'.[5]

13.8.1 Stainless steels

Stainless steels are iron-based, with 10–30% of chromium, 0–22% of nickel, and minor amounts of carbon, columbium, copper, molybdenum, selenium, tantalum and titanium. Several are produced in wrought and cast form and widely used in the process industries, although they generally lack resistance to hydrochloric acid and reducing acids. There are three groups of stainless steels.

- *Martensitic alloys* contain 12–20% of chromium with other additives and are used in mildly corrosive environments. Welding is by manual arc, with preheating to 200°C, and post-weld treatment at 650–750°C to prevent cold cracking.
- *Ferritic alloys* contain 15–30% Cr with low carbon content. Their corrosion resistance is good except against reducing acids such as HCl, although they are sensitive to intercrystalline corrosion. Welding requirements are similar to martensitic alloys although they tend to enbrittlement on welding.
- *Austenitic stainless steels* which contain 16–22% Cr, 6–22% Ni, with low carbon are the most corrosion resistant, although their resistance to chloride ions and reducing acids is generally poor. Some have been developed for high-temperature service (e.g. tubes of ethylene-cracking furnaces), and others containing 2.5–3.5% Mo for very corrosive duties at lower temperatures. They are readily welded by manual arc, tungsten inert gas (TIG) and metal inert gas (MIG). Ti, Nb or Ta are added to prevent intergranular corrosion of field-welded joints.

13.8.2 Other ferrous alloys

There are two groups of 'ferrous' alloys (in which iron is not a major constituent), with higher corrosion resistance than stainless steels: *medium alloys* and *high alloys*.

- *Medium alloys* contain 30–44% Ni+Co, 13–30% Cr and other elements. They include Carpenter 20, Incoloys which resist sulphuric acid over a wide range of concentrations and temperatures, and Hastelloy G which is used in wet-process phosphoric acid where fluorides are present. They have limited resistance against hydrochloric acid.
- *High alloys* contain 55–80% Ni and various amounts of Cr, Mo, Fe and other elements. They include Hastelloy B which resists hydrochloric acid over a range of concentrations and temperatures but lacks resistance against nitric acid and oxidising salts. High-nickel alloys are sensitive to contamination during welding, especially from S, Pb and Zn. They should be stress-free before welding, which is similar to that of stainless steels, preferably with argon shielding.

13.8.3 Non-ferrous metals and alloys

Nickel is used to handle and concentrate alkaline solutions. Monel 400 (67% Ni, 30% Cu) resists alkalis and hydrofluoric acid, and moderately oxidising and reducing environments.

Aluminium resists atmospheric conditions, industrial fumes, fresh and brackish waters, concentrated nitric and sulphuric acids, but corrodes rapidly in caustic solutions.

Copper resists sea-water, alkalis and solvents but is rapidly attacked by oxidising acids. Brasses, though used for domestic water fittings, are little used for chemical duties. Bronzes, especially aluminium and silicon bronzes, have better resistance. Cupronickels (10–30 Ni) are the most resistant of copper alloys and used in heat-exchangers on sea-water duties. Copper, however, is a deleterious contaminant in many processes. This limits its use and that of its alloys.

Lead, often with the addition of antimony or tellurium, was once widely used in corrosive environments containing sulphate, carbonate and phosphate ions, which form thin protective coatings on its surface. Its toxicity, low mechanical strength and poor temperature resistance now greatly limit its use in the process industries.

Titanium is strong and resists nitric acid better than stainless steels, and it also resists sea-water, wet (but not dry) chlorine, oxidising solutions hot or cold, and hypochlorites. It does not resist aqueous hydrogen fluoride, fluorine, moderately concentrated sulphuric and hydrochloric acids with no oxidising agent present, oxalic, formic and anhydrous acetic acids, boiling calcium chloride >55%, hot concentrated alkalis and dilute alkalis containing hydrogen peroxide. Small additions (0.1%) of palladium and platinum greatly improve its resistance against hydrochloric and sulphuric acids.

Titanium reacts with both atmospheric oxygen and nitrogen above 600°C. TIG welding with careful argon protection is needed and fabrication can be difficult.

Zirconium has similar although generally superior corrosion resistance to titanium. It resists:

- alkalis of all concentrations up to their boiling point, and fused alkalis;
- hydrochloric acid of all concentrations up to the boiling point;
- nitric acid of all concentrations up to the boiling point;
- sulphuric acid <70%, boiling;
- phosphoric acid <55%, boiling;
- boiling formic, acetic, lactic, and citric acids.

Zirconium does not resist oxidising metal chlorides (e.g. $FeCl_3$, $CuCl_2$), hydrofluoric and fluosilicic acids, wet chlorine, oxygen, nitrogen and hydrogen at elevated temperatures, aqua regia and boiling trichloracetic and oxalic acids. TIG welding is used with argon shielding of all zirconium surfaces above 400°C.

Tantalum has even better resistance than zirconium to oxidising and reducing acids and it also resists wet or dry chlorine, aqua regia, oxidising metal chlorides hot or cold, and lactic, oxalic and acetic acid. It does not resist alkalis, hydrofluoric acid and fluorides, fuming sulphuric acid, oxygen, nitrogen and hydrogen at elevated temperatures, and methanol solutions of hydrogen chloride. Welding of tantalum is similar to that of zirconium although otherwise it is easily fabricated.

13.9 Examples of industrial corrosion problems

Corrosion is a complex subject and, without thorough knowledge of the resistance of metals and alloys under operating conditions, oil and chemical companies quickly run into trouble. The following examples are based on first-hand experience.

13.9.1 A mild steel sulphuric acid tank

Mild steel is corroded quickly by sulphuric acid in the concentration range 5–60% but becomes passive at higher concentrations, and it is used to store 90–98% sulphuric acid.

A water hose was used to clean down the conical roof of a tank containing 95% acid, without it being realised that the roof had corroded internally and had at least one hole. Next morning the tank was in two pieces with a clean cut running round it, just above the acid level.

Water had trickled down inside the wall and formed a layer above the denser acid. Diffusion then produced a narrow zone of hot, dilute and highly corrosive acid. The incident drew attention to the corrosion of the roof of this and similar tanks by dilute acid formed by condensation of acid fumes underneath the roof and absorption of atmospheric water vapour by the condensed acid.

13.9.2 An overseas herbicide plant

A herbicide plant built in the 1960s corroded so fast during commissioning that it could not be handed over to the client in running order. This was due to the use of stainless steel for handling dilute hydrochloric acid.

The more expensive alloys Hastelloy B or titanium with 0.1% Pd could have been used for heat-exchange surfaces and difficult parts, while less expensive non-metals, chemical glassware, polypropylene or ABS (acrylonitrile-butadiene-styrene polymer), or steel lined with rubber, glass or PTFE could have been used for other items.

13.9.3 A new petrochemical process

A plant was built by Shell in the late 1960s in southern France, using a new and revolutionary process based on iodine for the dehydrogenation of butane to butadiene. Many millions of pounds had been spent on development. In spite of this the plant could not be commissioned, apparently because of intractable corrosion problems. The project was abandoned and the large investment written off.

13.9.4 Overhead condensers of crude oil distillation columns

Rapid corrosion was occurring in the overhead condensers and associated pumps, pipework and vessels of the primary flash and atmospheric columns of the main crude oil distillation units at Abadan refinery in the 1940s. One serious fire followed a pipe leak caused by this corrosion.

The primary flash columns operated under moderate pressure and removed LPG gases and low-boiling liquid hydrocarbons. They had tubular water-cooled condensers with mild steel shells and 'Admiralty' tubes.

The atmospheric columns operated at atmospheric pressure with light naphtha as overhead product, and various sidestreams. They had direct-contact condensers made of mild steel in which the vapour was condensed by contact with cooling water in short columns packed with ceramic rings.

A simplified flow-scheme of the distillation units and their crude oil supply is shown in Fig. 13.7.

The corrosion of the condensers of the primary columns was caused mainly by dilute hydrochloric acid, which was sometimes found in them. This was traced to hydrolysis in the columns of inorganic chlorides present in the crude oil. These were only found in the oil from two of ten wells in one of the three oilfields which supplied the refinery by pipeline. They only occurred when the production rates from these wells exceeded a critical figure. The problem was solved by:

1. setting maximum production rates for each well, checked by regular analyses of the crude oil for inorganic chlorides;
2. injecting a small controlled amount of ammonia into each primary column to maintain the pH of water separated from the condensed liquid at about 8. Excess of ammonia which would have attacked the copper alloy condenser tubes had to be avoided.

The corrosion of the overhead systems of the atmospheric columns was found to have different causes and resulted from the combined action of:

1. dissolved oxygen in the cooling water;
2. small amounts of hydrogen sulphide formed by 'cracking' organic sulphur compounds present in the crude oil in the fired heaters of the columns.

As a result, a layer of elemental sulphur was found inside the domed tops of the direct-contact condensers.

The problem was solved by lining the shells of the overhead condensers and water separators with a corrosion-resistant cement, and replacing mild steel parts which could not be lined with stainless steel.

13.10 Notes on 'corrosion' of non-metals

As much care is needed in choosing and maintaining non-metallic constructional materials as with metals. The deterioration of plastics and other non-metallic materials which are susceptible to swelling, crazing, cracking, softening, etc. is essentially physiochemical rather than electrochemical.

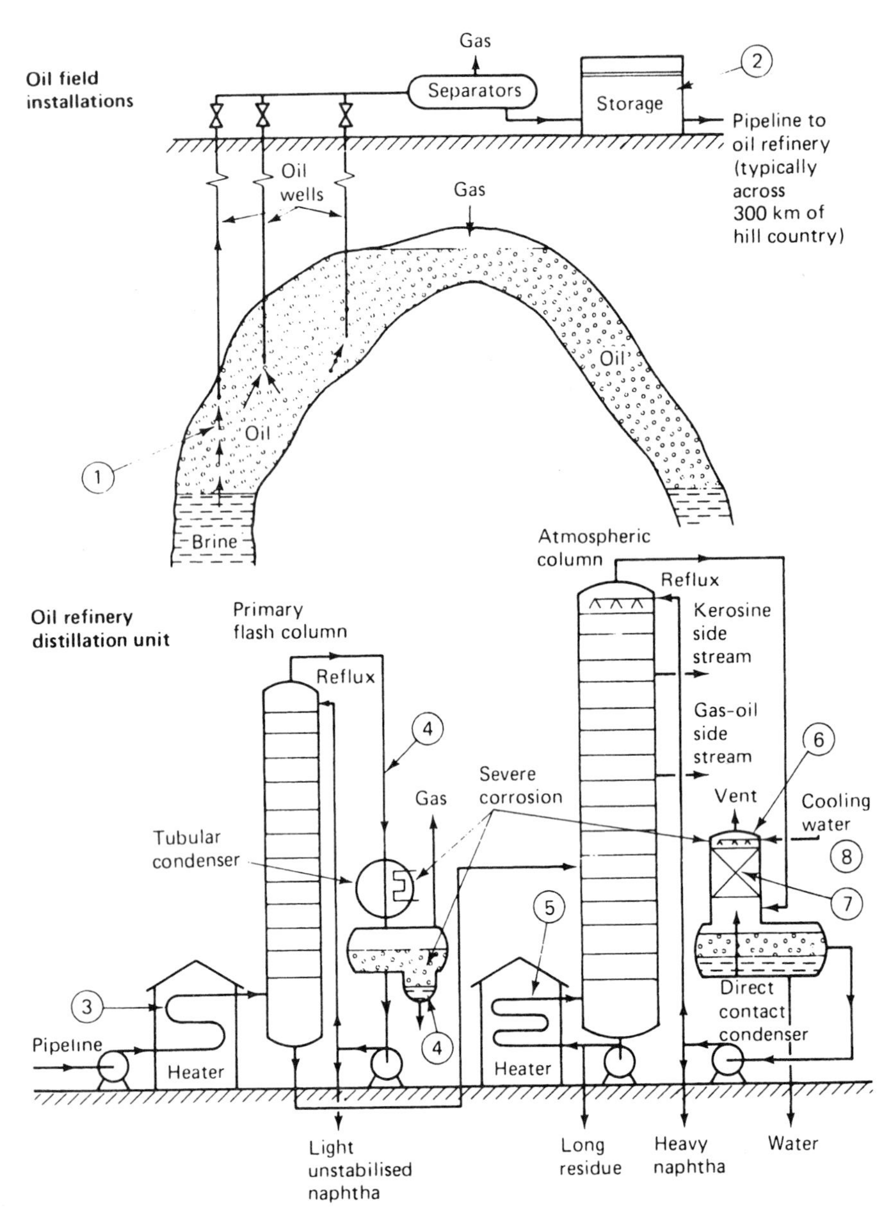

Causes of condenser corrosion

1. Brine drawn up into oil well at high oil production rate.
2. Oil always contains sulphur compounds, sometimes inorganic chlorides and water.
3. $MgCl_2$ hydrolysed here.
4. HCl found here.
5. H_2S formed here (from sulphur compounds in oil).
6. Sulphur found here.
7. $H_2S + O_2$.
8. Dissolved oxygen in cooling water.

Fig. 13.7 Simplified flow scheme of crude oil production and distillation

Thus it is seldom possible to evaluate the chemical resistance of non-metals by weight loss or surface penetration alone.

A range of coatings and linings have been developed to protect steels from corrosion by media ranging from rainwater to concentrated hydrochloric acid. These are often used in conjunction with cathodic protection. Electrical methods of detecting pinholes and other flaws in coatings and linings are widely used. The failure of a lining is then apparent well before the lined shell or pipe fails.

Some problems of non-metallic plant constructional materials follow.

13.10.1 Concrete foundations

These need to be protected both against salts and bacteria in the soil and the effects of spilt process fluids, particularly acids. In the mid-1950s the structure of a synthetic alcohol plant in which sulphuric acid was used nearly collapsed because the foundations had disintegrated through the action of acid which had spilt over a period of years from leaking pump glands. This could have been avoided had acid-resistant flooring and drains been provided in the area at risk.

13.10.2 Flange gaskets and jointing materials

A surprising number of leaks occur through the gaskets of flanged pipe joints of operating plant. Because of this, specialist contractors provide a service of sealing leaks from the outside, without shutting down the plant. Yet a wide range of jointing materials are available from which ones with good resistance to almost any process fluid can be chosen. The use of spirally wound gaskets reduces the danger of a sector of gasket material between two bolts being blown out.

The main causes of such leaks are probably mechanical, e.g. stresses in the pipework and joints (particularly due to differential thermal expansion), badly aligned flanges, improperly tensioned bolts, and the use of unsuitable flanges.

13.10.3 Rubber and plastic hoses

Serious accidents occur through the failure of flexible hoses of various kinds used to transfer fluids between storage tanks and vessels and transport vehicles, and for various unplanned and emergency operations. Many of these accidents result from mechanical causes — improperly secured hose connections and damaged hoses. Some are due to chemical deterioration of the hose, and the use of rubbers and plastics for fluids to which they are not resistant. It is therefore important:

- to provide secure storage for hoses which are not in immediate use;
- to check all hoses regularly;
- to have a system for identifying hoses of different materials;

- to have clear instructions on which types of hoses may and may not be used for the various fluids present.

There is often a tendency in an emergency operation to grab the nearest hose to empty a process vessel, etc. without considering whether it is resistant to the fluid. This may work at the time, but may cause the hose to deteriorate seriously before it is required for its normal duty.

13.10.4 Chemical glassware

Chemical glassware is widely used for small industrial-scale operations and for pilot plant and bench-scale work. Its good corrosion resistance and transparency give it a special appeal. It rarely fails through corrosion since its limitations against hydrogen fluoride and strong alkalis are well known, and early signs of deterioration are usually visible. Unfortunately it is all too easy to overlook its fragility and low tensile strength. In the only serious chemical accident for which the writer felt some personal responsibility, a friend and junior colleague (who since died), was badly burnt in a fire following the escape of a few litres of benzene from a pilot plant distillation unit constructed of chemical glassware. Hazards arising from its fragility always need to be carefully considered before it is chosen and used.

References

1. BSI, PD6493, *Assessment of Critical Crack Sizes*, HMSO, London
2. King, R. W. and Magid, J., *Industrial Hazard and Safety Handbook*, Butterworths, London (1980)
3. Uhlig, H. H., and Revie, W. R., *Corrosion and Corrosion Control*, 3rd edn, Wiley, New York (1985)
4. HSE, Guidance Note PM 13, *Zinc embrittlement of austenitic stainless steel*, HMSO, London (1976)
5. Perry, R. H. and Chilton, C. H., *Chemical Engineers' Handbook*, 5th edn, McGraw-Hill, New York (1973)

14 Reliability and risk assessment

14.1 Introduction

Some 21 years ago the Flixborough Report[1] (Paragraph 217) said, 'it was clear on the evidence that no-one concerned in the design or construction of the plant envisaged the possibility of a major disaster happening instantaneously.' And yet many major disasters had occurred before Flixborough. Indeed only the year before in the annual report of the Factory Inspectorate the then Chief Inspector was warning that such a disaster could take place. Moreover it was only a few years earlier that the Robens Report had also envisaged the possibility that major disasters could occur within the UK. These people were not clairvoyant but they had realised that reliability and risk assessment are intimately linked and by applying a common-sense approach they could foresee the possibility of disaster. Both the Chief Inspector of Factories and Lord Robens had proposed remedies that would reduce the likelihood of major hazards occurring but these had not come about at the time of the Flixborough disaster.

The Flixborough Inquiry (Paragraph 201) said that:

> At no point in the inquiry was there any evidence the chemical industry was not conscious of its responsibilities relative to safety. On the contrary there were indications that conscious, positive steps were continually taken with this objective in mind.

Although it is true that no evidence was taken at the inquiry regarding the subject of reliability in the chemical industry, there was a general notion within the UK (and much of the developed world) that major hazards always happened in somebody else's back yard and that because they had not happened here in the past they probably would not happen here in the future.

As the Flixborough inquiry was drawing to a conclusion, the Major Hazard Advisory Committee was starting its work. It produced three reports.[2] The first one at paragraph 64 says that 'the integrity of pressure systems is of the highest importance' and then at paragraph 70 recommends 'that existing installations should be brought up to an acceptable standard'. This was an echo of the considerable concern about the *laissez-faire* standard of plant integrity within the UK. Within the USA, and Europe there were strict laws and regulations with authorised bodies and mandatory standards which had to apply to process plant. Such regulation was considerably lacking within the UK. By the time the Major Hazard Advisory Committee had come to its final report it said at page 77 that 'it acknowledges that evidence of errors in design or construction may have been more significant causes of serious incidents than had so far been anticipated'. Although this statement should have rocked the process industry it had very little impact.

In the 1980s there were international calls for central registration of major disasters, coming from international seminars, articles in technical journals and debates within learned societies. But this has not come about. Understandably insurance companies and operators are very reluctant to admit liability or blame because of the insurance pay-out implications. Several studies have been published which attempt to collate the largest losses in the process industry. Marsh and Mclennan[3] have reviewed accidents over a 30-year period, examining 100 of the largest property damages in the hydrocarbon chemical industry. The Safety and Reliability Directorate, part of the UK Atomic Energy Authority, has also undertaken extensive international searches on failures of process plant, particularly with regard to missile damages. Mr Eddie Crookes of the HSE has attempted to analyse failures at process plant, and the trend he has ascertained is worrying. In the years 1956 to 1965 there were 13 losses, 1966 to 1975 29 losses and over the next ten years 61 losses; this shows that the failures in the process industry far from abating, are doubling every ten years. Mr Crookes suggests that refineries account for 50% of the failures, petrochemical plant for 32% and gas-processing plant for 5%.[4].

By Mr Crooke's analysis mechanical failure accounts for 63% of the known causes, with operating error, process upset and natural hazards accounting for the remainder. Upon further analysis it is realised that operator error and process upsets should have been prevented by the mechanical engineering solutions of adequate design, construction, maintenance and the provision of safety devices. The most important reasons for a process plant failure are mechanical failure of hardware, and failure of mechanical engineering software and protection devices.

Mr Crooke goes on to conclude that 64% of all accidents occurred to pipework, 14% to storage tanks and 14% to process towers and reactors. From the investigation of previous incidents it is obvious that there were no new startling technological secrets. Plant in the process industry is unreliable and increasingly so not because of total ignorance, but because the information, technology and know-how that does exist is not being used or not being used at its right source.

Attempts to make process plant more reliable follow [15.2, 15.3] but first let us consider some definitions.

14.2 Hazards

The term hazard particularly in the process industries relates to the danger that can occur either potentially or blatantly. Such hazards may cause asphyxiation, toxic poisonings, long-term dust inhalations, fire and explosions, burns from release of hot substances, or caustic or acid burns, or burns due to cold temperatures. Other hazards within the process industry will occur jointly in all industries and cover such things as electricity, noise, trips, slips or falls, and injury to human bodies due to repetitive strains, and manually handling awkward or heavy lifts. The hazards can affect on-site employees or off-site bystanders.

14.3 Risks

Risk is the chance of a hazard occurring. There are four reasons why people take risks in a given situation:

- they do not realise that they are taking risks because they are unaware of the inherent hazards of the situation — many accidents with children come into this category;
- they recognise the hazard in the abstract, but do not feel personally at risk — due perhaps to a belief in their own ability to control the situation;
- they recognise the hazard and the potential risk involved but adopt a somewhat fatalistic attitude — 'the preordination of fate';
- they act to keep the level of risk constant — there is a suggestion that, should there be a reduction in objective risk (e.g. through some engineering change) people will take more risks to restore the perceived risk to its original level.

It is obvious that all situations at home, at travel, or in work cannot be made absolutely safe. Sometimes the advantages of undertaking a very hazardous activity need to be considered. For example chlorine is extremely hazardous, but through its use in sterilising and bleaching it has become an essential part of modern, advanced culture. Man usually adopts a process of combining hazard and risk so that when the hazard is very high the risk is very low and vice versa. British safety law only ensures that process plant is safe so far as reasonably practicable. It does not have to be absolutely safe, for in reality nothing can be absolutely safe.

14.3.1 Quantified risk

There has grown up a desire to attempt to quantify risk. For example a certainty would have a probability of one. Table 14.1 attempts to collate visual descriptions with numerical assessments.

14.3.2 Historical risk assessment

Within the desire to have risk assessment numerated, organisations such as the Safety and Reliability Directorate or the UK Atomic Energy Authority have collected statistics on well-known pieces of equipment and articles, and then calculated a failure rate in numerical terms. For example large boroughs having many light bulbs in their street lighting complexes will have an accurate knowledge of the number of bulbs, the number of bulbs that fail per year, and this could be held over maybe a 50-year period. The failure probability of bulbs used in street lighting can therefore be assessed in a historical context. Such information is valuable not only for safety, but also for production and planned maintenance services.

The Atomic Energy Authority are able to supply failure rates of many components, but only when large numbers of components are in use. The HSE carry out hazard and risk assessments, and these can give ideas of whether a certain factory or industry is performing well in comparison with other industries or factories. For example it is known how many people are employed in the construction industry, and how many people are fatally injured each year, so at the beginning of a large contract such as the building of the second Severn Crossing it would be possible to carry out a statistical evaluation of the potential number of fatal accidents that may occur. If the actual number of fatalities is lower than that predicted people can feel satisfied with the safety measures taken; if the fatalities exceed the anticipated number then the same satisfaction will not prevail.

To the individual any fatality is a disaster but society cannot avoid disasters at any cost. Society accepts that there will be many more killings on the roads of the UK than there will be in process plant. The acceptability of risk and hazard has often been driven by public debate and the media. For example a single fatality at a fairground will attract media coverage for a

Table 14.1

Descriptions	*Frequency per year*
Extremely unlikely	less than 10^{-6}
Very unlikely	10^{-6} to 10^{-5}
Unlikely	10^{-5} to 10^{-4}
Quite unlikely	10^{-4} to 10^{-3}
Somewhat unlikely	10^{-3} to 10^{-2}
Fairly probable	10^{-2} to 10^{-1}
Probable	More than 10^{-1}

sustained time, whereas multiple fatalities on a building site many be of local interest only.

14.3.3 Subjective risk assessment

Much process plant is unique or only occurs in small numbers. For example there may be only ten ammonia storage spheres within the UK. Each piece of pipework has its own special features, and the failure modes and combination of failure modes, although only small, will not always prevail.

Within the UK the boiler inspection companies had carried out accident counting exercises so that they were able to give a failure rate of for example a steam boiler, steam receiver or air receiver in any one year. This of course helped them to assess their insurance cover needs, and also helped their engineering surveyors to highlight particular weak features when assessing boilers or steam receivers. The information was based upon a large number of boilers and many years experience of examining them. It should not be assumed that the reasons for steam boiler failures can be directly applied to other pressure systems. It certainly could not be applied to pipework or storage tanks, by far the largest causes of failure within the process industries. An in-depth analysis of steam boiler failures will reveal unique circumstances such as low-water failure, failure at welds of the header to furnace tube, failure of water-treatment processes, that would not occur in the majority of process plant. On the other hand plant that is liable to operate at low temperature, e.g. ammonia, LPG, chlorine or carbon dioxide storage vessels, will fail in a brittle manner due to fast fracture, and will be totally different from the hot failures usually associated with steam boilers.

It is therefore fairly obvious that with small numbers of unique plant, risk assessment based on analysis of historical data will not be useful. It therefore relies on management, insurance companies, inspection bodies, health and safety specialists devising systems of subjective analysis of how plant fails, and then going on to underpin safety by carrying out remedial actions.

The Flixborough Report almost let the process industry off the hook by saying that no-one involved was aware of how the failure could occur. In reality there were professional experts available who could have ascertained how failures could occur and then gone on to say if they would occur at that site before the disaster happened. Such assessments are regarded as subjective risk assessment analysis. They are based upon strength of material, fracture mechanics, stress analysis, design and fabrication assessment and maintenance and inspection know-how.

14.3.4 Legal implications of risk assessment

Chapter 2 deals with UK Law and shows that risk assessment must be undertaken before a disaster occurs. The regulations which apply particularly to process industries are CIMAH[5], COSH[6] and Management at Work.[7]

14.4 HAZOP

Hazard and operability (HAZOP) studies are the process engineer's concept of risk assessment. However, they have considerable shortcomings when used to assess the reliability and safety of major hazard process plant. As shown in 12.1 the only time a major hazard will arise is when there is a loss of containment. Loss of containment is always due to mechanical engineering failure at the design, fabrication or testing stage or to failure to keep the plant within its safe operating limits. A HAZOP study starts with the assumption that the plant is adequately designed, tested, installed and maintained. It does not question the mechanical integrity of the plant and therefore is not a good tool for assessing reliability. HAZOP studies, had they been carried out, would have failed to predict many of the catastrophic failures which have taken place. They are discussed more fully in Chapters 4 and 5.

HAZOP is described in some detail by Kletz[8] and Lawley[9] and only a brief account is given of it here. It is based on a team study of the P&I diagram as this reaches its final stage. The need for and the thoroughness required of a HAZOP study depend much on the degree of hazard of the process (e.g. as given by the Dow Index [11.1]). Since it is also an operability study, it may be combined with the writing of the operating manual [20.2.3]. Although a HAZOP study is undoubtedly expensive, it frequently saves much greater expenditure later when the plant may have to be modified because of some problem which would have come to light in a HAZOP study.

A team of four or five people each provided with copies of the PFD and the P&ID sit round a table in sessions of two to three hours and study the P&ID in a formal and systematic way two or three times per week. Because of the degree of attention required, and the need to incubate ideas which the study may release, longer and more frequent sessions are not recommended. Gibson[10] in 1976 estimated about 200 man-hours per million pounds of capital for these studies, i.e. about eight sessions. Typical members of the team would be:

- the process engineer responsible for the P&ID;
- the project engineer responsible for the mechanical design;
- the commissioning engineer responsible for start-up;
- the plant manager who will be responsible for operating the plant;
- the instrument engineer responsible for plant instrumentation;
- a research chemist (specially needed if new chemistry is involved);
- the computer programmer (for computer-controlled plant);
- a hazard analyst who acts as independent chairman. He records the ground covered at each meeting and issues reports of each meeting with agreed action items and responsibilities before the next meeting.

If the plant has been designed by a contractor, a mixed team of contractor's and client's staff is needed, and some functions may have to be duplicated.

Not all of these team members listed above need to be present at every meeting, but the process engineer, the commissioning engineer and/or the plant manager as well as the chairman should be present throughout.

The team studies each pipe and vessel in turn using a series of expressions to stimulate thought as to what would happen if the fluid in the pipe were to deviate from the design intention. The expressions recommended for a continuous process are:[9]

NONE	MORE OF	LESS OF
PART OF	MORE THAN	OTHER THAN

These are then applied to each possible variable in the conditions in the pipe or vessel, e.g. flow, temperature and pressure in all cases, with level, reaction, mixing, etc. added for vessels. The causes and consequences of all these variations are then considered and, where necessary, rough estimates of their probability are made.

Start-up, turndown, shutdown and other special operations such as catalyst regeneration should be considered as well as normal operation, and the study may be extended to utility units.

Batch operations are studied in a similar way by listing the sequence of operations and applying these expressions to each step. On computer-controlled plants the instructions to the computer should be studied as well as the P&IDs.

While a P&ID provides a good general plant model for HAZOP studies, its limitations must be recognised. It is purely a diagram, is not to scale, and does not always show equipment items in their proper vertical relationship to each other. It does not show all mechanical and instrument details, nor whether drains for removing water from low points in a distillation column are all correctly situated. What may appear to be a short straight pipe in a P&ID may in fact be long and tortuous, and vice versa. The chairman should have the ability to draw out useful information half-buried in the minds of team members. Drawings, manuals and a three-dimensional plant model (if one exists) should be available to the study team as needed.

The usual outcome of HAZOP studies is a number of (hopefully minor) design modifications, and for example, the provision of special communication channels and lighting, changes to the programme of a computer-controlled plant and to the analytical programme for quality and hazard control. The design modifications might include changing some materials of construction, adding, subtracting or altering the type or position of valves (particularly drains and vents) and vessels, modifying pipework, altering the instrumentation, particularly emergency trips and alarms, steam tracing and/or insulating lines, and improving access to particular valves or equipment. Since these modifications sometimes bring fresh and unsuspected hazards, the exercise must be repeated to the modified design.

While HAZOP studies may extend over weeks or months, they have to be fitted into the relatively short interval between the advanced drafting of P&IDs and other drawings and their freezing and approval for construction. Many of the points that emerge from a HAZOP study could have been avoided if the engineers had been more experienced, better informed, or had checked their own work more thoroughly.

A HAZOP study may also reveal that it would have been difficult or impossible to have carried out certain steps necessary during operation or maintenance on the plant as originally designed. This could have been anticipated if those writing the plant-operating manual had done their homework more thoroughly, and had 'gone through all the motions' on paper of opening and closing valves and starting and stopping motors, etc. as needed for all operations.

14.5 Worst credible event

Under the CIMAH Regulations[5] Schedule 6 for every major hazard site a report has to be prepared and sent to the appropriate authorities. Paragraph 5a of Schedule 6 calls for 'a description of the potential sources of a major accident and the conditions or events which could be significant in bringing one about'. It is therefore incumbent upon management of major hazard sites to carry out a worst credible event scenario. It would also be sensible for management of other process plant to carry out the same type of assessment. When the worst event and the conditions which could bring it about are considered then plans can be made for prevention, control or minimisation.

There is a certain reluctance within industry to admit what the worst credible event could be. It is difficult to persuade a waterworks using chlorine as a sterilising agent to accept that the worst credible scenario is the total catastrophic failure of the chlorine storage tank which may contain up to 30 tonnes of chlorine. It is possible, considering the weather conditions and the topography of the site, to model mathematically the distance that a released cloud of chlorine could travel. It is thought that a release of 15 tonnes of chlorine could travel up to 14 miles in very still air or during an inversion. The harm to humanity, animals and plant life would be devastating. And so it is a logical step to see how that condition could arise.

The reluctance to accept that a worst case scenario could arise, leads to them being dismissed as impossible. From first principles we know that catastrophic failure of the chlorine vessel can occur only as a result of over-stress, over-strain, creep, fatigue, brittle fracture or stress-corrosion cracking. It has the potential to fail due to over-stress, particularly in the saddle support area where the local stresses are very high; it has the potential to fail from brittle fracture because chlorine can be stored and delivered at subzero temperatures; it has the potential to suffer from fatigue because the vessel is emptied and filled on numerous occasions each year; and the highest points of stress will be caused by the loading of the chlorine inside the vessel, rather than the pressure fluctuations. If the chlorine was delivered in a wet state, or water were allowed to enter the vessel, it could suffer from rapid corrosion which could result in over-stress, fatigue or brittle fracture. It is also common practice on a chlorine storage vessel to have block valves before the pressure-relieving devices, which can in certain circumstances make them inoperative; this could lead to over-stress or brittle fracture. It is also known that chlorine is transferred from delivering road tankers into the vessel by use of padding or pressurised air thereby forcing liquid chlorine without the use of a pump. This has great advantages in as much as the glands are not a source of leakage, but it can introduce air into a vessel that is designed for the vapour pressure of chlorine. The addition of air in a chlorine vessel will increase the pressure, because the vapour above the liquid is non-condensible. This could lead to over-stress, fatigue or brittle fracture. Having analysed how a major hazard could come about, or the conditions or events which could be significant in bringing one about, it is then possible to carry out the necessary safeguards, e.g. design the vessel to sustain the maximum pressure, design the saddle supports so that they do not suffer from fatigue, design the vessel so that it will actually sustain a cold temperature. The pressure-relieving devices can be twinned so that each relieving device can be isolated in turn but still leaving its partner or twin on-line. It is only by carrying out an informed subjective study of the worst credible event that worthwhile reliability can be assured.

14.6 Cost of accidents

The HSE have produced a booklet entitled *Cost of Accidents at Work*, based on surveys carried out by HSE's Accident Prevention Advisory Unit over a period going as far back as 1978. Accidents, including major accidents at process plant, will have obvious cost effects such as insurance claims on behalf of injured persons and to cover damage to property. But financial costs, loss of productivity because of poor working conditions, absenteeism, and general debility can be hidden costs which can break an organisation.

Very large major hazards can cost so much money through loss of production that Government may suffer significant loss of tax revenue. This happened to the UK Government after the disaster in Pembroke in 1994 [13.3].

14.7 The accident triangle

A spin-off from the research into cost of accidents has been to produce the so-called 'accident triangle'. The principle has been to consider the worst case scenario, and relate it to other major injuries, loss-of-time injuries, minor injuries and non-injury accidents. For example in a case study done at a National Health hospital there was one three-day-injury accident to every ten minor accidents and this was in relation to 195 non-injury accidents. The accident triangle for an oil platform or a creamery might represent one three-day-injury accident to four or five minor injuries to 140 non-injury accidents. Figure 14.1 shows a typical accident triangle. However, it is difficult to attach reliable figures to each category, and make these meaningful in the context of the process industry. Although the principle of the accident triangle has been shown to be statistically valid, if it is applied to major incidents (such as Piper Alpha, Flixborough, Feyzin, Mexico City, Bhopal, BP Grangemouth or Texaco Wales) it will be seen that it is no longer relevant in the case of complete disasters.

Major disasters — due to unforeseen loss of mechanical engineering integrity — can happen even in well-organised, well-resourced safety-conscious companies if they have not applied adequate technology. For example a safety department can only prevent a major disaster if its staff are able to undertake reliability and risk assessments themselves or (recognising their limitations) are willing to call in outside expertise.

14.8 Use of outside expertise

A large refinery in South Wales had a safety team that was well aware that pressure-relieving pipes leading from knockout drums to the flare stack were considerably corroded.

But they did not have the expertise to see how vulnerable the plant was to total catastrophic failure; nor did they appreciate their own limitations and seek outside expertise. The flare stack lines were rarely used but on one fateful day flammable gas was passed along. This caused the corroded pipe to fail due to over-stress. The contents were discharged leading to a fire which destroyed a section of the refinery.

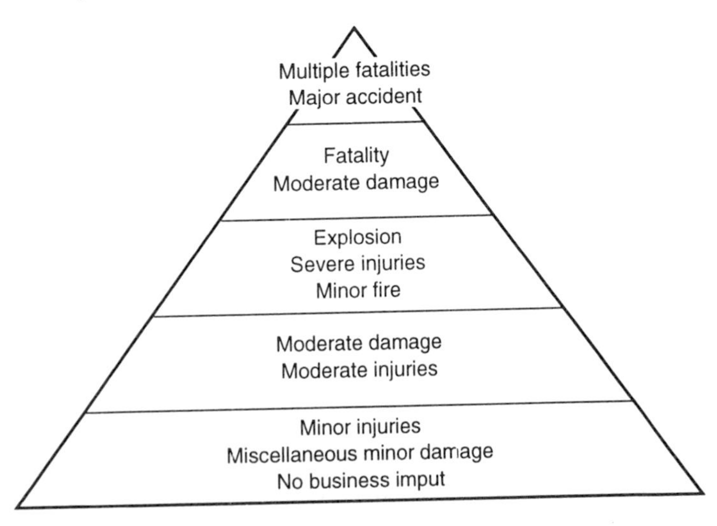

Fig. 14.1 The accident triangle

14.9 Fault trees

This section can only serve as an introduction to a complex subject which holds many pitfalls for the unwary novice. Fault tree analysis needs undivided attention and often expert guidance.

Before attempting a fault tree analysis, we should be thoroughly familiar with the system analysed, and have a suitable model in front of us. This should show all items and their interconnections whose faults (basic events) could affect the top event under study. For this a P&ID is a good starting point, though we may require more detail as we move down the tree.

Starting with a particular type of failure termed the 'top event', we work back to a first level of faults which, acting singly or in combination, could have led to it. Next we look for all faults which could cause the first-level faults and continue down the tree until we reach failures for which data are available (or can be invented!). These failures are called basic events.

Both failure rates (dimensions time^{-1}) and probabilities (dimensionless) that an item is in a failed state (sometimes referred to as fractional deadtime) are used in building fault trees. Confusion can readily arise between them. The failure rate is usually expressed as so many times per year (often less than one) that the item is expected to fail. This is usually assumed to be constant. The probability that the item is in a failed state is a function both of its failure rate and the period during which the unit of which the item is a part continues to operate while the item is in a failed state.

If the unit does not operate continuously and has to be shut down to repair (or replace) the item, as well as for general inspection and maintenance, the period is simply that between the occurrence of the failure and the shutdown of the unit. If the unit operates for 2000 hours per year and the average time lapse between discovery of the failure and shutdown of the unit is 10 hours, and the item fails, on average, three times per year, then the probability of the item being in a failed state while the unit is operating is $3 \times 10/2000 = 0.015$. Provided the failure rate of the item remains the same after repair or replacement as before it, and the failure and repair do not affect the number of hours per year when the unit operates, the probability of an item being in a failed state is given by:

$$\frac{(\text{Failure rate, years}^{-1}) \times (\text{hours between failure and shutdown})}{\text{Hours of operation per year}}$$

If the item is tested, and if necessary, repaired or replaced while the unit is in operation, the calculation of the probability of the item being in a failed state or out of commission while the unit is running becomes more complicated.

Lines from lower-level faults shown on a fault tree can join at one of two kinds of 'gates' (similar to those used in Boolean algebra) from which a single line passes to a higher fault.

Figure 14.2 shows the principal symbols used in drawing fault trees. At an OR gate, any of the lower events is sufficient to propagate the fault, and their probabilities or failure rates are added. At an AND gate, all the lower events leading to it must have happened in order to propagate the fault. While probabilities can be multiplied at an AND gate, two or more failure rates cannot be multiplied, since this would make the figures for the higher fault dimensionally inconsistent with the failure rates for the lower ones, and the result would be rubbish. *At an AND gate a rate can only be multiplied by one or more probabilities. Great care must be taken to keep the numbers used in fault tree analysis dimensionally consistent.* If the top event of a fault tree analysis is to be given as a rate, we need to know the failure rates of most items and the probabilities of others being in a failed state or out of commission. If the top event of a fault tree is to be given as a probability, we need to know the probabilities of all items corresponding to low-level events being in a failed state.

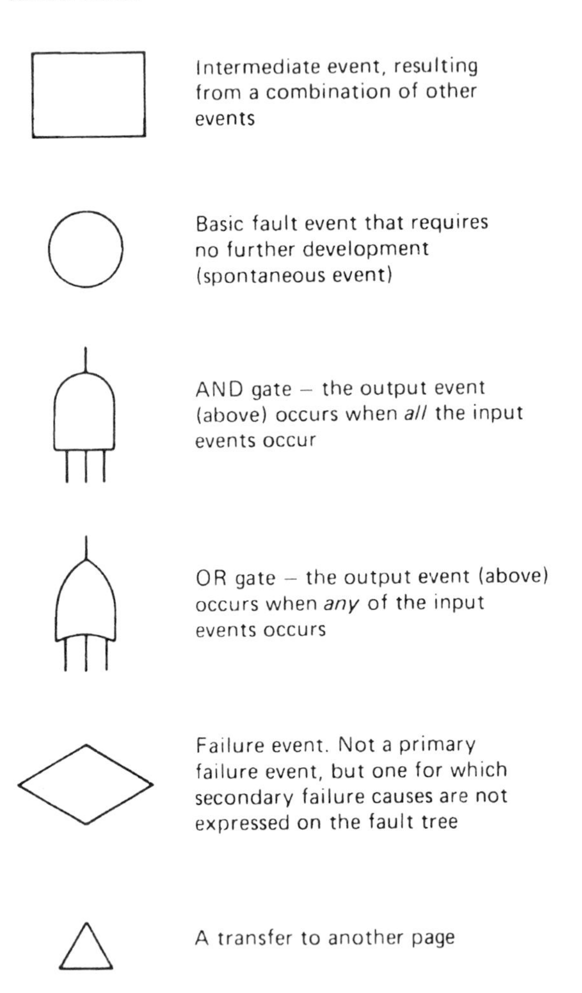

Fig. 14.2 Some symbols used in fault trees

14.9.1 Example of fault tree

The method is best illustrated by an example, for which I use a simplified model of my 30-year-old domestic gas central heating system which also provides me with hot water. A P&I diagram of the system is shown in Figure 14.3 (which should on no account be taken as a model for a safe central heating system!). The pilot burner is lit and the mains gas valve open for 340 days per year (i.e. except when the system is being serviced or the house is unoccupied for several days).

The top event considered for this tree is the flow of gas from the main burner into the combustion chamber of the so-called boiler for at least 10 minutes without igniting. We want to know how often this is liable to happen. (Although technically incorrect, the term boiler is used here following general usage.) This stipulation that the failure lasts at least 10 minutes excludes some shorter temporary failures which would rectify themselves automatically within that period. The basic events which could lead up to this top event are considered to be failures of the gas-control system, human error and interruption of the gas supply. This analysis does not include:

- the entry of unburnt gas from the pilot burner, which is considered a minor hazard;
- leakage of gas into the boiler-room which could result from a leak in the piping system, or from unignited gas in the boiler.

Gas from the mains passes through isolation valve V1, gas governor GG and meter M to the boiler, where it splits. A small flow passes through valve V3 to the pilot burner PB. This is lit manually and V1 is never closed while the boiler is in service. The main flow passes through valve V2 and solenoid-operated valve SV to the main burner MB. SV closes when the solenoid is de-energised, and is operated by current from a thermostat TC1 on the boiler. This current is cut off by switch S which is actuated by the flame-failure device FFD when PB is unlit. FFD consists of a fluid-filled bulb connected to S by a copper capillary.

Water is circulated through the boiler, pipework and radiators by pump P whose switch is actuated by a time-switch and room thermostat TC2. On starting the pump the water temperature in the jacket falls, causing TC1 to switch on current to open SV, provided S allows it. MB is lit from PB and continues burning until the water temperature rises and TC1 closes SV again. So the cycle is repeated. Any loss of water from the system is automatically made up from an elevated head tank, and the boiler is protected against overpressure by a safety valve which discharges into the space below the roof of the house.

On starting to draw the fault tree (Fig.14.4) from the top event, two alternate cases or intermediate events are considered. An OR gate is drawn below the top event, with lines leading to it from the two intermediate events:

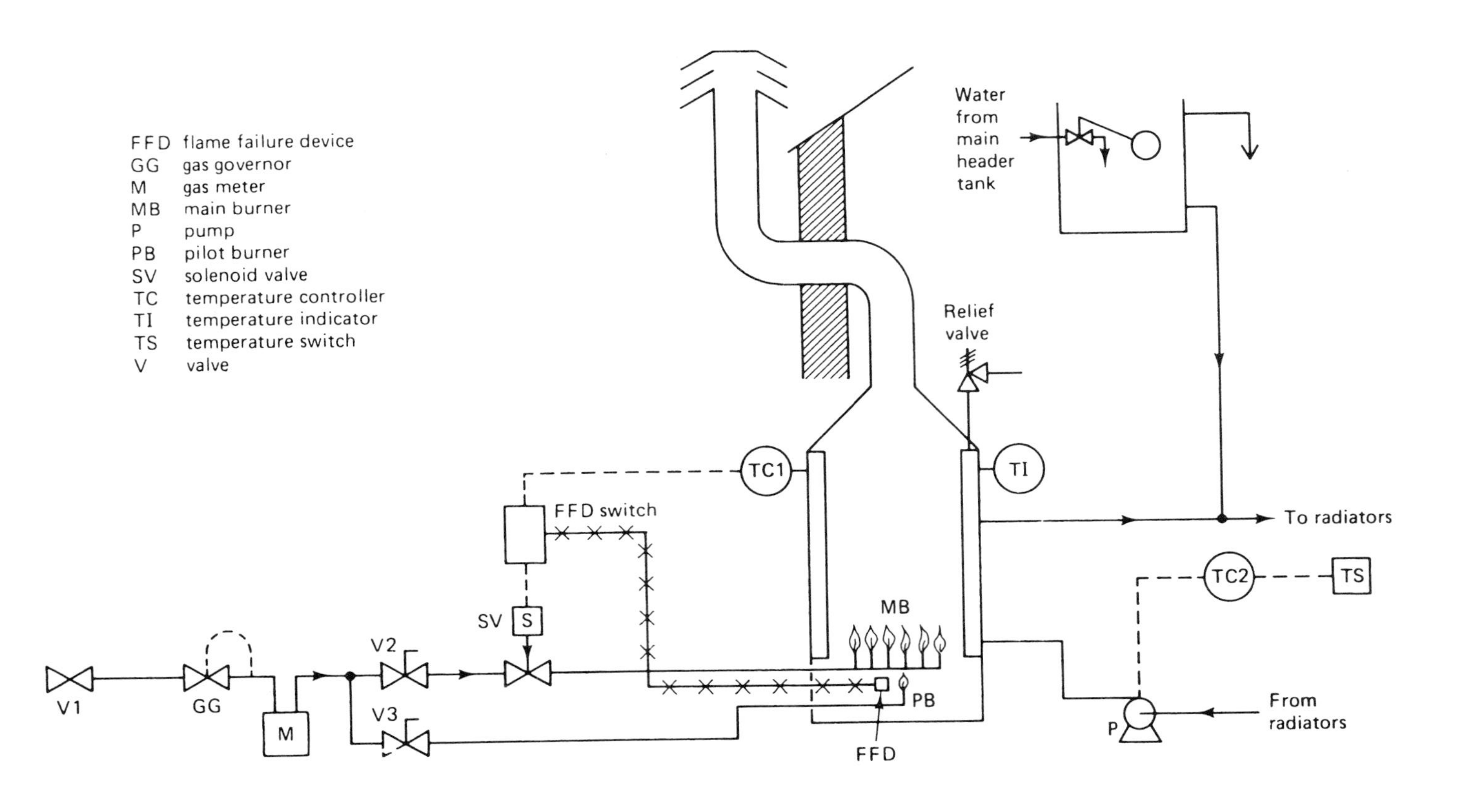

Fig. 14.3 P&I diagram of domestic central heating system

1. pilot burner unlit and solenoid valve fails to close on demand;
2. pilot burner lit, solenoid valve open, but main burner fails to light.

As (1) requires two conditions to be met simultaneously, an AND gate is drawn below it, under which the two conditions are shown side by side and connected by lines to the AND gate. As a rate is wanted for the top event of the fault tree, we need one of the conditions below the AND gate to be shown as a rate or frequency and the other as a probability. Frequency will be used for the first condition, the PB being unlit when V3 is open and probability will be used for the second condition, failure of SV to close on demand.

As we identify four possible alternate causes (A to D) for the first condition, 'PB unlit', we draw an OR gate under it. These causes are basic events which are shown on the fault tree under this OR gate and connected to it by lines. The frequencies assigned to them are given in Table 14.2. Both A and B result from human faults and their frequencies depend on the occupants of the house. C depends on wind and other variables and D on the gas supply.

We also identify three possible alternate causes E, F and G for the second condition, 'SV fails to close on demand', again drawing an OR gate under it. They are mechanical faults to which we can assign failure rates. We have to convert these to probabilities by the equation given in 14.4 in order to use them in the fault tree. If SV is open while PB is unlit, the boiler temperature and pressure will rise, causing the relief valve to blow, discharging steam and hot water. We assume that this will be noticed within a day or two when someone will investigate, ring up the 24-hour emergency service of the gas supply company and be advised to shut gas valves V2 and V3 pending the arrival of an engineer. We estimate the average period between the occurrence of the failure and the closing of V2 as 15 hours. Dividing this

Table 14.2 Assumed probabilities of basic events in fault tree

Basic event or condition	*Comment*	*Rate times/year*	*Probability* $\times 10^{-4}$
A V1 closed and reopened without relighting PB	Gross human error	2	
B. V3 opened without lighting PB	Same, more frequent than A	3	
C PB light goes out	Draughts	0.25	
D Transient failure of gas supply	Householder unaware of event	0.25	
E SV stuck open		0.2	7.4
F S stuck closed		0.5	18.5
G Fault in FFD	Broken capillary	0.1	3.7
H MB does not ignite because of soot	Boiler badly needs service	0.002	
J MB does not ignite because PB out of place	Poor design or installation	0.002	

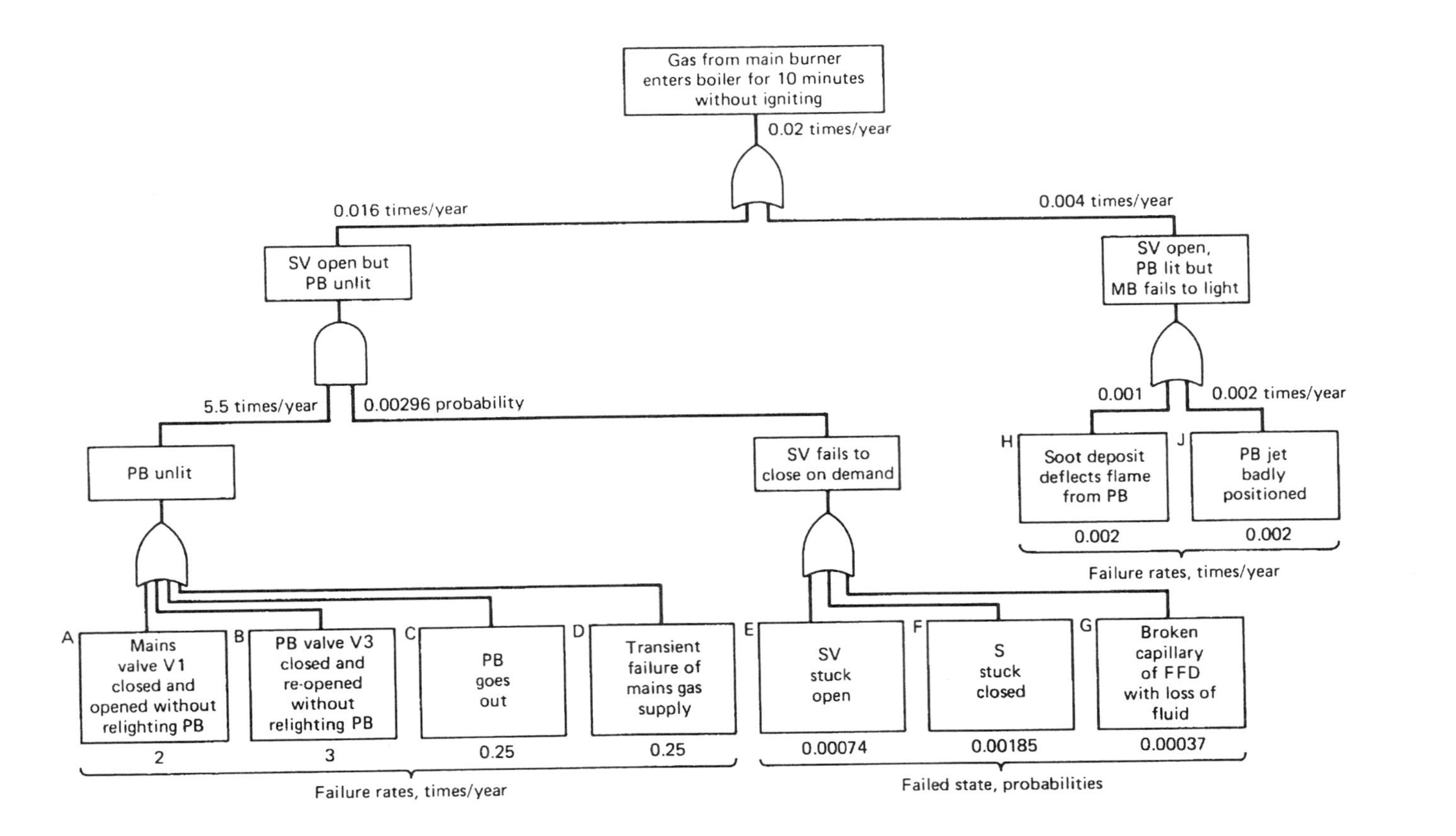

Fig. 14.4 A fault tree for the central heating system shown in Figure 14.3

by (340 × 24) gives the factor 3.7×10^{-3}, by which the failure rates must be multiplied to give failed state probabilities. These three causes with their rates and probabilities are also given in Table 14.2.

For (2), we identify two possible alternate causes H and J which we are prepared to treat as basic events. They are given with the rates assigned to them in Table 14.2 and marked on the fault tree under an OR gate.

The resulting frequency of unburnt gas flowing through the main burner into the combustion chamber for at least 10 minutes without igniting is thus 0.02 times per year.

If we consider this is excessive, the analysis can suggest ways of improving the situation. Thus the gas company might reduce the frequencies of the human errors A and B by posting warning notices to the householder with the gas bill, and of events E to J by providing a free annual service of the installation.

14.9.2 Comment on fault trees

It should be apparent that the construction of a fault tree and the assignment of frequencies to various base events and probabilities to various conditions is as much an art as a science, and much depends on the experience, judgement and mood of the person doing it. One person might immediately recognise an important branch of a fault tree which another had missed entirely. There is nothing absolute about the probabilities assigned to the various base events. The main value of fault tree analysis may be that it stimulates our curiosity about the various ways in which a system can fail, so that we learn more about it and understand it better. It also puts these failure modes into some perspective, and concentrates our attention on the more serious ones. The educational effect of the exercise can be more important than its numerical results.

Many fault tree analyses include recommendations that one or more safety gadgets be added to the system. The writer has yet to see one which showed how the risk could be reduced by simplifying the system and throwing out superfluous items!

The main drawback of fault tree analysis is the large amount of time which it takes to investigate a single top event of even a simple system. Here the writer has a suggestion. Shift operators of process plants usually have time on their hands (especially at night) when the plant is running smoothly. The could be encouraged (say, through a works safety committee) to use some of this time in making their own fault tree analyses on selected sections of the plant, for particular top events. Fault tree competitions could be organised between different shifts on the same plant. Provided care was taken that operators did not become so engrossed in their fault trees that they neglected their normal duties, they could learn a great deal about their plants in this way, and thus make a real contribution to their safety. Judging of the fault tree analyses submitted by the different teams could also set in motion a useful dialogue between the teams and the judges, during which misconceptions would be exposed and discarded, and new and useful information would come to light.

Finally, while fault trees are generally used as an aid in studying how a potentially disastrous event might arise in an existing plant or one under design, more use might perhaps be made of fault trees in investigating causes of disasters. Accident inquiry reports, after weighing up the merits of different theories, often base their conclusions about the causes on 'a balance of probabilities'. Why, then, not present the various theories in fault tree form? After identifying and allocating probabilities to the different possible base events on which each theory is based, a fault tree could be drawn for each theory to show its intermediate events and provide an estimate of its probability. This could help to rationalise accident investigation even though it does not solve the more difficult problem of recognising base events. Nevertheless, if objective fault tree analysis shows that all the theories hitherto put forward have low probabilities, it is a fair bet that some base event or vital link in the chain of causation has been missed!

14.10 Event tree risk assessment

14.10.1 Introduction

If HAZOP studies are based on process engineering, and fault tree analysis considers how faults in a process may lead to catastrophic failure, neither can carry out an assessment of whether a catastrophic failure is likely to occur due to the shortcoming of the mechanical engineer. There is a need to consider events rather than faults.

The UK Atomic Energy Authority has developed the rule of event tree analysis in the study of safety of reactor accident analysis. Their booklet SRD128[11] describes and analyses the role of event trees. An event tree is a diagrammatic representation of the sequence from initial events through secondary events or conditions to the final end effect. The forward-thinking inherent in an event tree distinguishes it from the backward thinking of a fault tree. The secondary events should be independent of each other. Each has a path of success or failure to which probabilities are assigned. It is usual to allocate probabilities between 1 and 0.1, 1 being certain and 0.1 being uncertain or extremely remote.

14.10.2 Predicting loss of containment

It is possible to study the mechanical engineering aspects of a pressure vessel, pipework, heat-exchanger, boiler, distillation column, storage tank etc. considering all the logical failure routes: stability, over-stress, plastic collapse, fatigue, brittle fracture, stress-corrosion cracking and creep. By considering each potential failure mode and assigning failure probabilities between 0.9 and 0.1 for each an event tree can be established. The risk or chance of a final failure (either a leak or a catastrophic failure) is arrived at by multiplying together each assigned risk assessment. Figure 14.5 shows an event tree for a 60-tonne propane vessel. Considering each stage in its potential failure and using expertise in stress analysis, failure analysis and metallurgy, it is possible to assign risks at each stage. For example: Instability is unlikely to occur — allocate a factor of 1

Over-stress may well occur in this vessel, because it has peculiar stress intensifications where the saddle support is welded directly onto the shell, and, contrary to pressure vessel codes, the flanges are let into the vessel rather than attached with stand-off branches. It is also noted that the flanges have blind tapped holes for the studs that are used to hold on valves etc. The stress engineer may attribute a factor of 2 to stress intensification. As the design stress was two-thirds of the yield stress, and there was a stress intensification of 2, there would be local stresses well past the yield point, making this vessel susceptible to failure by over-stress — allocate a failure event factor of 0.9.

Plastic collapse is thought to be unlikely except in the saddle support area and a failure factor of 0.1 to 0.9 is allocated.

This vessel is not considered likely to fail due to fatigue — allocate a factor of 1.

Brittle fracture is a strong possibility because: (a) there are likely to be defects up to 2 mm deep within its surface; (b) the material used for its fabrication was not resistant to low temperature; (c) there were local stress raisers that could induce brittle fracture; (d) the ambient temperature can be –5°C, and the withdraw temperature can be sub-zero. PD6493[12] would suggest that the critical crack size for the vessel would be 2 mm deep, and it is extremely difficult to find cracks so small, and therefore they would be assumed to be pre-existing. The failure rate due to brittle fracture would be high. Assign a factor of 0.9.

By multiplying the assigned risk factors on each branch of the event tree together a final risk number is obtained. The example shown in Fig. 14.5 would suggest that there is a 0.009 chance of the vessel failing by leakage, i.e. that failure by leakage was quite unlikely. The failure from brittle fracture when operating under cold conditions on the other hand is arrived at by multiplying each of the assigned failure rates on that branch together, and arriving at a failure rate of 0.7, i.e. the probability is high that it will fail by brittle fracture. It must be remembered that brittle fracture will actually produce a fast running defect that would lead to total catastrophic failure and in this instance that vessel would be totally unsafe.

The remedial action that could be undertaken, i.e. not to operate the vessel at sub-zero temperatures, is not an acceptable option, and therefore the vessel should not be used. There are no safety devices that are available that could be fitted to this vessel to stop it operating at sub-zero temperatures. Therefore this vessel is not economically viable.

Using the above mechanical integrity, safety and hazard assessment probability scheme, it is possible to analyse past failures. We can ask whether the No. 5 Reactor at Flixborough would have failed by cracking

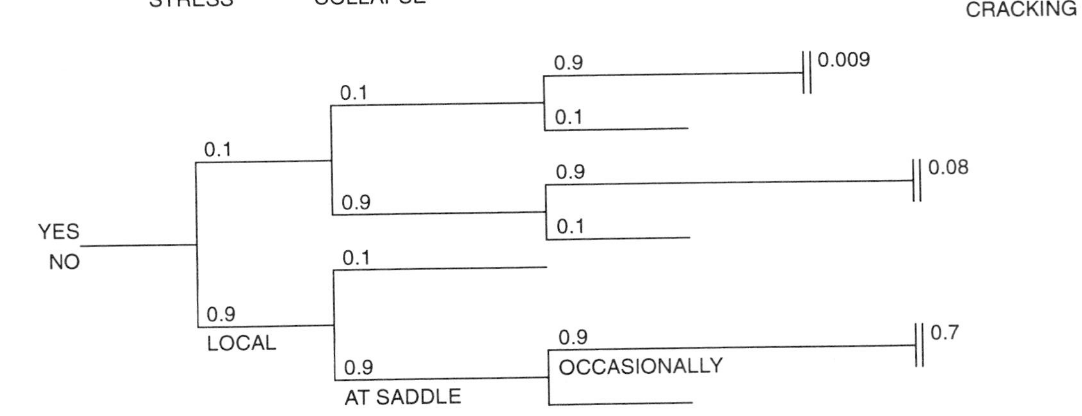

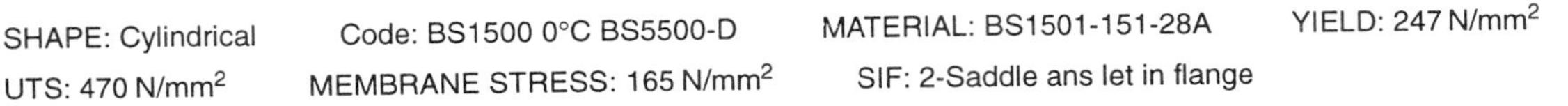

SHAPE: Cylindrical Code: BS1500 0°C BS5500-D MATERIAL: BS1501-151-28A YIELD: 247 N/mm^2

UTS: 470 N/mm^2 MEMBRANE STRESS: 165 N/mm^2 SIF: 2-Saddle ans let in flange

CRACK LIKE DEFECT: Possible up to 2mm FABRICATION: Suspect FATIGUE: Possible (see SRD 314)

SCC: N/A PROCESS: Storage MEASURED: Not measured ASSUMED: N/A SCC

TEMP: –5°C THICKNESS: 15mm 20mm Flange PRESSURE: 40 to 70 Psi LOADS: 60 Tonnes Propane

VACUUM: N/A OTHER RELEVANT INFO: Saddles welded on, no stress relief. Assumed impact test and 20°C, critical crack size 2 mm

Fig. 14.5 Event tree for 60t propane vessel

and leaking or by total disintegration. It is also possible to analyse the 20-inch bypass pipe at Flixborough to see whether it would fail by leaking or by catastrophic failure.

The event tree has proved itself to be an acceptable and powerful aid to risk assessment — considering either the potential catastrophic failure of existing plant or the potential for newly-designed plant to fail.

14.11 Change of reliability with time

Even when conditions remains roughly constant, the reliability of any item changes during its working life. From early reliability studies, mainly on electronic components which then depended on thermionic valves with short working lives, it was concluded that the working life of most items has three phases.

1. The early failure period, beginning at some stated time and during which the failure rate decreases rapidly. This is a sort of 'running-in' period when manufacturing flaws (e.g. in materials or assembly) and installation faults cause a small proportion of the items tested to fail. In process plant many early failures occur during the commissioning period, when the failed items are repaired or replaced.

2. The 'constant failure rate period' during which some of the surviving items fail apparently at random at a low but approximately uniform rate. This defines the working life of the item under the conditions specified. Despite the random nature of these failures, they all have definite causes which can usually be diagnosed. For non-repairable items in process plant, the constant failure rate period should exceed the anticipated life-time of the plant.

3. The 'wear-out' period during which the surviving items (usually the great majority) fail at an increasing rate due to deterioration (e.g. as a result of wear, corrosion or fatigue).

Since the early days of computers which depended on thermionic valves the reliability of electronic equipment has increased spectacularly, so that modern integrated circuits on microchips might outlast the Pyramids!

A plot of failure rate against time (of failure-prone items) gives the classic 'bath-tub curve' of Fig. 14.6(a). Many components of process equipment show this failure pattern.

An easily visualised example of an apparent 'constant failure rate' is provided by an underground lighting system consisting of a large number of identical bulbs wired in parallel, each of which is replaced as soon as it fails. When the system was new, most bulbs would have lasted for several months and then failed one by one within a shorter period. As time went on and all bulbs had been replaced several times, bulbs would fail more or less at random, giving a 'pseudo-constant failure rate'. The appearance of a constant failure rate is in fact due to a random and roughly constant replacement rate. If records of individual bulbs had been kept, it might have been found that no bulbs failed for the first three months of service although most failed in the fourth. Their true failure rate would then not have shown the first two phases of the bath-tub curve at all. but would have been better represented by a 'ski curve' as shown in Fig. 14.6 (b).

The same appearance of a constant failure rate is commonly found with the instruments of a process plant which has been in service for several years, where a policy of repair maintenance was adopted (for the instruments).

Carter and other reliability engineers in the mechanical field have long questioned the universal validity of the bath-tub curve, particularly its implication that they are dependent on components which are liable to fail at random at a low but steady rate for most of their working lives.[13] This fatalistic view would make air-travel rather unsafe. It would also make long production runs on complex single-stream continuous plant almost impossible, and there would be little point in scheduled preventative maintenance.

With improvements in engineering, manufacturing techniques and quality control it is now generally possible to produce components in which the first

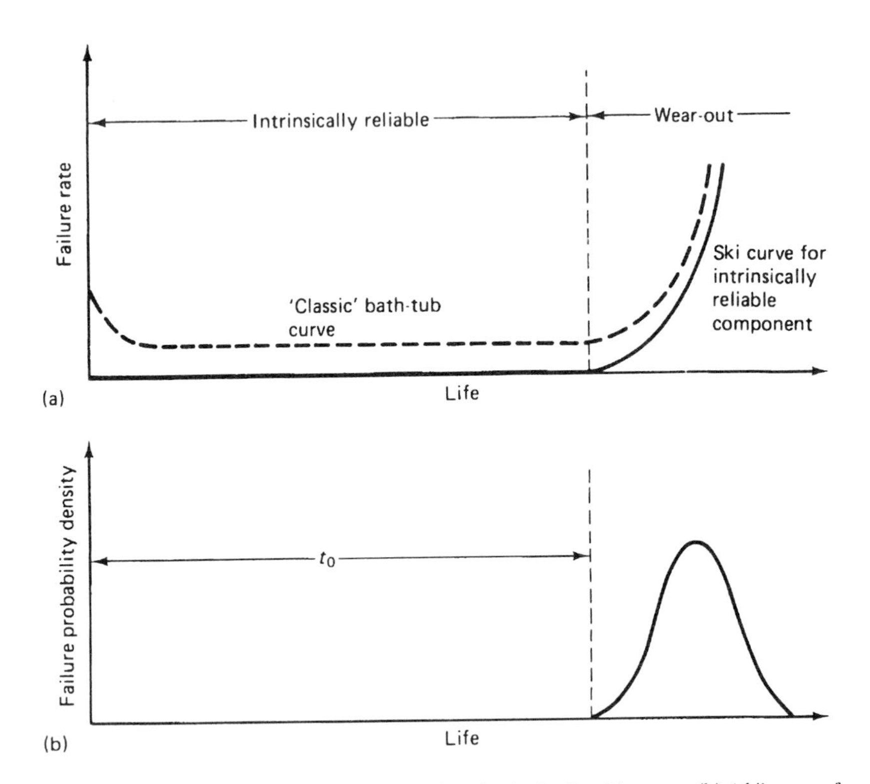

Fig. 14.6 Failure-rate versus age curves (a) the classic 'bath tub' curve; (b) 'ski' curve for intrinsically reliable component

two phases of the bath-tub curve are replaced by a single one which is essentially free from failures, provided the loading and other conditions to which the component is subjected remain within specified limits. This is achieved for mass-produced components without incurring economic penalties. Indeed in most cases it is more profitable to produce equipment (like a quartz crystal watch) which can be guaranteed to remain failure-free for, say, a year's use, than lower-quality items which fail at random. The bath-tub curve is now becoming outmoded. Although the ski curve cannot always be achieved, it is the target for which most engineers strive. Three things should be noted about it:

1. The early failure period of the bath-tub curve can be virtually eliminated by improved quality control, proof testing and/or a period of test running before the item leaves the works or factory.

2. By providing a margin of safety in design which is adequate for all anticipated variations in loading and other conditions, and in the strength of the items themselves, the constant failure rate period can be replaced by a 'zero failure rate period' during which the item is said to be 'intrinsically reliable'. This is not done by applying large and arbitrarily selected margins of safety, but it requires detailed study of possible load and strength variations, wear and corrosion rates and mechanisms, and careful attention to all possible modes of failure.

3. Wear, corrosion and other forms of deterioration still impose a limit on the period of use during which components remain intrinsically reliable. This (like the learners' skis of the *ski évolutif* system) can be long, medium or short. At the end of this period the items are worn but by no means all worn out. Although now past their guarantee periods, and subject to occasional failures, they may continue to give useful service for much longer. The shape of this part of the curve can vary, as shown in Fig. 14.7, and is often typical of the mode of failure, e.g. stress-rupture, fatigue or corrosion.

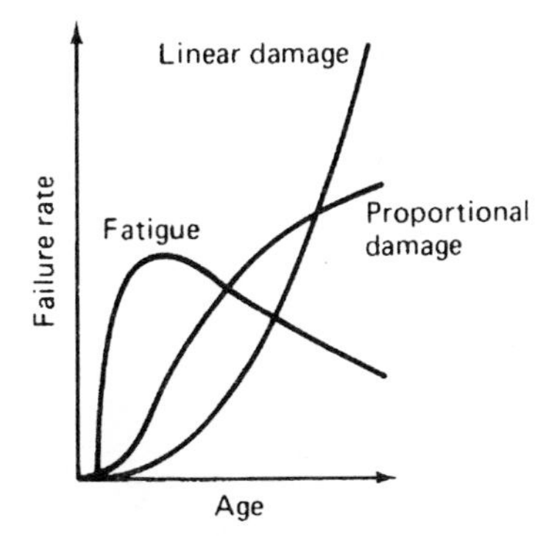

Fig. 14.7 Failure-rate versus age curves for different failure modes during wear-out period

The policy to be adopted at the end of the intrinsically reliable period depends on considerations both of safety and of economics. It is sometimes recommended, as in the case of factory lighting, to scrap and replace all bulbs or other short-life items at the end of this period. While there may be an occasional failure within this period, it is usually difficult to attach any meaningful statistic to it.

The obvious advantages to the user of intrinsically reliable items have resulted in their becoming the norm for most machine components and many domestic articles. Such items are generally sold under guarantee for a certain period. Because of this, premature failures are quickly reported to the manufacturer, who takes the necessary action to prevent their recurrence.

References

1. Flixborough Report HMSO 1975
2. HSC, Reports *on Major Hazards Reports 1, 2, 3* HMSO, London
3. Marsh and Mclennan, 100 Worst Accidents in the Process Industries, 1988
4. Crookes, E. Property losses in the process industries; an engineering problem — I, *Proceedings Management of in-Service Inspections of Pressure Systems*, Institution of Mechanical Engineers (1993)
5. S.I. 1984, No. 1902, *Control of Industrial Major Accident Hazards Regulations 1984*, HMSO, London
6. S.I. 1988, No. 1657, *Control of Substances Hazardous to Health Regulations 1988*, HMSO, London
7. S.I.1992, No. 2051, *Management of Health and Safety at Work Regulations 1992*, HMSO, London
8. Kletz, T. A., *HAZOP and HAZAN — Notes on the identification and assessment of hazards*. The Institution of Chemical Engineers, Rugby (1985)
9. Lawley, H. G,. 'Operability studies and hazard analysis', in A. I. Chem. E. Symposium, *Loss Prevention in the Chemical Industry*, New York, (November 1973)
10. Gibson, F. B., 'The design of new chemical plants using hazard analysis', in I. Chem. E. Symposium Series No. 47, *Process Industry Hazards, Accidental Release, Assessment, Containment and Control*, Rugby (1976)
11. UKAEA, SRD128
12. PD6493, *Guidance on Methods for Assessing the Acceptance of Flows in Fusion Welded Structures* (London: BFI, 1991)
13. Carter, A.D.S. *Mechanical Reliability*, 2nd edn, Macmillan, London

15 Design for safety

15.1 Introduction

No major hazard accident has occurred or is likely to occur unless there has been a loss of containment. Safety in the process industries is largely dependent on the elimination of losses through good design and maintenance.

This chapter does not attempt to change the safety specialist, or anyone else into an expert on the safe design of process plant. That should always be delegated to experts. It is intended rather to highlight past deficiencies in design that have led to losses of containment, and suggest ways of making plant safe as far as reasonably practicable.

Whether the intention is to provide a service, or to produce a profit, whether it is to be a turnkey project or the expansion of an existing plant, the first step will normally be a process or chemical engineering design. The raw materials needed, the end product and by-products, as well as air temperatures, flow-rates, weight, dynamic activities, etc. will need to be considered, together with the feasibility and cost-effectiveness of the design. At this stage health and safety will not be a factor.

Once the functional design and geometry of the plant has been laid down by the process experts, the concepts will be passed on to mechanical engineers who will design the hardware, instrumentation, and software for the process operation. Any faults in the hardware design or instrumentation could lead to disaster. Should anything go wrong, even the most toxic, explosive or flammable material, provided it is confined or contained in adequately designed equipment, will not bring about a disaster so long as that containment is maintained throughout the lifetime of the plant.

At the turn of the century there were 300 boiler explosions per annum. The carnage and injury to people and property could not be tolerated, and mechanical engineers put their minds first to considering why the boilers were failing and then to designing boilers that were unlikely to fail.

Process plant is only likely to fail catastrophically because of the six reasons given in Chapter 12, i.e. over-stress, over-strain, metal fatigue, brittle fracture, creep and stress-corrosion cracking.

15.2 Design errors

The Flixborough Report[1] at Paragraph 209 suggests 'the disaster was caused by the introduction into a well designed and constructed plant of a modification which destroyed its integrity'. That suggestion had no detrimental effect on the report but others have seized upon it to suggest that a majority of chemical plants were well-designed and well-constructed. Running parallel to the Flixborough Report was an advisory committee which produced three separate reports on major hazards.[2] The first report at Paragraph 64 says that the integrity of pressure systems is of the highest importance and at Paragraph 70 recommends that existing installations (i.e. existing in 1976) should be brought up to acceptable standards. In this there is an implication that not all plant in existence at that time was designed to acceptable standards. The final report at Paragraph 77 acknowledges evidence that errors in design or construction may be more significant causes of serious incidents than has so far been accepted.

The design of process plant is not easy and there is no room for complacency. Chapter 14 deals with reliability and it shows that major failures of process plant are continuing at an alarming rate. The majority of these failures are correctly attributed to failure to contain the substances safely. Therefore the design has not been adequate.[3]

The Pressure System Regulations[4] require safe operating limits (SOL) to be set for process plant pressure systems. But it is good engineering practice to set such limits for storage vessels and pipework which do not come under the regulations.

- Over-stress is caused by pressure, weight, expansion and wind.
- Over-strain is caused by wind forces, vacuum and expansion.
- Creep comes about because of loads (stresses), high temperature and time.
- Fatigue comes about because of stress reversals (i.e. the degree of stress and the number of cycles the process plant undergoes).
- Brittle fracture comes about because of stress, temperature and stress raisers.
- Stress-corrosion cracking comes about due to stress and the environment.

From the above list we can see the issues to be addressed when setting SOL. For example it may be obvious that creep will not be a failure mode in a static storage tank operating at ambient temperatures. But it is still worth going through the safe operating checklist to see which conditions do apply. It is quite often thought that pressure vessels fail because of over-pressure.

In fact, failure by excessive pressure is extremely rare;[1] it is more likely that combinations of other factors such as expansion loads, weight, wind-induced eddies and temperature may give rise to catastrophic failures. All loading conditions should be used as design criteria. The design of process plant will certainly include stress analysis, metallurgical strength and behaviour, and will also deal with the geometry of the structure.

Through long years of trial and error, and by laboratory testing and experimentation, stress analysis has achieved a reliability and confidence that is unshakeable. Modern UK, US and Continental standards deal with the stress analysis of pressure systems, pipework, storage tanks, heat-exchangers, distillation columns and boilers. The same standards will give recommendations on welding profiles, methods of fabrication, and methods of laying down welds.

15.3 Stress raisers

The stress analysis of a perfectly homogeneous structure may be relatively simple and straightforward. Once local deformations, notches, sharp edges, weld profiles or crevices caused by corrosion or poor workmanship are introduced then there may well be localised stress intensifications.

For example theoretical calculations that have been checked by strain gauge testing have confirmed that very large spheres have local areas where the stresses are three times those anticipated for a perfectly homogeneous sphere. If the design engineer has anticipated using a design stress of 2/3 of the yield stress of the material, but there are localised stress intensifications of three, then in very localised pockets there will be stresses which are twice the yield strength. This means that there will be very local yielding or local deformation. If that local deformation causes a structure to shape itself into a better profile all is well. But some structures, such as a homogeneous sphere, cannot be improved. If there is a local deformation then stress intensification in that area will cause it to grow worse, leading to progressive deformation and failure through plastic collapse. A proper design analysis will either reduce the stress raisers by producing good geometric designs, and ensuring that fabrication keeps to them, or reduce the allowable stresses so that even when stress intensifications are inadvertently experienced the material is still within its SOL.

15.4 Critical crack sizes

The object of a design is to produce a structure that will not fail catastrophically. Leaks, provided they remain small, may well be tolerated in process plant but catastrophic failure must be avoided at all costs. There is a mathematical modelling technique which is based upon considerable experience issued by the BSI as PD6403 and called *Critical Crack Sizes*.[5] This standard analyses the load, the crack size and the stresses within a structure so as to predict whether the structure would leak, or catastrophically fail should the considered crack grow to a postulated size.

15.5 Designing for vapour pressures

According to Charles' Law a perfect gas will have a vapour pressure which is proportional to its temperature. In some quarters it has been common practice for the design of pressure vessels for storing liquefied gases to be based purely on their physical vapour pressures. Such gases would be propane, butane, chlorine and ammonia. Experiments looking at the temperature and pressure of large storage vessels containing liquefied gases show that there is a vast discrepancy between theory and practice (see Fig. 15.1). This has come about due to liquid being drawn off from the static storage tanks, and the remaining liquid boiling so as to fill the vapour space. This boiling action has super-cooled the liquid. But at the same time sunlight may be warming the vapour space, so that the gas in the upper portion of the containment vessel may become relatively warmer than the super-cooled liquid. Many storage vessels for liquefied petroleum gas have not been designed for this phenomenon.

Some old standards that were used for designing liquefied petroleum or chlorine vessels did not adequately cover the concept of brittle fracture. When it is realised that such vessels may be operating in a brittle fracture regime it is necessary to revalidate the SOL using modern-day stress analysis, critical crack size analysis, brittle fracture analysis accepting as well as that vapour pressure/temperature curves may not necessarily indicate the worst conditions that the containment vessel will sustain.

The HSE has in recent years produced guidance entitled *The Assessment of Pressure Vessels Operating at Low Temperatures*.[6] Within this, it acknowledges that there are some existing pressure vessels that were originally designed to standards that do not comply with later design methods. Some of them may well be unfit for their intended use. The HSE goes on to acknowledge the usefulness of the modern-day UK Pressure Vessel Standard BS 5500 and recommends that the safety experts revalidate old plant.[7]

15.6 Revalidation of design

New process plant naturally must be designed and fabricated to appropriate standards. For new concepts design may have to be done from first principles. As the plant deteriorates [16], or when modern standards readdress design principles, plant will have to be redesigned or revalidated.[6]

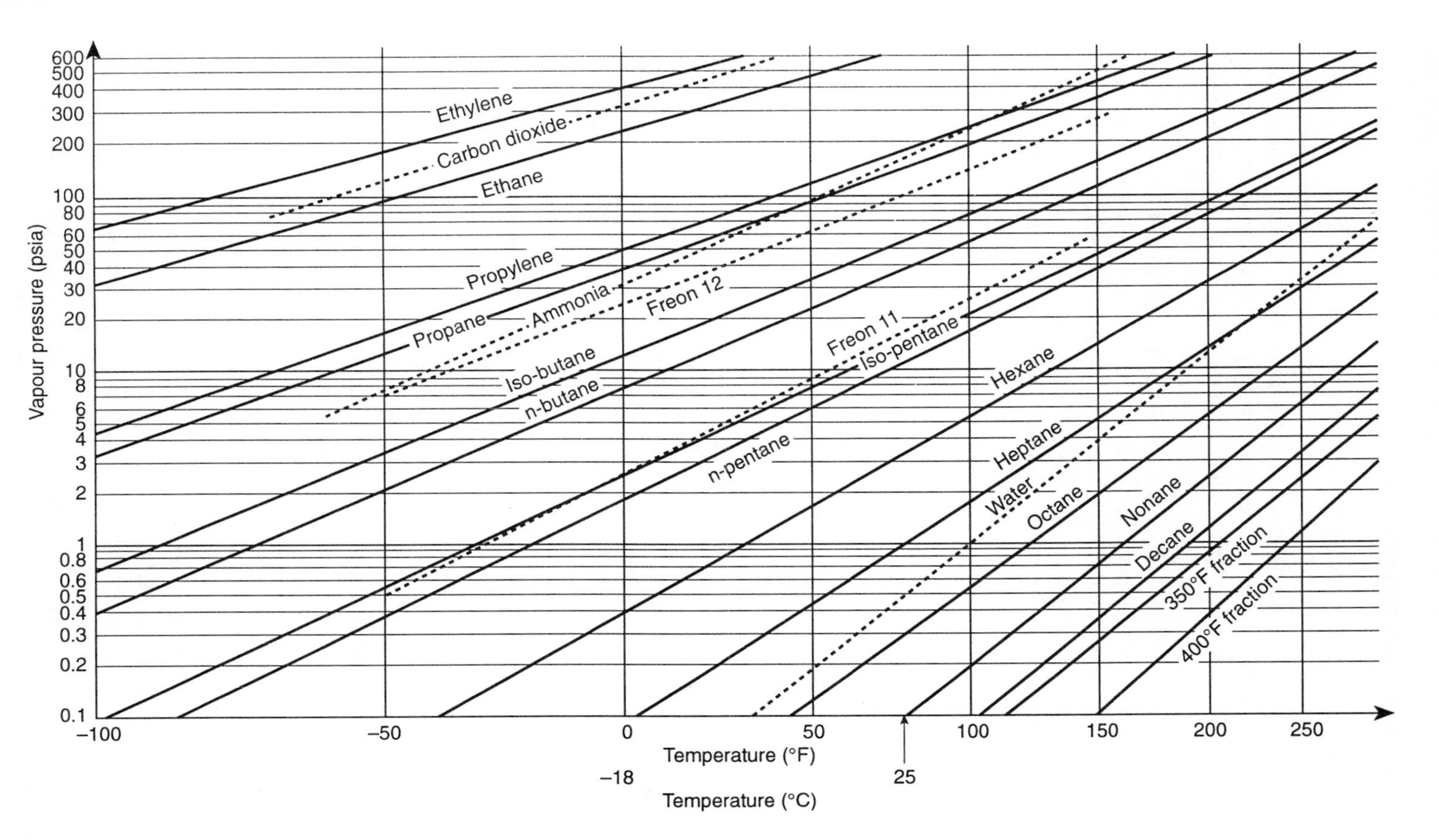

Fig. 15.1 Vapour pressures

15.7 Quality assurance

The design, fabrication and testing of process plant needs to be double-checked if the hazard is high or the risk of failure needs to be very low. That is not to say that all process plant has to be treated with the same degree of sophistication. It may be that to ensure product reliability and avoid loss of production, users may need high-integrity plant. But here we are concerned only with designing for safety. There are organisations approved by BSI and the Institution of Mechanical Engineers — such as various insurance companies and the leading inspection bodies — that will carry out quality assurance checks on design features (stress analysis, critical crack size analysis) and on fabrication. In this way the procurers of process plant can be assured that the design and fabrication reaches them in first class condition. BS 5500[7] calls for third party independent quality assurance assessments. But there are no UK mandatory prescribed rules.

15.8 Design myths

After the Flixborough disaster it was thought that process plant control rooms should be designed and fabricated to withstand an over-pressure up to 3 psi. But in fact explosions such as at Flixborough developed over-pressures vastly in excess of 3 psi, and so building new control rooms or rebuilding existing ones to that standard (if it had been undertaken) would have been useless. Moreover had the control room at Flixborough been designed to sustain an over-pressure in excess of 3 psi, people could well have been trapped within it, with access possible only after life had been extinguished. It is a far better use of resources to attempt to design or revalidate plant so that it is capable of working within safe operating limits and then to ensure that it does so [16, 17].

There has been and still is a concept that process plant can be protected by water deluge systems should a major fire develop. Almost all process plant within the UK which contains flammable substances is protected with some sort of emergency deluge device. When major fires or explosions have occurred it is known that deluge systems, pipework hydrants and even underground pipework has been severely disrupted and the systems have not worked. It is also realised that once flame is impinging on a vessel or storage tank and water is applied afterwards the thermal shock of cold water dousing hot metal can cause thermal contractions and shocks that have caused vessels to split and intensified the already devastating fire. It is also realised that when water is played onto very hot surfaces there is flash boiling whereby the water on the surface of the metal instantly vaporises with little cooling of the metal, and for these reasons very few water deluge systems have ever done anything to preclude a major disaster. It would be far better to invest money in the revalidation of old vessels, and the assurance that new vessels are designed, fabricated and tested to adequate and acceptable standards.

The law requires that emergency arrangements are made in the largest major hazard plants. It should be realised that emergency services such as fire brigades and ambulances cannot contain an escalating major hazard. They can only deal with some of the casualties. There are some hazards which are so disastrous that only a cosmetic emergency services response is possible. For example where such toxic substances as hydrogen fluoride are made and stored, should there be a loss of containment of the main storage vessels, the hydrogen fluoride is likely to drift in a cloud which would not be dispersed for 5–6 km. This type of disaster would cause a large number of fatalities. The general emergency advice is for people to go indoors, shut their doors and windows and if necessary place wet towels above their heads. When it is realised that the dispersion of some toxic gases may take several hours and the normal air changes through gaps in windows and floor boards etc. would be two per hour, there would in reality be very little hope for anyone attempting to shelter in this makeshift manner. The only sensible way to cope with such a situation would be to do all that was humanly possible to avoid a disaster, i.e. by designing or revalidating the structure so as to be assured that catastrophic loss of containment cannot take place.

15.9 Process plant units of measurement

Much of the petrochemical industry takes its lead from the USA where imperial units are still in use. Many of the other process units have imported technology from the Continent where units have been metricated. The UK has attempted to use so-called 'SI units', and this has often caused problems and difficulties. Units of length, area and volume cause very little difficulty either in imperial or metric units. Units of weight — 1 kg is equal to 2.2 pounds or 1 tonne is equal to 0.98 tons — are within everyday working knowledge. The unit of pressure probably causes the most confusion.

Process personnel tend to use the term psi standing for pounds per square inch: UK British stress personnel tend to use the term lbf/in^2; the modern unit of pressure is the 'bar', with 1 bar equal to 14.5 lbf/in^2. The SI unit of pressure is the newton per square metre and one mega newton per square metre is equal to 145 lbf/in^2.

The unit of force used to be pounds but is now newtons, with 1 newton equal to 0.2248 lbf.

At any process plant there should be a consistent system of units.

15.10 Process engineering

Process engineering is a skilled profession which operates at the nerve centre of process development and plant design. While vital to safety, it is often misunderstood. Most process engineers are chemical engineers, but only a minority of chemical engineers work as process engineers.

There is constant competition between processes for making the same end product. This is particularly felt in the drive to reduce consumptions of raw materials, power, steam, fuel and cooling water and to utilise cheaper raw materials. It has led to the development of new catalysts which give higher yields and reduce operating temperatures and pressures, and to the extensive use of power and heat recovery by the use of waste-heat boilers, gas turbines, low-temperature expansion turbines (which also provide refrigeration) and heat exchange between outgoing and incoming streams. Many cracking, reforming, dehydrogenation, ammonia, gas separation, hydrocarbon oxidation and even polymerisation plants now have many resemblances to power stations. This is not always an unmixed blessing for safety, particularly when the process design becomes so 'tight' that operators have little room for manoeuvre if things go wrong. This situation can usually be improved by the provision of recycle loops, particularly round distillation columns, which allow production to be halted while vital equipment is kept 'ticking over' ready to come on-stream again as soon as the problem is sorted out.

The wastes and effluents from a process are also determined by its process design [15.10.2] and process engineering is involved in the design of their treatment facilities.

15.10.1 The role of the process engineer

The highest skills are needed by process engineers who design new processes from the results of small-scale experiments. They require a good knowledge of chemistry (particularly of the process in question), chemical engineering and process hazard evaluation and a working knowledge of economics, metallurgy, corrosion engineering, computer programming and mechanical and other branches of engineering, a varied operating experience and sometimes a foreign language or two.

Some companies distinguish between *process design* and *process engineering design*. For them, *process design* is oriented towards generating the process information given in a licensor's process package or a feasibility study. This includes process flow-schemes, material and heat balances and many other items. They regard *process engineering design* as part of the contractor's function and oriented more towards the preparation of piping and instrument diagrams, equipment data sheets and specifications, and information required by mechanical, civil, electrical and other engineering departments. This division of the subject into two functions needs to be borne in mind, but is not adhered to here.

Hazards such as those discussed earlier in this book creep readily into design in these early stages, and are often difficult to recognise and eradicate later before they cause mischief.

Ludwig[8] summarises the responsibilities of the average process engineer thus:

1. preparing studies of process cycles and systems for various product production or improvements or changes in existing production units;
2. preparing economic studies associated with process performance;
3. designing and/or specifying items of equipment required to define the process flowsheet or flow system;
4. evaluating competitive bids for equipment;
5. evaluating operating data for existing or test equipment;
6. guiding flowsheet draughtsmen in detailed flowsheet preparation.

The process engineer must understand the interrelationship between research, engineering, purchasing, construction, operation and safety, and appreciate how any of these may affect process engineering decisions.

The scope of process design includes:

1. process material and heat balances;
2. correlation of physical data, and data from research, pilot plants and test runs on other plant;
3. material, heat and energy balances for power, water, steam and other auxiliary services;
4. development, detailing and completion of flowsheets;
5. specifying conditions and performance of equipment shown on flowsheet consistent with mechanical practicality;
6. specifying instrumentation required for process requirements and safety and interpreting this to instrument specialists;
7. interpretation of process needs to all other engineering departments involved in the project.

Process engineers are generally expected to record all their calculations and sources of information in a design book so that these can be checked. Most process engineers, if asked to check the calculations of a particular design, prefer to start with a clean sheet and see what conclusions they reach, before checking someone else's calculations. This may take longer but it provides a better check on the data and methods used and on the assumptions made.

Process engineers are consulted on plot plans and plant layout (both in plan and elevation), process drains and effluents and usually on most other matters which form part of a design. To advise on these they need, *inter alia*, to evaluate the fire, explosion and toxic hazards of the process, e.g. as determined by the Dow or Mond hazard indices [11]. They may also need to modify their designs to reduce their hazard potential.

15.10.2 Flowsheets

Flowsheets are maps of a process and define the responsibilities of other engineers (mechanical, electrical, etc.) who are not expected to understand the process itself. Most flowsheets show the process steps in sequence pictorially, using standard symbols for different types of equipment, valves, pipes and instruments. Austin gives sets of drawing symbols according to British and American standards,[9] but most contractors appear to follow American practice and use the symbols given by Sherwood and Whistance.[10]

Many flowsheets are drawn and printed from left to right on long strips of paper, of the same height (generally A4) as used for office documents, into which they can be bound and folded, to be extended concertina-wise when required. These are more convenient to use than large drawings, although they cannot provide as much detail. Separate flow diagrams are used for each process and utility unit of which the plant is composed. Flowsheets include block diagrams, process flow diagrams (PFDs) and piping and instrument diagrams (P&IDs). Material balances are prepared and used in conjunction with process flow diagrams and sometimes printed on the same strip, above or below them.

Block diagrams (Fig. 15.2) are the simplest form of flowsheet and set the basic process concept. They show only what each step has to achieve but not how it is done.

15.10.3 Process flow diagrams (PFDs) and material balance sheets

Process flow diagrams (Fig. 15.3) show all major equipment items with descriptions and interconnecting flow lines for process materials. Most show no valves or instruments. Some indicate only control valves (automatic or manual). Important flowlines are usually 'flagged' or otherwise marked with the stream number, and the temperature, pressure and design flow quantity. Design flow rates, temperatures, pressures and stream compositions are also shown.

The stream number refers to a material balance sheet which accompanies or forms part of the diagram. This sheet contains descriptions, quantities and components of each numbered stream, with its composition. It is useful if it also shows the physical state, enthalpy and important physical properties such as specific gravity and viscosity of the flowing material. Stream data given on material balance sheets are, in the case of a new process, calculated theoretically from basic data and principles, and from experimental

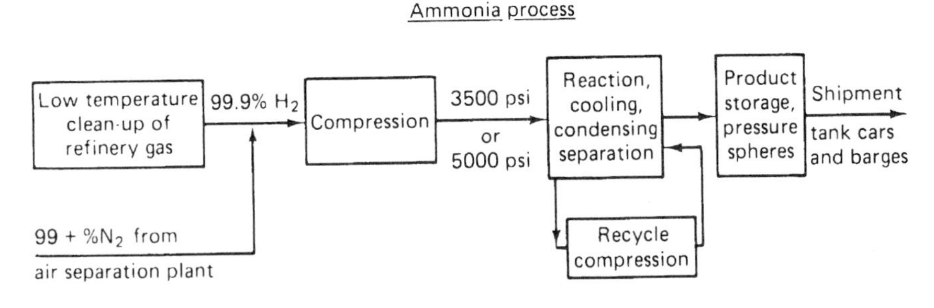

Fig. 15.2 Process block diagram (from Ludgwig,[8] courtesy Gulf Publishing Co., Houston)

results. The preparation of material balances is an important part of the process engineer's job, and being often very time-consuming does not always receive the care and attention needed. Thus whenever a process stream containing several components enters a separation step from which two or more streams leave, one needs to know what proportion of each component is present in each of the outgoing streams. The problem becomes more difficult in process steps in which reactions occur. Material balance data are, where possible, checked against information from similar plants or from pilot plants or laboratory data. They often have to be checked by measurement and analysis when the plant which is being designed comes on-stream. Computer programs are developed for many of these calculations. These may have to be repeated several times with different inputs to select the optimum design and operating conditions.

The material balance sheet should also show the nature and quantity of all solid wastes and liquid and gaseous effluents from the process. This information is vitally important in order to comply with anti-pollution requirements and to design the treatment plant required. This may be an important cost element in the process and has to be considered when choosing between different processes and raw materials.

The 'information' (hopefully, not misinformation!) developed in the course of preparing PFDs and material balance sheets is used to specify design duties of equipment, pipes and valves, as well as in calculating how much steam, electricity and other ancillaries are needed. These calculations are important to the technical and economic success of the project as well as to its safety.

15.10.4 Piping and instrumentation diagrams[10] (P&IDs)

The object of a P&ID (Fig. 15.4) in to indicate all service lines, instruments and controls and data necessary for the design groups. The PFD is the principal source of information for developing the P&ID. This should define piping, equipment and instrumentation well enough for cost estimation and for subsequent design, construction, operation and modification of the process. Material balance data, flow rates, temperatures, pressures, etc. are not shown, nor are mechanical piping details except for permanently installed 'spectacle plates', pipe flanges which may require to be broken to insert line blinds, and removable spools needed for isolation. P&IDs show all process pipes and valves (including drains and vents) with numbers and sizes which refer to pipe and valve lists, all pressure relief valves and bursting discs. (Not all P&IDs show valve numbers, although these are particularly needed on P&IDs used to illustrate operating manuals and procedures [20.2.1].) P&IDs are laid out in the same way as PFDs, but with greater horizontal spacing between equipment, and the process relationship of equipment should correspond exactly. P&IDs are not to scale, but, where possible, equipment is drawn in the correct vertical proportions, and critical

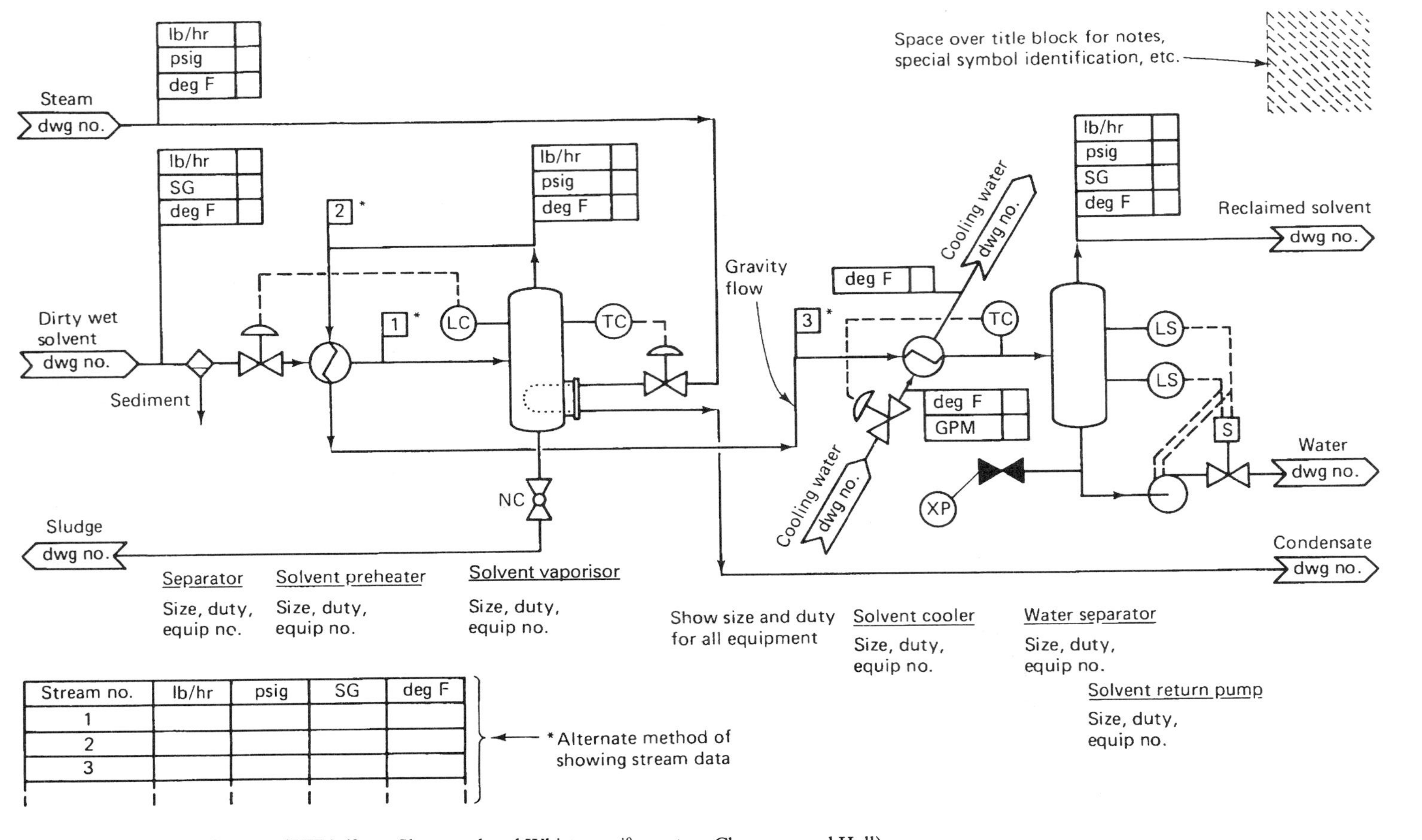

Stream no.	lb/hr	psig	SG	deg F
1				
2				
3				

Fig. 15.3 Process flow diagram (PFD) (from Sherwood and Whistance,[10] courtesy Chapman and Hall)

elevations are noted. P&IDs show symbolically all instruments and instrument loops (including trips and alarms) and control valves, with conventions to indicate whether they are located in the control room or on the plant. They show impulse and power supply lines (pneumatic, electric or hydraulic) to control instruments and valves, and the mode of valves if power fails (open, shut or same position). Separate PFDs and P&IDs indicate the pressure relief systems with headers, catchpots and flares, etc.

From these P&IDs (or sometimes replacing them) more detailed drawings (engineering line and general arrangement drawings) on which equipment is shown to scale in correct elevation are prepared. On these, piping, valves and instrumentation are developed in greater detail.

Utility flow and piping and instrument diagrams (UFDs and UP&IDs), similar to PFDs and P&IDs, are prepared for steam, water, compressed air, vent and relief systems, effluents and fire water. Electrical 'single-line' diagrams are prepared for power and lighting. Both PFDs and P&IDs are used in the preparation of operating manuals, and also in HAZOP studies which are best done at the same time. These are discussed in [14.4].

15.10.5 Simplification of PFDs and P&IDs

Once PFDs and P&IDs have been drafted two or three times, process engineers should try to simplify them (an exercise in which Boolean algebra can be helpful). For this they need to reconsider all the possible needs for pipes and valves during start-up (including the use of recycle circuits), normal operation, turndown, normal shutdown, emergency shutdowns in various circumstances, emptying and flushing. This done, they should systematically work out the simplest network of pipes and valves which allows each of these operations to be carried out safely, and make notes on how these are to be done. This generally results in improved and simplified PFDs and P&IDs. The notes written at this time form a basis for the plant-operating manual, when the PFDs and P&IDs are finalised. This manual is best written by a small working party which includes start-up and operating personnel as well as designers. It is best combined with the hazard and operability (HAZOP) study discussed in 14.4. There is not, of course, always time for such exhaustive reviews of the PFDs and P&IDs, but where they can be done they generally result in significant reductions in the plant costs which more than pay for the additional design hours, as well as improving operability and reliability.

15.10.6 Gaps in knowledge of process chemistry and trace components

Such gaps sometimes arise with a newly developed process and lead to the unsuspected formation and accumulation of a hazardous compound at some point in the plant. The process engineer may have considered all the

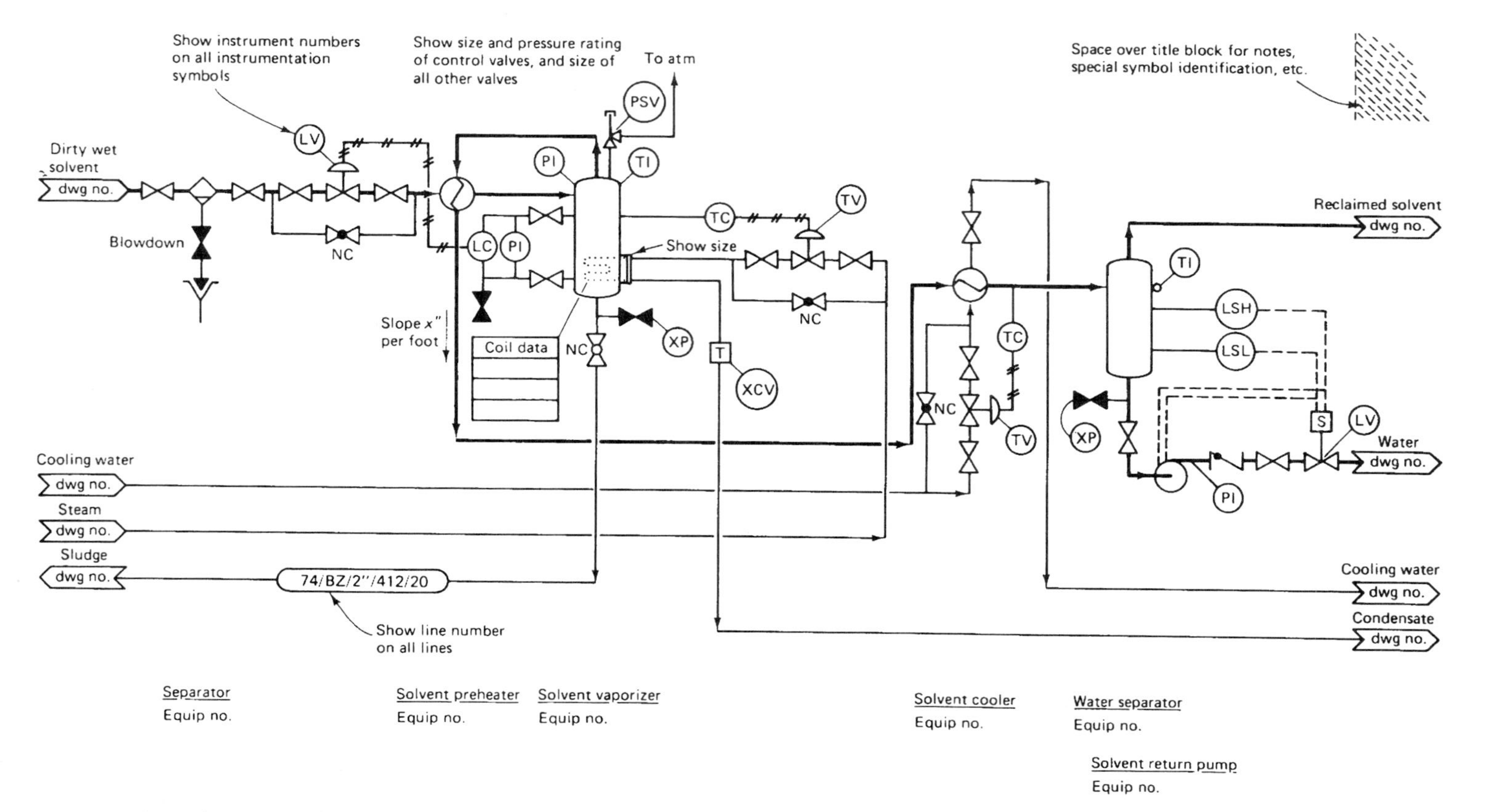

Fig. 15.4 Piping and instrument diagram (P&ID)(from Sherwood and Whistance,[10] courtesy Chapman and Hall)

known by-products of a process which has been developed on a small scale, and designed the plant to take care of them. He may be unaware of an unidentified pimple on the chart of the GLC analysis of the crude reaction product. This was not identified during experimental development because it was not found in any other analyses, and it disappeared when attempts were made to concentrate it. On building and starting up a continuous commercial plant with a capacity of 40 000 t/a, what was a small pimple on the GLC was now found to be an unidentified by-product formed at the rate of several kilograms per hour, although there may still have been no sign of it in any of the product streams. It may in fact have been trapped inside the plant like a *jinee* in a bottle from which it cannot escape. This happens in distillation columns if the compound is more volatile than the 'bottoms' product and less volatile than the 'top' one, and is quite common when the compounds present are of different types. So it simply accumulates somewhere inside the column until it reaches a high concentration. The material may be harmless, but it may be explosive, highly toxic or corrosive.

One example was an air-separation plant built in the 1950s in Manchester (UK) where there was a trace of acetylene in the air intake. One day the plant suddenly exploded without warning. Something similar happened with vinyl acetylene in the purification train of a butadiene plant, which also exploded[11] [9.8.2].

Other examples discussed in this book where unrecognised by-products caused trouble were:

- dioxin in trichlorophenol production [5.2];
- elemental sulphur in the condensers of a crude oil distillation unit [13.9.4].

The HAZOP type of study discussed in 14.4 helps the study team to face up to such a problem provided one of its members was dimly aware of it in the first place. It is hard to see how the study can help here if none of the team has the slightest inkling of the problem. One would need to run courses for process engineers on 'how to develop a sixth sense'.

Checks and double-checks of their calculations and design assumptions may be needed on hazardous plants. Sources of help for such checks include recently retired technical staff with appropriate backgrounds.

References

1. Flixborough Disaster Report, HMSO, London (1975)
2. HIC, *Advisory Committee on Major Hazards Reports 1,2,3*, HMSO, London
3. Crookes, E., Property losses in the process industries: an engineering problem — I, Proceedings Management of in-service Inspections of Pressure Systems, Institution of Mechanical Engineers (1993)
4. SI. 1986, No. 896, *Pressure Vessels (Verification) Regulations*, 1988, HMSO, London
5. BSI, PD6403, *Critical Crack Sizes*, HMSO, London

6. HSE HS(G) 1993, *The Assessment of Pressure Vessels Operating at Low Temperatures*, HMSO, London
7. *BSI* BS 5500, *Unfired Pressure Vessels.*
8. Ludwig, E. E., *Applied Process Design for Chemical and Petrochemical Plants, Volume I*, 2nd edn, Gulf Publishing Company, Houston, Texas (1977)
9. Austin, D. G., *Chemical Engineering Drawing Symbols*, George Godwin, London (1979)
10. Sherwood, D. S. and Whistance, D. J., *The 'Piping Guide'*, Chapman and Hall, London (1979)
11. Jarvis, H. C., *Chemical Engineering Programs*, **67**(6), 41–44 (1971)

16 Safety by inspection and maintenance

16.1 Introduction

Inspection and maintenance of process plant should only be undertaken by personnel who are technically competent and adequately resourced. This chapter is not intended to give the lay reader the in-depth knowledge necessary to achieve plant which is safe so far as reasonably practicable, but to give sufficient knowledge and instruction so that intelligent questions may be asked and more informed decisions made. There should be no complacency. Many failures have led to fatal and serious injuries and loss of production, even to the point where national Governments suffer balance of payment crises.

It is essential that process plant is thoroughly inspected and adequately maintained. Poor inspection and maintenance can lead to all the main causes of process plant failure, i.e. over-stress, over-strain, creep, fatigue, brittle failure, stress-corrosion cracking [13].

16.2 Inspection practice

Inspection can be carried out during design and after completion, but usually this type of double-checking is considered to be quality assurance. All UK standards, and those of many other countries, concerned with the design of pressure vessels, boilers, piping, heat exchanges and storage tanks, call for one organisation to carry out the stress analysis and geometric design, and another to check those calculations and designs before items are fabricated.

Inspection is also undertaken periodically to investigate whether deterioration has taken place, and in order to revalidate the safe operating limits of pressurised and storage plant.

16.3 Competent persons

The term 'competent persons' really only applies within the UK. Most other countries have defined third party, state-approved or legally-sanctioned organisations or bodies who carry out inspection, quality assurance and/or revalidation certification.

Within the UK the concept of competent persons has been used in health and safety legislation, particularly regarding boilers, steam receivers, air receivers and lifting equipment. To some it is surprising that the term is not legally defined, particularly as the functions of the competent persons were clearly prescribed by law. The term competent persons has come to encompass an individual or a body corporate whose formal education and training coupled with their experience qualifies them to undertake a certain function. For example, the competent person who certifies forged, welded chains, can be the blacksmith who 'knows about such things'; on the other hand the competent person who carries out the examination of a steam boiler has to be a member of a boiler-inspecting organisation. But this term is vague too; usually within the UK it has come to mean somebody from one of the major insurance companies that has engineering expertise.

The Pressure System Regulations are very restrictive, only dealing with certain pressurised plant, but within this legislation, which certainly does not cover all process equipment, there are guidance notes published by the HSE[1,2,3] defining competent persons as either self-employed, or a corporate body having such depth of knowledge that they can competently carry out the desired functions.[3]

In order to bring the UK into line with European, US and world-wide practice, there is now a Government-approved scheme which is run in conjunction with the Institution of Mechanical Engineers and the HSE whereby organisations can submit themselves and be approved. This organisation is called the Pressure Vessel Quality Assurance Board. It is acknowledged by European Government safety agencies, but it may not be universally accepted, particularly in the USA.

Organisations can be approved in certain categories, e.g. fair-ground rides, power presses, lifting equipment and particularly in process plant. The approved list is not static, and the latest list of organisations that are approved, and remain approved, can be inspected at the premises of the Institution of Mechanical Engineers, 1 Birdcage Walk, London. At present the types of organisation approved for process plant inspections cover the design, fabrication, inspection and installation, and are made up generally of the engineering departments of the large insurance companies, the engineering departments of the large chemical and petrochemical manufacturers, and a very few independent inspection bodies.

16.4 Process plant quality assurance

Many industrial nations had Government-sponsored or Government-controlled organisations that carried out both quality assurance on design, fabrication and testing and periodic re-examination and revalidation. Certain manufacturers and certain users were not confident in and did not want to rely upon state organisations, and they ran their own inspection bodies; on top of this, insurance companies that were covering actuary risks also insisted upon carrying out validation of plant; and this could be coupled with the main designer, the main fabricator and any subcontractors. Within the UK in the past it was not uncommon to find seven inspection bodies carrying out similar if not identical safety checks on pressurised systems before they were installed. This was naturally time-consuming and not economical.

Often such quality assurance provided no more than an illusion of respectability. For example it is well known that the No. 5 Reactor that failed at Flixborough had undergone design to the then current pressure vessel standards, had been fabricated and designed by a well-known manufacturer and had undergone an examination and tests by leading insurance companies; and yet despite all this, there were fundamental design faults at its 28-inch connection, which led to the initial six-foot long crack developing.

Modern UK Pressure Vessel Standards call for the design to be undertaken by one organisation, which could be approved by the Institution of Mechanical Engineers, and for that design to be checked by another organisation, which could also be approved by the Institution of Mechanical Engineers. Both organisations, whether approved or not, would have to sign a certificate confirming the work they had carried out. There is now an obligation within the Pressure Systems Regulations for both the designer and the inspecting authority to aim for safety so far as reasonably practicable and so there is a legal obligation and penalties if they do not do their job correctly.

Undoubtedly technology and fabrication techniques have developed to the point where process plant can be safe so far as reasonably practicable. Unfortunately process plant continues to fail, and the carnage in injury to personnel and loss of production continues.

16.5 Reasonably practicable inspection

There is within the USA and mainland Europe a concept that all process plant will be safe if fabricated and designed to the national standard, and overseen by the national inspection organisation. This has developed into a system of world-wide safety technology.

Within the UK process plant has to be safe so far as reasonably practicable. Plant in which failure has more serious consequences should be designed to higher safety standards than low hazard plant or plant where failure has very low consequences. It could be argued within UK legislation, that if say, a cold water tank failed catastrophically it would not be likely to cause harm, and so there would be no health and safety obligation to carry out elaborate inspection and maintenance checks. On the other hand, if a large sphere contained LPG, the consequences of failure would be the catastrophic loss of containment leading to fire and explosion, damaging life and property over a large area. In the same way, the failure of a large chlorine storage vessel could affect people over large distances (it is considered that chlorine could travel 13 km from a 30-tonne storage bullet). The maintenance and inspection of these vessels would need to be of the highest order.

16.6 Inspection techniques

There is an arsenal of inspection techniques available to the competent person. It is doubtful if any one person is capable of undertaking all the available techniques or is able to interpret the findings. This is why the corporate body is considered to be the competent person. As in all things, knowing your limit of expertise is essential.

Basically inspection techniques devolve into two camps; either destructive testing, whereby the component may fail; or non-destructive testing, where techniques are used which do not cause failure.

16.7 Destructive testing

Inspection techniques such as pressure testing, over-filling storage vessels, or overloading pipework, cranes, lifting equipment, conveyors or heating equipment, can lead to catastrophic failure, and for this reason such testing techniques are considered to be destructive.

If an engineer was perfectly confident that the component would never fail there would be no need to test it. Tests are only undertaken to prove or produce forensic evidence that the chance of failure is remote. It is essential that it is assumed that all destructive testing may result in failure, and precautions should be taken to reduce the hazard from failure to a minimum.

16.7.1 Pressure testing

Overload testing has often resulted in cranes and lifting apparatus failing catastrophically to considerable embarrassment. Pressure vessels have been heated, and boilers have been set on fire to prove their safety, only to have the reverse proved and they have failed at the first application.

A very large boiler drum destined for use in Scotland had been designed, fabricated and quality-tested by a very large boiler-fabricating organisation. The outside drum thickness was 150 mm, and it was pressure-tested filled with water. The water temperature was about 4°C, and during the initial pressurisation regime, the vessel failed catastrophically and a large 'chunk' of metal the size of a saloon car fell from the structure. Naturally the contents of water cascaded out. There was no loss of life, but the repair costs were considerable. Investigations found that there were small welding defects near to a nozzle, and this stress intensification had produced brittle fracture due to the low temperature of the water. This illustration is used to show that:

- the inspection techniques had not been adequate in finding the small welding defects;
- the testing procedure of using cold water was intrinsically unsafe as it could produce fast or cold brittle fracture;
- the saving grace was that using a test medium of water reduced the hazard considerably.

A defect of this size would not have led to failure at the normal operating temperature of the vessel.

Pressure testing of itself does not produce a safe item. A system may sustain an adequate safe pressure but in sustaining that pressure the vessel or pipework may have been damaged, and that damage may only materialise at a later date. For example it may be safe to bend a paper clip several times, but you can never be sure that the next time you bend it it will not snap. There are records of large storage spheres being filled and successfully tested with water, but where this testing with water has subjected the sphere to a weight far greater than it would experience when being filled with LPG, because the specific gravity of LPG is approximately half that of water. The pressure test combined with the weight has caused local cracking, and local deformation in the vessel skin, which progressed when in service to the point of catastrophic failure.

There can be a false sense of complacency when a component has been pressure tested. For example the bypass pipe at Flixborough had been successfully fabricated, filled with water and pressurised on the shop floor. It did not fail, the water was emptied and then the pipe was taken to be erected between the two reaction vessels where it subsequently did fail. No-one at the site appreciated the considerable end-thrusts to which the pipe was subjected during operation, that would not be simulated by a mere overpressure test.

In 12.3.2 it is explained that the old concept of design was based upon using a safety factor of four applied to the ultimate strength of the material. In these circumstances it would have been safe to have the design pressure multiplied by two or even safer to multiply by one and a half. It is not uncommon even to see modern pressure vessels pressure tested to one and a half times their design pressure, and this in itself is foolhardy. For example if the design stress is two-thirds of the yield stress, and it was tested to one and a half times the design stress, the stress at maximum pressure would exceed the yield stress and the vessel, pipework, boiler or heat-exchanger would change shape or fail.

Pressure testing is still undertaken on pressure vessels, pipework, boilers etc. but the modern concept is to use only one and a quarter times the design pressure, in order to minimise the maximum pressure so that any component is exposed to 90% of the proof stress.

In South Africa a large cylindrical ammonia vessel was found to have a defect adjacent to a weld, and this was repaired. The vessel was subjected to hydraulic pressure testing, the liquid was emptied, the test found to have been successful, and then ammonia was fed to the cylinder. As the pressure was increased in the vessel there was a catastrophic failure and a hole more than $2m^2$ was blown from the end discharging the complete 30-tonne contents. Fatalities occurred up to five miles away. In all probability the pressure test had been undertaken at a pressure that caused local yielding and induced stresses into the vessel shell that were latent until cold ammonia was applied and fast brittle fracture occurred. If the weld had been adequately inspected using a technique other than over-pressure testing, in all probability the welding defect would have been recognised and remedial action could have been taken. As it was, there was a false sense of security induced by thinking that the vessel had safely sustained a hydraulic pressure test, when in fact that pressure test had damaged the vessel beyond its safety regime.

16.7.2 Safety when pressure testing

When carrying out the design of pressurised equipment, it is usual to concentrate on stresses, as these will give the key to potential failures be it due to over-stress, strain, creep, fatigue, brittle fracture or stress-corrosion cracking. The same stress will be applied whether the test media is hydraulic, e.g. water or oil, is a compressible gas, e.g., steam etc.

When work is done upon a gas to compress it, that energy is stored, and will be given up on decompression. It is a simple matter to calculate the energy stored within a compressed gas. The energy is proportional to the pressure and the volume contained. Energy calculations used to be undertaken in Imperial units in foot pounds and of course these are still used in the USA. In continental Europe the usual unit of energy was the calorie, and now under SI units the unit of energy is the joule. The explosive energy is often related to pounds of TNT. For comparison 1.5×10^6 lbf of stored energy is equal to one pound of TNT. The nomograph (Fig. 16.1) compares volume, pressure and pounds of TNT energy stored and gives potential damage after explosions have occurred. Thus it is possible to estimate the structural damage caused by an over-pressure blast wave at a given distance from the release of that energy. Figure 16.1 also demonstrates this function.

Water and liquids on the other hand are not compressible at normal pressures, and hence they contain no energy; and so should a containment vessel fail whilst being pressure tested, the resultant damage will not cause a blast wave.

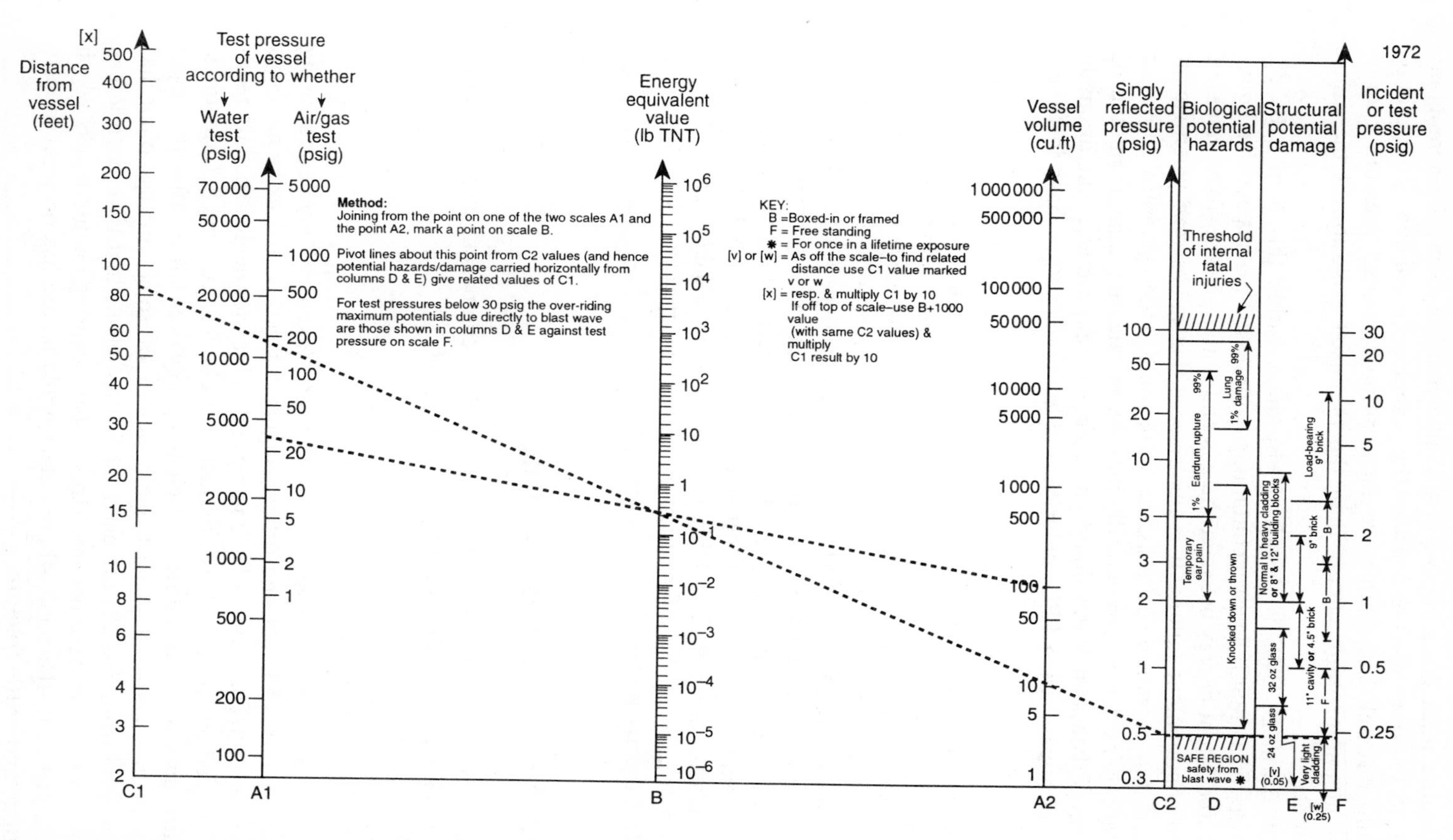

Fig. 16.1 Safetygram – pressure test explosions — approximate damage potentials

Because most gases behave in a similar manner to air, the nomogram (Fig. 16.1) can be used to consider the potential hazard due to the energy within a vessel. The blast-wave hazard due to stored pressure energy can amount to more energy than the premixed explosion of the same contents.

The term BLEVE is described elsewhere within this book [10.2.3]. Suffice to say that if a vessel containing compressed gas has become liquefied, the latent energy between the gas and liquid phases will be released upon containment failure. It is this release that is observed when steam boilers fail catastrophically as the water held under pressure above its boiling point flashes dramatically into steam releasing a large amount of energy. Likewise it is this liquid phase change that produces the dramatic energy release when LPG, chlorine and ammonia vessels have lost their containment.

In summary, pressure testing should always be undertaken with the least hazardous material in case a failure should occur. For this reason it should always be undertaken using water or other non-compressible fluids. If there are circumstances where pressure testing has to be undertaken using air, steam or compressible gases then adequate precautions should be taken. The area where gas pressure testing takes place should be cleared of all personnel, and preferably the vessel, pipework or component should be housed in a blast-proof container. For a temporary measure this may be made up from sand bags; or if regular testing of the component should be done then in a blast-proof cell. The HSE has produced a valuable guidance note[4] which gives adequate information on precautions necessary when pressure testing is carried out.

16.8 Non-destructive testing

The term non-destructive testing applies where components are not expected and not planned to fail during the test regime.

16.8.1 Visual examination

Visual examination covers measurements taken by hand and using micrometers, callipers and other gauging devices. Visual examination is also useful to check overall geometry and to look for defects that are readily noticed by eye, and these will include weld defects, local deformations, inadequate supports, fabrication cleats and devices left on after the vessel has been completed. Catastrophic failure of some process plant has emanated from casually attached fabrication cleats that have been left in place, and these have become stress raisers causing minute cracks which have grown to the point where catastrophic failure has occurred due to fatigue and brittle fracture.

16.8.2 X-ray

A very powerful tool used in non-destructive testing is X-ray examination of welds and metal profiles. X-rays can detect weld inclusions and defects which are sub-surface and are very useful to encourage welders to carry out adequate quality assurance. The function of X-raying welds should not be over-stated, because X-raying cannot find surface-breaking defects such as cracks or fissures, and it is these surface-breaking defects that are notorious in causing fatigue, and brittle fracture.

For example within the UK and much of the developed world very large spheres containing ammonia will have had quality assurance including X-raying of welds undertaken when fabricated. Users of such large spheres have been actively encouraged to carry out revalidation and re-examination looking for surface-breaking defects that could lead to catastrophic failure due to fatigue or brittle fracture. It is no good relying on the original X-rays, because these did not look at the surface. But when surfaces are examined cracks numbering many hundreds may be found.

It is common within pressure system standards to call for 10% of the main seam welds to be X-rayed, probably to encourage quality assurance rather than to ensure safety. But it has been found that a very large pressurised natural gas cylinder with a 9-inch diameter opening had surface-breaking cracks measuring 19 mm in a 29 mm thick wall. X-raying was not carried out and so this defect was not found on initial fabrication but only in service. If X-rays had been taken upon initial fabrication, there would have been a benchmark to see how much this cracking defect had grown because, although X-rays cannot detect surface-breaking defects up to about 2 mm deep, there is no doubt that even an X-ray would pick up a crack that was 19 mm deep in a 29 mm thick shell.

16.8.3 Ultrasonic examination

Ultrasonic examination works by passing high-frequency sound through material, and picking up the echo as the sound wave is bounced against the opposite surface. Electronic instrumentation measures the time taken for the sound waves to travel through the material, and in its crudest form can be used for thickness checking. If there was an internal defect, the bounce or travel of the sound wave would be reduced, and this could be plotted by skilled technicians, giving an indication of the depth and shape and position of an interior defect.

Generally ultrasonic scans and X-rays are interchangeable, but like X-ray techniques, ultrasonics are incapable of finding defects less than 2 mm from the surface. It is also extremely difficult to use ultrasonic measuring techniques either for thickness testing or for defect testing in certain materials, and it is impossible in stainless steel.

A very large contract to test welds ultrasonically for internal defects was carried out on a large cross-country gas pipeline. Despite adequate records being taken of the scan, it was realised at a later date that the technician applying the technique had 'cheated' and supplied the same identical records on many pieces of weld. Fortunately this slovenly work did not lead to failure but it highlights how people can cheat the system.

Another large cross-country pipeline in Canada which in all probability had been either X-rayed or ultrasonically tested, developed a through-wall thickness crack that led to fast fracture. Once the crack had reached a critical level it probably travelled at the speed of sound (the speed of sound in that particular metal which is not the same as the speed of sound in air); and that crack travelled unabated for over seven miles.

16.8.4 Magnetic particles

It has been said on numerous occasions that surface-breaking defects have the most potential to cause catastrophic failure due to over-stress, over-strain, creep, stress-corrosion cracking, brittle fracture and fatigue. It is most important that surface-breaking defects are considered both at the new fabrication stage and as plant is periodically examined throughout its working life. Even modern pressure vessel standards do not call for all surfaces to be checked for surface-breaking defects; this would have to be a special requirement called for by the purchaser and the fabricator.

One successful method of looking for surface-breaking defects is to use magnetic particle inspection techniques. The surface is cleaned and coated with a base colour such as white powder, magnetic particles are sprinkled onto the surface in the region where cracks may occur (e.g. near to welds, and changes of material thickness) and then powerful magnets are placed onto the surface which cause the magnetic particles to align themselves, and show up any surface crack. The magnetic particles are often suspended in a fluid, which enhances their mobility, and this fluid can be stained and can contain ultraviolet dyes to fluoresce and show up the crack more positively. Magnetic particle crack detection has the ability to show up very fine shallow cracks as well as deep cracks. Naturally it has its limitations and cannot be used on non-magnetic surfaces such as stainless steel, aluminium, copper and cast iron. Normally no permanent record is taken but it would be possible to grid-reference a containment vessel surface and photograph particular crack-like defects, but of course this would be most expensive.

16.8.5 Dye penetrance

For non-magnetic surfaces it is possible to carry out a form of surface-breaking crack detection, by cleaning the surface and then spraying it with a very penetrating or searching fluid that will by capillary action sink deep into any cracks or crevices. This surface is then cleaned and sprayed with a precipitated white chalk suspension, that rapidly dries, and the penetrating fluid inside the defects or cracks gently permeates out showing a surface outline of where the defect lies. Here again the fluid that it used for penetrating can be dyed, usually bright red, green or fluorescent to enhance the crack detection. Naturally this is not so sensitive as magnetic particle crack detection, but has the great advantage of being able to be used on stainless steel and aluminium as well as copper.

16.8.6 Acoustic emission

When metal undergoes stressing, the individual molecules rub together and produce very minute sounds. It is possible by putting very sensitive microphones onto the surface of a structure to hear these individual sounds. By putting several microphones in a grid system, it is possible to ascertain where particularly large sounds are emanating from, and this may indicate either large stresses or the formation of cracks. In principle there is a great advantage in using this block type of testing, particularly on large structures such as distillation columns, spheres and boilers. In theory once a defect has been indicated by acoustic emission, then detailed investigation in the area of the potential crack could be carried out, e.g. magnetic particle, dye penetrance or ultrasonic testing. Very large undertakings have been commissioned, but often the acoustic emission testing has either proved too sensitive (even picking up rain drops), or has been desensitised to the point where even large cracks have been missed. The process of scaffolding a large vessel, and putting microphones in many places across its surface, and the sophisticated recording and decoding instrumentation using highly skilled operators, can be expensive, and has not always proved worthwhile.

16.8.7 Conditioned monitoring

The term conditioned monitoring covers a wide variety of subjects. It can be applied while machinery is running using acoustic emission to listen for defects in bearings, indicated by a change in sound. This has proved most effective, allowing testing to take place without the necessity of taking the plant off-line and dismantling it prematurely.

16.8.8 Metallurgical examination

Metallurgical examination is a specialism of its own. At its very simplest it will consist of doing hardness testing on the surface of a material, and then using charts to give an indication of the allowable or safe or design stresses that can be used for the design and revalidation of that material.

On manufacture it is common to put an extra piece of material onto the structure and carry out the same welding techniques on this as on the parent material. This is designated a ‘coupon plate’. The coupon plate is removed from the parent metal, and can be taken to the laboratory for tensile testing, metallurgical examination and general welding quality assurance control and can be used to carry out corrosion-resistance testing. The customer is offered the coupon plates from a new plant to keep with the general pressure vessel or pipework records, so that in future years reference can be made back to the original weld construction.

Metallurgical examination can be undertaken using spectrometry, whereby the actual metal constituents of a particular plate can be ascertained to check for quality assurance that the exact material which was ordered is being used.

Random selection has been undertaken by large quality-assurance organisations in well-equipped pressure-system workshops being run by reputable organisations, and it has been found that up to 7% of the material being used to produce high-quality pressure vessels has not complied with the original orders. That is not to say that all the plant will be understrength or potentially dangerous, but it actually highlights how metallurgical examination can be used to gain assurances that the correct material has been used.

16.9 Maintenance

In common language maintenance suggests changing oil, adjusting the tightness of some screws, doing some decorating and general titivation. In law the term maintenance has a much stricter definition, and it implies that the component (be it a pressure system, crane, centrifuge, distillation column or boiler) has been safely maintained in the same safe condition as it was when it was first put into use. This is a most arduous burden to place upon a user, but one that is essential for the health, safety and well-being of employees, and those outside the plant premises who could be affected by a catastrophic loss of containment.

16.10 Revalidation

Old UK law on pressure vessels, e.g. boilers, steam receivers and air receivers, as well as cranes and lifting equipment, had set periods when items had to be examined and safety certificates issued.

Modern pressure systems legislation calls for the safe operating limits to be set by the user of the plant, and these must be valid at every point in time. In other words as plant deteriorates SOL may well change. It is therefore essential that from time to time plant is thoroughly examined, and the results used to calculate the present state of the SOL.

It has often been the case that very casual examinations of pressurised systems, storage tanks and pipework have been undertaken, and very casual revalidation certificates have been issued. Terms such as ‘in order as seen’ are commonplace. If the plant is hazardous or is a major hazard, then there is an even greater obligation to revalidate the plant adequately from time to time.

The periodicity of revalidation should be set before the plant is first used, and then the plant should be revalidated using all the necessary inspection techniques above.[2,3] After the plant has been examined, it can be revalidated stating the current safe operating limits, and at the same time the competent person should set the time for the next revalidation to take place.

To carry out a full range of inspection testing and revalidation on a large pressure vessel such as an LPG sphere could cost in the order of £100 000. On top of this the loss of production and down-time of the plant must be considered. There is a concept that the whole revalidation exercise could be undertaken as a sophisticated exercise on paper carrying out both hazard and risk assessments. For example if it can be safely argued that the only way the sphere could fail would be through a minor leakage, and there were no possibilities of total catastrophic failure, it may therefore be said that it is not reasonably practicable to take the plant out of use to carry out what would be cosmetic checks. For example if by using sophisticated crack analysis methods it is thought that surface-breaking defects less than 0.5 mm cannot be found, then there is no point looking for them. If it is also known that defects less than 0.5 mm are critical then it is obvious that such plant should not be used; it should be decommissioned, thereby removing the hazard. If on the other hand it is calculated that a vessel or containment structure can tolerate a through-thickness crack breaking right the way through the wall, and it will remain as a minor leak, then there is very little point in looking for that defect, unless the contents are particularly lethal, because it is known that leaks can be safely dealt with in the majority of major hazard plant. It can also be successfully argued that if the rate of crack growth is likely to necessitate periodic examination being undertaken at certain intervals then it would not be reasonably practicable to reduce the interval.

16.11 Certification

Operators, owners, and people who could be affected, e.g. nearby neighbours, will need to know that major hazard plant, and other plant, so far as is reasonably practicable, has been certified for its safe operating limits. That certification should be undertaken by one organisation, and checked or quality assured by another, if the hazard is high and the risk of failure cannot be tolerated.

It is realised that some old pressure system codes were inadequate when dealing with some very important subjects, including catastrophic failure by brittle fracture. It is also realised that some existing plant will have been designed to those old standards, and the original quality assurance may leave a lot to be desired.

It is also realised that some operators rely on vapour pressures to set the design conditions of storage tanks containing such things as chlorine, propane, butane and ammonia, see Fig. 16.1. Modern research has indicated that such vessels can operate either colder or at much higher pressures

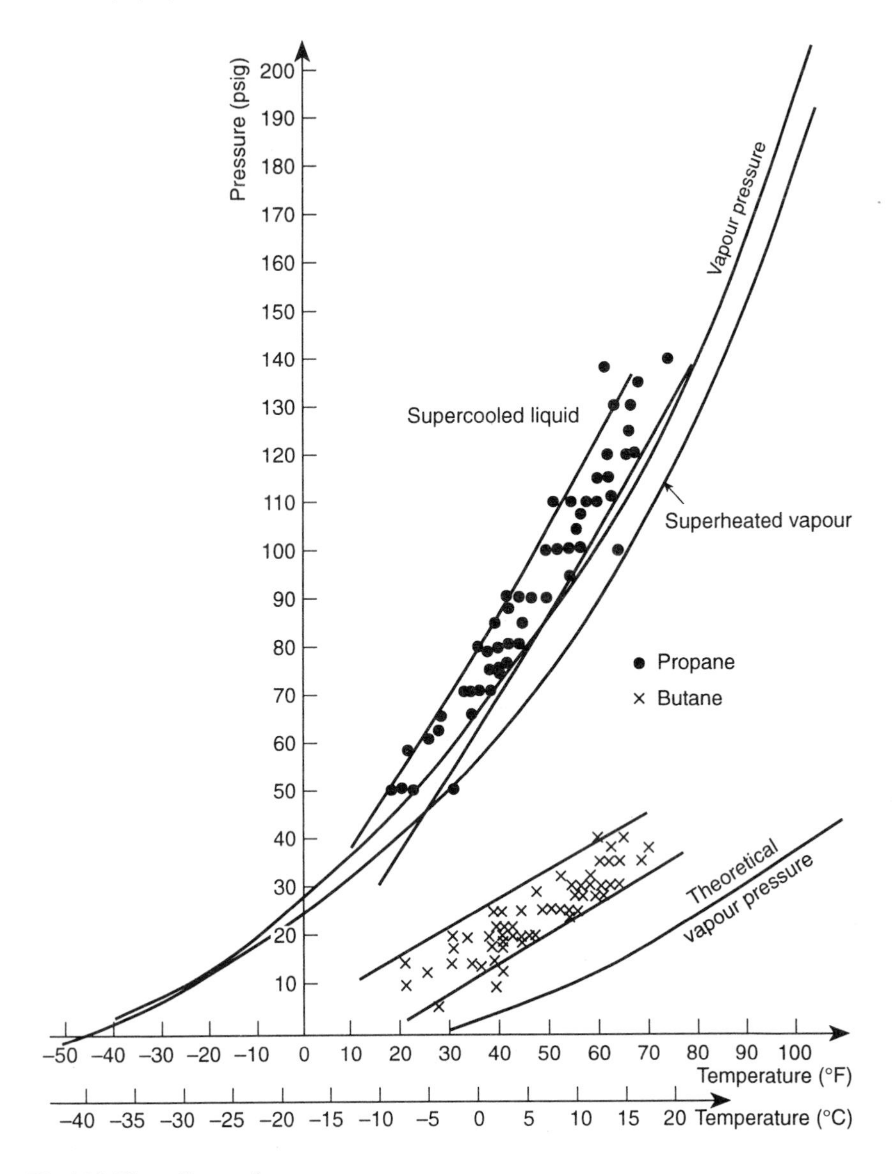

Fig. 16.2 Phase diagram for propane

than would be anticipated by purely considering the vapour pressure, see Fig. 16.2. The difference in operating conditions would be brought about by drawing off liquid, and the remaining liquid evaporating to fill the void space, or the vessel being heated up or cooled down by sunlight or cold nights. The reassessment of existing pressure vessels that may be designed to inadequate old standards, or may not comply with the standard in all important respects, or where the operating conditions may have been optimistically, but incorrectly, chosen has taxed both industry and the HSE. The HSE has now produced a worthwhile booklet entitled *The Assessment of Pressure*

Vessels Operating at Low Temperatures under its designation HS(G)93[5] and although this is only specifically aimed at low-temperature vessels, the principles contained within it would be quite valid for all pressure vessels and storage vessels as well as pipework.

References

1. HSC, *Safety of Pressure Systems, Pressure System and Transportable Gas Containers Regulations 1989*, ACOP, HMSO, London
2. HSE, *A Guide to the Pressure Systems and Transportable Gas Container Regulations 1989*, HMSO, London
3. S.I. 1989, No. 2169, *Health and Safety — the Pressure Systems and Transportable Gas Containers Regulations 1989*, HMSO, London
4. HSE, *GN GS4, Safety in Pressure Testing*, HMSO, London
5. HSE, *HS (G) 93, The Assessment of Pressure Vessels Operating at Low Temperatures*, HMSO, London

17 Active protection systems and instrumentation

17.1 Introduction

This chapter discusses both active protective systems that are intended to keep process plant within its safe operating limits and the type of process instrumentation needed to keep the plant in everyday working order. The active protective systems are often referred to as safety systems, whereas the instrumentations etc. is regarded as process systems. The process is kept within its SOL for two reasons: first to meet production targets and maintain financial profitability; and second to ensure the safety and welfare of the operators and innocent bystanders who might be exposed to hazardous materials.

17.2 Fail safe

We have already considered that process plant should be designed, fabricated and tested with the view of its failing safely. The same objective should be pursued when considering protective systems. Some protective systems such as pressure-relieving devices can be made to fail safe, whereas other devices such as boiler blow-down systems have not as yet been made capable of fail-safe design. Generally the simpler the design the more likely it is to fail safe; the advantages of mechanical systems are well tried and proven. Protective systems which rely on electrical, pneumatic or hydraulic systems can invariably be designed to shut-down, or to go on to a fail-safe condition. Such systems are discussed in BS 5304 *Safety of Machinery*.[1]

Modern-day process plant is often controlled by computers, and it is known from process engineers' experience that such computers can fail within either software or their hardware designs. Even for a system that has not failed, a computer can have so many variables within its software that it is not possible to test all combinations, and it must therefore be assumed that PES systems will fail to danger. For this reason computer-controlled PES systems are actively used for process control, but are not used for safety protective systems because they are assumed to fail to danger. All safety protective systems should be hard-wired direct from the sensing device to the protection device without going via a computer. The HSE has worked with EEMUA to publicise this concept in public seminars and training sessions. A booklet on this subject, *Safety Related Instrument Systems for Process Industries (including PES)* has been published by EEMUA (Publications Number 160.[2])

17.3 Misconceptions

'If it has not failed yet it is unlikely to fail in the future', is a dramatic misconception where process plant is concerned. The reality is that all disasters have been waiting to happen, until the circumstances for disasters have occurred. Protective devices of themselves cannot ward off failure, which is inevitable if the plant has not been designed, fabricated or tested taking into account all its failure modes. It is essential to consider first how the plant may fail, e.g. by over-stress, over-strain, creep, fatigue, brittle fracture or stress-corrosion cracking, and then go on to the types of situations which would cause breaches of the SOL.

The assumption that vapour pressure is dependent only upon temperature in such liquefied substances as propane, butane, ammonia, or chlorine is a basic misconception. For example, when liquefied gases are drawn from a vessel the volume taken out has to be filled by gas, and that gas is generated by the liquid boiling, which in turn supercools the liquid and the vapour above it. This may happen in the summer time, when sunlight on the top of the vessel will heat the vessel shell, but the liquid at the bottom of the vessel will make the temperature of the skin dramatically lower. In Chapter 16, Fig. 16.2 shows typical storage conditions for LPG (propane and butane) as compared with the phase diagram and vapour pressure curves. For propane for the same temperature, say 10°C, there can be a pressure range of 1.7 bar or conversely for the same pressure, say 6 bar, there can be a temperature difference of 8°C. This difference between theoretical superheated vapour pressure, and the actual temperatures under working conditions leads to difficulty in providing adequate safety protective systems.

17.4 Safe design

There are some adverse operating conditions encountered in process plant that are not readily solved by the use of protective systems. Steam hammer and water hammer are caused by pulsating liquid and vapour flowing within the same pipe. Water hammer is often experienced in domestic premises where air is contained within a water system; when the valve is opened the rush of water is cushioned by an air pocket, but then the air pocket shifts and causes a banging or hammering vibration. Steam hammer produces similar effects to water hammer but is caused by slugs of condensed water propelled by travelling steam, which often occur when valves are opened or closed. Such hammer hazards are not solved by protective systems, and it is therefore essential to design the system so that it is capable of sustaining the stresses, strains and fatigue conditions that will occur. The remedy will be to use material that is ductile and which does not suffer from brittle fracture, i.e. avoid cast iron, wrought iron and non-ductile steels. It will also be a sensible precaution to provide steam traps to remove condensation and to carry out adequate maintenance to see that those steam traps are functioning; but steam traps themselves do fail to danger, and can never be relied on to remove all traces of condensed water.

Cavitation occurs in pipes, pumps and vessels when air or other gases are drawn into the system and are injected into flowing liquids. This can cause erosion and corrosion and, as no safety system is known, prevention is better than cure. This can be achieved by making sure all pipe flange gaskets are adequately sealed, that stuffing boxes and glands in pumps are properly maintained, and by the effective use of sparges or close attention to gas injection methods into vapour condensing spaces. These of course are really functions of design and maintenance rather than protective devices.

17.4.1 Safety devices to protect SOL

The prime process safety concern is adequate containment. Containment will only be maintained provided the SOL have not been breached. The SOL will include pressure (both positive and vacuum), temperature (both high and low), weight, time or number of cycles, and environmental conditions that could lead to stress corrosion cracking (see Fig. 17.1).

UK legislation calls for steam boilers, steam receivers and air receivers to have safety valves. This concept goes back to the turn of the century, and has laid the foundations for the idea that the main type of protective system is a safety valve. For boilers and air receivers this is merely an overpressure-release device. The process industry has generally called such valves pressure-release devices, which are accepted to be the prime safety device. Mr Alan Smith of the HSE has carried out a survey of pressure-relief valves and published his findings in the Institution of Mechanical Engineers Conference in March 1993.[3] The concept is that a spring-loaded pressure-relief device will fail to safety, i.e. if the spring should prematurely fail then the pressure inside the system would be released. But Mr Smith's survey showed that this has not been the case. He found that 39% of the safety valves that were checked would not have operated as required in the event of an emergency. This could show that there have been very few emergencies, or that the usefulness of a safety valve is over-estimated. Very few

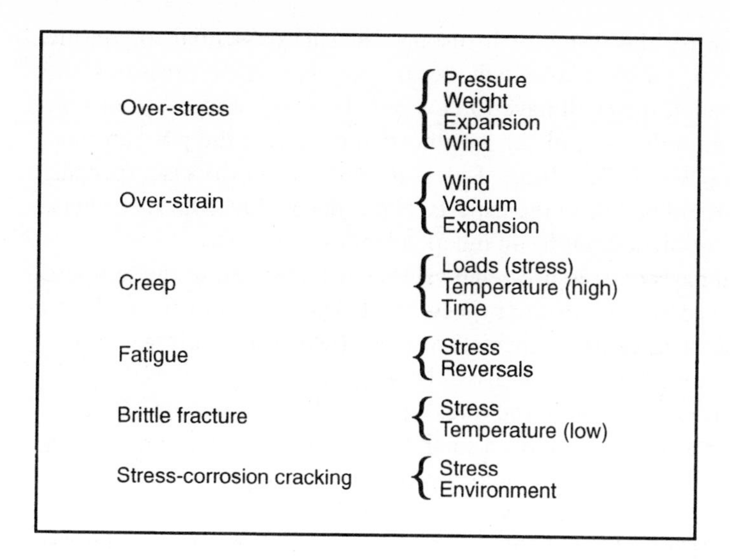

Fig. 17.1 Safe operating limits

process plants have failed because of straight overpressure. There is one instance of a steam boiler that failed catastrophically late one Saturday evening when no-one was on the premises, but detailed investigations suggested that the safety valve had been 'doctored', gagged or tied down and deliberate sabotage was suspected; in any case the insurance claim was never paid out.

The reality is that process plants need more than overpressure relief devices to protect them from operating outside their SOL.

17.4.2 Overpressure-relief devices

Failure in pressure vessels can occur due to both positive overpressure and negative or vacuum conditions. Within the food-processing industries, stainless steel vessels and pipework are sterilised by using high-temperature steam, which condenses when cooled down, forming a vacuum. Unfortunately this gentle vacuum formation has caused very large vessels to suck in or collapse causing considerable process and financial disarray; but it has rarely led to injury. Nevertheless it would be a simple process to use a spring-loaded vacuum release device, so that if the safe operating vacuum conditions of a vessel are correctly established and a safety device is fitted, the embarrassing situation of a failed vessel will not occur.

Pressure can also occur due to static head of liquid, or the increased vapour pressure due to heating liquid until it boils, e.g. when water boils, or when chlorine or LPG evaporates. Pressure can be increased by displacement pumping or by use of compressors. The design of overpressure-relief devices should be carried out in two stages. First of all the process plant hardware and its failure modes must be considered to ascertain the maximum pressure the system will sustain. Next it is essential to consider what theoretical pressure rise will occur, because when pressure-relief devices are needed they will have to be sized. The rate of increase of pressure will give an indication of the area needed to relieve the pressure once the device has opened. The sizing of pressure-relieving devices is a specialist subject, and should be left to the experts. Naturally quality assurance checks should always be carried out by an independent or third party.

The internal parts of pressure-relief valves must be suited to the temperature range and corrosivity of the process fluid. The temperature at which a pressure-relief valve can be expected to open (which may differ considerably from its temperature when it is relieving) must be considered in relation to its set pressure and the pressure at which it reseats. Types of pressure-relief valves which allow a small leak of process fluid to the atmosphere when the valve opens should not be used for highly toxic fluids, or flammable fluids when the relief valve is in a building.

'Bursting discs' which are held between pipe flanges are defined by API as 'rupture disc devices'. They are used instead of pressure-relief valves where the pressure rise may be so rapid as virtually to constitute an explosion, and upstream of pressure-relief valves in cases where minute leakage cannot be tolerated, or where blockage or corrosion might render a relief valve ineffective. Bursting discs are available in various metals, plastics, combinations of both, and carbon. Metal bursting discs are usually very thin, and domed. They are secured between special flanges and supported on the high-pressure side by a perforated concave metal plate to prevent flexing of the disc if the pressure difference is reversed, e.g. when the equipment is evacuated.[4] When fitted in series with a pressure-relief valve it is essential to prevent debris from the burst disc from interfering with the operation of the valve, and to provide a device to indicate leakage into the space between the discs and the valve (e.g. a pressure gauge). When used alone, bursting discs have the obvious disadvantage of allowing most of the contents of the equipment to escape. When used in conjunction with a pressure-relief valve, the process should be shut down when a bursting disc has ruptured. The cause of the rupture should be satisfactorily accounted for before replacing the disc and restarting the process.

If it is neccessary to fit a valve on either side of a pressure-relief device, e.g. to isolate either the equipment protected by the device or the device itself for maintenance, it must be secured (e.g. by lock or interlock) in the open position while the pressure-relief device is in active service. On difficult duties two pressure-relief devices are sometimes installed in parallel, with isolation valves upstream and, where necessary, downstream, so that one device is always functional while the other is available for maintenance or cleaning. Three-way isolation valves with locks or interlocks [17.7.1] should then be chosen which ensure that one pressure-relief device is always on duty.

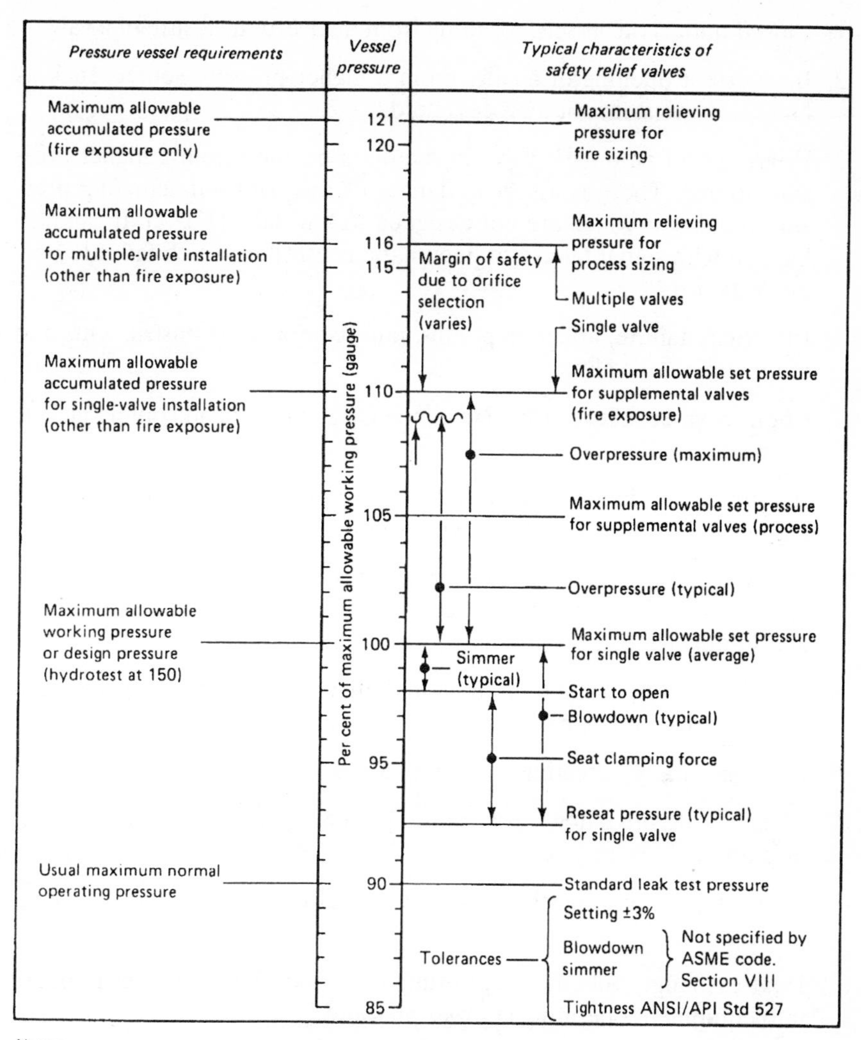

Notes:

1. The operating pressure may be any lower pressure required.
2. The set pressure and all other values related to it may be moved downward if the operating pressure permits.
3. This figure conforms with the requirements of the *ASME Boiler and Pressure Vessel Code,* Section VIII, 'Pressure vessels', Division 1.
4. The pressure conditions shown are for safety relief valves installed on a pressure vessel (vapour phase).

Fig. 17.2 Pressure levels referred to in connection with pressure relief devices (courtesy API).

17.4.3 Causes of overpressure

Overpressure in process equipment, pipelines and boilers can result from a variety of causes. Those given here are mainly based on the API guide[5] for oil refineries. While these include most causes to be found in the process industries, the list is for guidance only and is not claimed to be complete.

1. Closed outlets on vessels, resulting from inadvertent closure of a valve.
2. Inadvertent opening of a valve from a higher pressure source, such as high-pressure steam or a process fluid.
3. Failure of an automatic valve in the open position from a higher pressure source. There is a special danger of this with self-actuating pressure regulators which are not designed to 'fail safe' [17.7.2]. It can also happen with signal failure to pneumatic controllers which fail safe only on air failure.
4. Electricity failure, affecting pumps, fans, compressors, instruments and motor-operated valves.
5. Cooling-water failure, affecting condensers, coolers and cooling jackets on machines.
6. Instrument air failure, affecting transmitters, controllers, process-regulating valves, alarm and shutdown systems.
7. Steam failure, affecting steam-driven machines (pumps, compressors, generators, etc.), steam-heated equipment, steam injection to processes and steam ejectors.
8. Fuel failure (gas, oil, etc.), affecting boilers, fired heaters, engine- and turbine-driven machines.
9. Inert gas failure, affecting seals and purges.
10. Mechanical breakdown of fans, pumps, compressors and other machines.
11. Instrument failure, particularly of temperature, pressure and flow controllers.
12. Heat exchanger tube failure, leading to flow of fluid at higher pressure into equipment operating at lower pressure.
13. Failure of tube in fired heater for process stream, causing increased combustion in furnace and excessive heat input to waste heat boilers, etc.
14. Thermal expansion of liquids in liquid-filled pipes and equipment (especially heat exchangers), on which valves have been closed ('boxed-in equipment'). This can be a serious problem with long pipes carrying liquids which are exposed to the sun.
15. Plant fires, causing expansion and vaporisation of fluids in equipment and pipes. Pressure-relief systems for plant fires are usually designed on the assumption that external thermal insulation on equipment remains intact, but if it is destroyed or removed, e.g. by an explosion, the pressure-relief system may be unable to cope. Metals are also weakened when heated by a fire, so that equipment may fail at normal operating

pressures with the pressure-relief valves closed. To protect against fire, pressure-relief systems need to be backed up by other forms of protection, including depressurisation, to reduce the danger of a vessel bursting.

16. Chemical reactions, particularly exothermic ones [8.6]. These are normally controlled by other means to prevent overpressure, but where they fail, the pressure-relief device should be able to save the situation, provided the reaction is not extremely fast. The danger of very fast run-away reactions can sometimes be averted by carrying out the reaction in a boiling solvent under a reflux condenser, which returns condensed solvent to the reactor. This removes the heat and controls the temperature.
17. Presence of unexpectedly volatile material in feed to distillation column caused, for example, by failure of reboiler heat in immediately preceding column.

17.4.4 Excessive load

Profitability and production control often demand that the throughput, weight or loading within process plant are increased, and there are some occasions when safety has been adversely affected. Bulk silos and hoppers have failed because their content has risen too high and the bulk load has been more than the structure could sustain. In such cases where overload can occur and where it may lead to disaster it is essential to set load cells that measure the weight, and are automatically coupled with shut-off valves which inhibit entry into the hopper or vessel.

Because chlorine is both toxic and corrosive when wet, it is usual not to supply dipsticks, site glasses or level gauges, but rather to rely on storage vessels being mounted on load cells so that the contents can be measured by weighing. This is a sensible precaution but care must be taken when fitting the load cells. Inexperienced persons have been known to put them in the centre of the vessel, thus making the saddle supports on which the total weight of the vessel rests non-symmetrical. This in turn has increased the stresses in the saddle support directly above the middle saddle. This excess stressing has the potential to cause catastrophic vessel failure. Considerable financial costs have been incurred in modifying existing installations to reduce the hazard caused by an ill-conceived design.

17.4.5 Time-dependent safety

The setting of safe operating limits may indicate that there are time-dependent safety factors based upon fatigue or vibration considerations, or stress reversals, which are time- or temperature-dependent. In these circumstances it is essential that safety devices are used to count the number of reversals, hours under operating conditions, cycle reversals, rotation reversals, etc. Having done this an adequate management system must be adopted to review the number of reversals and to take adequate steps to ensure that safety has not been affected.

Hydrofluoric acid is an extremely corrosive substance which attacks the phosphorus and calcium in flesh and bones. The processing starts with the roasting of naturally occurring felspar in large revolving drums which are heated, similar to cement kilns. The weight of the natural felspar rock, the high temperatures and the revolving of drums induces fatigue, stress corrosion and creep conditions, which have led to cracking in the support regions where the revolving drum rests on rollers. Cracks have occurred which have led to minor leakages, which were adequately dealt with by water sprays. The cracking can potentially cause the complete catastrophic failure of these 30 m long by 3 m diameter drums, discharging their hydrogen fluoride gas, and at the same time causing physical damage by their impact on the surrounding process plant. Of course ideally the original design would have taken into account all the failure modes that could apply and adequate maintenance would have been carried out. But safe monitoring of the number of cycles and the temperature of the plant would have warned management so that the plant should be taken out of service. This would lead to considerable cost savings as planned maintenance is far more economical than breakdown maintenance.

17.4.6 Temperature controls

The SOL, including the temperature of process plant, should be ascertained both at design and by periodic examination. Over-temperature can result in the weakening or softening of metal which will lead to premature failure, but very high over-temperature will lead to creep conditions where the material stretches progressively to the point of failure. Creep conditions will be both time- and temperature-dependent, and will only occur in temperature over 450°C. Low temperature conditions can bring about fast fracture due to the brittle condition of the material with instant catastrophic loss of containment. It is essential that once the safe operating limits have been established, automatic safety devices should be provided to monitor the temperature and then actuate safety systems such as pressure release or cooling should it become excessive; or, if the temperature becomes dangerously low, there should be devices to relieve the pressure and to increase the temperature of the vessel pipework or storage tank.

Generally speaking within steam boiler and steam receiver technology the temperature of the steam is directly related to the pressure. If the pressure does not rise the temperature does not rise, and therefore a pressure-relieving device is adequate for safety. The same conditions can occur when other liquids are being boiled in distillation plant and so temperature can be directly related to pressure and it is far easier to maintain safety by controlling pressure than by controlling temperature. Unfortunately this safety system does not always operate. For example when a boiler is working properly there is an adequate level of liquid, e.g. water, within it, that effectively cools the furnace tubes. Should that water be lost, then the furnace tubes will not be cooled and they will be heated by the fuel within them. UK safety legislation under the Factories Act Section 32 states that boilers should be provided with a fusible plug or an efficient low-water alarm device. The fusible plug was mounted within the furnace tube, and when water was lost within the boiler it became overheated and was blown out from the furnace tube. This was never intended to relieve the pressure, but just to warn the furnaceman or the stoker to increase the water level so as to prevent total failure by steam boiler explosion. When a fusible plug dropped out of a furnace tube a great deal of water and steam poured onto the fire, which to some extent cooled it and at the same time was a clear warning. In reality this legislation has become superfluous because modern-day boilers are usually gas- or oil-fired, and stokers are not in attendance. Likewise the low-water alarm device which would have been operated by a fireman when the water dropped low produced a whistle to warn the furnaceman. In modern-day use this is inoperative because boilers are rarely manned 24 hours a day. Although not a legal requirement it would be feasible to fit a temperature recorder onto a furnace tube, and when the temperature of the furnace tube was increased (because of low water) then the fuel to the boiler would be shut down.

17.4.7 Low temperature

Low temperature conditions can result in the most dramatic instantaneous loss of containment as happened at the ammonia tank in South Africa, LPG storage tanks in Asia, and carbon dioxide storage vessels in Eastern Europe. Cold temperature can come about due to atmospheric conditions, deliberate refrigeration, inadvertent refrigeration when liquid or vapour is withdrawn from the vessel containing a liquefied gas, or inadvertent local refrigeration when liquefied gas leaks from a vessel via a crack, and the crack area is locally supercooled.

An ammonia cylinder had been filled by drawing off surplus ammonia in a refrigerated section of a brewery in South Wales. The cylinder was then placed outside in the sunlight and, due to a slight temperature increase, the pressure in the vapour space rose to a level that it had probably never reached before, and this caused a local corrosion defect to become overstressed and crack. Because of the quality of the cylinder material, its thickness and the shape of the crack, if the leak had been at ambient temperatures the crack would not have grown, and there would not have been a discharge of ammonia from the cylinder. Unfortunately as the ammonia gas escaped via the crack there was a Joule Thompson refrigerating cooling effect (as ammonia expands when passing the crack, it cools). This dropped the temperature of the materials locally down to the point where fast fracture or brittle fracture occurred and the cylinder disintegrated into three pieces discharging its contents instantly.

It is probably far better to design process plant to sustain the minimum temperature that can be experienced, e.g. propane –40°C, butane –20°C, rather than attempt to fit safety devices that will preclude the skin temperature reaching a point where it becomes susceptible to brittle fracture.

It is the possibility of local cooling, leading to brittle fracture, which makes low temperature plant difficult to protect by supplying safety protective systems. Theoretically it would be possible to measure the skin temperature of the metal and, as it approached a temperature low enough to produce susceptibility to brittle fracture, to warm the contents or prevent any further drop in temperature. There are proposals to spike butane vessels with propane and this may prevent problems arising due to low temperature. But there would be other safety problems when the ambient temperature rose, because the propane would have a higher vapour pressure than the butane and this might cause the lifting of the pressure-relieving device or some other form of failure.

The discharge from pressure-relieving devices relieving liquefied gases such as propane, chlorine or ammonia, or the weeping of stop-valves, can supercool the area, leading to a build-up of ice. A carbon dioxide storage vessel in a Scottish brewery had a leaking stop-valve, and the issuing of carbon dioxide supercooled the stem of the valve on which ice built up, thus causing the valve to be inoperative. The valve was passing carbon dioxide to the main evaporator, and it was then dumped back into the vessel in a closed loop. Process operators noticed the pressure was rising and attempted to lift the easing gear on the safety valve so as to discharge the carbon dioxide and so make the system safe. Unfortunately they were not able to move this easing gear, and a decision was taken to open the main filling line which allowed carbon dioxide to vent directly into the room. As this valve was opened it discharged liquid carbon dioxide, and that valve became jammed due to ice build-up, and so gradually the whole contents of the vessel were released into the brewery, resulting in asphyxiation of the operator who opened the valve.

17.4.8 Fire impingement

Steam boilers and other boilers are designed to have one side of the furnace tube with flame impingement while the other side of the furnace tube is water- or other liquid-cooled. The skin of the furnace tube in a steam boiler probably never gets warmer than 360°C.

When a fire occurs, the skin of storage vessels, pressure vessels or pipework will be kept cool by any liquid inside, but if that liquid is not flowing, or is covered by a gas, then the skin of the containment vessel or pipework will become increasingly hot either by radiation or by direct flame impingement. The material will become weaker as it becomes hotter and may fail in a ductile manner opening up a longitudinal crack.

It is common practice to supply LPG vessels, and much process plant containing flammable substances, with water sprays. This can be considered a protective system, as the intention is to detect fire and then cool the skin of the containment vessel or pipework by use of cold water. Such systems are notorious for failure. If there has been flame impingement on metal surfaces for more than a few minutes a layer of steam may form between the water and the metal, and cooling will be reduced. Road tankers that have been engulfed in fire have been sprayed with cold water by the fire brigade, and the thermal shock of cooling part of the vessel has produced local stressing which has led to cracking due to excessive thermal contraction. Such a situation happened near Exeter in Devon, on the hard stand of the motorway, but fortunately the skin of the road tanker was sufficiently warm to allow ductile cracking to occur without the cold conditions of fast fracture.

A road tanker in Jersey was discharging its contents when there was a leak from a hose which resulted in a fire engulfing the tanker. It was not possible to apply cooling water and the complete contents of the small tanker discharged but without a catastrophic failure. In all probability the dowsing of that tank with cold water would have resulted in thermal shock and total catastrophic failure.

A road tanker in Holland was discharging LPG to bulk storage vessels at a petrol filling station when once again a hose developed a leak, which spread into a fire and engulfed the tanker. Unfortunately there was a stress raiser where the manway was welded onto the tanker, and that caused local over-stressing in the heat conditions which produced a small crack. LPG escaping through the crack caused local cooling, which led to brittle fracture. The tanker failed catastrophically, releasing the total contents, which burned in a rising fire ball.

An LPG static storage vessel in a site near Glasgow was subjected to flame impingement which lasted for several hours before firefighting personnel could bring the fire under control. Although the vessel was discoloured due to fire, the skin had stayed intact and no catastrophic failure occurred, probably because there were no stress raisers due to design or manufacturing defects. Water sprays were not applied, and the fire was allowed to burn itself out. It is suspected that if water had been applied this may have cracked the vessel.

There are numerous records of water sprays failing within the first few minutes of the fire occurring. This can happen because the water within the cooling system boils and the pressure rise is too great for the ring main supplying the cooling water, or because explosions can shatter the pipework as happened at the Flixborough disaster when the cooling water pipework was totally destroyed in the initial blast. The LPG storage spheres in Feyzin, Texas City, Mexico City and Asia were all supplied with water sprays but in none of these situations were they effective in stopping total catastrophic failure.

There have been numerous instances where during road transport and rail transport tanks containing pressurised liquefied gas have failed violently or BLEVED. This can happen when the vessels are carrying flammable materials or inert materials such as carbon dioxide or chlorine. The system of failure due to BLEVES is discussed in 10.2.3. In this chapter it is prudent to suggest that water spraying probably would not have protected the vessels from fire-induced BLEVEing. There are proprietary fire insulating materials that could be applied to the outside of the tanks which could in general be described as safety protective systems.

17.5 Revalidation

In Chapter 16 there is a discussion on how plant should be periodically retested, inspected and revalidated. At this stage vessels, pipework, heat-exchangers, storage tanks, distillation columns etc. should have their SOL restated. It may be that the safe operating limits after a periodic examination have to be reduced due to local or general deterioration. Once new SOL have been set it will be necessary to reset or adjust the existing safety protective systems described above.

17.6 Disposal of released fluids

Three principal methods are used for the disposal of fluids released by pressure release devices:

1. by discharge to the atmosphere;
2. by burning;
3. by disposal to a lower-pressure system. This includes the use of separators to collect released liquids and special chemical scrubbers to prevent the release of highly toxic gases and vapours to the atmosphere.

The API guide[5] deals with (1) and (2) in detail. The design of systems for dealing with specific toxic gases follows normal chemical engineering practice.[6]

Factors to be considered when deciding on the method of disposal include toxicity, odour, smoke, particulate matter, noise, heat, reliability and ease of maintenance

Refinery gases consisting of combustibles with small amounts of hydrogen sulphide are an interesting case. Hydrogen sulphide has an OEL of 10 ppm, which is higher than that of sulphur dioxide (OEL 5 ppm), its main combustion product. So far as the toxic hazard is concerned, there is at first sight little to choose between discharging the gas at high level and at high velocity or burning in a flare. In the latter case the sulphur dioxide will be carried to a higher level by thermal convection, and its concentration at ground-level will be low. A quantitative assessment of all the factors involved needs to be made when deciding on the disposal method.

Released-fluid disposal systems, for which a high degree of reliability is required, unfortunately tend to be the 'Cinderellas' of plant design. Serious deficiencies in the discharge arrangements of the pressure-relief system of a new LPG installation which the writer encountered in a Third World country are described in 23.5.1.

The pipework of disposal systems must be designed and supported to withstand the thermal strains resulting from the entry of hot and cold fluids, and the shock loading resulting from the sudden release of compressible fluids and slugs of liquid. If compressed liquefied gases are relieved, flash evaporation will occur at the release device subjecting the pipework to very low temperatures. It is then essential to choose an alloy which does not suffer embrittlement at the lowest temperature which may be reached, and to provide adequate bends in the pipework to cater for contraction. If solid materials (such as rubbers and polymers) are liable to be released into or formed in the pipework, adequate flanged joints must be provided to allow it to be cleaned and inspected internally. Means must also be provided (e.g. locks on valves or line blinds [18.6.1]) for leak testing the pipework after installation and maintenance. (This is often overlooked.)

If liquid is liable to be present in the discharge, the piping should be self-draining into an adequately sized knockout pot before the vent stack, flare or scrubber, and effective means of removing liquid collected must be provided.

17.6.1 Discharge to the atmosphere

This is usually the cheapest method and is commonly used not only for air and steam and other gases of low toxicity but also for flammable gases and vapours. Individual vent pipes near the equipment relieved are generally used. The levels of noise and atmospheric toxicity to which nearby workers and members of the public are liable to be subjected by the discharge should be evaluated and be within accepted limits for the chosen location. Mist is generally only a problem with steam discharges since most saturated vapours superheat on passing through a relief valve.

In the case of flammable gases, these must leave the vent at a sufficiently high velocity to entrain enough air to reduce its concentration to below the lower explosive limit before the energy of the jet has dissipated. Hydrocarbon gases and vapours are generally diluted to below their lower flammability limit [10.1] at a distance from the vent exit equal to 120 times its diameter. When the cross-sectional area of the vent exit is equal to that of the full valve opening, this condition is usually met so long as the valve is passing gas only and is at least 25% open. This applies to 'pop-open' relief valves which close positively when the pressure in the equipment relieved falls below the set pressure. The vent should be located on the top of a tall structure and so directed that the jet does not impinge on any solid object which would cause a stable flame if ignition occurred. The vent should preferably point upwards, and be adequately supported against reaction from the jet discharge. A small drain hole is needed at the bottom of the vent pipe to ensure drainage of rain-water. This should be arranged so that a little air is sucked into the vent pipe by the momentum of the gas rather than allowing gas to escape through it at low level.

There is always some risk of ignition when flammable gases are discharged to the atmosphere, especially if the relief device opens as a result of fire. This happened in the Fézin disaster of 1966,[7] when the gas released ignited, causing a huge flame which radiated heat back onto the LPG spheres and was a major factor in causing their disastrous rupture. The consequences of radiation from burning gas at the vent exit must therefore be considered when deciding its position. Possible causes of ignition include lighting, static electricity and iron oxide particles. The risk of ignition increases if the gas contains molecular hydrogen (Table 10.1). The exit of the vent can can be designed to reduce the chance of static ignition when hydrogen-containing gases are vented, and also to ensure that any flame which tends to form is blown out by the jet. There should be no risk of forming an explosive vapour cloud with a properly designed vent discharge since the concentration of flammable gas is only momentarily within explosive limits in the jet itself. Considerations of jet dispersion favour separate elevated vents for each pressure-relief valve rather than manifolding several discharge lines together into a common stack. For similar reasons, process vents of flammable gases should be kept separate from pressure-relief vents.

Wells[8] recommends that the following criteria should be met before venting to atmosphere under emergency conditions is permitted:

1. molecular weight of vented material <61;
2. liquids and solids are absent (including liquid condensate in lines and liquid droplets in the discharge resulting from condensation);
3. the vapour release velocity is adequate to ensure that in the case of flammable gases and vapours the expanding jet entrains sufficient air for the mixture to be non-flammable by the time it has ceased to move under the influence of the jet;
4. concentrations of toxic and noxious materials in the air meet industrial hygiene specifications;
5. harmful airborne materials cannot enter buildings, enclosed areas or other workplaces to a dangerous extent;
6. the discharge tip is >25 m above grade and >3 m above the highest working level or roof within a radius of 12 m;
7. the discharge of any flammable material should be more than 30 m horizontally from heaters, air intakes and sources of ignition;
8. should the material ignite, the radiant heat flux should not exceed 4700 J/m^2s on any neighbouring access floor.

17.6.2 Disposal by burning

Disposal of combustible gases by burning is generally accomplished in elevated flares which can be a nuisance to local communities, especially at night-time. Flare stacks and burners are best designed by, obtained from and installed by specialist firms. This is because of the difficult problems of ensuring positive pilot ignition, flame stability and acceptable levels of noise, thermal radiation and luminosity. Smokeless operation generally requires considerable quantities of steam, high-pressure waste gas or forced-draught air. Carbon deposited on the lip of the burner presents an ignition hazard, since red-hot particles may be detached and travel considerable distances in the air, with the risk of igniting pockets of flammable vapour, e.g. above floating roof tanks. Another potential hazard of flare stacks is that of explosion within the stack resulting from air entering it when no gas is flowing through it. Care should be taken to avoid holes in the piping, knockout vessel and stack through which air could enter, and a small continuous flow of inert gas through the system is recommended to prevent back diffusion of air through the burner.

If space permits, a ground-level flare has the advantage of facilitating maintenance and allowing a light shield to be constructed which obscures visible flames under all conditions other than major releases. When continuous flaring is necessary, smokeless combustion can be achieved through the use of a number of small burners at ground-level without the need for steam or forced-draught air. Combustible liquids released by pressure-relief valves can where space is available be discharged to a burning pit in which a fire is maintained continuously.

17.6.3 Disposal to a lower pressure system

For this method to be safe the lower-pressure system must have adequate capacity to take the maximum amount of fluid liable to be discharged from the higher-pressure system without itself becoming overpressurised. This method might be used in an LPG storage terminal with spheres and a considerable amount of pipework which requires thermal expansion relief for liquid LPG trapped between closed valves. Here the small amount of liquid LPG relieved could generally be routed to a storage sphere provided all possible hazards are checked.

17.6.4 Treatment of toxic gas and hot fluid discharges

Properly engineered systems for dealing with releases of highly toxic and hot fluids are essential. Highly toxic gases must be treated in adequately sized towers with liquid absorbent or solid adsorbent, or catalytically to form non-toxic compounds. Companies and industries which have these problems usually develop their own methods and codes for dealing with them, such as BCISC's chlorine code.[9]

Even where an adequate system is installed, the fact that it may not be needed for a long time may induce managements to question its need and allow it to lapse into a state of disrepair, as appeared to happen at Bhopal [5.3]. The rate of discharge which such systems may have to deal with in an emergency must also be properly appreciated when they are designed. The high cost of an adequate system is a temptation to provide a 'cosmetic' but inadequate solution.

17.7 Instrumentation for control and safety[6,10–12]

The importance of instrumentation, which typically accounts for 5% of the capital cost of process plant, can hardly be exaggerated. It has considerably reduced the number of operators required and increased the capital investment per employee. It has enabled more complex and efficient plants with lower material inventories to be built. Such plants may incorporate several reaction and separation stages, internal recycle streams, and integrated fuel and power supply combined with waste-heat recovery. If the same process had to be designed for manual operation without control instruments, it would have to be split into several much simpler units, with considerable intermediate storage between them and a large team of operators. Today many processes are controlled by programmable electronic systems (PESs), with only a handful of highly trained operators in supervisory and monitoring roles. While the term PES includes a single monolithic computer such as those installed in control rooms in the 1960s and 1970s, it also covers more modern systems consisting of networks of individual electronic devices linked together with coaxial cables or fibre optics. These, like single computers, can select the optimum conditions to give the required product output and specifications at minimum cost.

The design of plant instrumentation and the selection of instruments is, even more than pressure-relief systems, a specialised engineering field. It cannot, however, be left entirely to instrument engineers, but requires close co-operation with process engineers who design the process, operating staff who will run it, and safety and reliability specialists. While these people cannot all be instrument experts, they need a general appreciation of instrument and control theory and of the instrument symbols used on P&I diagrams. These generally follow the Instrument Society of America's standard S5.1.[13] A selection of symbols is shown in Fig. 17.3.

The reliability of process instruments is discussed in 14.1 and again under hazard and operability studies in 14.4. While complexity is inevitable in modern plants, every effort should be made to reduce the instrumentation to its essentials, and keep it as simple as possible. The measuring elements of control instruments should be as close as possible to the point where control is most needed, and response times should be minimal.

Instruments are used both for normal plant control and for special safety functions. Many of those used for plant control incorporate such safety

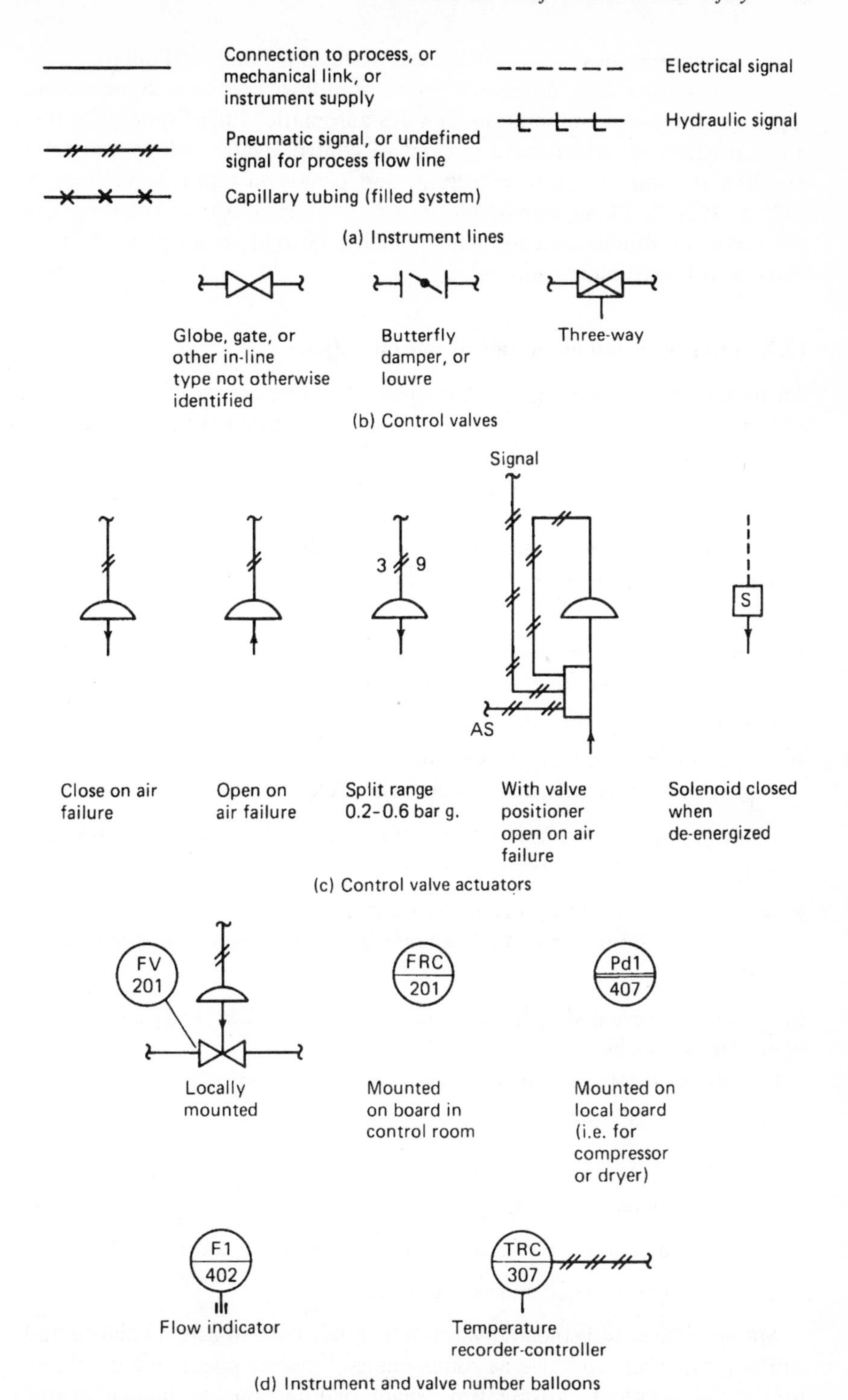

Fig. 17.3 Instrument symbols used on P&I diagrams

features as interlocks [17.8.1], 'fail-safe' systems [17.8.2], and microswitch-operated alarms and automatic plant shutdown systems. Special and separate safety instrumentation includes automatic 'dump' systems to dispose of reactions which have got out of control, water deluge and total flooding systems for fire protection, and explosion-suppression systems [10.3.5, 10.4.2]. These only come into play when a critical situation has been reached which demands action which is beyond the scope of the normal control instrumentation.

17.7.1 Interlocks and other aids to safe operation

An interlock is something which connects two or more adjustable devices and prevents one or more of them being set in a particular mode (e.g. open or shut) which would be hazardous while others are also in particular modes. The first interlocks were mechanical and applied to the hand-operated levers used to switch railway signals and points. Interlocks today may be electrical, electronic, pneumatic and hydraulic as well as mechanical, or may involve combinations of these signalling or actuating media. They are very widely used and a few examples follow:

- in starting plant or machinery, interlocks are used to ensure that all prestart conditions (e.g. adequate lubricating oil pressure) are met and that the correct sequence is followed;
- interlocks are used to prevent unauthorised entry to electrical switch-rooms, process vessels, and cubicles where explosives are tested;
- isolating valves fitted in series with pressure-relief devices have interlocks to prevent all of them being closed at the same time;
- interlocks are used to prevent control instruments from being decommissioned for calibration or maintenance unless certain safe conditions are met.

Many hazards revealed by hazard and operability studies [14.4] can be eliminated by interlocks.

Features of a good interlock system as stated by Lees[7] are that it:

1. controls operations positively
2. cannot be defeated
3. is simple, robust and inexpensive
4. can be readily and securely attached to the devices associated with it
5. can be regularly tested and maintained.

Microswitches and solenoid valves are widely used to actuate alarms and shutdown devices, and also as components of electro-pneumatic interlocks to prevent accidents arising from plant malfunctions or human errors. Richmond describes several examples, including the fitting of microswitches to the levers of cocks on the oleum charge lines to three sulphonation reactors, and a method of protecting a chemical reactor from the consequences of agitator failure.[14]

17.7.2 'Fail-safe' design

This philosophy which is found in all branches of engineering applies particularly to automatic control systems which contain several subsystems, any one of which may fail. Claims that a particular system or instrument can only 'fail safe' need critical appraisal. The term 'fail-safe' is something of a euphemism, and cannot disguise the fact that failures have occurred, causing loss and sometimes hazard. Thus it is important to concentrate on choosing reliable instruments and instrument systems with little likelihood of failure.

In opting for one mode of failure as 'safer' than another, the choice may be between the devil and the sea. With control systems the main criterion for fail-safe is that the valve or other control element should assume one of three positions if failure occurs: open, closed and the last position of the valve before failure. Possible failures affecting the valve include:

1. the power supply to the valve
2. measuring and sensing elements
3. the control system itself, which may depend both on electric and pneumatic power sources.

While a valve can be arranged to fail safe in case (1), this may be difficult or impossible for the other two since the valve has no way of knowing whether such failure has occurred. The extent to which 'fail-safe' applies is therefore limited.

With powered control valves with spring return mechanisms, particularly those with pneumatic actuators, the designer can choose between an actuator which opens or closes the valve on power failure. With a pneumatic actuator he can also arrange for 'last position failure' by the addition of a solenoid valve in the air line to the actuator. Thus it is important for the process engineer to study which is the least dangerous position of each control valve on loss of control, and specify this when ordering and installing the valve.

Richmond[14] has shown through the analysis of several types of self-actuating pressure and temperature regulators that all types of pressure regulator failed in the unsafe condition and that only one type of self-acting temperature regulator would 'fail-safe'. This type, however, is operated by a fluid under vacuum which has limited power to move the valve stem.

The consequences of all possible modes of failure of critical instruments need to be studied. For modes in which it is not possible to secure fail-safe operation, other means of protection should be provided, e.g. a pressure-relief valve or a high-temperature trip switch.

17.7.3 Process stability

Some processes are inherently stable or self-adjusting, so that any deviation from the required value of an important variable produces some effect in the process which tries to restore the variable to its original value. Others are inherently unstable, in that a change from the desired value of the variable produces some effect which increases and accelerates the change. Sometimes the same process has both a stable and an unstable regime. A pilot cracking plant which the writer designed and which was fired by its own cracked gases, a mixture of hydrocarbons and hydrogen, was an example of a very stable process. If (due to some upset) the cracking temperature increased, the calorific value of the cracked gas (and its Wobbe index [17.8.4]) fell, so that with other conditions unchanged the heat input to the furnace fell, thus restoring the original cracking temperature. This resulted in a very constant cracking temperature without any temperature control instrument. An example of a process with both stable and unstable regimes is a water-cooled exothermic reactor for which there is critical temperature below which the temperature naturally falls to a stable value, and above which it accelerates upwards [8.2].

When planning a new process and how to control it, the stability of different process variants and control methods should be explored. An inherently stable natural control system is likely to be safe and simple. The characteristics of the system adopted should be also studied to discover any unstable control regimes and delineate their boundaries. Precautions are needed to prevent crossing such boundaries while special protective instrumentation may be needed in case an unstable control regime is reached.

17.8 Component features of instrumentation

Some or all of the following features appear in the instrumentation of all process plant:

1. a control room
2. one or more power sources
3. basic control systems
4. measurement and sensing of process variables
5. receivers, i.e. indicators, recorders, controllers and alarms, etc.
6. final control elements
7. signal and power transmission systems
8. PES control

Within each of these there are various alternatives to choose from. In doing this, reliability and safety are key issues.

17.8.1 Control rooms

Most process plants have a control room as their nerve centre. This typically houses a control desk or console with visual display units (VDUs), one or more control panels, sometimes a computer, and operating personnel. On the panel is mounted a network of instruments (known as receivers) which receive information from the plant. These receivers increasingly tend to incorporate PES devices which extend their functions and allow plant data to be displayed on VDUs built into the console instead of on panel-mounted instruments. Receivers continuously receive pneumatic or electronic signals which relate to variables such as temperature and pressure and are transmitted by measuring instruments in the plant or elsewhere. From the control room pneumatic and/or electric signals are automatically sent to valves, motors, etc. to control the variable as required. No hazardous materials should be allowed to enter control rooms via pipes or tubes (e.g. to reach meters or gauges).

Control rooms are usually located close enough to their plants for operators to be able to move quickly from the control room to inspect plant items or open and close valves, etc. Good communications (e.g. by radio-telephone) are also essential between those working in the control room and those on the plant. Factors which affect operator morale and instrument performance such as the temperature, humidity, air movement, noise, vibration, illumination, interior decoration and furniture in a control room should be considered.

On oil and chemical plants where flammable gases and vapours are present, the control room may have to be located in an area classed as 'hazardous' and in which special precautions against electrical ignition are required [10.5]. It is usual here to 'pressurise' the control room with air supplied with a fan from a 'non-hazardous' area. The only admittance to the control room is then through an air lock with self-closing doors which allow a slight positive air pressure to be maintained inside it. This allows standard industrial electrical equipment to be used in the control room.

Since the Flixborough disaster [4] in which 18 men were killed in a control room, most control rooms built near plant where there is an explosion risk have been designed as separate blast-resistant single-storey buildings[7,15]. Windows, if fitted, contain only small panels of toughened glass.

17.8.2 Power sources

Most instrument systems rely on pneumatic or electrical power for signal transmission and control and electricity for chart drives and illumination. Pneumatic systems should create no ignition risks in electrically 'hazardous' areas, while pneumatic valve actuators are generally cheaper and simpler than electric ones. Pneumatic signal transmission is, however, relatively slow, thus limiting transmission distances to about 100 m. Low d.c.

intrinsically safe [10.5] electric current signal transmission and electronic instruments in the control room are often combined with pneumatic valve actuation.

All instrument power sources must be completely reliable. While failures of plant electrical power, steam, fuel gas and water usually cause an emergency shutdown, sudden and complete loss of control instrumentation is even more serious, since without it, it is far more difficult to shut the plant down safely, particularly if one of the other services fails at the same time.

On a large or medium-sized works, a reliable supply of clean, dry, compressed air is conveniently provided by an electrically-driven air compressor, with a diesel-driven stand-by compressor, and a large compressed air reservoir capable of meeting all likely demands until the stand-by unit is operating in the event of electrical failure. Instrument air should be distributed by a ring main, to give two alternative supply routes to every plant. Branches from the main should be provided with excess flow valves or other means of protection against escape of air and loss of pressure if a branch pipe is broken. The instrument air supply should be kept entirely separate from compressed air used for pneumatic tools and other purposes. Local compressed air reservoirs and sometimes even separate air compressors may be justified for large or hazardous process units. The compressed air is reduced in control rooms to the pressures required by the instruments.

Fitt[16] made the point that some instrument rooms are fed from a single leg of the air main. Failure of this leg alone without failure of the air supply to the control valves could lead to valves assuming 'unsafe' positions. The consequences of partial failure of instrument air can thus be more serious than total failure.

A reliable supply of high-quality electrical power, with emergency supply to critical instruments from batteries and/or a standby generator, is needed for electronic instrumentation. The instrument circuits should be kept entirely separate from the normal power and lighting circuits, and must meet the relevant flameproofing or intrinsically safe requirements.

17.8.3 Basic control systems[7,10,12]

The main control systems considered here are of analogue type employing continuous feedback, and are based either on pneumatic signals in the pressure range 0.2–1.0 bar g. or d.c. electronic signals in the current range 4–20 mA. The various control modes and other features considered here apply to both pneumatic and electronic systems despite their other differences. Pneumatic systems employ small mechanical devices with moving parts, e.g. baffle-nozzle amplifiers and metal bellows. Electronic systems use electronic amplifiers, switches and relays as well as electro-mechanically balanced potentiometers and Wheatstone bridges. Pneumatic signals interface directly with mechanical control devices. Electronic signals generally interface via pneumatics with mechanical control devices.

Also considered briefly here is 'Power Fluidics',[17] a control system with no moving parts, and self-actuating controllers which depend for their power on the moving process fluid. PES control which today is mostly of the digital type is considered later [17.8.8].

Instrument loops and degrees of freedom

The feedback control loop is basic to process control. It includes the process, the measuring element, the controller which receives its signal from it and the final control element (generally a valve) which manipulates a process variable and receives its signal from the controller. It is a 'closed loop' when controlling automatically (Fig. 17.4) but an 'open loop' when the control element is actuated manually or not at all. (*Warning*: Some writers refer to a loop as closed when a human operates the control valve!) In flow control loops the variable manipulated (by a valve) is generally the same (flowing fluid) as the one being controlled, but in most other control loops, one variable is manipulated in order to control another, e.g. the flow of steam may be manipulated to control the temperature of a continuous liquid mixer (Fig. 17.5).

Most controllers have an adjustable 'set point' which can be set at the required value of the variable under control, and they measure the deviation between the set point and the measured value of the variable. Their signal to the final control element aims to correct deviations which result from process disturbances (known as 'upsets'). When steady conditions are re-established, and depending on the control mode, the deviation will have been eliminated, or a residual deviation known as the 'offset' will remain (see 'Proportional control' later).

With feedback systems and their components, there is a time lag referred to as 'dead time' between any stimulus and the response of the system or component to it. This sometimes causes serious problems.

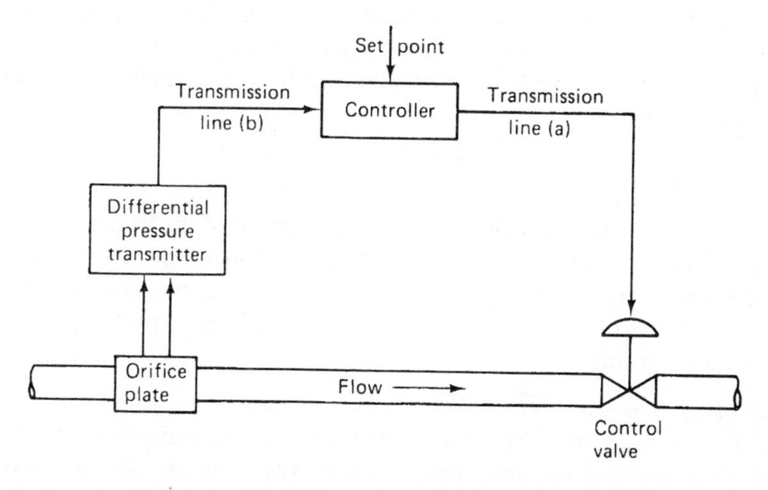

Fig. 17.4 Process-flow control loop (from Perry and Chilton,[6] courtesy McGraw-Hill)

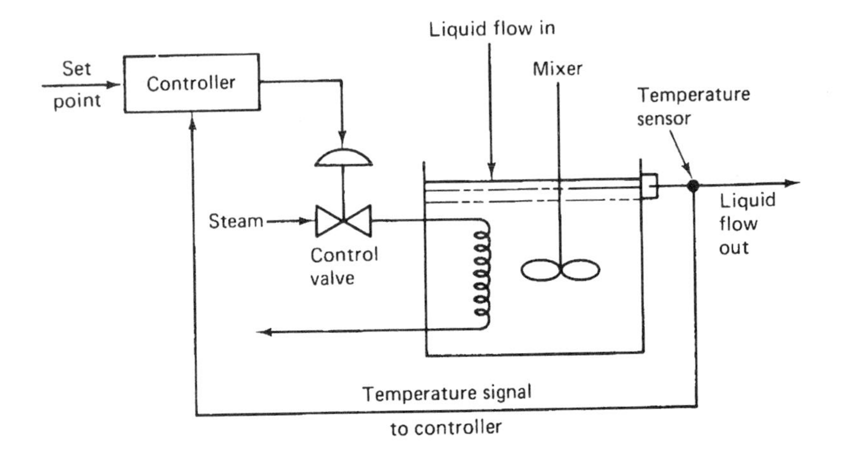

Fig. 17.5 Temperature control loop for liquid mixer (from Perry and Chilton,[6] courtesy McGraw-Hill)

The number of closed control loops needed for any process equals the number of independent variables or degrees of freedom (e.g. temperature, pressure, flow rate, level, concentration) which have to be controlled. But if too many control loops are provided, they will fight each other for control and the system will not work. This can happen because some variable was mistakenly thought to be independent whereas it was in fact dependent on and uniquely determined by other variables. As a simple example, consider two liquids which flow into a vessel, mix there and flow out as one stream. This gives three degrees of freedom, the flows of each stream into the vessel and the liquid level in it. We can control these three independent variables but we cannot at the same time control the flow of mixed liquid from the vessel.

Most cases are, of course, more complicated than this, but the same principles apply. In complex cases Boolean algebra can help to determine the irreducible minimum number of independent variables which need to be controlled, thereby simplifying the plant design and avoiding superfluous instrumentation.

Control modes

Analogue controllers may incorporate one or more of the following characteristic modes, the choice of which depends, *inter alia*, on the inertia of the process, the various time lags and any hysteresis in the instrument loop. The different modes of any controller have to be tuned for its particular process application (see later).

1. *On-off or 'bang-bang' control*
This is a simple, accurate but rather drastic method which is widely used (e.g. for refrigerator thermostats) where there are no process constraints (e.g. water hammer) which prohibit sudden control action. It is

used where the controlled variable changes slowly in relation to changes in the manipulated variable. Where a constant value of the controlled variable is needed a single set point is used, but when control is not critical, e.g. the level of liquid in a supply tank, two set points may be used, between which the level oscillates slowly.

On-off control is used more in batch processes than in continuous ones. Microswitches used in safety instruments to protect plants by flaring, venting or dumping process materials, when the temperature, etc. exceeds a critical value are a form of on-off control.

2. *Proportional control, single mode (P)*

In proportional control the difference in output of the controller from its mid-point is proportional to the deviation of the input signal from the set point. The relationship between the ranges of the output and input signals is called the proportional band and is expressed as a percentage. Thus a 20% band is narrow and gives sensitive control because 100% output change is produced by a 20% change of the input signal (i.e. the measurement scale).

The controlled variable only steadies out at the set point when the output signal is at 50% of its range. For other outputs the controlled variable will steady out at a value which differs from the set point by the offset. The maximum and minimum values of the offset for any proportional band would be plus or minus half the proportional band. If the proportional band is too wide, the offset will be large and the control insensitive. If it is too narrow, although the offset will be small, the loop will cycle and the controlled variable may never reach a steady value. Responses of a purely proportional controller to a process upset with too high, too low and correct proportional band settings are shown in Fig. 17.6. The best setting of the proportional band for the range of process conditions likely to be encountered is generally found by trial and error.

Proportional control works best when the controlled variable responds rapidly to changes in the manipulated variable, and where the value of the controlled variable is not critical, e.g. in some temperature and level-control loops.

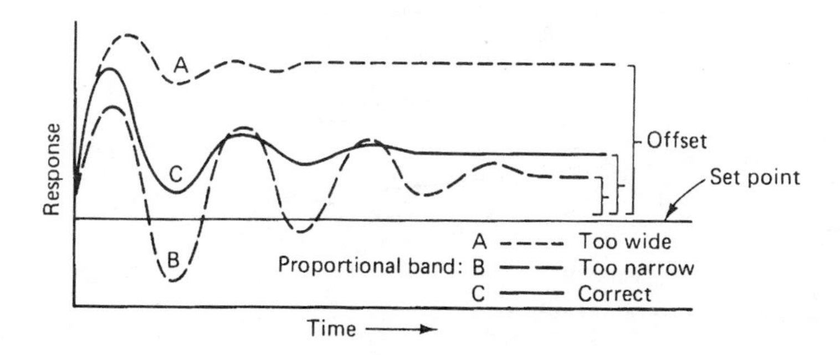

Fig. 17.6 Responses of proportional controller to a process upset (courtesy The Foxboro Company)

3. *Proportional plus integral (reset) control, two-mode (PI)*
Integral action, formerly known as 'reset', is a means of correcting the offset when proportional control is used. It produces a change in the output which is proportional to the integral of the deviation with time. It is combined with proportional control on critical loops where no offset can be tolerated, and on loops requiring such a wide proportional band that the amount of offset would be unacceptable. Integral action is expressed as 'minutes per repeat', the time needed for the integral mode to repeat the response of the proportional mode for a sudden upset. If the integral time is too long, the offset will persist for a considerable time, and if it is too short, the system will tend to cycle continuously. Figure 17.7 shows the PI system response to a process upset with different integral times. The addition of integral to proportional control, while eliminating the offset, increases the initial response of the loop to an upset and causes some tendency to cycle.

A problem of integral action which is specially found with batch processes is known as 'integral wind-up'. Here the controller (say, of a reactor temperature by a valve on the steam supply to the jacket) may start to work while the controlled variable is well below the set-point (and perhaps below the temperature scale of the controller). So much integral action may then have been stored up that the steam valve opens fully and remains so until the reactor temperature passes the set point. The temperature continues rising for some time before the output of the controller falls and the steam valve starts to close.

To overcome this problem, a 'batch switch' may be fitted to a controller with integral action. This cancels the integral action when the offset exceeds a preselected limit. Beyond this limit only proportional control operates, but the integral action restarts once the offset has been reduced to within this limit. Batch switches are used mainly in discontinuous processes, and in continuous ones where speed of response is vital to equipment protection, e.g. on anti-surge flow controllers for compressors. Tuning of a PI controller with a batch switch may be very critical.

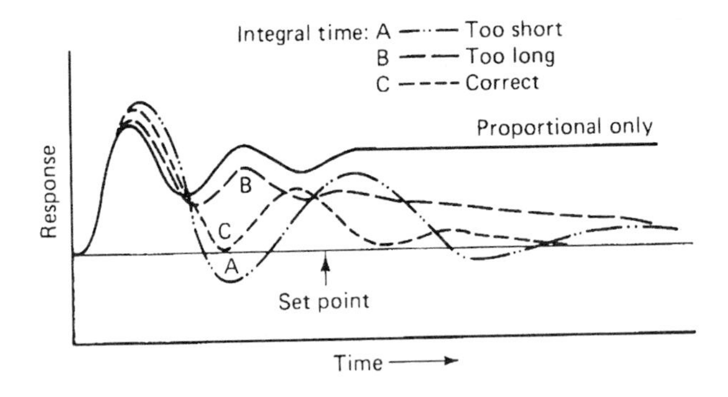

Fig. 17.7 PI system response to process upset with different integral times (courtesy The Foxboro Company)

4. *Proportional plus integral plus derivative control, three-mode (PID)*
Derivative action is a means of speeding up the control action when both proportional and integral modes are used. (It can also be used with proportional control only.) This measures the rate of deviation of the controlled variable from the set point and applies a control action proportional to this rate. It is usually expressed in minutes which represent the reduction in response time which the derivative action gives when added to a proportional controller. The action adds stability to the loop and is useful in cases where the overall lag is rather large or where upsets tend to be large and rapid. It is seldom used in 'noisy' control loops where random fast-moving signals are superimposed on the measured variable. Derivative and integral action are in many ways complementary. Figure 17.8 compares the responses of loops with PI and PID control to a process upset.

Feedforward

Another control system, known as feedforward, reduces the time lag. Here a process variable which would affect the variable to be controlled is itself measured, and changes in it are compensated for without waiting for changes in the controlled variable. If the temperature of a continuous mixer is the main one to be controlled, feedforward control might be applied to compensate for changes in the flow and temperature of its main feed stream (Fig. 17.9). The temperature of the main feed stream but not that of the reactor would then be part of a closed loop. This system requires a computational model of the process which can be pneumatic or electronic and which calculates and applies the correct compensation. Feedforward control works faster than feedback control, but is less accurate if there are other process variables whose changes are not allowed for by the model.

Cascade control

In cascade control a master controller which has to keep a particular variable constant operates by resetting the set point of a slave controller which

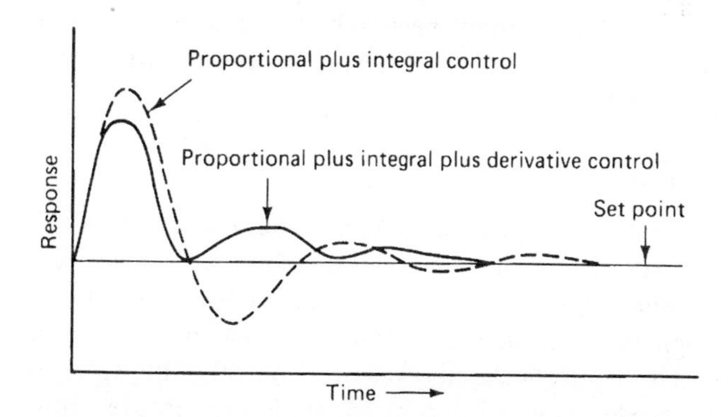

Fig. 17.8 Comparison of system response to a process upset with PI control and PID control (courtesy The Foxboro Company)

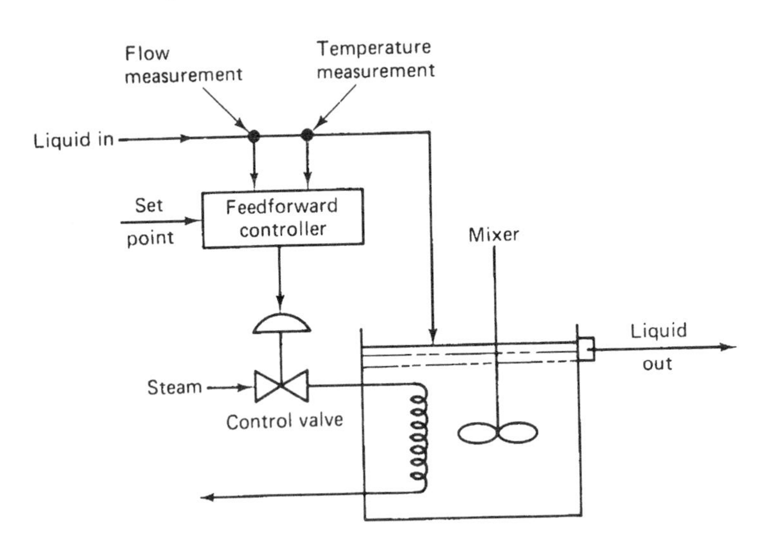

Fig. 17.9 Feedforward control of a continuous mixer (from Perry and Chilton,[6] courtesy McGraw-Hill)

controls another, faster, variable which affects the first one. The feedforward control system just discussed could, for instance, be converted to cascade control by installing a master controller to control the mixer temperature by adjusting the set point of the feedforward controller (Fig. 17.10). Cascade control might, for example, be used to control the (slowly changing) composition at the mid-point of a large distillation column operating with a constant reflux rate and top pressure by a master controller which adjusts the set point of a steam pressure (slave) controller for the column reboiler (Fig. 17.11). Cascade control can produce significant improvements where upsets in the supply variable which affects the main controlled one are large and frequent, where the main variable responds slowly and where the supply variable can be controlled in a rapidly responding loop. Where integral action is included (as is usual) in the primary controller, or in the slave controller, special precautions have to be taken in the link-up of the two controllers to counter the effects of 'integral wind-up', which can otherwise cause serious trouble.

Other compatible control systems[12]

The 'auto-selector' control system is used to keep several variables which are affected by one control valve within safe operating limits. Each variable has an associated controller whose output goes to a device which selects the controller with the highest or the lowest output, provided it is above or below a certain value. This controller then takes over control of the valve. It might, for example, be used on a pipeline pumping station to throttle a normally open valve if the pump suction pressure falls too far, if the motor load gets too high or if the pipeline pressure rises too high. Although somewhat

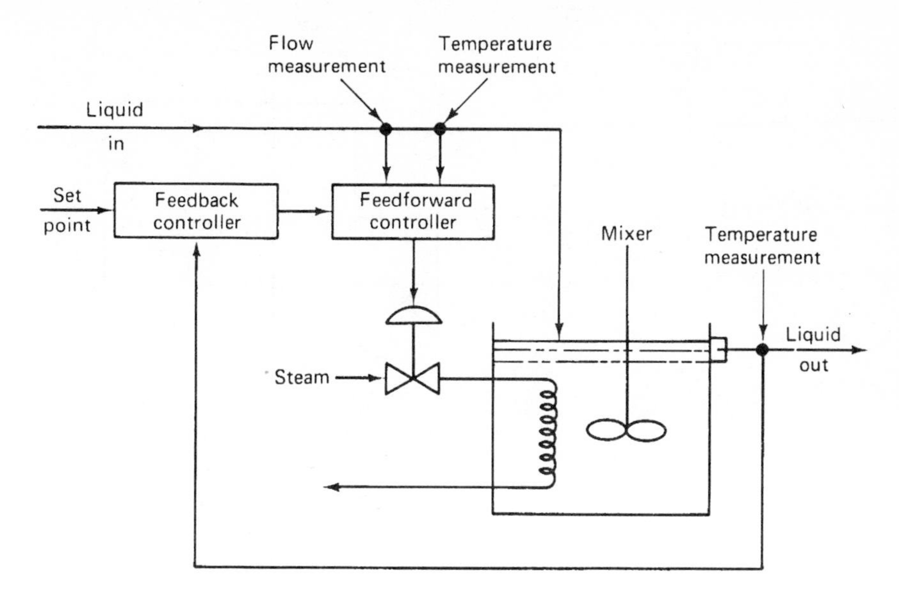

Fig. 17.10 Cascade control of a continuous mixer (from Perry and Chilton,[6] courtesy McGraw-Hill)

complicated, the system can form a useful first line of defence and prevent a complete shutdown if the trouble is a minor one. If, in spite of it, one variable goes further into an unsafe operating area, the use of microswitches may be needed to actuate complete automatic shutdown.

Ratio control is used for blending two flowing fluids where the flow of one fluid is measured by adjusting the set point of a flow controller for the second fluid.

Tuning control loops

The width of the proportional band and the speeds of integral and derivative action need to be tuned finely enough for each control loop to respond adequately over all likely combinations of plant conditions without 'cycling'. The resulting oscillations in the controlled and manipulated variables may lead in complex plant to worse cycling of other variables. This can be a serious hazard, causing pumps to lose suction and pressure-relief valves to blow. Tuning should be done only by trained and authorised persons, and a record kept of the adjustments made. Tuning a proportional controller is largely a matter of trial and error. An empirical method of tuning three- and two-mode controllers is given by Foxboro[12] and summarised as follows.

1. *Tuning three-mode controllers.*
   Step 1: Set the integral time of the controller at maximum and the derivative time at minimum (giving only proportional control). Reduce the proportional band until cycling starts. Measure the natural period (between two successive crests or troughs, Fig. 17.6).

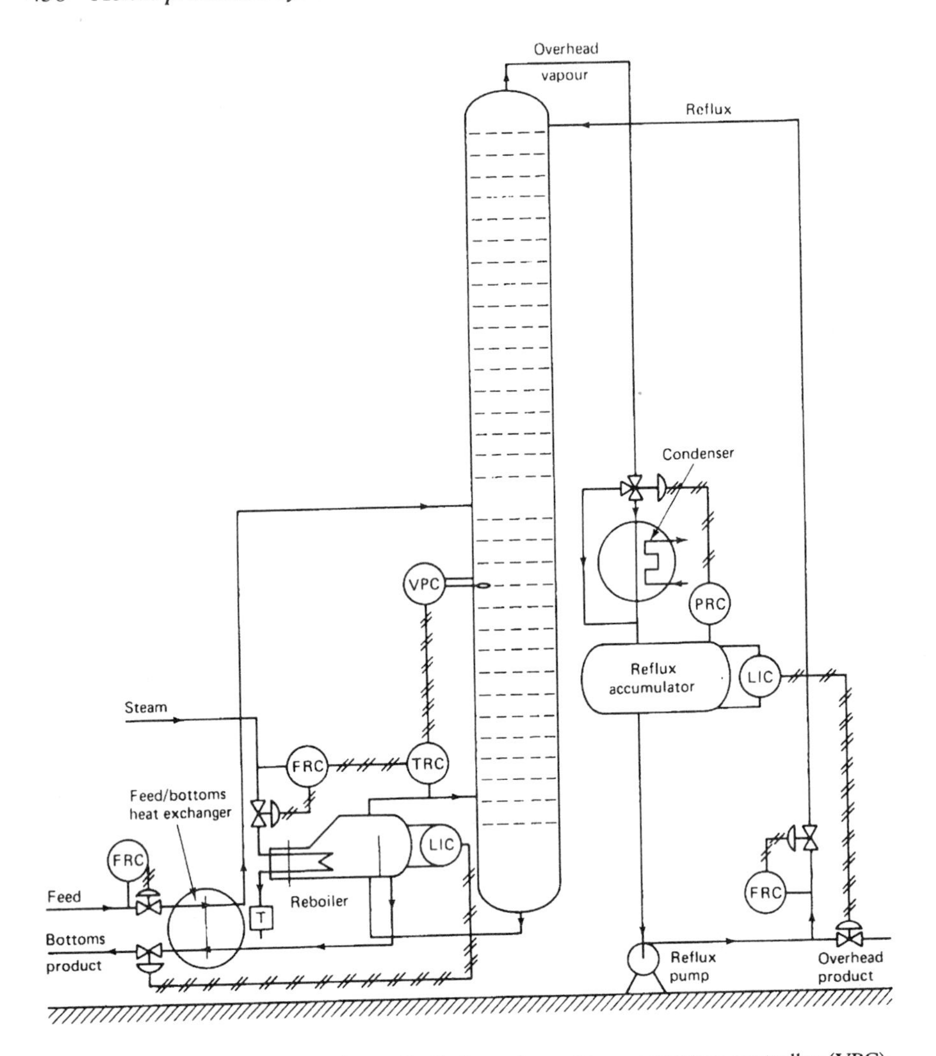

Fig. 17.11 Cascade control of a large distillation column. Vapour pressure controller (VPC) measures difference between vapour pressure of reference substance and column pressure at a sensitive point in the column, and resets bottom vapour temperature controller, which resets stream flow control

Step 2: Set the derivative time at 0.15 times the natural period and the integral time at 0.4 times the natural period and observe the new period. If it is less than 75% of the natural period, reduce the derivative time, but if it is longer, increase it.
Step 3: Finally, adjust the proportional band to give the required degree of damping.

2. *Adjusting two-mode controllers (proportional-plus-integral).*
   Step 1: Same as step (1) for three-mode controller.
   Step 2: Set the integral time to the natural period and observe the new

period which should be 140% of the natural one. If it is longer, increase the integral time.
Step 3: Same as step (1) for three-mode controller. (Adding integral always increases the proportional band required for stable control.)

More sophisticated methods of tuning are sometimes used.[6,10]

When correctly tuned, a disturbance in the process causes the value of the controlled variable to somewhat overshoot its final value at which it should settle after three or four damped oscillations (Fig. 17.8).

A loop which has been in satisfactory operation for some time may start to cycle because a control valve is sticking, or because backlash has developed in an instrument mechanism. Regular lubrication of valve spindles, adjustment of glands and maintenance and checking of all instruments are essential. If despite proper tuning the control is still unsatisfactory, the use of cascade control and/or a valve positioner [17.8.6] should be considered.

In plants with conventional instrumentation, the tuning should be checked under various process conditions so as to give stable operation under the worst combination of flows, temperatures, etc. which may occur. This may result in rather poor responses under normal conditions. In plant where the controllers are incorporated into a PES, it is possible to arrange for the tuning constants to be changed automatically to correspond to the current process conditions.

Remote manual control

Most control instruments can be switched by the operator from automatic to (remote) manual control. Limits should be imposed on the number of control instruments which may be put on manual control at any one time, and on the length of time during which a controller may remain on manual control. The compatibility of various combinations of automatic and remote manual control should be checked. The switching of one control instrument from automatic to manual control may upset the operation of other control instruments. This can cause special problems in PES-controlled plant.

For critical variables (especially flows of dangerous fluids) which may have to be controlled in an emergency, duplicate and well-separated remote manual control stations in well-chosen locations are often needed so that one can function if the other is knocked out.

Self-actuating controllers

Most self-actuating controllers are small, self-contained mechanical devices which incorporate a measuring element, a simple adjustable control system and a valve whose stem is actuated by the force of the process fluid balanced against a spring or weight. They do not usually show the value of the controlled variable. These controllers are often used on simple plants for control of pressure, temperature, flow and level, and on more complex plant in auxiliary control functions (e.g. P&V valves for nitrogen blanketing of vessels). They are mostly of on-off or proportional type. It is usually difficult to incorporate 'fail-safe' features into them [17.7.2].

Power fluidics[17]

One common feature of the 'power fluidic' devices which have been developed for control and other applications (which include pumping, ventilation, precipitation and phase contacting) is that they have no moving parts and should hence be more reliable and require less maintenance than conventional devices. Their main applications to date appear to be in the processing of nuclear fuels and other very hazardous materials. Their control applications include fluid diversion, flow restriction and pulsing. The principle of a power fluidic diverter is shown in Fig. 17.12. The main limitation of power fluidics as applied to control valves is that it is seldom possible to achieve a positive shut-off without employing some moving part.

17.8.4 Measuring and sensing process variables

Process variables include the physical temperatures, pressure, flow and level, physical properties such as density, viscosity, thermal conductivity, refractive index, calorific value and Wobbe index of fuel gases, vapour pressure and boiling point of liquids, and chemical composition. Each of these can be measured by several different methods, the choice of which needs careful study. Important considerations include simplicity, speed of response, accuracy, reliability, the consequences of failure (e.g. safe or unsafe), cost, standardisation and familiarity by the operating organisation, ease of installation and servicing. In the writer's experience, more problems were caused by faulty installation and servicing and by inappropriate choice of instrument than by failure of instruments themselves.

Clear thinking is needed about what should be measured and how this relates to other measurements. One should consider the principle of the

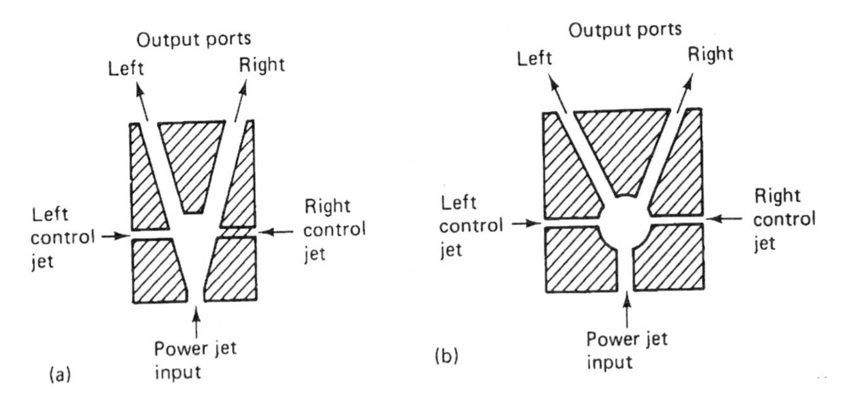

Fig. 17.12 (a) 'On-off and (b) proportional fluidic controllers (from Perry and Chilton,[6] courtesy McGraw-Hill)

meter and whether it measures the required variable directly, or whether it measures some related variable and converts it. Thus some thermometers measure the vapour pressure of a liquid and convert this to a linear temperature scale, whereas the vapour pressure itself may be a better control parameter than the temperature. Some flow meters measure volume flow (L^3T^{-1}), some measure mass flow (MT^{-1}), while variable head and variable area meters measure volume flow divided by the square root of the fluid density ($L^{1½}M^{-½}T^{-1}$). If the fluid density is liable to change and the figure required is either mass flow or volume flow, a continuous density meter and a computing device would be needed as well if a variable head or variable area meter were used. The flow of a fuel gas measured by such a meter is, however, directly compatible with its Wobbe index and not its calorific value (J/m^3), since

$$\text{Wobbe index} = (\text{calorific value})/(\text{gas density})^{½}$$

Continuous measurement of calorific value of a gas involves measuring its Wobbe index and continously compensating this for changes in its density. Thus to control the heat output of a large burner by controlling the volumetric flow of the fuel gas by an orifice meter, as well as its calorific value, means putting opposing compensators for gas density into each meter so that their corrections balance out.

Measurement of the physical variable is next discussed followed by a very brief discussion of the measurement of physical properties and chemical composition.

Temperature measurement[18]

Thermocouples and electrical resistance thermometers are commonly used for direct intrinsically safe cable transmission to receivers. Instruments with fluid-filled bulbs and bimetallic strips used for local temperature indication are simple and generally reliable. Fluid-filled bulbs with long flexible metal capillary tubes (which are easily damaged) are used for remote control and recording, usually on small plants with little instrumentation. Radiation pyrometers of various types are used for measuring temperatures of hot visible objects. Temperature-sensitive pigments are available as paints and self-adhesive labels which change colour reversibly or irreversibly at known temperature up to 350°C in steps of 5°C. They are useful for monitoring the temperature of bearings, etc. and showing the temperature at various levels of a burning oil tank.

Thermocouples give a small electromotive force (e.m.f.) between the working junction in the plant and the cold junction in the receiver. The receiver includes a device which compensates for variations in the cold junction temperature. Several combinations of thermocouple wires are available for temperatures from –250 to 2600°C. The combination must be suited to the corrosivity and temperature range of its working environment. Only couples whose e.m.f. increases continuously with the temperature difference between the cold and hot junction over its working temperature range can be used. With strip recorders the e.m.f. is measured by a potentiometer, but for temperature indicators a galvanometer or an electronic amplifier is often used. A failure in the thermocouple circuit (e.g. caused by breakage of the welded hot junction) causes a drop in e.m.f. which results in a lower apparent temperature than the real one. Thus it fails in the unsafe mode.

Since most thermocouples are fragile and liable to chemical attack, their working junction is generally enclosed in a metal sheath, or inserted in a thermowell on the plant, or both. These reduce the speed and accuracy of the measurement, particularly of gas temperatures. A bare hot junction may reach thermal equilibrium in less than 30 seconds, but may take several minutes to do so in a thermowell. If the closed end of the thermowell is its lowest point and the temperature is not too high, the time lag may be reduced by the use of a heat transfer liquid in the well.

Platinum **resistance thermometers** are available for temperatures from 15K to 800°C and nickel ones for the range –200 to 350°C. Their resistance increases with temperature, so that any failure of the resistance circuit gives too high a temperature reading and their failure mode is safe.

Thermistors with a negative temperature coefficient are available for measuring temperatures over five ranges between –100 and 300°C. A break in the resistance gives an unsafe failure mode. Thermistors with a positive temperature coefficient are also available and used to protect the electrical windings of motors and transformers.

Resistance thermometers are protected in the same way as thermocouples, and the same problems apply.

Pressure measurement[19]

Pressures are measured as absolute, relative to a perfect vacuum, or as gauge, relative to the surrounding atmosphere. While devices measuring gauge pressure are simpler than those measuring absolute pressure, gauge pressures are often converted to and quoted as absolute pressures [3.1.1].

There are three main types of pressure-measuring devices: those based on the distortion of an elastic element, electrical sensing devices and manometers. Measurement of low absolute pressures is not discussed here.

All pressure-measuring devices except manometers need regular checking and calibration, generally against a 'dead-weight' tester. Their performance varies widely, not only as a result of their basic design and materials of construction but also because of their service conditions. Care should be taken to ensure that the element can withstand the maximum possible process pressure. Errors can arise from hysteresis, corrosion, changes in temperature, friction, backlash and pulsations. An isolating valve is generally needed between the pressure-measuring device and the process. Elements should not be mounted where they are subject to vibration or extremes of temperature. Pulsations in the process pressure should be damped by a restriction between the process and the measuring device. Any pressure element has its own natural frequencies of vibration, and if one of these coincides with the frequency of a process pulsation, it is likely to be quickly damaged.

Elastic elements include Bourdon tubes, spiral and helical elements (Fig. 17.13), bellows and diaphragms. The effects of temperature changes are minimised by using elements made from an alloy with a low coefficient of thermal expansion. Elements from which the process fluid is excluded by a thin diaphragm are available for corrosive and fouling process fluids.

Locally mounted Bourdon tube gauges are widely used for pressure and vacuum. Bellows are widely used for transmitters and receivers of pneumatic signals. Diaphragm elements are of two main types, those which depend on the elastic properties of the diaphragm and those which are opposed by a spring.

Electrical sensing elements include strain gauges and piezoelectric transducers, all of which lend themselves to electrical transmission of the signal by intrinsically safe circuits to potentiometer-type receiving instruments.

An electrical strain gauge uses a fine electrical resistance wire or a semiconductive wafer which is attached to an elastic element and stretches with it, its resistance increasing with the strain. If the electrical circuit of a strain gauge is damaged or broken it would indicate a higher pressure than the true one and fail safe.

Piezoelectric transducers which generate a potential difference proportional to a pressure-generated stress are used for measuring rapidly changing pressures but are not usually suitable for measuring fairly static ones.

Manometers are based on the height of a liquid column, usually mercury, water or an organic liquid of low volatility. They are simple and reliable and mainly used for measuring small differences in pressure, up to a maximum of about 2 bars. There is a heath risk with mercury manometers but although once common, they are now little used in industrial instruments.

Flow measurement[20]

The term 'flow' usually refers to the volume of fluid passing in unit time, the term 'mass flow' being used for mass of material passing in unit time. Here we are mainly concerned with the flow of a single liquid phase or a gas.

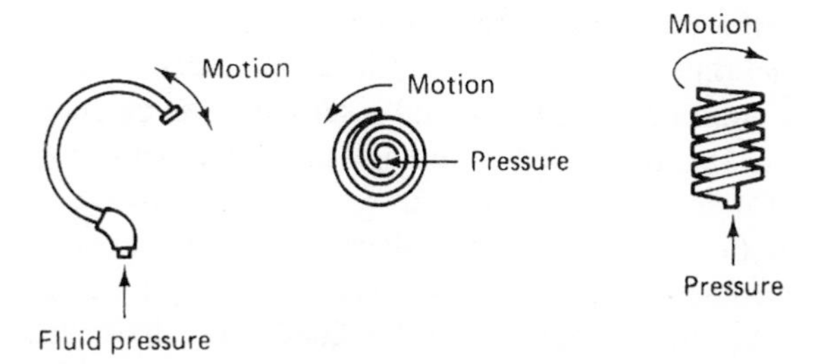

Fig. 17.13 Bourdon tube, spiral and helical pressure elements (Foxboro, Figure 24)

The main classes of flow-measuring devices used in the process industries are variable-head, variable-area, positive displacement, turbine, electronic flow and mass flow meters, and weirs and flumes for flow in channels. Variable-head and variable-area devices are commonly used for clean liquids of low viscosity and gases flowing in pipes. Corrections have to be applied for changes in fluid density and sometimes viscosity. For this reason, the gas density needs to be controlled at the constriction producing the differential head, while only small temperature changes can be tolerated at this point.

A **variable-head** device has a fixed restriction with upstream and downstream pressure tappings which allow the differential head to be measured with a differential pressure (d.p.) cell and transmitted to a receiver. As the flow is proportional to the square root of the differential pressure, these instruments show a maximum sensivity at their full-scale reading while their sensivity at 10% of maximum flow is low. 'Square root extractors' are often used to convert the differential pressure into a signal proportional to the flow before transmission. Orifice plates are the most commonly used form of restriction while Venturi and Dall tubes have lower energy losses. Errors arise through deposits of solids on either side of the restriction and through the presence of gas or a second liquid phase in the lines between the pressure tappings and the d.p. cell.

The most common form of **variable-area** meter is the rotameter, a tapered vertical tube with a linear scale and a plumb-bob spinner (with a higher density than the fluid) which moves up and down the tube as the flow increases or decreases (Fig. 17.14). The area between the spinner and the wall of the tube is proportional to the height of the spinner in the tube, and the flow is nearly proportional to this height, giving an almost linear scale. Transparent glass tubes are used for local metering of small flows, and metal tubes with electromagnetic sensing of the position of the spinner are used for larger flows and dangerous fluids, and allow easy signal transmission.

Several types of **positive displacement** meter are used to give the total volume of fluid which has flowed since the last reading. They are mostly of rotary type, of high accuracy, and are little affected by pulsations in flow and by changes in fluid density or viscosity. They are mainly used for accounting purposes and dispensing known quantities of liquids and not for flow control.

A **turbine** meter consists of a tube with an axially mounted turbine wheel rotating between almost frictionless bearings. The rotational speed of the wheel is proportional to the volumetric flow rate and is sensed by an electric pick-up coil outside the tube. It can be used as the measuring element in a flow control loop as well as for accounting purposes.

Several types of **electronic flow** meters have been developed over the past two decades. Of these, electromagnetic flowmeters have special uses, being able to measure dirty liquids, slurries and pastes provided these have an electrical resistivity greater than 1 microhm/cm. Their accuracy is largely unaffected by changes in temperatures, viscosity, density or conductivity.

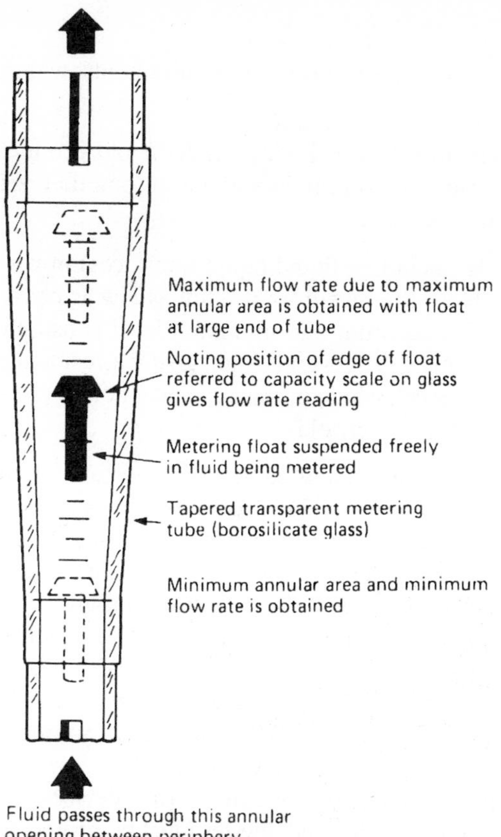

Fig. 17.14 Variable area meter (Rotameter) (courtesy Fischer and Porter Ltd)

Ultrasonic flowmeters based on the Doppler effect and on the transmission of an ultrasonic pulse through the flowstream also have considerable promise, as do those based on the frequency of vortex shedding from a bluff body in the flowstream.

Mass flow meters are of two types: true mass flow meters which measure the mass flow directly, and inferential mass flow meters which measure both a combined function of the mass flow and the density itself and compute the mass flow from the two measurements. Several rather sophisticated types of true mass flow meter have been developed, including ones based on angular momentum in the fluid and on the Coriolis effect.

Weirs depend on the flow of liquid through a shaped notch to a lower level. The shape of the notch may be triangular, rectangular, trapezoidal, parabolic, etc. **Flumes** are used where there is not enough liquid head for a weir or where the stream carries much suspended matter.

Level measurement[21]

There are two circumstances with rather different requirements where levels are measured or sensed:

- Storage of liquids and particulate solids including the feed and product tanks of processes. Here accurate measurement without automatic level control is needed for accounting purposes.
- For the control of liquid levels (including liquid/liquid interfaces) in vessels of continuous processes. Here the inventory is often as low as possible, the difference between the maximum and minimum level is usually small and the flow of liquid through the vessel is high, giving a residence time of only a few minutes. The level, which in this case may not be critical, is measured or sensed in order to control it.

In both cases high- and low-level alarms and sometimes trips are often needed to prevent accidents. In the case of level instruments one can rarely speak of fail-safe or unsafe since any failure is often unsafe.

Methods of liquid level measurement include dipsticks and tapes, gauge glasses, float-actuated devices, displacer devices, head devices and electrical methods based on electrical conductivity and dielectric constant. Radioactive level gauges are used to some extent for measuring the depth of solids in silos. While these are adequately shielded in use, they may augment fire risks since nobody usually knows what has happened to the radioactive source if the silo is involved in a fire, and firefighting is thereby inhibited.

Some useful guidelines on the selection of level measuring devices are given by Sydenham.[21] The floats of float-operated devices must be protected against the forces of moving liquids. Backlash and leakage must be avoided with the glands of lever and shaft mechanisms. Head devices have the advantages of no moving parts in contact with the liquid and of giving the weight of liquid in a tank without correcting for its density. Specially designed head device systems are needed for boiling liquids.

For critical applications it is sound practice to use a device based on one principle for level measurement and/or control and one based on a different principle for safety alarms and/or trips.

For measurement of liquids in storage, the traditional **dipstick or tape** is still widely used, since it is simple, accurate and usually reliable. For light petroleum fractions there is here a risk of discharging static electricity and igniting a vapour/air mixture in the tank if appropriate precautions [10.5.1] are not taken. There may also be other hazards to the dipper such as toxic vapours, slipping on a tank roof on a windy night and even falling into the tank (as happened to a man whom the writer knew).

Although special **gauge glasses** are used for quite high pressures and temperatures, they should never be improvised and only complete assemblies from reputable suppliers should be used, within their design specifications for temperature, pressure and process liquid. All gauge glasses should have isolation valves or cocks and drain valves, and those used under pressure or for dangerous liquids should have safety devices which will cut off the flow if the glass breaks. It is good safety practice to avoid gauge glasses entirely on vessels containing flammable and other hazardous liquids.

Physical properties and chemical composition

It is impossible to discuss here the many different methods and automatic measuring instruments available. It should be stressed that for most control purposes, and especially for safety, simplicity, speed of response and reliability are more important than absolute accuracy. Measurement of a physical property is usually simpler than chemical analysis and for process streams consisting essentially of two components, the composition can be inferred from an appropriate physical property such as density, viscosity, refractive index or thermal conductivity.

Continuous sampling of process streams can present serious problems, e.g. in securing a representative sample, in reducing time lag, in removing interfering materials such as water from the sample and in disposing of toxic or flammable sample streams. According to Giles,[22] 'Analytical instruments are out of commission more frequently due to troubles in the sampling system than to any other cause.' Whereas the instrument itself may have been developed over many years, the sampling system is often designed on a 'one-off' basis by someone who only understands part of the problem. A complete system contains the following five elements: the sample probe, the sample transport system, sample conditioning equipment, the analyser itself and sample disposal. The subject is treated in detail by Cornish.[23]

Monitoring for safety

Continuous analytical control is sometimes vital to safety. An example is the continuous monitoring of the vinyl acetylene content in a butadiene-purification column. Another is the monitoring of the oxygen content of the mixed feed gases to a direct hydrocarbon oxidation process. Many process analysers whose main function is to monitor the purity or quality of the product also provide a warning when something goes dangerously wrong with the process.

Other continuous instruments detect fires and flammable and toxic gases in the atmosphere. Several principles are used for automatic fire detection, i.e. detection of flame (by IR and UV methods), of heat and of smoke [10.6]. The choice of method and location of detectors require case-by-case study, not only to ensure rapid response but also to avoid false alarms. Most flammable gas detectors are based on detecting the rise in temperature of an electrically heated capsule of a combustion catalyst (a 'pellistor') in contact with air containing the flammable gas. Designers of such detection equipment have to overcome two problems with conflicting requirements, (1) to ensure that the detector does not become a source of ignition and (2) to make certain that it responds fast enough for effective action to be taken [10.6].

Errors in measurement

Some measuring instruments are prone to errors which may develop gradually while the plant is operating and which cannot be corrected until it is shut down. These may be due, for example, to deposition of carbon or polymer on a thermocouple pocket or on the moving vanes of a turbine-flow meter. Much ingenuity has been spent in finding ways of living with and allowing for this problem. Computer programs have been developed for reconciling the simultaneous outputs of all relevant plant instruments, e.g. by means of mass balances, showing which ones are most likely to be in error and suggesting correction factors to be used. This 'solution' is most applicable on plants under PES control.

Instrument errors also result from condensation and deposition of solids in connections and short lines between plant and transmitting instruments, and in the instruments themselves, and in compressed air signal lines. To avoid the former, inert gas purges, diaphragms and steam or electrical tracing are sometimes needed, and to avoid the latter, precautions are required to ensure that the compressed air remains clean and dry.

17.8.5 Receivers

Controllers and their tuning were discussed in 17.8.3.

Marking of instrument scales

Normal (safe) and unsafe operating ranges should, where possible, be marked prominently on instrument scales. There is a distinct danger with modern recorders which simply have a linear scale with graduations of 0–100, of operators, supervisors and even plant managers thinking of these as mere numbers, without being clear what temperatures, pressures, etc. they represent, or knowing, for example, whether the figure for a pressure being controlled is in gauge or absolute units. Conversion factors with unambiguous units should be clearly marked on all receivers.

Ergonomic considerations[24]

Instruments on a control panel should be arranged logically for the operator and the most critical instruments should be well within his or her field of view. 'Mimic diagrams' (basically enlarged flowsheets [15.10.2]) on the control panel, with colour-coded lines and lights to show the condition of the process, can greatly assist operation. (Similar but more versatile and interchangeable diagrams are shown on VDU screens at the console control stations used with PES control.)

The accuracy of instrument reading should be appropriate to the need. Too fine subdivisions increase reading errors. These are less likely with digital displays than with pointers and dials, but the latter are better for rough checks and for assessing rate of change. The best arrangement for reading accuracy is a moving dial with a fixed pointer. The next best is a circular or semi-circular scale with a moving pointer. Horizontal and vertical scales are most prone to reading errors.

Instruments should react quickly enough to show expected movements and be sensitive enough to show the smallest meaningful change in the variable measured.

Where the readings of several dials have to be checked, e.g. the temperatures or flows of several parallel streams, all pointers should point in the same direction for the same plant condition. Identical instruments and scales should be used for identical duties on parallel processes.

Dials and indicators should be logically laid out and related to the control involved. If the dials and their controls cannot be placed side by side, they should be arranged in a pattern where they relate to each other.

Controls should be grouped according to their function and to the part of the plant which they affect. Those which have to be operated in sequence should be placed near each other and, if possible, in their order of operation, even though this leads to an asymmetrical layout. Different types of control knob should, if possible, be used for different functions.

The fact that most of an operator's bodily movements result from habit, and from subconscious rather than from conscious thought, must be appreciated by the designer. Most manual skills depend on the development of reflexes. Thus the arrangements of manually operated valves, switches, and gauges for similar equipment should follow similar patterns. Errors readily creep in when there is 'an odd one out'. Habits die hard, and most of us become aware of the extent to which we are creatures of habit when a small change in our domestic arrangements (such as the position of a light switch) interferes with a long-established routine.

Alarms and trips

All critical points of operation in a plant are normally protected by alarms and/or shutdown devices which are actuated by microswitches triggered by high or low pressures, flows, levels, temperatures, etc. An audible/visual alarm is actuated when the variable deviates from normal and reaches a certain figure, to allow the operator to take corrective action. If this is unsuccessful and the variable deviates further to reach another figure, a shutdown device may be actuated, which, by means of solenoid valves in the appropriate instrument air lines, shuts down one or more sections of the plant. The alarm consists of a horn or other audible device and lights which appear at labelled positions on an annunciator board mounted on the instrument panel.

The operator can usually stop the audible device, but cannot switch off the light completely, although he or she may be able to alter it in some way (e.g. change from flashing to continuous) so that it is not confused with a new warning signal from another point in the plant which appears on the board. All warning lights continue to show until the variables which set them off have returned to their normal range. More sophisticated alarm display and cancelling arrangements are available with PESs.

Trouble shooting and checking

When some malfunction, perhaps of an instrument, develops suddenly on a highly interactive plant, it generally actuates an alarm for a particular part of the plant. It is then often surprisingly difficult to pinpoint the source of the malfunction and distinguish between cause and effect. Misconceptions readily arise.[9]. The plant has the same propensity to disguise its ailments as the human body, where even doctors can be misled, and such dictums have arisen as 'for pain in the knee, treat the spine'. Confusion is compounded by the simultaneous blowing of a pressure relief valve, with an ear-splitting roar. The near-disaster on the Three Mile Island atomic power plant in the USA was a classic example of this kind of difficulty in fault diagnosis.[25]

On PES-controlled plant it is possible to provide software programs to help operators in rapid fault diagnosis. These programs are designed by experts who have carefully studied the various faults which might develop and how to diagnose and handle them.[26] Even here the operators generally need special training. For this special process simulators with similar VDU- and keyboard-equipped control stations are being increasingly used[27] [21.3.5].

To avert complete plant shutdowns because of minor malfunctions, the panel instruments should allow operators to take some holding action, such as putting the plant on total recycle or columns on total reflux, while the problem is being investigated.

Procedures must be established between operating and instrument department personnel for the calibration, zeroing and other necessary testing of instruments carried out while the plant is running. This generally involves temporarily putting each control instrument in turn on remote manual control and similarly de-activating alarms and trips. Clearly marked tags or notices should be displayed on any panel instruments which are out of service. Verbal and written warnings of this must be given to incoming shifts.

17.8.6 Final control devices[6,10,12]

The final control device consists of two parts, an actuator and a valve or other mechanism (such as a variable-output pump) which adjusts a flow or other manipulated variable. The actuator translates a signal, usually pneumatic, electric or hydraulic, into a force which operates the valve, etc. Pneumatic signals usually lie in the range 0.2 to 1.0 bar g. and are transmitted through 6 mm tube. With pneumatic actuators the air pressure acts on a diaphragm attached to a stem and is opposed by a spring. Two versions are available, one which opens and the other which closes the valve with increasing air pressure (Fig. 17.15). Ideally, the actuator should respond quickly and assume a position proportional to the signal pressure. In practice the action may be sluggish and the position reached may deviate from linearity. The delay is caused by the time taken for sufficient air to travel through the transmission tube to pressurise and enlarge the space behind the diaphragm in order to move it. This delay increases with the volume of the diaphragm chamber and the length of the transmission line, and typically amounts to 5 seconds for a line length of 100 m and a terminal volume of 3 litres. Lack of linearity may be caused by the force of the process fluid on the valve plug and friction in the valve gland.

To overcome these problems valve positioners and signal boosters are often used. The pneumatic valve positioner uses a separate air supply to operate the valve and adjust its position to correspond exactly to the signal pressure.

Direct-current electrical signals are generally converted to variable air pressures by electro-pneumatic transducers mounted on the control device. A typical electro-pneumatic actuator is a combination of a current-to-pressure transducer, a feedback positioner and a pneumatic spring-diaphragm actuator.

Hydraulic actuators employ a fluid acting on one or both sides of a piston. In the first case the force of the fluid is opposed by a spring and the operation is similar to a pneumatic actuator.

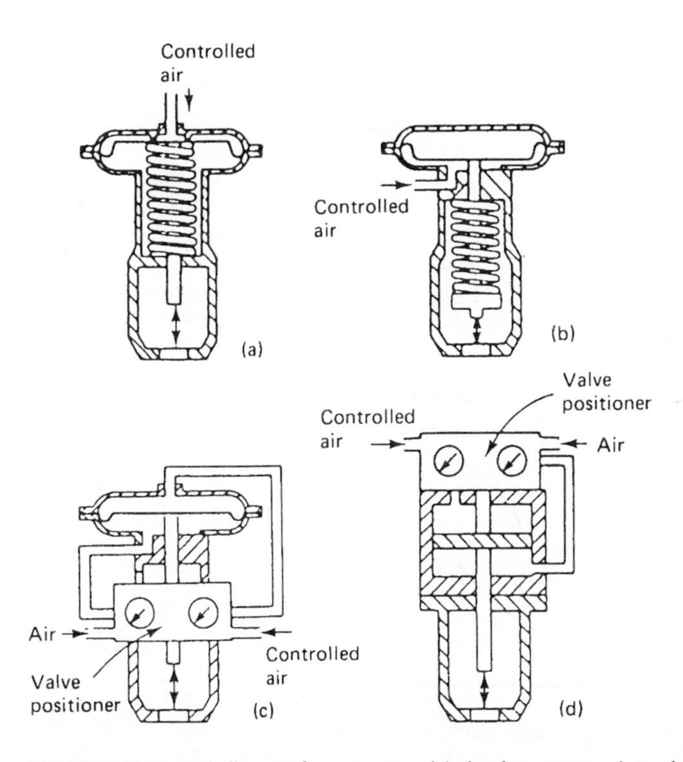

Fig. 17.15 Pneumatic linear valve actuators: (a) simple actuator, air to close; (b) simple actuator, air to open; (c) actuator with positioner, air to close; (d) actuator with positioner, air to open (from Perry and Chilton,[6] courtesy McGraw-Hill)

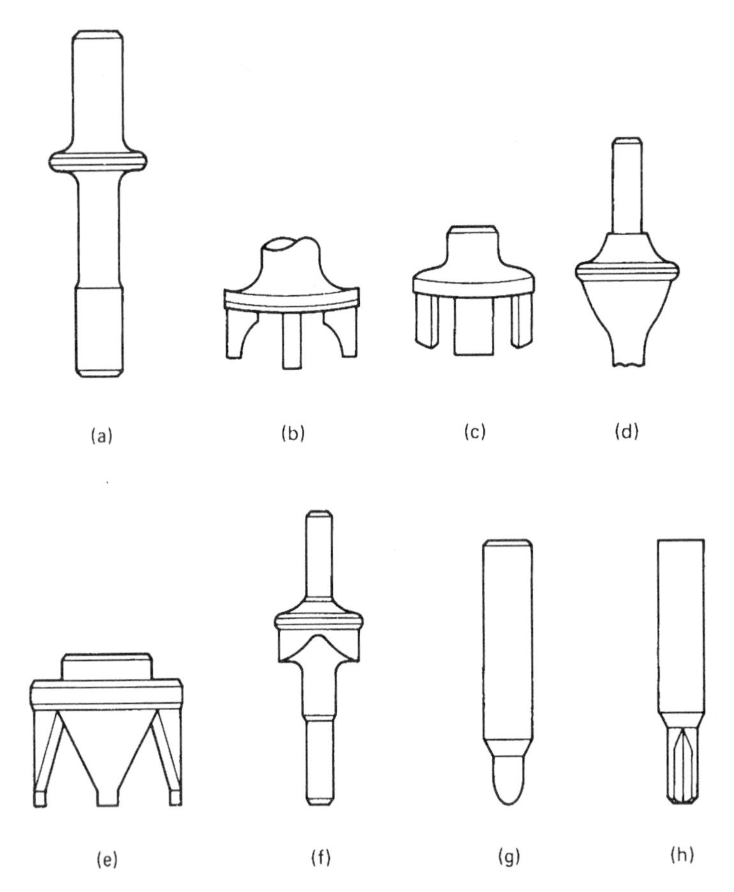

Fig. 17.16 Shapes of typical valve plugs. (a) Top and bottom guided single-port quick-opening; (b) port-guided quick-opening; (c) rectangular (linear) port; (d) throttle plug (modified linear); (e) V-port (modified linear); (f) equal-percentage V-port; (g) miniature throttle plug (equal percentage); (h) miniature fluted plug (equal percentage) (from Perry and Chilton,[6] courtesy McGraw-Hill)

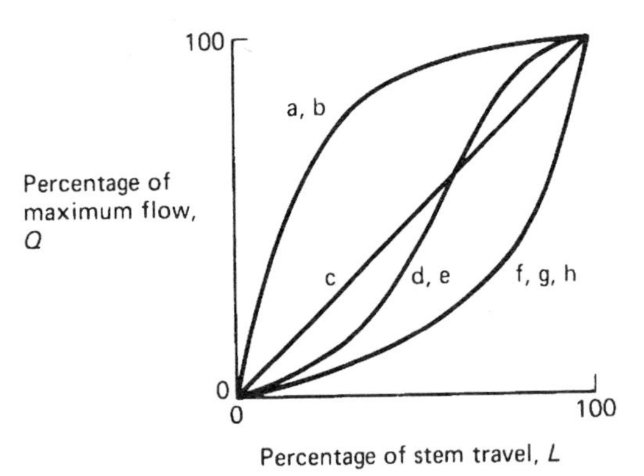

Fig. 17.17 Flow characteristics of valve plugs shown in Figure 17.16 (from Perry and Chilton,[6] courtesy McGraw-Hill)

Purely electrical actuators are also used in which an electric motor or solenoid provides the driving force. These can easily be used to open or close a valve completely, but those which adjust their position to correspond to an electrical signal are more complicated.

A wide choice of valves and valve plugs, some of which are illustrated in Fig. 17.16, is available for operation by pneumatic and other types of actuator. These give a choice in the flow characteristics of the valve which are shown in Fig. 17.17 as a plot of the percentage of maximum flow against the percentage of stem travel. Some control valves do not give a positive shut-off of the process fluid unless this is specified, when they can usually be provided with soft seats. For good control the size of a control valve is usually considerably smaller than the pipe where it is installed. Valve manufacturers will advise on the best type and size of control valve for different applications.

Actuators can also be used to control the speed of variable-speed motors which are used, for instance, with solids-metering valves and for variable-output pumps used for slurries and metering small flows (e.g. chemical additives and catalysts). Most of these are either constant-volume pumps operated by variable-speed drives or variable-volume pumps operated by constant-speed motors.

Formerly a hand-operated by-pass valve was usually installed in parallel with an an automatic control valve for use while the controller was being adjusted or the control valve was being serviced. This requires a local indicator of the controlled variable within view of the human by-pass operator. A manual by-pass valve might also be installed to assist in start-up or to reduce the sensitivity of the control valve.

A by-pass valve can, however, cause hazards, particularly if the control valve 'fails-safe' in the closed position. Such by-pass valves can generally be avoided, e.g. by a better choice of control valve. The installation of a manual by-pass valve round any control valve thus requires special justification, e.g. by a simple hazard analysis [14.4].

Isolation (block) valves on one or both sides of an automatic control valve are more often needed, particularly where the control valve does not give a positive shut-off. For very critical duties the duplication of the entire control loop, valve included, with block valves on either side of each control valve, may be justified on grounds of reliability.

17.8.7 Signal-transmission systems

Most telemetering systems used in process control involve fixed electrical or pneumatic conductors. Apart from thermocouples whose millivolt output is transmitted directly by wires to the receiver, most instrument outputs are transformed into electrical or pneumatic signals, which in the case of receiver instruments, are reconverted to a numerical value representative of the measured variable.

Analogue signals are used exclusively with pneumatic and often with electrical transmission, while digital signals are also used with electrical transmission. Analogue electrical signals which are now usually of d.c. current type with the range 4–20 mA are carried by wires, as are voltage signals.

Digital signal transmission within a PES can be by frequency change or pulses, and sent by coaxial cable, optical fibres or short-wave radio. Complex information can be sent rapidly in these ways. Binary codes are generally used, with electronic coders and decoders.

Electrical 'noise', that is, unwanted signal, can be a serious problem with electrical signal transmission, particularly with digital signals. A false or distorted signal in a PES-controlled plant can result in the wrong opening or closing of valves, with resultant hazards. The main sources of electrical noise are electrostatic and electromagnetic fields, instruments and sensors, and grounding problems. The subject lies outside the scope of this book and requires expert attention during plant and system design.

17.8.8 PES control

PESs may be used as an aid to the operator, e.g. by logging data and performing calculations, or for automatic process control. Here there are two main alternatives, both supervised by the operator:

1. using the PES to adjust the set points of conventional controllers;
2. by direct PE control, where all controllers are part of a PES.

With PESs, the panel indicators, recorders and controllers used with conventional instrumentation are replaced by VDUs and keyboards at the control station. Instead of scanning a row of instruments on a panel the operator sees a more logical representation of present and past values of process variables on a VDU. By touching selected spots on the screen he or she can call up other 'windows' onto the screen which provide more detailed and appropriate information which is needed about the state of the process. Instead of tweaking a knob on a conventional controller to change its set point, the operator types abbreviated commands on the keyboard, with the assistance of 'help' menus on the VDU screen in case he has forgotten the procedure laid down in the operating manual [20.2.2]. At a simple command the PES will alter the set point for him gradually over a period of an hour or so, saving him the tedium of altering the set point a little at a time every few minutes. The PES will also arrange the 'bumpless' introduction of cascade or manual control. But if something goes wrong with a screen and his 'window' onto the plant is lost, he may be in for serious trouble. This is when he is in most need of traditional operating skills.

The advantages claimed for PES control are mainly economic and apply particularly to large-scale process plant and complexes.

A hierarchy of control systems exist when PESs are used, those at a lower level receiving instructions from the one above. Systems at the lowest level control unit operations involving individual plant items such as heat-exchangers and dryers. Next comes the control of individual process units each containing several plant items whose operations must be co-ordinated to meet specific objectives such as product purity, yields on raw materials and production rate. The third level, plant control, involves the co-ordination of all process units in the plant to achieve the required performance at minimum cost. The PES will adjust the control parameters of other units to cater for upsets in the operation of any particular one. On a fourth level is the control at a single location of several interdependent plants (such as a petrochemical complex) to optimise the yields and qualities of several products within the constraints imposed by the plants, processes, raw materials and utilities available. Above these levels come department and company control. Although computers are used here to assist decision making, such decisions fall outside automatic process control and are made by humans.

One problem with PES control is that the variables controlled at higher levels include chemical compositions of process streams as measured by on-line analytical instruments which have higher failure rates than most other types of instrument. In normal operation the analytical controller may reset the set points of other controllers in cascade fashion, but if it develops a fault the operator must take the analytical controller 'off-line' and reset the set points of the slave controllers manually in the light of the results of laboratory analyses.

PES control improves the performance of unit operations hitherto controlled by conventional instrumentation in cases where uncontrollable disturbances in the process occur without compensation, and where the variables controlled are of only indirect interest. With PES control systems it is possible to calculate, specify and control performance variables which previously could be neither measured nor controlled. A further advantage with direct PES control is that it can automatically adjust the tuning of individual controllers [17.8.3]. The PES generally allows the number of uncontrolled variables or degrees of freedom to be considerably reduced.

PESs used for control are now almost exclusively of digital rather than analogue type. They are used for on-line control of both batch and continuous processes, for process simulation, as a diagnostic aid to human operation, for logging, and monitoring operating variables and for actuating special protective features to safeguard the plant in particular emergencies. They range from single microprocessors to mainframe computers. Although they can greatly increase control efficiency and safety, they pose additional and different safety problems compared with non-electronic control and protection equipment. By distributing programmable electronic devices and personal control stations and incorporating a degree of redundancy among them, the consequences of any single failure can be reduced and fault location can be made easier. A high standard of safety and reliability, which should be at least as high as for non-PES systems, is needed when equipment is controlled or protected by a PES.

HSE has published guidelines on PES in safety-related applications[28] in two parts, a general introduction for the non-specialist and technical guidelines which are mainly intended for specialists. The Engineering Equipment Material Users Association (EEMUA) has also published 'companion guidelines'.[29]

17.9 Features of PES systems used for control and safety[25]

The first feature which strikes the novice is the strange, new vocabulary used by the initiated. Most readers will be familiar with the terms 'hardware' and 'software'. Hardware refers to all the physical components of the PES system, the chips, bits of wire, keyboards, VDU screens, printers, etc. Software refers to the instructions which the system follows. Some software is incorporated into the PES, while other software consists of programs written in electronic language on hard and 'floppy' discs, magnetic tape, etc. The software can be introduced into the 'programmable electronics' (PE) of the machine which is, in effect, its brain. The PE communicates with the installation which is being controlled, and with its human operators, through input and output units (Fig. 17.18). Input units include plant sensors, e.g. conventional temperature- and pressure-measuring instruments and transmitters which send coded messages to the PE unit, and control devices which enable operators to give instructions to it in the same way. The PE analyses the data thus received and transmits instructions and information to the output units. These include plant actuators (e.g. pneumatic control valves), information-storage devices and communication devices (e.g. VDU screens, printers). The input and output units are linked to the PE by electronic interfaces. Information is stored in two types of electronic memory, Read-Only Memories (ROMs) which are mostly built into the machine and Random Access Memories (RAMs) which are used for the

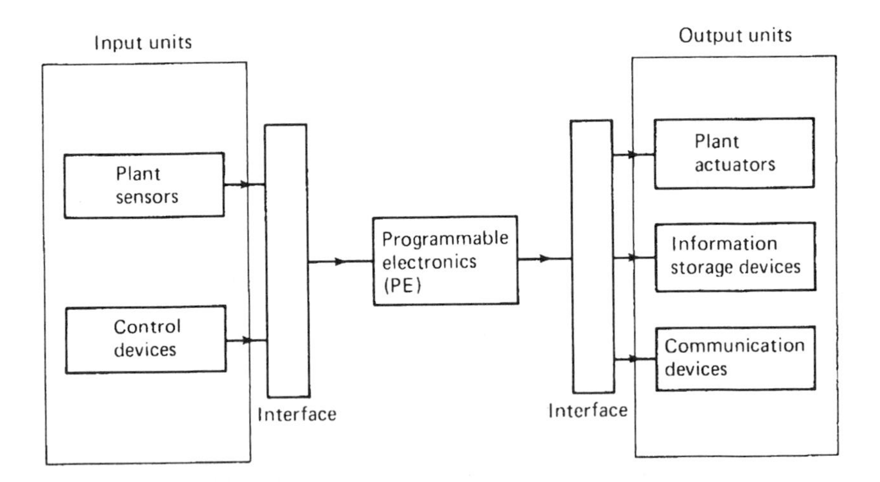

Fig. 17.18 Structure of a programmable electronic system (PES) (courtesy HSE)

temporary storage of information which can be read and used later by the machine. PESs may control process plant continuously through sensors and actuators, or act only in emergencies as revealed by similar sensors, to protect the plant by opening and shutting valves, stopping motors, etc. Possible failures of such PESs are of two types, *random hardware failures* and *systematic failures*.

17.9.1 Random hardware failures

These are caused by the wear-out of one of the large numbers of components in the PES which are assumed to be unpredictable and to occur at random [14]. Redundant back-up components or systems which continue working if one breaks down and control of the physical environment in which the PES operates are used to increase reliability. The PES should then have the means of identifying the failed part so that it can be replaced or repaired promptly to restore the reliability of the system protected by a redundant item.

17.9.2 Systematic failures

These happen every time a particular set of conditions occurs and are due to mistakes in the specification, design, construction or operation of a system. Such failures may lie dormant until the particular circumstances arise and then lead to accidents. Three types of error which can lead to systematic failures are:

1. *specification errors* made when the system was planned;
2. *equipment errors* in design, construction, installation, etc.;
3. *software errors* left in the original program or introduced into it during modification.

To ensure that the safety of the plant or equipment controlled is adequate, five steps should be taken:

1. a hazard analysis to identify the hazards and how they could arise [14.4];
2. identification of the systems on which the safety of the installation depends;
3. determination of the required safety level;
4. design of the systems identified in (2);
5. analysis (e.g. by a check questionnaire) to ensure that the installation meets the safety requirements.

There are three fundamental aspects of an installation which require expert examination and may need improvement: *reliability* [14], *configuration* and *overall quality*.

Configuration means the way in which the hardware and software components of a PES are organised and the ways in which PESs and non-programmable equipment are linked to make up a complete installation. Overall quality depends both on the quality of manufacture (e.g. whether the PE was manufactured using an established quality-assurance system) and on the competence of the designers.

Control and protection systems should be completely separated, with their own sensors and actuators (Fig. 17.19). Although some control systems include protective features, these may fail if the control system fails.

PESs must be protected from accidental damage (by position), from environmental hazards (temperature, humidity, dust and corrosive vapours), from fire, electrical interference and electrostatic effects.

17.9.3 Concerns about the use of PESs for process control and protection

Many of HSE's concerns about the application of PESs to control and protect process plant were voiced by Dr Jones.[30] Failures can occur in both the hardware and the software.

Large companies usually have specialists who can match the process parameters to the control and/or protective system. Medium- and smaller-sized companies who lack such specialists tend to buy control and protective packages 'off the shelf'. Such companies need to call in specialist help to match the process chemistry to the electronics.

Sound systems for the validation of software are not yet generally available. Even in high-quality software, errors can exist which may lie dormant until triggered by an unusual combination of process parameters.

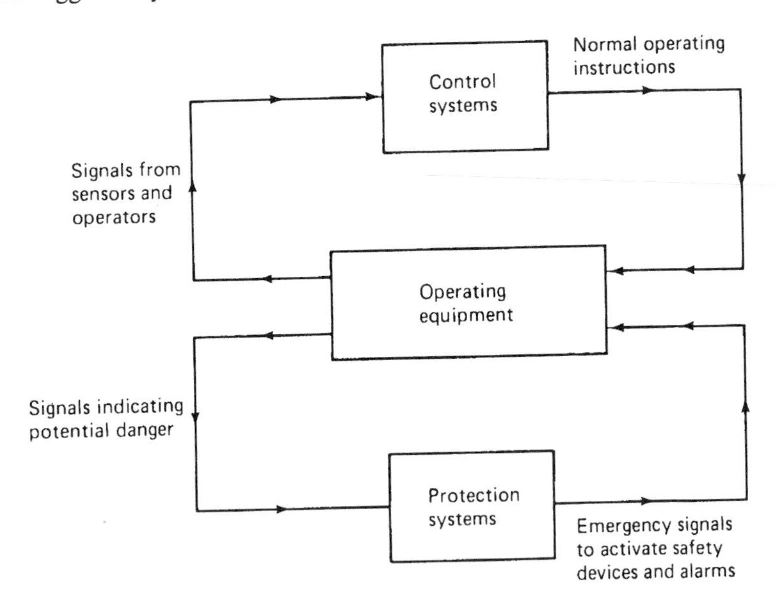

Fig. 17.19 Separation of control and protection systems (courtesy HSE)

The sensing elements and actuators (valves, etc.) on the plant form an essential link in any system. Their state and that of instrument housing, wiring runs and installation workmanship are often below standard, particularly on plants where the atmosphere is dusty or contains acid fumes.

P&IDs do not always correspond to what actually exists on the plant Sometimes vital protective items have not been fitted and sometimes they have been removed or by-passed during plant maintenance or modification. Plant managers therefore need to check process instrumentation thoroughly and ensure that it corresponds to design drawings during plant audits.

The protection of the plant control system prompts four questions:

- can the equipment be easily damaged by accident or by tampering?
- are the instruments and controls appropriate to the atmospheric environment?
- could electrical interference or electrostatic discharges affect the operation of PESs?
- could the safety back-up system(s) be knocked out by a fire or flood?

The security of PESs is also critical and access to software should be restricted to key personnel. (Visits by school and college parties can spell danger to unprotected software.)

Appropriate staff training is needed when PESs are introduced into a process plant to ensure that they are used and maintained safely.

Exact procedures for operating PES-controlled plant need to be spelt out (e.g. in the plant operating manual [20.2.2]) and must be followed religiously. Critical points in these procedures can be shown as 'menus' on one of the VDU screens.

The importance of electrical and process isolation before maintenance must be stressed [18.6], and a check should be made after maintenance or modifications to ensure that the software has not been corrupted.

HSE are also concerned about the use of 'expert systems' [19.4] for online process control, although agreeing that they may be a valuable diagnostic aid in tracing the causes of faults arising during the operation of complex plants. There is still truth in the statement that 'the best expert system for process control is the well-trained, experienced and dedicated plant operator'.

17.10 Hazards of instrument maintenance and modifications

Instruments, particularly control valves, require scheduled preventative maintenance, and lubrication in the same way as mechanical and process equipment [16.10, 16.11], although prompt repair maintenance is also essential to deal with faults in operation.

The ranges of transmitters and receiving instruments are mostly readily changed. If this is done without operating staff being aware of the change and its implications, the results can lead to serious accidents. An operator may think he had a certain flow through a line or level in a tank, whereas it was double that figure. Such modifications to instruments, despite their apparent minor nature and the ease with which they can be made, require proper scrutiny and authorisation before they are carried out, positive notification of all concerned when they are undertaken, and old conversion factors replaced by new ones. Similar precautions are needed when instrument loops themselves are modified, which should only be done after thorough discussion with operating and safety personnel.

References

1. BSI 5304, *Safety of Machinery*
2. EEMUA, No. 160, *Safety Related Instrument Systems for Process Industries (including PES)*
3. Smith, A., Proceedings Institution of Mechanical Engineers, March 1993
4. BS 2915: 1984, *Specification for bursting discs and bursting disc devices*
5. API, *RP.521 Guide for pressure-relieving and depressurising systems*, 2nd edn, *ibid.* (1982)
6. Perry, R. H. and Chilton, C. H., *Chemical Engineers' Handbook*, 5th edn, McGraw-Hill, New York (1973)
7. Lees, F. P., *Loss Prevention in the Process Industries*, Butterworths, London (1980)
8. Wells, G. L., *Safety in Process Design*, George Godwin, London (1980)
9. BCISC, *Codes of Practice for Chemicals with Major Hazards: Chlorine*, British Chemical Industry Safety Council, London (1975)
10. Anderson, N. A., *Instrumentation for Process Measurement and Control*, 3rd edn, Chilton, London (1980)
11. Noltingk, B. E., *Instrumentation Reference Book*, Butterworths, London (1988)
12. Foxboro, *Introduction to Process Control*, The Foxboro Company, Foxboro, Mass., USA (frequently revised)
13. ISA Standard S5.1, 'Instrument symbols and identification', The Instrument Society of America (1985)
14. Richmond, D., 'Instrumentation for Safe Operation', in *Safety and Accident Prevention in Chemical Operations*, edited by Fawcett, H. H. and Wood, W. S., 2nd edn, Wiley-Interscience, New York (1982)
15. Forbes, D. J., 'Design of blast-resistant buildings in petroleum and chemical plants', *ibid.*
16. Fitt, J. S., 'The process engineering of pressure relief and blowndown systems' in *Loss Prevention and Safety Promotion in the Process Industries*, edited by Buschmann, C. H., Elsevier, Amsterdam (1974)
17. UKAEA, 'Power Fluidics — Control without movement', *Loss Prevention Bulletin 040*, The Institution of Chemical Engineers, Rugby (1981)
18. Hagart-Alexander, G., 'Temperature measurement' in ref. 16
19. Higham, E. H., 'Measurement of pressure', *ibid.*
20. Fowler, G., 'Measurement of flow', *ibid.*
21. Sydenham, P. H., 'Measurement of level and volume', *ibid.*
22. Giles, J. G., 'Sampling', *ibid.*
23. Cornish, D. C. *et al. Sampling Systems for Process Analysers*, Butterworths, London (1981)
24. Grandjean, E., *Fitting the Task to the Man: an ergonomic approach*, 2nd edn, Taylor and Francis, London (1980)

25. Marshall, V. C., 'What happened at Harrisburgh?', The Chemical Engineer, *346,* 479 (1979)
26. Forsyth, R. (ed.), *Expert systems*, Chapman and Hall, London (1984)
27. Pathe, D. G., 'Simulator a key to successful plant start-up', *Oil and Gas Journal*, 7 April (1986)
28. HSE, *PES — Programmable electronic systems in safety related applications: 1 An introductory guide, 2 General technical guidelines,* HMSO, London (1987)
29. EEMUA *Guidance notes on programmable electronic systems*, Engineering Equipment Material Users Association, 14 Belgrave Square, London SW1 (1988)
30. Jones, P. G., 'Some areas of HSE concern about safety of computer control systems', in symposium *The Safety and Reliability of Computerised Process Control Systems*, NW Branch of the Institution of Chemical Engineers, 24 March (1988)

Part IV
Management, production and related topics

18 Safe work permits

This chapter owes much to Professor Trevor Kletz,[1] who gives details and underlying principles of the permit system used at ICI's Wilton works; to Bill Sampson, safety officer of Dow Chemical Company's King's Lynn works, who kindly provided information about the permit system[2] developed and used there; and to HSE's guidance note *Entry into confined spaces*.[3] Kletz also gives numerous examples of real accidents which could have been avoided had an effective permit system been in force. Probably the one best known to British readers was the fire and explosions on the Piper Alpha platform in 1988 [5.4].[4]

18.1 Why permits are needed

Maintenance is essential for the safety and integrity of process plant. Yet many plant accidents have occurred during or following it because of misunderstandings and neglect of essential precautions when plant was handed over from production to maintenance workers, and vice versa. The maintenance workers may be company employees in a particular section of the engineering department, e.g. instruments, electrical, general mechanical, rotating machinery, thermal insulation, drains, painting, or they may be employed by an outside contractor. The possibilities of misunderstandings between operating and maintenance personnel are aggravated by shift work and by the use of outside contractors.

Production workers usually play only minor roles in maintenance e.g. lubrication and the replacement of filter-elements, while maintenance personnel play no direct part in production. Serious injuries to maintenance workers have occurred because some flammable, corrosive, toxic, hot or asphyxiating material had been left in equipment which they were required to work on, or which had entered the equipment while they were working on or in it, or because an electric motor driving a pump, compressor or stirrer had been accidentally started while they were working.

Typical accidents involving production personnel have occurred because the wrong pipe flanges were disconnected by a maintenance worker, or because a motor was wrongly connected, so that when started it ran backwards, or because a joint which had been disconnected during maintenance had not been tightened and tested before restarting production. The writer still vividly remembers his narrow escape over 40 years ago from a 30-foot fountain of concentrated sulphuric acid, which was caused by a fitter removing a pipe bend from the discharge line of the wrong pump.

Errors in instrument maintenance are particularly insidious since instruments form an extension to the eyes, mind and hands of the operator. Thus the connection of the wrong thermocouple wires to the temperature controller of a reactor may lead to a runaway reaction and an emergency shutdown or a serious accident.

To prevent such errors and the accidents to which they readily give rise, most enterprises in the process industries, particularly oil and chemicals, have found it necessary to use formalised procedures whenever work is done on a plant by personnel other than those who are normally in charge of it. There are also such potentially dangerous activities as entry into pits and confined spaces (where the atmosphere may be unsafe to breathe), digging holes in the ground (with the risk of cutting a buried pipe or cable) or working at a height without guard rails. For these a written permit system which incorporates the necessary safeguards is also needed.

In many countries there are legal requirements for a written permit to be issued by a competent person before certain things are done. In the UK this applies to entry into various types of confined spaces under the following legislation:

- Chemical Works Regulations 1922
- Kiers Regulations 1938
- Shipbuilding and Ship Repairing Regulations 1960
- Construction (General Provisions) Regulations 1960
- Factories Act 1961 — sections 30 and 34.

Official forms are available for such permits which are dealt with in HSE's Guidance Note, *Entry into confined spaces* [8.7].[3] Other UK legislation which requires permits-to-work includes:

- The Electricity at Work Regulations 1989 (safe use of electricity);
- Factories Act 1961 — sections 27(7) (precautions needed to prevent persons being struck by overhead travelling cranes);
- Factories Act 1961 — section 31 (precautions regarding explosive or flammable materials);

- The Ionising Radiations Regulations 1985 (safe use of radioactive sources);
- Health and Safety at Work etc. Act 1974 — section 2(2)(a) (a general duty for safe systems of work).

18.2 Principles of permit systems

A written permit-to-work should identify the precise item on which work is to be done, record the hazards anticipated and detail the precautions needed before work starts. Bamber[5] gives the following principles for operating a permit-to-work system.

1. The information given in the permit must be precise, detailed and accurate.
2. The work to be done — and who will supervise and undertake it — must be clearly stated.
3. It must specify which apparatus or plant has been made safe and should outline the safety precautions already undertaken.
4. It should also specify the precautions still to be taken by the employees prior to commencement of the work (for example, fixing locking-off devices, siting of warning and danger boards, etc.).
5. The permit should specify the time at which it comes into effect and for how long it remains in effect. A re-issue should take place if the work is not completed within the allocated time.
6. The permit should be regarded as the master instruction which — until it is cancelled — overrides all other instructions.
7. Work must not be undertaken in an area not covered by the permit.
8. No work other than that specified should be undertaken. If a change is considered necessary to the work programme, a new permit should be issued by the authorised person who issued the original permit.
9. The authorised person who is to issue any permit must, before signing it, assure himself — and the persons undertaking the work — that all the precautions specified as necessary to make the plant and environment safe for the operation in question have in fact been taken.
10. The person who accepts the permit — i.e. the person who is to supervise or undertake the work — becomes responsible for ensuring that all specified safety precautions continue in being, and that only permitted work is undertaken within the area specified on the permit.
11. A copy of the permit should be clearly displayed in the work area.
12. All persons not involved in the work should be kept well away from the defined area.

13. Where relevant, regular environmental monitoring should be under taken throughout the time the permit is operative.
14. The precautions to be followed for cancelling the permit should be clearly stated so that a smooth hand-over of plant or machinery occurs.

To this list should be added the need, which may occasionally arise, to inform people outside the works about intended activities which may affect their own operations or indeed their safety. This could involve for example a neighbouring works, a highway or river authority, or, if roads are to be closed, the fire brigade and the police. A copy of the permit should be taken to an appointed representative, to whom the work to be undertaken can be explained. A signature should then be obtained on the permit copy. Where several works belonging to the same company are located in one area there will often be a central safety office. Copies of all permits can be sent to this office where a check will be made to ensure that work on one site will not adversely affect another site. Also the office will be alerted to the need to visit plant where potentially hazardous work is to be done.

The correct functioning of what can be a complicated procedure depends on the knowledge and experience of everyone who is involved. When work is being done by a contractor in a Third World country, his employees may be completely unaware of the importance of the permit system. This problem is discussed at the end of this chapter [18.9.6].

18.3 Permits for maintenance

In the hand-over of single plant items, or even complete units, from production to maintenance personnel the main points of information to be transmitted by the production personnel are:

- the identity of the item or items on which maintenance is needed;
- a clear description of the work to be done;
- a statement and guarantee that all normal and necessary precautions for the work as specified have been taken, e.g. complete isolation of the item or items from process materials, electricity and service fluids, as well as the elimination of process materials, fuels and heat transfer fluids from the items isolated;
- a statement of the remaining hazards facing the maintenance worker(s), and any special protective clothing, equipment and precautions which he or they should use.

It should be clear at this point that the nature of the work to be done largely determines the precautions which have to be taken. Thus the changing of an external bearing on the shaft of a stirrer in a vessel may require little more than that the process in the vessel and the electrical drive be isolated. But if it is found that an internal bearing has to be changed, far more careful precautions will be needed, e.g. complete isolation of the item concerned, removal of all process materials from it, and purging. If workers have to enter the item to replace the bearing, further precautions, including testing of the atmosphere within the item, will be needed.

A permit system should not require every simple and straightforward maintenance job to be referred to senior management level for authorisation. This would be enormously time-wasting and would blunt priorities. The system should act as a filter which allows straightforward jobs of low hazard potential to be dealt with at senior operator or shift supervisor level, but catches more potentially hazardous tasks and refers them for review and authorisation to appropriate staff of higher seniority.

Where the maintenance job is a routine one, the precautions needed are usually clear at the outset. But where the work appears to be a little unusual, discussion is necessary between the production and maintenance personnel to decide how to tackle it, and what precautions to take. If it is discovered after the work has been started that the job is more complicated than was foreseen, and requires additional preparations by production personnel, then the maintenance work must be stopped. The work to be done and the precautions needed must be redefined, and any further ones necessary must be taken before the maintenance continues. In such cases it often saves time to assume the worst initially and take the more stringent precautions at the outset.

In some companies a form requesting a permit is completed before an appropriate permit is prepared. The form is used by the supervisor of the affected plant to indicate the potential hazards and to give instructions about the precautions which must be taken before the work can be started. The permit itself will then certify that this preliminary work has been completed (and will also specify the precautions which must be taken during the performance of the requested work). The Dow Safety Planning Certificate (Figs 18.1 and 18.2) has a similar function but is used only for specified work which could be unusually hazardous. Many companies continue to use the terms 'Requesting Authority', or 'Area Authority', or 'Owner' for the plant management, and the people who undertake the work are frequently called the 'Performing Authority'.

Although the principles of permit-to-work systems are well established in the process industries,[6] there is no universally agreed system of permits for maintenance and other work and there is even ambiguity about the use of the terms 'permit' and 'certificate' in this connection. Most large companies appear to have developed their own systems for their own internal use. The permit system as used at ICI's Wilton works have been described at length by Kletz.[1] In this the preparations and precautions needed for most maintenance work are covered by one general 'permit-to-work' (also called a 'clearance certificate'!). Special permits are used for purposes not covered by the general permit, e.g. hand-over of new equipment by contractors, entry into vessels and other confined spaces, excavations, use of portable

DOW KING'S LYNN **PERMIT TO WORK** No: 109608

1. **AREA/TANK/VESSEL/EQUIPMENT/PIPELINE**
   EXACT LOCATION:

2. **WORK TO BE DONE:**

COMPANY:	No. of Men:
Man in Charge:	
Attendant for Line Breaking is:	

3. **SAFETY PRECAUTIONS:**
   a) **Gas Tests** (when applicable) (Results and initials in boxes)
   b) **Protective Equipment to be worn** (ring items which apply)

Time							
Flammable Vapours							
Toxic Gases							
Oxygen							

1 Safety Helmet	2 Safety Spectacles	3 Chemical Goggles	4 Face Shield	5 Updraft Helmet	6. Air Hood	7 Self-contained C A B A	8 Compressed Airline B.A	9 Dust Mask	10 Ear Muffs
11. Gloves P.V.C. General Special	12 Rubber Boots (Steel Toecaps)	13 Protective Footwear	14 Boilersuit	15 PVC Suit	16 Neoprene Suit	17. Disposable Suit	18. Plant Overalls Rubber Overboots Rubber Gloves	19 Safety Harness	

c) Other Precautions

4. **STATE OF ISOLATION**

No. of Lines	Depressurised and Drained	Positive Isolation	Tagged Off	Valve only	Initials as applicable	Not Isolated	N/A
					Steam		
					Gas/Vapour		
					Liquid		
					Solids		
					Air		
					Nitrogen		

a) Although the job may be isolated and depressurised, small residual quantities of hazardous chemicals may still be present **so wear protective clothing suitable for the risk.**

b) All Motive Power has been isolated and any logic control interrupted. (LOCK/TAG/TRY). | Yes | No | N/A | Signed . . . . Approved Signator

c) I have placed my lock. Signed . . . . Person undertaking work

d) Electrical fuses have been withdrawn, all circuits are dead. Signed . . . . Electrician

e) Electrical circuits are live for 'Troubleshooting' only. Signed . . . . Electrician/Inst.

5. I certify that a Safety Planning Certificate is **not** required because the work does not involve Projects, Plant Changes, Confined spaces, Hot work in Zone 1 or 2, Open flame, Critical line breaking, Asbestos, Excavations, Mobile Cranes, Rootwork or heights >5m, H.V. Electricity.

   Signature: . . . . Permit Signator

6. a) **CONFINED SPACE ENTRY** (Cancelled if Site Alert (pips) sounds)
   In accordance with Regulation 7 of the Chemical Works' Regulations, 1922, and Section 30 of the Factories Act, 1961, I have inspected the above confined space, it has been tested, is fully isolated, has been safely prepared according to the precautions above and on Safety Planning Certificate No: . . . . and is, therefore, safe to enter from . . . . to . . . .
   on . . . . Signed: . . . . Approved Signator
   Name of competent Attendant outside vessel: . . . .

   b) [illegible]

   c) **OTHER HAZARDOUS WORK** (See Safety Planning Certificate)
   I have inspected the above job which has been safely prepared according to the precautions outlined above and on Safety Planning Certificate No: . . . . , therefore work may start from . . . . to . . . . on . . . .
   Signed: . . . . Approved Signator

7. **APPROVAL OF PERMIT TO WORK**
   I am satisfied that this permit is properly authorised and that safe access is provided and that no work is taking place above or below this job. Work may proceed
   from . . . . to . . . . on . . . . Date. Signed . . . . Permit Signator

8. **ACCEPTANCE OF PERMIT TO WORK**
   I have read and understood the above precautions and agree that for our/my protection we/I will observe tham. I confirm that all our/my Power Tools and Equipment have been registered and inspected as required by Dow Standards and that we/I understand the Site and Area Emergency Plans.
   Signature: . . . .

9. **COMPLETION OF PERMIT TO WORK**
   I certify that this job is complete/incomplete (ring appropriate word), that all guards have been replaced and secured in position, that all Tools and Equipment have been removed and the Job Site has been left clean and tidy.
   Signature . . . . Time: . . . . Date: . . . .

RENEWAL OF PERMIT TO WORK (CONSECUTIVE SHIFTS ONLY)

10. Approved until . . . . Time/Date . . . . Permit Signator . . . .
    Approved until . . . . Time/Date . . . . Permit Signator . . . .

NOTE: When a job is finished this Permit must be signed off in Section 9 and returned. Should the job not be completed by the time specified this Permit must be renewed. This Permit is cancelled if Area Alarm (warble) sounds.

Fig. 18.1 Example of permit to work (courtesy Dow Chemical Co.)

SAFETY PLANNING CERTIFICATE

KING'S LYNN

(A PERMIT TO WORK IS REQUIRED BEFORE WORK STARTS) CERTIFICATE No: 6953

1. FOR WORK INVOLVING:	Projects	Confined Spaces	Open Flame	Asbestos	Mobile Crane	H.V.Electricity	Other:
	Plant Change	Hot work in Zone 1 or 2	Critical Line Breaking	Excavations	Roofwork and Heights > 5m		

2. **CERTIFICATE APPLIED FOR BY:** Department/Contractor:

Area/Tank/Vessel/Equipment/Pipeline:

Exact Location:

WORK TO BE DONE:

TOOLS TO BE USED:	Welding Cutting Equipment	Gas	Mobile Crane	Mobile Pump		Compressor	M/Vehicle	Cold Tools only
		Arc	Excavator	Temporary Lights	110V / 24V	Electric Power Tools	Other	

3. **USE OF A MOBILE CRANE:** I have inspected this job and it may proceed subject to the following precautions: N/A

Signed: Approved Crane Supervisor. Date:

4. **EXCAVATIONS:** I have inspected this job and it may proceed subject to the following precautions: N/A

Signed: Approved Construction Signator. Date:

Signed: Approved Electrical Signator. Date:

5. **ROOFWORK – WORKING AT HEIGHTS AND ASBESTOS:** I have inspected this job and it may proceed subject to the following precautions: N/A

Signed: Approved Construction Signator. Date:

6. **H.V. ELECTRICITY:** I have inspected this job and it may proceed subject to the following precautions: N/A

It will be switched by ....

Signed: Approved H.V. Electrical Signator. Date:

7. I confirm that the Area/Tank/Vessel/Equipment/Pipeline, as described above, will be safe for the proposed work provided the precautions listed above, together with those ringed on the check list opposite, are taken.

Additional precautions: (if none, write none)

A permit-to-Work must be obtained from: before work starts. This Safety Planning Certificate is valid from hours on to hours on

Section 6(C) on the Permit to Work may be signed by ....

Signed: Approved Safety Planning Certificate Signator. Time: Date:

8. **HOT WORK IN ZONE 1 OR ZONE 2 AREAS OR H.V. ELECTRICAL WORK OR ANY PLANT CHANGE:** N/A

I confirm that the above work may take place provided all the stated conditions are satisfied.

Signed: Authorised Signator; Time: Date:

9. **RENEWAL:** Subject to the provisions and precautions stated above and opposite this certificate is further valid.

Renewed from hours on to hours on Signature (approved SPC):

Renewed from hours on to hours on Signature (approved SPC):

Renewed from hours on to hours on Signature (approved SPC):

NOTE: (a) A separate signature is required for Section 8. (b) This Certificate is not valid until all necessary signatures have been obtained.

Fig. 18.2 Example of safety planning certificate – continued on next page (courtesy Dow Chemical Co.)

SAFETY PLANNING CERTIFICATE CHECK LIST (ALL REQUIRED PRECAUTIONS TO BE RINGED)

SIGN

GENERAL PRECAUTIONS

001. All power tools and equipment (including steps and ladders) must be registered with valid label affixed.
002. All power tools must be 110 volts maximum.
003. Ensure that power supply cables to transformers and welding sets above 110 volts are less than six feet long.
004. Suitable steps or ladders to be used.
005. Scaffolding to be erected and inspected by competent persons and notice fixed before use (mobile or fixed).
006. Provide life-line.
007. Use inertia fall arrestor (e.g. Sala Block).
008. Cordon off work area, above and below.
009. Notify adjacent plants/areas.
010. Check that all holes, excavations, work areas where covers or drains are removed are barricaded off and warning notices affixed. At night any such hazards must be adequately lit.
011. Isolate all power driven equipment before work starts – LOCK, TAG and TRY.
012. Check showers and eye bath units before work starts.
013. Instigate safe procedures for materials containing asbestos to comply with King's Lynn Site Standard No. 20 – Asbestos.
014.
015.

PROTECTIVE CLOTHING

100. Protection required:

1. Chemical Goggles	2. Face Shield	3. Updraft Helmet	4. Air Hood
5 Self-contained C.A.B A.	6. Compressed Air line B.A.	7 Dust Mask	8 Ear Muffs
9 Gloves PVC Gen. Spec.	10. Rubber Boots (Steel Toecaps)	11 Protective Footwear	12. PVC Suit
13. Neoprene Suit	14. Disposable Suit	15 Plant Overalls Rubber Overboots Rubber Gloves	16 Safety Harness

ATMOSPHERE TESTING

200. Test for flammable vapours (explosimeter) BEFORE WORK STARTS/REPEAT EVERY . . . . HOURS/MONITOR CONTINUOUSLY.

201. Test for oxygen BEFORE WORK STARTS/REPEAT EVERY . . . . HOURS/ MONITOR CONTINUOUSLY

202. Test for toxic gas BEFORE WORK STARTS/REPEAT EVERY . . . . HOURS/ MONITOR CONTINUOUSLY.

LINE BREAKING

300. Positively identify by tagging, taping or painting.
301. Before cutting into a pipeline a 'test' hole should be drilled in the pipe.
302. Process operator to 'stand by' (protected to same standard as craftsman).
303. Check pipeline suspension.
304. Drain and isolate line, lock off pump(s).
305. Provide scaffolding – fitter should work at waist height.
306. Blank off open ends of pipelines.
307. Flush area with water after job to ensure no spillage left.
308. Decontaminate tools, protective clothing and boots, gloves, face and eye protection (keep goggles on until last and then remove in safe area wearing clean or disposable gloves).
309.
310.

HOT WORK

400. Guard against falling sparks and slag.
401. Keep work area and below wet with running water.
402. Instigate fire watch.
403. Check area 30 minutes after cessation of work.
404. Check work area every .... minutes.
405. Run out fire hose.
406. Provide fire extinguisher, Type
407. Clear all combustible materials from work area.
408. Remove all full and empty drums from area.
409. Use only approved welding set, see Safety Standard No.17.

SIGN

HOT WORK . . . . . continued

410. Check welding cables are in good condition and where they must cross pipelines a suitable insulating bridging must be used to prevent possible contact. Weld return routing via installed equipment is prohibited.
411. Site gas cylinders so as to be clear of sparks and slag.
412. Check detachable cylinder key in situ.
413. Check compressed gas cylinders are used in metal wheeled trolley (not free standing or fixed to a structure).
414. Test all compressed gas connections using soap solution before work starts.
415. Check that oxygen and fuel gases have flash-back arrestors fitted between regulators and supply hose and that non-return valves are fitted between torch and supply hoses.
416. Check that all hoses are in good condition and located away from traffic. They should not present a tripping hazard to personnel.
417. Erect screens to safeguard personnel from U.V. radiation.
418. Site diesel driven D.C. generating sets in open air to prevent fumes accumulating in work area.
419. Check that smoke detectors are isolated.
420.
421.

ENTRY INTO CONFINED SPACES

500. All pipelines must be isolated, either by removing spool pieces and blanking off live ends or by inserting spade in lines.
501. Isolate agitator by removal of fuses, followed by LOCK, TAG and TRY.
502. Trained attendant to stand by outside vessel (must be named on Permit-to-Work).
503. Use mini-winch with life-line and full hoister-type safety harness.
504. Check vessel is cool enough to enter (< 35°C).
505. Use air mover or fan (must be grounded).
506. Use 24 volt lamp.
507. Check adequacy of means of vessel entry/exit.
508. Provide portable alarm for attendant.
509. Provide two sets of breathing apparatus outside vessel.
510. Compressed gas cylinders must be kept out of confined spaces.
511.
512.

MOBILE CRANES

600. Simple lift - banksman to be named on Work Permit(3c)
601. Qualified Dow representative in control- Name . . . . . .
. . . .
602. Critical lift – check list completed –Construction Supervisor or Owner's Representative (mech.) in control.
603.

EXCAVATIONS

700. Over 1.2 metres deep - Construction Department in control.
701. Hand dig only.
702. Sides of excavation made secure.
703. Test ground water for contamination.
704.
705.

ROOF WORK & HEIGHTS GREATER THAN 5 METRES WHERE THERE IS NO PERMANENT ACCESS

800. Crawling boards must be used.
801. Working method and safety devices to be approved and recorded by Construction Signator.
802. Provide working platform with handrail and toe boards.
803.
804.

Fig. 18.2 (continued)

radiation producing equipment, work on pipelines connecting units under the control of different supervisors, and work on equipment sent outside the plant.

The scope of a general permit-to-work can be extended to cover special cases by the use of certificates. Such certificates are not in themselves permits to do anything and can only be used in conjunction with a permit. For example, a vessel entry permit might allow a person wearing breathing apparatus to enter a process vessel, subject to certain conditions. It would not, however, allow him to enter it without breathing apparatus unless the permit is backed up by a valid gas test certificate, signed by the person who tested the atmosphere inside the vessel, and certifying that it is safe to breathe, again subject to certain conditions.

The Dow Chemical Company uses a two-tier system for its UK operations. This was developed by discussion with the personnel concerned and consensus over a period of 18 months at their King's Lynn works. It was instigated by Dow's safety officer, Bill Sampson, after studying permit systems in use by other UK companies. It attempts to match the hierarchy of hazards inherent in different maintenance and other tasks with a hierarchy of professional experience and expertise among those who authorise and supervise the work. It has been used without serious incident since 1982.

18.4 Outline of the Dow system

This uses a single multipurpose permit-to-work (Fig. 18.1) which can be authorised by a single designated signatory for most maintenance work, but which requires to be backed up by a 'Safety Planning Certificate' (Fig. 18.2) authorised by higher-level signatories for several specified jobs of greater hazard.

A permit-to-work is not required for work covered by 'Job safety analyses' [21.2.1] or 'Safe operating procedures', which is done by people within their own departments but is required for any site work involving construction, installation, alteration, dismantling and maintenance.

In addition to a permit-to-work, a safety planning certificate is required for the following types of work:

1. projects (for which capital has been authorised);
2. any plant change (affecting its design or integrity);
3. work in confined spaces (defined in a company standard);
4. hot work in electrical zones (1) or (2) capable of igniting a flammable vapour or dust; [10.5.2, 10.5.3]
5. use of an open flame anywhere on site capable of igniting combustible liquids and solids;

6. critical line breaking (where a reasonably sized leak could lead to a major accident) — critical lines are defined as those carrying certain hazardous fluids and are designated on boards close to plant control rooms;
7. asbestos (refers to a company standard);
8. excavating or digging (refers to a site standard);
9. use of a mobile crane (refers to a site standard);
10. roofwork and work higher than 5 m where permanent access is not provided, and excluding scaffold erection;
11. high-voltage (>1000 V) electrical work.

18.4.1 Safety planning certificates

Except in an emergency or exceptional circumstances, the safety planning certificate must be raised at least two days before work is planned to start. The person initiating the work must determine if it needs a safety planning certificate and if so, which sections of it will apply. He then fills in sections (1) and (2), thus indicating which other sections should apply.

All persons authorised by management to approve certain sections of the safety planning certificate should have appropriate job and safety training and minimum periods of service with the company (five to ten years) and of experience in the job or area concerned (three months to two years, depending on the job). Two to four signatories have been approved for each function and their names and duties are made known to all personnel. They are as follows.

- *Approved crane supervisor* must inspect the job, check ground and overhead conditions, specify the precautions needed and sign section (3) of the safety planning certificate.
- *Approved construction signatory* is responsible for inspection and precautions for any work involving excavations, roofs, working at heights and asbestos. He liaises with the approved electrical signatory regarding buried power cables and is responsible for signing sections (4) and (5) of the safety planning certificate.
- *Approved electrical signatory* checks for underground power cables prior to any excavations.
- *Approved high-voltage electrical signatory* is responsible for safety on work involving over 1000 volts. He will name the electrician who will do the power switching and sign section (8) of the safety planning certificate.
- *Approved safety planning certificate signatory* is responsible for the overall safety planning aspects of the job, reviewing, liaising and specifying any additional safety precautions needed with particular emphasis on process hazards and for signing section (7) of the safety planning certificate.

- *Approved signatory for section (8)* is generally the plant superintendent or senior process engineer and must confirm that the work specified in section (8) of the safety planning certificate may proceed by signing it. While an approved safety planning certificate initiator may initiate a safety planning certificate as well as signing section (7), sections (7) and (8) must always be signed by different signatories.

The original safety planning certificate is given to the person responsible for carrying out the work, the first copy is filed in the plant control room or area where the permit-for-work is normally issued and the second copy is retained by the initiator. The safety planning certificate may be issued for an initial period of up to two weeks. If more time is needed it may be renewed up to three times by an approved signatory. *Before work can start a permit-to-work is required.*

18.4.2 Permits-to-work

Approved signatories for permits-to-work require training and experience broadly similar to those of safety planning certificate signatories. Their authority to sign should be limited to their normal areas of work. Persons undertaking the work, and electricians, etc. who sign section (4), should have had appropriate training and a minimum of two years' appropriate work experience, at least three months of which should generally have been spent in the present job and in the same plant or area.

To initiate a permit-to-work *the person undertaking it* reports to the plant/area and contacts the permit signatory. If the work requires a safety planning certificate, the person undertaking the work must take this with him whenever he requests a permit-to-work, which can only be issued within the times specified on the safety planning certificate.

The permit signatory completes sections (1) and (2) naming the person in charge of the job, the company (where relevant), the name of the attendant if line breaking is involved and specifying the number of people working. He then arranges for any gas tests to be carried out and the results entered in section 3(a), rings the items of protective clothing to be worn and specifies any other precautions. He also arranges for the required isolation of pipelines and electricity to be carried out.

The person undertaking the work will place his own lock (on the power switch), try the equipment to ensure that it is isolated and sign section 4(c). Where an electrician or instrument has to 'trouble shoot' on live equipment, he will sign section 4(e).

The authorisation of a permit-to-work depends on whether or not a safety planning certificate is involved. If it is, an approved safety planning certificate signatory will inspect the job, sign the appropriate parts of section (6) and stipulate time constraints. The permit signatory then authorises work to start by signing section (7). He will assume responsibility in an emergency.

If no safety planning certificate is involved the permit signatory will sign both sections (5) and (7) of the permit-to-work.

The individual undertaking the work signs section (8) of the permit thereby accepting it before starting work.

The permit-to-work may be renewed by the signatory for up to two shifts, providing the work is continuous and that he is satisfied that all stipulated safety conditions are still in force and that it is safe to continue the work. If a safety planning certificate is involved, any time constraints stipulated on it should still be valid.

On completion of the job or period of work, the person responsible for it signs off his permit-to-work in section (9), returns it to the designated control area and attaches it to the plant copy.

Permits become void when the works evacuation alarm sounds and can only be renewed or reissued when the incident is over.

18.5 Precautions before issuing a permit

The following recommendations are due to Kletz.[1] Before a permit is issued, the person who issues it should go with the person who will be doing the job, or his supervisor, to the job-site and there discuss with him or her the work to be done and the precautions to be taken, as detailed on the permit. If the item to be maintained has no number, a tag should be attached to it which bears the same number as that used on the permit. The issuer of the permit should point out all valves which have been closed and locked, all electrical isolation switches which have been locked, and all fuses which have been removed as part of the isolation procedure. These should be verified on the spot by the person accepting the permit. Tags should also be attached to all unnumbered joints on pipes and equipment which will have to be disconnected. The working copy of the permit should be placed in a transparent cover and hung in a prominent position near the job while it is in force so that it can be readily referred to and signed as required by those doing the job.

Precautions before entry into confined spaces are discussed later [18.7].

18.5.1 Hot work

Operations classed as 'hot work' include welding, burning, the use of industrial (non-flameproof) electrical equipment, the entry or use of vehicles and plant with internal combustion engines, and the use of pneumatic chippers, hammers and rock drills. Three hazards must be recognised:

- those due to the presence of flammable materials in the item on which hot work has to be done;
- those due to the presence of combustible materials in the item which burn or give off flammable vapours when heated as a result of the work being done;

- those due to the presence of flammable gases and vapours in the surrounding atmosphere.

The first hazard will have been removed if the isolation and purging discussed in 18.6 have been carried out effectively, and confirmed by a negative test for flammable gases [18.7.1].

The second hazard is more difficult to eliminate, and has caused several explosions. Where hot work has to be done on the outside of vessels, etc. which may still contain combustible materials in inaccessible places, the vessel, etc. should either be filled with water, purged with inert gas, or filled with a stable foam containing only inert gas before starting hot work. Where possible, cold methods of work should be used rather than hot ones.

Where welding has to be done on pipes which may contain residues or polymers, there should be at least two openings for fumes to escape, fire extinguishers should be available, and the welder should have a clear escape route.

The third of these hazards arises when work has to be done near running oil and chemical plants. The area within which hot work is to be carried out should be cordoned off and warning notices displayed. The atmosphere in this area should be tested for flammable gases, and no hot work should be allowed if the flammable gas content exceeds 10% of the lower explosive limit. Portable combustible gas detector alarms should be placed nearby, mainly upwind of the hot work operations, and hot work should cease at once if an alarm sounds. Supervisors of nearby plants from which flammable materials might escape should be consulted. They should be required to give loud audible warnings of any escape of flammable materials from their plants, or of any abnormal plant conditions which might herald such escapes.

The hazard of flammable vapours reaching the hot work area from nearby drains should be checked by flammable gas tests near the drains and by warning those liable to discharge flammable materials into the drains.

Many works have internal roads where vehicles, plant and cranes are allowed without a permit and where the risk of flammable vapour is low. Before vehicles, etc. are allowed to leave these roads, a permit may be needed.

Many hot work permits include a section which must be completed if the presence of one or more firefighters (sometimes called 'firewatchers'), or the Works Fire Brigade, is requested. The equipment that they will need, possibly including breathing apparatus, is also indicated.

18.5.2 Excavations, etc.

Excavations have a wide range of hazards including the striking of buried cables, pipes and other buried objects (even unexploded bombs!), those of persons and vehicles falling into them, the collapse of their sides onto people working, flooding, and the accumulation of dangerous heavier-than-air gases such as propane, carbon dioxide and chlorine. Their constructional hazards are covered in the UK by The Construction (General Provisions)

Regulations 1961. These, among other things, require them to be inspected at least once every day by a competent person while people are working in them (Regulation 9).

Besides excavations, the operations of levelling ground and driving piles, poles and stakes often carry the risk of striking a buried cable or pipe, and should also require a permit. Before any such work is started, a responsible person should study the site plans and check whether buried cables or pipelines are present within 1 m of the place of work or not. Problems arise when records are incomplete or inaccurate, and in such cases cable- and pipe-locating devices should be used.[7]

Where there is a risk of hitting a buried cable or pipe, the person issuing the permit should clearly mark the limits of the excavation, etc. on the ground with paint or pegs. The permit should state whether the area is clear of buried cables and pipes, whether machinery may be used, and within what boundaries. If cables or pipes are present within 1 m of the work, only hand tools should be allowed, taking care not to disturb the cable or pipe which should, if possible, be isolated before the permit is issued. The permit should also state whether the cable is alive or the pipe is in service.

18.5.3 Ionising radiation

Before portable equipment producing ionising radiation (as used for radiographic weld inspection) is introduced into a plant, all necessary precautions (e.g. barrier fences round radiation areas) should be in force before the permit is signed.

When instruments (such as level and density gauges) which contain sealed radioactive sources are present, these may have to be removed by a 'competent person' before a permit-to-work is issued. Plants which contain unsealed radioactive substances (e.g. catalysts containing uranium) also require a competent person to be satisfied that all necessary precautions and legal requirements have been met before a maintenance permit is issued. (See The (UK) Ionizing Radiations Regulations 1985.)

18.5.4 Work on live electrical circuits

Occasionally work has to be done on live electrical circuits to prevent a plant or factory coming to a complete standstill. The permit should be authorised by the chief electrical engineer, and only certain named electricians should be allowed to do it, under close supervision. Special training, techniques and equipment are needed. (See The (UK) Electricity at Work Regulations 1989.)

18.6 Practical preparations for maintenance

As people belatedly discover to their cost, maintenance starts in the design stage. Wherever possible, the future plant manager and maintenance engineer should be members of the design team. The team should ensure that all likely maintenance work can be done expeditiously and that every item needing maintenance, etc. while the plant is running can be isolated, and made safe for the work. It is, however, as important to avoid unnecessary valves and fittings as it is to install enough of them. Every one represents a potential leak and will itself require inspection, maintenance and access.

18.6.1 Equipment isolation

Few valves can be guaranteed leak-free after a period in service. For all critical isolation duties, particularly when equipment has to be entered, spectacle plates should be turned or line blinds fitted, or a spool pipe section removed between the closed valve and the item being isolated and a blank flange fitted after the closed valve.

Valves used for isolation should be capable of being locked shut, whether a blank or blind is fitted or not. The commonest procedure is to close and lock all isolating valves first, then remove residual material from the item, and only then break joints between the closed valve and the item to fit blinds or disconnect pipe sections.

Blank flanges, spectacle plates and line blinds should be of as high engineering standards as the flanges to or between which they are fitted. Permanently installed spectacle plates should be used on rigid lines. These only have to be turned after loosening the bolts and renewing the gaskets. Line blinds should only be used on flexible lines. Blank flanges should be used when a pipe spool section adjacent to a closed valve has been removed if a leak through the closed valve could present a hazard.

Approved line blind valves (Fig. 18.3) past which leaks are a proven impossibility offer an alternative to the above methods. Although dearer, they save much time in use and are the cheapest solution when frequent isolation is needed.

Entire plant or plant units are usually isolated at their battery limits by spectacle plates on all incoming and outgoing lines, thus avoiding the need to isolate individual plant items.

Special care is needed when isolating plant items from relief and vent headers. The relief or vent valve from an item being isolated may discharge into a valve-free common relief header, since some codes and organisations do not allow valves to be placed downstream of a relief or vent valve or bursting disc. It is then dangerous to disconnect a joint on the discharge side of the vent valve or relief device to fit a blind when there may be pressure in the header (Fig. 18.4). Unless a blind can be fitted between the item and the closed relief device, or unless the relief device can be relied on completely for isolation, the item should not be isolated alone. Maintenance on it should only be done when all items connected to the common relief header have been shut down and depressurised.

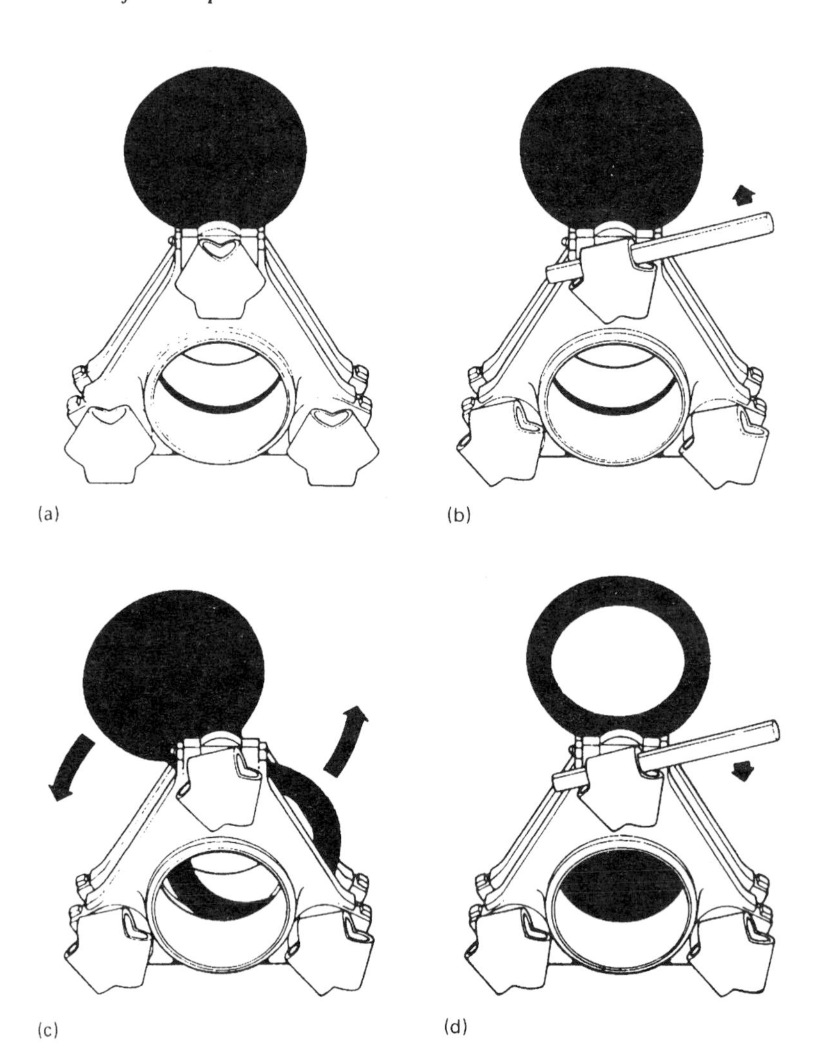

Fig. 18.3 Operation of line blind: (a) line open; (b) bolts slackened, line spread; (c) spectacle plate reversed; (d) bolts re-tightened (courtesy Hindle)

Vent valves and lines used for emptying, cleaning and purging should not be isolated or disconnected until this work is complete.

Fuses should be removed as part of the electrical isolation procedure. (Locked switches do not always prevent live circuits.)

In situations where very reliable isolation is frequently needed, it is usual to employ a double block and bleed arrangement: two valves are shut and the short length of pipe between them is bled by opening a third valve (the released material being discharged safety, possibly to a flare). Under no circumstances should a single-valve isolation be accepted on any plant from which a hazardous material can be released.

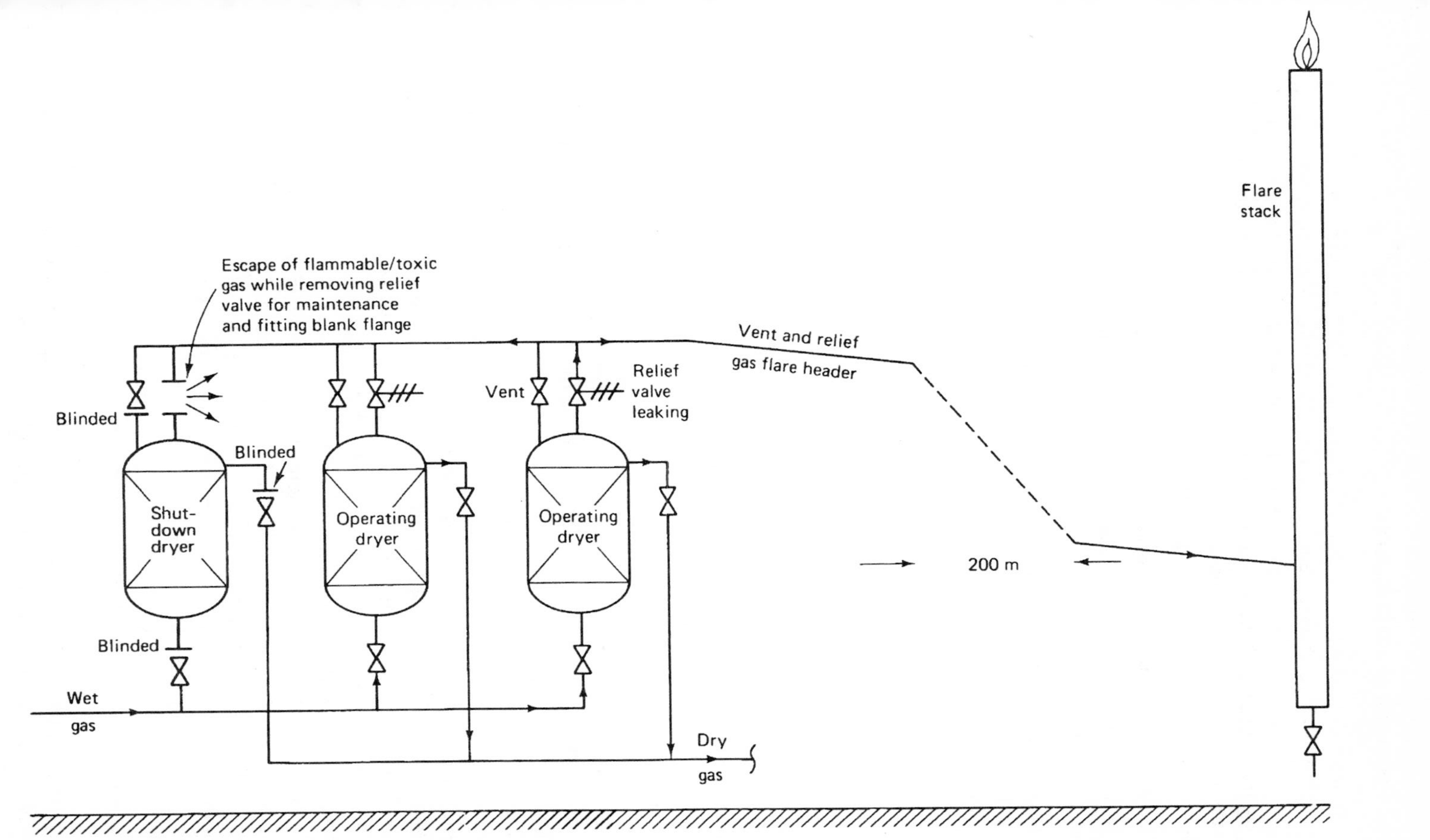

Fig. 18.4 Danger of blanking discharge flange of relief valve of shut-down process vessel

18.6.2 Removing residual hazards

Equipment should not be opened up for maintenance, etc. until risks to those doing the work have been removed. The worker is least exposed when all the work can be done from outside the equipment without applying heat. Exposure is greater if he has to enter the equipment, and greater still if he has to do 'hot work' while inside it.

It is impossible to lay down universal procedures because of:

1. the wide range of hazards which may be encountered;
2. the ease or difficulty of removing hazardous materials, and the many different methods which may have to be used;
3. the wide differences in the degree of worker exposure, depending on the job to be done and the method used.

Hazardous materials may be biologically active, carcinogenic, super-toxic, explosive, pyrophoric, highly reactive, flammable and/or corrosive. They may be gases, liquids and/or solids, including tars, resins, rubbers and fibres. The production department is generally responsible for removing them but the problems involved in this should have been studied and provided for during design.

For totally enclosed processes, the item or unit is first depressurised by venting gas or vapour through the normal disposal system (e.g. flare, scrubber, vent stack). If volatile liquids are present, they are generally next removed by draining to a safe place, displacing by or dilution with water, followed by blowing through with steam, air or inert gas, usually for several hours, until no trace of the liquid can be found in the steam or gas leaving the item. A special sampling point and condenser and some simple test may be needed to check this. Where steam is used the flow should be enough to ensure that the entire equipment is swept through. The item steamed out must be left vented until cool to prevent formation of a vacuum with risk of collapse. Most solids can be removed by high-pressure water jets. Their use requires training and personal protection including shields.

Difficulties will arise if there are no drain valves or connections at the lowest points of lines and plant items. These may have been omitted because of the risk of leakage of some very hazardous material such as carbon disulphide through a joint or valve.

The use of special solvents and chemicals to remove difficult materials is sometimes necessary, but should only be considered as a last resort. Most strong solvents are flammable, toxic or corrosive, and their final removal from the equipment and disposal may cause as many problems as that of the material they are intended to remove.

Chemical reactions with the surfaces of equipment which is being prepared for maintenance and entry can pose problems. When steel tanks which have contained concentrated sulphuric acid are washed out with water, the dilute acid formed attacks the steel and generates toxic gases derived from impurities in the metal, as well as hydrogen.

The inside surfaces of steel tanks, vessels, pipes, etc., even after cleaning, are sometimes still coated with iron oxide or sulphide, which react when air is admitted and reduce the oxygen content of the air inside. Such reactions with air also generate heat and the material may burn unless kept wet.

Water containing dissolved chlorides can damage stainless steel. It is thus best to use demineralised water or steam condensate for washing out stainless steel equipment.

The removal of residual hazards from process equipment before maintenance, and especially before entry or hot work, can give rise to unexpected problems (Fig. 18.5). It is not a task to be undertaken lightly or hurriedly and may require expert advice.

18.6.3 Staged maintenance jobs

Many maintenance jobs proceed in three stages.

1. disconnection of joints and fitting blinds/blank flanges, etc. as part of the isolation procedure;
2. the maintenance work proper, after residual hazards have been removed from the job;
3. removal of blinds, etc. and reconnection of joints. Before this is done, the production team must have satisfied themselves that the job has been completed satisfactorily, and that the item is clean and in a fit state to bring back into production. (There are endless anecdotes of fitters' cleaning rags and even sandwiches being left inside equipment when production was resumed.) In most cases a leak test is required, and in some processes it is critically important to remove all water from equipment before process materials are admitted. A suitable inspection procedure and checklist for handing equipment back to production is often necessary.

Fig. 18.5 Hidden hazard to welder inside process vessel (courtesy T. A. Kletz and John Wiley & Sons)

In this three-stage process, the maintenance worker is more exposed to process materials in the first and last stages than in the second, when the real maintenance work is being done. It is therefore recommended that separate permits be made out for each stage. The personal protection (gloves, goggles, etc.) specified for stages (1) and (3) will be of a standard higher than that required for stage (2). The first permit should then have been signed off, and all work called for under it completed, before the next permit is issued.

Some straightforward jobs such as renewing a gland seal in a valve or pump can often be done in less time and with less hazard exposure of maintenance workers than is involved in disconnecting and reconnecting joints, etc. in stages (1) and (3). In these cases it is safer and more practical for the work to be done under a single permit, using shut and locked valves and electrical isolation switches as the only form of isolation. Here the person doing the job must wear appropriate protection against leaks of the process material. To assist persons issuing permits to decide when this simpler procedure may be used, a list of such jobs should be discussed and agreed between senior members of the production, engineering and safety departments, and given for guidance to those authorised to issue permits.

18.7 Entry into confined spaces

Section 30 of the UK Factories Act (1961) refers to 'any chamber, tank, vat, pipe, flue or similar confined space', and Chemical Works Regulation 7 (1922) refers to 'any absorber, boiler, culvert, drain, flue, gas purifier, sewer, still, tank, tower, vitriol chamber or other place where there is reason to apprehend the presence of dangerous gas or fume'. HSE's guidance note GS 5[3] refers to 'reaction vessels, closed tanks, large ducts, sewers and enclosed drains', as well as 'open topped tanks and vats, closed and unventilated rooms, and medium-sized and large furnaces and ovens'. Kletz[1] recommends that excavations more than 1 m deep be treated as confined spaces and that entry permits be required for them unless their width at their widest point is more than twice their depth.

Before deciding to enter a confined space, all reasonable possibilities of doing the work from the outside should have been explored.

The requirements of the Factories Act are summarised thus in HSE's GS 5:[3]

> *Atmospheres in which dangerous fumes are liable to be present*
> No-one may enter or remain for any purpose in a confined space which has at any time contained or is likely to contain fumes liable to cause a person to be overcome, unless:
>
> (1) He is wearing approved breathing apparatus;
> (2) He has been authorised to enter by a responsible person;
> (3) Where practicable, he is wearing a belt with a rope securely attached;
> (4) A person keeping watch outside and capable of pulling him out is holding the free end of the rope.

Alternatively, a person may enter or work in a confined space without breathing apparatus provided that:

(1) Effective steps have been taken to avoid ingress of fumes;
(2) Sludge or other deposits liable to give off dangerous fumes have been removed;
(3) The space contains no other material liable to give off such fumes;
(4) The space has been adequately ventilated and tested for fumes;
(5) The space has been certified by a responsible person as being safe for entry for a specified period without breathing apparatus.

No-one should be allowed to enter a vessel or confined space in any circumstances unless the size of manhole or other opening is large enough to allow a person to enter and be rescued while wearing breathing apparatus. The Factories Act 1961 stipulates the minimum internal diameter of circular manholes as 18 inches for stationary vessels. Kletz recommends a suitable manhole diameter of 24 inches.[1]

The person issuing a permit to enter a confined apace must be satisfied that:

- ventilation (forced where necessary) is adequate not only for breathing and comfort but also to cope with possible changes in the atmosphere inside a vessel as a result of disturbing scale, burning, welding and painting;
- rescue facilities and persons trained in their use are available in case a person is injured or becomes ill while working inside a vessel;
- adequate illumination complying with electrical safety standards is provided inside the vessel, etc.;
- there is safe access to all parts within the vessel, etc. which may have to be reached.

The permit should only be issued after a gas test has been carried out and attested by a gas test certificate, or the signature of the tester on the permit itself (depending on the permit system in use). Special care is needed in removing hazards and making rescue plans from vessels containing baffles or other obstructions. A group of vessels with large interconnecting lines and no valves or obstructions between them which have been isolated together may be treated as a single vessel. Where parts of the inside of the vessel, etc. cannot be seen from the outside, the permit should first be issued on a provisional basis for inspection only. Only after this has been done should the permit be re-issued or endorsed as a working permit.

Kletz recommends three types or grades of vessel entry permits with a maximum validity of 24 hours from their time of issue.[1]

Type A permits are issued where a gas test certificate shows that the atmosphere within a vessel, etc. is fit to breathe.

Type B permits are issued when the atmosphere in the vessel, etc. is hazardous or unpleasant, but would not prove immediately fatal to a person breathing it. These require the following further precautions.

- Appropriate, well-fitting and adjusted respiratory protective equipment [22.7] and a harness and life-line should be worn by anyone entering. The harness and line need to be adjusted and worn so that the wearer can be drawn up head-first through any manhole or opening. An armlet attached to the life-line and fastened to the wrist or forearm of the wearer will facilitate this. Any lifting gear needed for rescue should be ready in position.
- Another person qualified in the use of the breathing apparatus and resuscitation equipment should remain at the entrance to the vessel, etc. as long as anyone is inside, and be in constant communication with him. He should have another set of breathing apparatus for himself, and resuscitation equipment with him, and should have the means of summoning a rescue team without leaving his post.
- A rescue team within easy reach should be available at short notice.
- Rescue plans should have been practised so that all know what to do in an emergency.

Type C permits are issued when the atmosphere in the vessel is so deficient in oxygen or contains so much toxic material as to present an immediate danger to life. This situation should be avoided if at all possible, but where it cannot, in addition to the precautions listed for type B, two trained rescue workers should be on duty at the entrance to the vessel, keeping the one inside continuously in view and in radio contact with rescue and medical services. Only breathing apparatus of a type approved for use in atmospheres immediately hazardous to life should be allowed. Further precautions may be needed depending on the hazards involved.

18.7.1 Gas tests

These are required as a condition for issuing permits for vessel entry and for hot work. They may be attested by special certificates and/or on the permit forms. Only trained and authorised gas testers should sign. Where the test is made by a laboratory technician, a separate certificate is usual. Most tests are made on the spot with portable apparatus. This is connected via a flexible tube to a long sample probe which can be inserted from outside the space to be tested to any point within it. The outlet of the apparatus is connected to an aspirator bulb or suction pump. At least two of the three following tests are usually required.

1. *Oxygen content.* For a normal vessel entry permit, this should lie between 20% and 22% by volume. An oxygen content below 15% presents an immediate danger to life.
2. *Flammable gas content.* For a hot work permit, this should give a maximum reading at any point in the space of less than 10% of the lower explosive limit in air. The apparatus should first have been calibrated with the same flammable vapour as that present in the space.

3. *Toxic gas or vapour.* Since tests for toxic gases and vapours are specific to particular compounds, one can only test for those likely to be present. Tests which depend on the extent of a colour change in an adsorbent-filled glass tube through which a known volume of the gas sample has been drawn[8] are available for many gases and vapours. For those for which no portable test apparatus is available, air samples should be taken and analysed in a laboratory. The analyses together with HSE's 'Occupational Exposure Limits'[9] and the toxicity data discussed in 7.4 provide guidance on the type of vessel entry permit which should be issued.

Gas test certificates should state:

- the time and exact place where the air was sampled;
- the oxygen content;
- the flammable gas or vapour content as a percentage of its lower explosive limit;
- the names of any toxic gases and vapours liable to be present and tested for, the concentrations found by testing and the degree of hazard which these represent;
- the signature of the tester;
- the period of validity of the test (usually not more than 24 hours from the time of sampling).

If the work for which the test was required is not complete by the time of expiry of the certificate, a further test is needed before it proceeds.

For work in an area where dangerous gases and vapours may intrude from sources other than the item which has been isolated, continuously monitoring gas detectors should, where possible, be placed round the work being done. Otherwise air samples should be taken at frequent intervals and tested promptly.

18.7.2 Hot work inside confined spaces

The hazards which can arise when hot work, especially oxy-fuel gas cutting and welding, has to be done inside vessels, etc, must be specially appreciated and the following precautions observed.

- Fuel and oxygen cylinders and their pressure regulators should not be brought into the vessel, etc.
- These gases should only be introduced into the vessel, etc. at reduced pressures via flexible hoses which are in good condition with secure connections.
- Forced ventilation should be used when hot work is done inside vessels.
- Workers must be warned of the dangers of gas escapes from damaged hoses, unlit burners, and of carbon monoxide poisoning from incomplete combustion. Carbon monoxide is liable to be produced when a large, cold, metal object is heated directly by a flame.

- Another worker should always be present outside the entrance to any vessel, etc. in which hot work is being done. He or she should be able to shut off fuel gas and oxygen instantly if needed.
- Continuous monitoring of the atmosphere in the vessel may be needed.

18.8 Other permits and certificates used

In addition to those used for maintenance, Kletz[1] discusses some other types used with process plant.

18.8.1 Permits to work on inter-plant pipelines

The normal responsibility for pipelines between plants may lie with two or more departments or plant managers, e.g. those at each end of the pipeline, and of any areas through which the pipeline passes. Close liaison between all of them is required. The general permit-to-work discussed in 18.4.2 might be used provided it has sufficient space for authorised persons in all plants and areas concerned with the pipeline to sign and write the precautions they have taken, and those which the maintenance team should take. The pipeline and those parts of it on which work is to be done should be clearly identified.

18.8.2 Certificates for used equipment sent outside plant

Before process equipment leaves a plant for repair, maintenance or scrapping, all hazardous material should have been removed from it. This is not always possible, and since accidents have resulted from this cause, a certificate should be used. The certificate should be issued by the head of the process department which is sending the equipment and should accompany it. The certificate should state that the equipment:

- is free of all hazardous materials, or
- contains certain named and potentially hazardous materials.

In the second case, the certificate should state what hazards may arise from the materials, under what circumstances this may happen, and what precautions should be taken. If necessary, someone from the department despatching the equipment should visit the organisation receiving it and explain the hazards, or even supervise the work done on it until all hazardous material has been removed.

18.8.3 Certificates of plant hand-over by contractors

Before new plant and equipment is handed over by contractors to an operating company and/or connected to operating plant, a formal hand-over procedure which includes inspections, test-runs and the issue of a certificate [20.1.2] should be followed. A normal permit procedure for all subsequent engineering work would then be applied. Connections from new to existing plant are preferably made by the engineering department of the operating company, under an appropriate permit, rather than by the contractor.

18.9 Pitfalls that must be avoided

Of the many pitfalls in maintenance work which the use of a good permit system helps to prevent, several common ones are mentioned.

18.9.1 Inadequate isolation

Several accidents have occurred through failure to isolate equipment from all lines connected to it, from reliance on leaking valves for isolation, and from failure to lock closed isolation valves to prevent them from being opened.

18.9.2 Faulty identification

Faulty identification by the maintenance person of the item to be maintained or the joint to be disconnected can have fatal consequences. The procedure of handing over and explaining the job to be done on the spot can be rendered ineffective by shift changes and other interruptions. It is essential for the permit issuer to write clearly and legibly, and to double-check against slips of the pen. Mistakes can also occur if the equipment numbering does not follow a consistent and logical pattern.

18.9.3 Changes in intent

Changes in the scope of work to be done once the permit has been issued and the work has started can also have serious consequences. It is important that the work requested be clearly detailed on the permit. If it is found that the job is more complex than appeared at the outset, this must be discussed with the issuer, and a fresh permit issued which states all precautions needed for the redefined job.

18.9.4 Inadequate communication

Lees[10] quotes a case of a maintenance fitter who left a job unfinished overnight intending to complete it the next day. The job was, however, finished and signed off by a night fitter. When the original fitter began work again the next day, the plant was no longer safe for work.

18.9.5 Unauthorised and unrecognised modifications

There are many temptations during maintenance to improve, simplify or streamline pipework, always with good intent and often with justification.

Sometimes, however, some important reason for apparently untidy or complicated pipework is totally missed by the maintenance worker. Kletz[1] quotes a case of the air supply used for breathing apparatus which came from a branch-pipe originally 'teed' in to the top of a horizontal compressed air main. During maintenance a fitter thought the pipework would be neater if the branch-pipe were teed into the main from below. When the next person used breathing apparatus fed from the branch-pipe for work inside a vessel, he received a faceful of water, fortunately without serious consequences.

18.9.6 Ignorance

A permit-to-work system will operate safely only when everyone who is involved fully understands the procedures. The following sequence of events has happened frequently in emerging countries and indeed is probably still happening. The plant, perhaps a refinery, has been installed by a European or American company. All of the documentation, including the permits-to-work, is written in the language of the company, and all top-management discussions are in that language. This is necessary because initially the plant will be supervised largely by foreigners, and all of the technical information about the process will be in the foreign language. This practice will continue even when the plant is eventually operated solely by the local people (who will probably have foreign degrees), and lessons in the foreign language will be provided for the operatives. The real problems start when local contractors are employed to do maintenance work. Suppose for example that a pipeline is to be replaced. The purpose and functioning of the permit system is explained to the contractor, and its observance forms part of the agreement with him. When everything is ready to start work, the contractor tells his foreman to go to the office and collect a piece of paper, which gives him permission to do the work. The foreman cannot read, or does not understand, what is written on the form, so he puts it in his pocket and gets on with the job.

One way of avoiding this potentially disastrous situation is to ensure that anyone who authorises, issues, or accepts a permit, carries an identity card which shows his photograph, his signature, and the actions concerning permits that he is permitted to take. The card is issued only after the man has attended a training course and has satisfactorily completed a written test. The test could comprise a number of multiple-choice questions which will assess his knowledge of the permit system, and of the language in which it is conducted. If simple yes/no questions are used, care must be taken to ensure that it is not possible to pass the test by randomly selecting the answers. General plant safety and the use of firefighting equipment can usefully be included in the course, and it is quite reasonable to make a contractor pay for the training received by his employees.

If it is agreed that only the foreman is to be trained in this way, then there should be a section on the permit form which the Issuing Authority signs to certify that he has discussed and explained the requirements of the permit with the foreman. The foreman will sign a similar section when he has instructed his men. Preferably, however, everyone involved in the work should receive some training [23].

References

1. Kletz, T. A., 'Hazards in chemical system maintenance — permits', in *Safety and Accident Prevention in Chemical Operations*, edited by Fawcett, H. H. and Wood, W. S., 2nd edn, Wiley-Interscience, New York (1982)
2. Dow Chemical (internal standard), *Permits to work and safety planning certificates*, Dow Chemical Company Ltd, King's Lynn, Norfolk, PE30 2JD
3. HSE, Guidance Note GS 5, *Entry into confined spaces*, HMSO, London (1980)
4. Lord Cullen, *The public inquiry into the Piper Alpha disaster*, Cm 1310, HMSO, London (1990)
5. Bamber, L., 'Techniques of accident prevention' in *Safety at Work*, edited by Ridley, J., 3rd edn, Butterworths (1990)
6. Chemical and Allied Products Industry Training Board, *Permit-to-work systems*, CAPITB, London (1977)
7. National Joint Utilities Group, *Cable locating devices* (available from any Area Electricity Board) (1980)
8. Lee, G. L., 'Sampling: principles, methods, apparatus, surveys', in *Occupational Hygiene*, edited by Waldon, H. A. and Harrington, J. M., Blackwell Scientific, London (1980)
9. HSE, Guidance Note EH 40, *Occupational Exposure Limits*, 1985, HMSO London (1985)
10. Lees, F. P., *Loss Prevention in the Process Industries*, Butterworths, London (1980)

19 Management for Health and Safety (HS)

By health and safety, abbreviated HS, we include here many matters and activities which are often classified under other headings such as 'loss prevention', 'reliability' and 'risk analysis', so long as they relate to or contain an element of unintended risk to human life or health. However, we exclude matters and activities which relate mainly to security and protection against deliberate acts of theft and destruction.

In discussing industrial management, it is hard to escape its economic role as put by Professor Drucker:[1]

> Management must always, in every decision and action, put economic performance first. It can only justify its existence and its authority by the economic results it produces.

The cost of major accidents such as those discussed in Chapters 4 and 5 can, however, be crippling to a company. The legal penalties in the UK for managerial failure to implement proper HS policies are also increasing, as in 1989 one large oil company was fined £500 000 for a repeated safety lapse, and manslaughter charges were brought against another company and some of its senior managers. Accidents also make for poor and costly relations between a company and both its employees and the general public. Thus even by Drucker's yardstick, managements need to include HS in the economic equation and pay special attention to protecting members of the public.

This leads to cost-benefit analysis, which attempts to balance the probabilities and costs of various types of accidents against the costs of preventing them (Fig. 19.1).[2] The total cost of accidents in any year is taken as the sum of two parts, A the direct costs of the accidents plus P the costs incurred in preventing them. As more money P is spent on preventing accidents, their number and direct cost A fall. When both A and P and their total are plotted against some index representing the degree of risk reduction achieved,

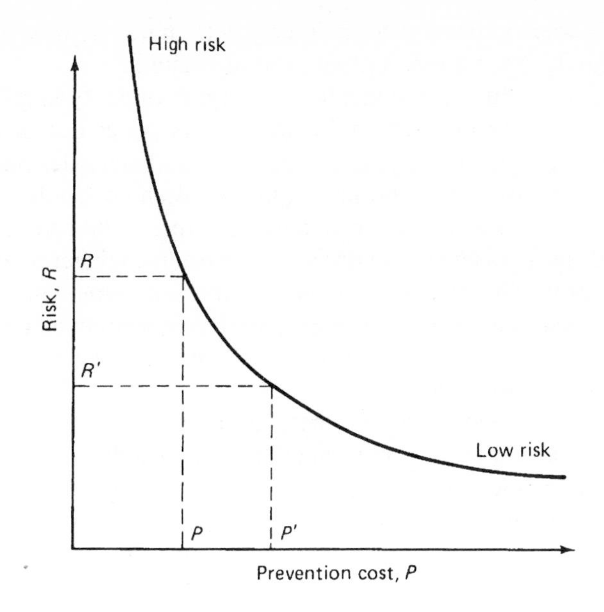

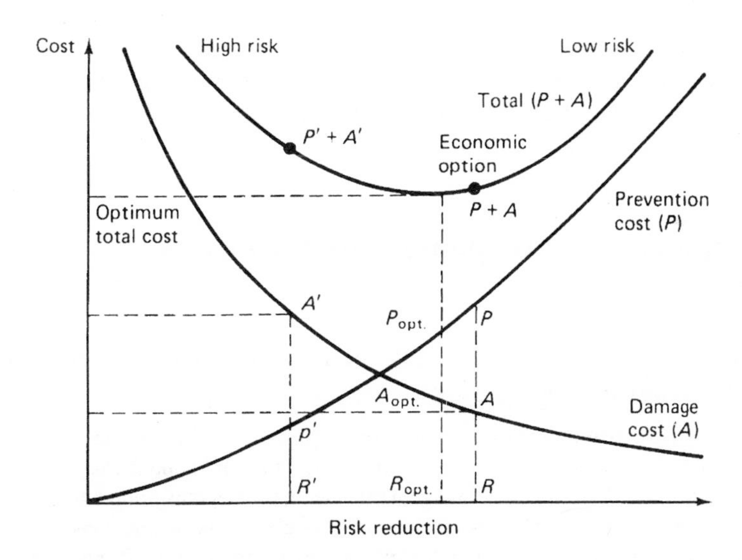

Fig. 19.1 Accident costs versus risk reduction

the total cost passes through a minimum. This approach has stimulated reliability studies and risk analysis and the development of hazard indices (e.g. Dow and Mond) [11].

Cost-benefit analysis of HS is still, however, at an early stage and has two main weaknesses. One is the difficulty of costing the benefits arising from any level of expenditure on HS, and the other is the reluctance felt by many people at equating human life in money terms.

The safety of modern process plant is largely determined by the competence of those who design, manage, operate and maintain it.

A pipe carrying a hot, highly flammable liquid under pressure may break because there was no stress engineer in the organisation which designed the plant. and the piping draughtsman had not provided sufficient flexibility to allow for thermal expansion. On another plant an austenitic stainless steel pipe may break because it was carrying an aqueous process fluid containing chloride ions, and there was no metallurgist or corrosion engineer to advise the piping design group that this would result in rapid corrosion. On a third plant a pipe may burst as a result of the explosive decomposition of a peroxide inside it, which could have been avoided if the process engineer had a better knowledge of peroxide chemistry.

Someone capable of recognising all these hazards before they caused damage would need a very broad technical education, including mechanical engineering, chemistry, metallurgy and corrosion engineering. To have been able to avert them he would have needed the authority to shut down an operating plant for inspection when he felt that something was amiss. A safety professional who combined all these qualities would be a rarity. The responsibility for such accidents can therefore lie only with the technical executives and their staff who caused the hazards in the first place. Safety in the process industries is thus everybody's business, especially management's. It requires that all employees, particularly those doing technical jobs, have the necessary training, experience and *hazard awareness* to avoid accidents arising from their work.

19.1 Management's responsibility for health and safety

Management's responsibility for the safe performance of its workers was expressed by Heinrich in 1931:[3]

> Management's responsibility for controlling the unsafe acts of employees exists chiefly because these unsafe acts occur in the course of employment that management creates and then directs. Management selects the persons upon whom it depends to carry on industrial work. It may, if it so elects, choose persons who are experienced, capable and willing to do this work, not only well, but also safely. Management must also train and instruct its employees, acquaint them with safe methods, and provide competent supervision. In following the principles of delegated authority, management, through its representatives in the supervisory staff, may set a safe example, establish standards for safe performance, and issue and enforce safety rules.

In the UK, after the recommendations of the Robens Report of 1972[4] had been embodied in HSWA 1974 [2.5]; management's key role in the safety of operations under its control was given legal emphasis. HS should now be treated as integral parts of the work processes. Managerial competence in HS must match the risks inherent in the undertaking and be no less than that needed to run the business successfully. HS at work requires methods, time and money similar to those devoted to other business objectives.

Management responsibility for HS does not apply only to top management or to shop-floor supervisors, but it must be recognised and accepted at all levels of line management, particularly by middle managers who may find themselves torn by conflicting requirements of production and HS. To resolve these they need help and support from senior managers. Senior managers, while motivating those at lower level in HS, should not, however, become too involved in detailed decisions which subordinates can take, given proper training and guidance.

The directors of companies involved in the process industries are responsible for ensuring that all staff participating in the design, operation and maintenance of the plant have an appropriately high degree of technical competence, hazard appreciation and personal responsibility. They also must ensure that adequate checks are made by hazard studies such as those discussed in 14.4.

The appointment of safety professionals whose roles are discussed later [19.1.7] cannot relieve line management of its safety responsibilities. The safety professional has no direct responsibilities for production or maintenance and the role is mainly advisory. The Accident Prevention Advisory Unit (APAU) of HM Factory Inspectorate (HMFI) provides the following key questions[5] which managers should ask themselves to assess their effectiveness in their HS performance.

1. Do we have a safety policy?
2. Is it up to date?
3. Do the subsidiary parts of our organisation have a policy?
4. Who is in charge of health and safety?
5. Are the technical problems of safety handled by competent persons?
6. Do we have a system to measure safety performance?
7. What is the worst disaster that could happen?
8. If the worst happened could we cope?
9. Would our workforce know how to react in an emergency?
10. What do our employees think of our safety standards?
11. What are we trying to achieve?
12. How much effort are we putting into safety?
13. Is the effort directed to the right place?
14. Is there an efficient system of checking that the duties are being carried out efficiently?
15. What are our long-term objectives?

19.1.1 Accident causes — technical and organisational

Industrial accidents have technical causes which need to be identified in order to prevent their repetition. Although the law has in the past been concerned to identify single 'proximate' causes for major accidents, careful analysis has shown that a combination of two or more technical causes was usually involved. But behind nearly all technical causes lie organisational weaknesses which allowed them to be present [3.3]. This can be illustrated by the simple example of a man falling from a ladder. The technical cause was identified as a defect in the ladder. Getting rid of the defective ladder will prevent it causing more accidents, but more questions need to be asked to discover the faults in the organisation, e.g.

1. Was a system of regular ladder inspection in force and who was responsible for it?
2. When was the defective ladder last inspected and why was the defect not then found and the ladder removed from service?
3. Did the injured employee's supervisor examine the job and the ladder before the accident? Why did the supervisor allow the ladder to be used?
4. Did the injured employee know that the ladder was defective and that he should not use it?
5. Was the injured employee properly trained in the use of ladders?

The answers to such questions might lead to the following organisational improvements:

1. a better ladder-inspection procedure
2. better training
3. clearer responsibilities
4. better job planning by supervisors.

The same types of question need to be asked and the same types of organisational corrections made time and again following the more complex accidents typical of the process industries. It is even more important for management to anticipate the hazards and create an organisation in which the accidents do not occur.

19.1.2 Policies and degrees of hazard

Since 1974 all undertakings employing five or more persons have been required under HSWA to have a written policy for HS. The policy, which is a statement of intent, will have limited value unless it is backed up by an effective organisation, adequate resources and motivated personnel. The statement and the degree to which it is implemented provide acid tests of managements' attitudes and commitment to HS.

Each policy should be unique to the special needs of the organisation for whom it is written. 'It cannot be bought or borrowed nor can it be written by outside inspectors or consultants.'[6] In large enterprises typical of the process industries, the most senior management specifies the overall objectives and top-level organisation for HS while each section of the enterprise amplifies its organisation and arrangements needed to meet these overall objectives. The policy should give a clear, unequivocal commitment to HS, be agreed by the board and be signed and dated by a director. It should be regularly reviewed, agreed with trade union representatives, brought to the attention of employees and should state that its operation will be monitored at workplace, divisional and group level. A 1980 review of the effectiveness of company policies for HS by the APAU of HMFI[6] contains a checklist of questions to probe the applicability, strengths and weaknesses of HS policy documents.

The effort and organisation which need to be devoted to HS is determined very largely by the magnitude and nature of the inherent hazards of the operations. In the process industries, which commonly employ flammable, reactive, toxic and corrosive substances from which to make useful products, the hazards may range from the mundane to the major. The HS resources needed in different cases have parallels with traffic control. Pedestrian precincts and shopping arcades from which motorised traffic is excluded are akin to safe processes such as solar evaporation of sea-water and salt crystallisation, which have only minor hazards. The accidental bodily contact of shoppers is quite common, but unless there is a sudden panic causing a stampede, injuries are rare. Normally there is little need for control.

On roads and motorways, which are akin to moderately hazardous processes, special protection has to be devised and procedures with formal safety rules have to be laid down by experts, enforced by police or inspectors.

In busy air-lanes near major airports, which correspond to highly hazardous processes, more sophisticated safety equipment and much stricter controls are needed. These must be continuously monitored by highly trained people.

19.1.3 HS goals

Several things have to be considered in setting HS goals. They should be practical, comprehensive, within the capabilities of the management and relevant to the conditions in the undertaking. It is even more important that every manager should understand his role in meeting them and that his progress in this can be and is monitored, preferably by his immediate superior. It will then be the latter's responsibility to check how far the junior manager is able to fulfil his HS goals (particularly if he has other urgent production tasks), to compliment him if he succeeds and to work out with him how to improve his performance if he fails.

All broad HS objectives of the organisation, both short-term and long-term, should first be listed by the most senior managers (assisted by

HS professionals), together with the names and functions of those next in line who will be responsible for implementing them. After the senior manager has discussed and agreed with each of his junior managers which of the broad goals apply to him, the latter should list his own goals in greater detail with the names and functions of those of his own subordinates responsible to him. In this way a complete, detailed and often unexpectedly long list of HS goals for the enterprise is built up. (The length of the list is less surprising when one considers that most chapters of this book include several different HS goals.). The list should then serve as the basis for HS planning, budgeting and costing. Some typical and rather broad goals follow by way of example.

- Provision of adequate resources both financial and in terms of man-hours of competent persons to meet the goals set. These include not only full-time HS professionals but a reasonable percentage of the time of line managers to meet their HS responsibilities.
- Maintenance of a sound organisational structure, with accepted job descriptions which include HS responsibilities.
- An adequate level of competent staff which leaves no gaps, especially in positions considered critical to safety and plant integrity.
- Achievement of lower-than-average accidental injury rates for the type of industry. This implies the reporting and analysis of all accidental injuries: better still, all accidents, whether causing injury or not, within the organisation. Here one needs to consider accident severity as well as frequency. Some authorities use a combined rate, the product of frequency and severity rate, as the best overall criterion [19.2.5]. One difficulty in judging progress toward this goal is that the figures for any one year may have little statistical significance unless large numbers of workers are involved. Changes in the type of manufacturing operation and in methods of accident reporting can be further difficulties.
- Compliance with relevant codes of practice, standards and regulations (such as COSHH), and the setting and monitoring of company HS systems (such as permits-to-work).
- Identification and elimination or reduction of specific hazards arising from the work. This includes careful investigation of accident causes.
- Adequate or improved HS training and commitment at all levels.
- Regular inspection, testing and and maintenance of protective systems for emergency use as well as the plant and machinery itself.
- Adequate protection of visitors, the public and the environment.
- Adequate monitoring of programmes in support of the HS goals.

Table 19.1, based on headings given in Dow Chemical's publication *Minimum Requirements*,[7] provides a list of items to be included in most lists of safety goals.

Table 19.1 Items to be covered in safety and loss-prevention goals

Safety	*Loss prevention*
Accident/incident investigation and reporting	Buildings and structure design
Audits	Capital project review
Confined space entry	Combustible dusts
Contractor safety	Electrical
De-energising and tag procedures	Emergency planning
Employee training and job-operating instructions	Equipment and piping
Government regulations	Fired equipment
Guarding and interlocking	Fire protection systems
Hot work and smoking	Firefighting capability
Job and process-operating procedures	Flammable liquids and gases
Ladders, scaffolding, work surfaces, etc.	Flexible joints in hazardous service
Line and equipment opening	Fragile devices in hazardous service
Personal protective clothing and equipment	Instrumentation
Safe operation of motor vehicles and motorised handling equipment	Leak and spill control/containment
Testing of emergency alarms and protective devices	Means of exit
	Pressure vessels
Related requirements	Process computers and data-handling equipment
Distribution emergency response	Reactive chemicals
Industrial hygiene and medical programme	Risk analysis
Material hazard identification	Rotating equipment
Product stewardship	Storage
	Technology centres

19.1.4 Management systems and accountability for HS

There are as many views on organisation and management systems as there are systems, whose structures vary from the nearly vertical to the nearly horizontal. From experience of several organisations, the writer prefers one which contains the following features:

1. a minimum of hierarchy;
2. structure oriented towards the project or activity;
3. decisions made at the lowest possible level with guidelines provided by higher levels;
4. supporting staff and techniques available where they are most needed;
5. communications across vertical lines of authority;
6. effective feedback of information.

Special problems of management including stress are discussed in 19.3. The position of safety professionals, particularly in organisations with hierarchical structures, can cause problems if their advice is ignored by their seniors [19.1.8].

Performance in HS should be included in staff assessments. Managers with successful HS records need encouragement. Those who fail must be made aware of where they have failed and the appropriate lessons discussed with them. Provision should be made to protect the 'whistle-blower' (who calls attention to particular hazards or dangerous practices) from being victimised.[8] Managers at all levels need to be convinced of the importance of HS goals, that the organisation intends to achieve them, and that they will be personally accountable for their part in it. The cue will be taken from the top.

19.1.5 Job descriptions and their HS content

The drawing up of job descriptions should go hand in hand with that of organisation charts. It is of immense importance in selecting applicants and determining the remuneration of various posts. No job description is complete unless it includes an agreed list of HS duties. The APAU of HMFI, which has seen many efforts at HS organisation within companies, gives the following advice about the issue of job descriptions for safety:[6]

1. The construction of a job description, a defined list of tasks for each manager, is a valuable exercise, so long as it is personally relevant to each person.
2. Since the person who will monitor the degree of success in meeting the tasks listed in the job description is the employee's superior officer, it is that officer who should first of all offer the suggested list of duties. He should then have an interview to discuss the job holder's view on his own work.
3. At the interview any uncertainties about the duties, on the part of the person for whom the job description is written, should be discussed and resolved.
4. Agreement between the two employees should then be followed by the issue of a final personal job description.
5. Monitoring by the senior employee should be against the job description, and reference should be made to it in assessing performance against the allotted and agreed tasks.
6. When either person changes, the exercise should be repeated, in order to establish the clearest possible common attitude to HS between the new parties.
7. The procedures listed above should be carried out progressively through the company, starting at senior management and finishing with the most junior manager in the organisation.

19.1.6 Resources for HS — time and money

A good deal has been written on the economics of loss prevention (see Lees[9]), but less has been said on what resources should be provided to meet companies' HS goals, particularly to support line managers. The full-time HS professional knows what he is being paid to do but the line manager, who among his other duties is directly responsible for HS performance, often has little idea of how much of his time is, should or may be devoted to HS matters. If he finds himself overstretched with other (e.g. production-related) duties which are given priority over HS, he may see all the paperwork about his HS goals and duties as merely a trap to make him a scapegoat if something goes wrong. If he is required by company accountants to record how he spends his working time under various cost centres, one or more of these should clearly relate to HS activities. This is a matter to be decided between the director who signed the HS policy statement and his chief accountant.

It is hardly enough to say that the time that the manager spends on HS matters should be charged to production activities to which they may relate. In this case nobody is any the wiser about how much time he actually spends or should reasonably spend on HS matters, and figures like 5% or 15% are simply plucked from the blue. Thus there generally seems to be a need for research on how much of their time line managers need to spend on discharging their HS duties. Top managers need to consider these findings and discuss them with their line managers. Individual records of time spent on particular HS matters should be kept, and managerial staffing levels should be adjusted to ensure that no line manager can plead that pressing production problems left him no time for his HS duties.

Likewise, when budgeting for capital expenditure, a special allowance should be made for safety items and special protective equipment needed over and above the normal protection incorporated into plant and equipment to meet the requirements of codes and standards. This is closely related to money spent on buying insurance. The application of the Dow and Mond hazard indices [11] sheds useful light on the subject even if it does not provide exact answers. The idea of financing a company's HS expenditure from a special fund rather than from its current account is raised in 19.3.6.

It is very important when choosing between different projects or process alternatives on which to invest capital that the differences in loss potential and safety/insurance expenditure are fully considered. Any project with a high accidental loss potential needs to show a correspondingly high potential for profit.

19.1.7 Motivation for HS

Motivation for HS is needed, particularly to change deeply ingrained habits and attitudes and to indoctrinate new employees. The key element is knowledge of work hazards and the effects they are likely to have if certain precautions are not taken. The dilemma for managers is how to put these facts over without creating undue anxiety and stress. As one worker interviewed by an industrial psychologist put it, 'If I knew all the hazards I faced at work, I'd never sleep at night'.[10] The relevant information needs to be presented in a way which assures the individual that the hazards have been assessed, provides him with the means of coping with them, and eliminates uncertainty. Sound knowledge is not, however, always enough to change bad habits, while the act of changing one bad habit can easily lead to another. A genuine desire to change must be there. If it does not or is only half-hearted, it must be fostered, preferably by objective discussion of the facts. But will alone is seldom enough to overcome subconscious impulses and something more is generally needed.

Management more than committees has the tools for the job. If it can organise people to achieve results in other fields, it should be able to do the same for HS. Its main tools are *communication, assignment of responsibility, granting of authority and fixing accountability*. Yet each individual will still make his or her own decisions. Management has to recognise those influences over which it has little or no control and extend its own influence in areas where it can. These include:

- group attitudes to safety
- selection and placement
- training
- supervision
- special emphasis programmes
- the media.

Studies on industrial motivation[11] suggest that the factors which motivate people are separate and distinct from those that cause dissatisfaction. Examples of both are:

Motivators	*Dissatisfiers*
Achievement	Company policies
Recognition	Supervision
Quality of work	Working conditions
Responsibility	Salary
Advancement	Relationships
	Status

Factors labelled as potential dissatisfiers need first to be brought up to the level of contentment. The next step is to work on the motivating factors.

19.1.8 HS professionals and safety organisation

Full-time HS professionals employed in industry include inspectors, doctors, nurses, industrial hygienists and fire officers as well as safety specialists. The last named act mainly as advisers. Their number and need depend both on the size of the undertaking and the nature of the hazards. The APAU of HMFI quote the following advantages in having a safety adviser[5].

1. He can keep abreast of HS developments and changes in legislation and provide line managers with such information as is relevant to their needs.
2. Training effort can be concentrated and specialist experience widened by seeing a range of problems.
3. He can advise whether the safety policy is being consistently implemented throughout the organisation's premises — particularly important in large undertakings such as local authorities or conglomerates having multiple premises.
4. Co-ordination of safety effort is simplified. He can avoid the duplication of effort that inevitably results from each location or department trying to resolve its own problems in isolation.

There is a clear role in the process industries for professionals who can advise and assist management in controlling the hazards common to most industries, and who are familiar with the law on safety. While they require some technical education, this need not always be to degree level. Otherwise they should be full members of the appropriate professional organisation, i.e. the Institution of Occupational Safety and Health. One prime requirement is that they should be good communicators. Hearn[12] has given a good description of their work and duties, of which an extract follows:

> In small or medium-sized units a safety officer may be appointed for duties which cover industrial accident and fire prevention, road safety, security, welfare, personnel, etc., but in large units a safety officer specialising in safety and hygiene is essential.
>
> He should advise on the formulation of a company's safety policy and guide all employees on the implementation of this policy, and he must be given the status necessary for him to carry out his duties *vis-à-vis* all levels of line management.
>
> *Duties of safety officer*
>
> The duties of the safety officer will depend to a great extent on the size and nature of the works. In general they will consist of advice on safety measures to all members of management, and to specialist employees such as architects, designers and purchasing agents, etc.
>
> He should ensure that basic safety principles are incorporated in the design stage of buildings, plant, processes, storage and distribution, not only to provide a safe working environment, but also to facilitate production. He should advise on the incorporation of safety measures in all operational procedures, machine usage, use of hoists and other lifting equipment; on the provision and usage of personal protective equipment; on the preparation of safety and emergency instructions; on reporting and investigating accidents and the preparation and analysis of records; on training, safety propaganda, safety incentive schemes, etc. He must at all times work in close harmony and collaboration with line management and the trade unions to ensure that no aspect of safety is neglected.

Such safety officers cannot, however, be expected to assume responsibility for a variety of special hazards in the process industries which they do not have the necessary technical training to recognise.

This still leaves a need for a safety professional at a higher technical level who may not be *au fait* with all the possible hazards which may arise in a process plant but can nevertheless discuss technical matters and hazards on equal terms with technical executives, audit their work, assess their own attitudes to safety, and strive to improve these where necessary.

Petersen[11] gives the following criteria on where safety professionals should be located in an organisation:

1. report to a boss with influence;
2. report to a boss who wants safety;
3. have a channel to the top;
4. perhaps — install safety under the executive in charge of the major activity.

In large organisations where several HS personnel in different fields are employed, the question often arises whether they would not be better organised in a separate department, responsible to a director or top manager. Arguments in favour are that it ensures a channel to the top, that all members of the department have a common objective and are not liable to find themselves reporting to hostile bosses. Against this it might be argued that a separate HS department can become so isolated from other professional staff that it defeats its own objectives.

An organisation chart for a large chemical company (making both toxic and flammable products) in which the HS functions are slotted in where there is most demand for their services is shown in Fig. 19.2. This contains three functions exclusively concerned with HS, i.e. health, safety, and fire brigade.

The HS organisation at Dow Chemical Company's King's Lynn site is shown in Fig. 19.3.[13] This includes both full-time HS professionals (the safety officer and the industrial hygienist) as well as departmental managers and other senior staff and workers' safety representatives. The works manager heads the organisation.

The safety council is the policy-making authority for site safety. It consists of the works manager (chairman) and the safety officer (secretary) as permanent members, with one member from each department having an accident-prevention committee. Meetings are held monthly and their minutes are circulated to departmental heads and displayed on the main notice boards.

The safety officer who is directly responsible to the works manager is responsible for the incident controller, one of several senior managers appointed in rotation to co-ordinate during an incident on or off the site.

The health and ecology council is the policy-making body for health and ecology and meets on a quarterly basis. The works manager is again

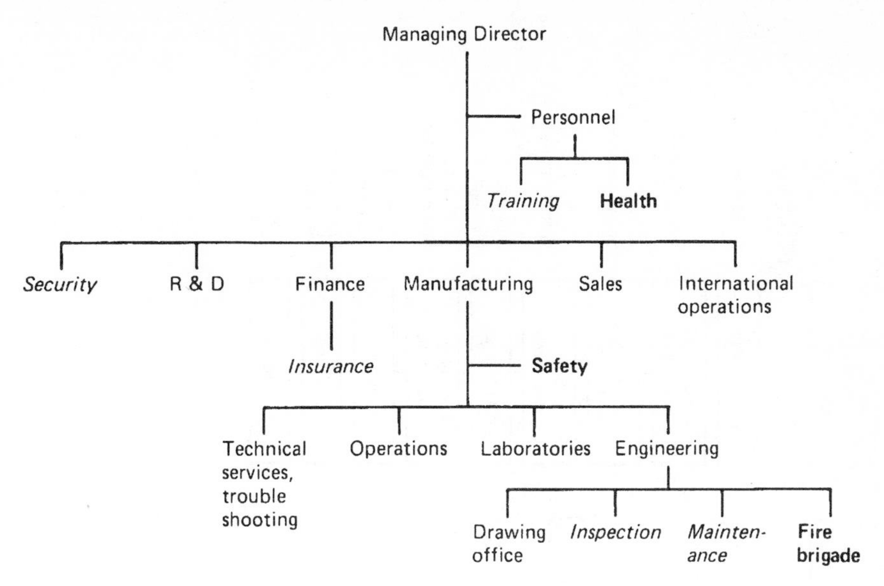

Fig. 19.2 Organisational chart for a chemical company making both toxic and flammable products

chairman, the industrial hygienist is secretary and the other members are plant managers and appropriate staff. The council deals with industrial hygiene, environmental health and waste disposal and the occupational physician reports to it.

There are also several part-time safety directors, usually senior staff members, who monitor specific areas of the safety programmes of which they have special knowledge. This work is very important since it would not be practical to employ a full-time safety specialist in each of these areas.

19.1.9 Safety committees

Safety committees may be set up either as a result of a management decision or at the request of two or more workers' safety representatives. In either case workers should be represented as well as managers and supervisors since they are the best people to ensure that hazards which may affect them are removed or contained. Small departmental committees with no more than 12 members each are more effective than a large organisation-wide committee. The departmental accident-prevention committees (APCs) at Dow Chemical's King's Lynn works shown in Fig. 19.3 provide a good example of this. Each departmental head is encouraged to form his or her own accident-prevention committee and may act as chairman. Small departments may amalgamate to form a single APC and rotate the chairmanship. Each APC is required to draw up and work to an annual safety programme which includes the following elements:

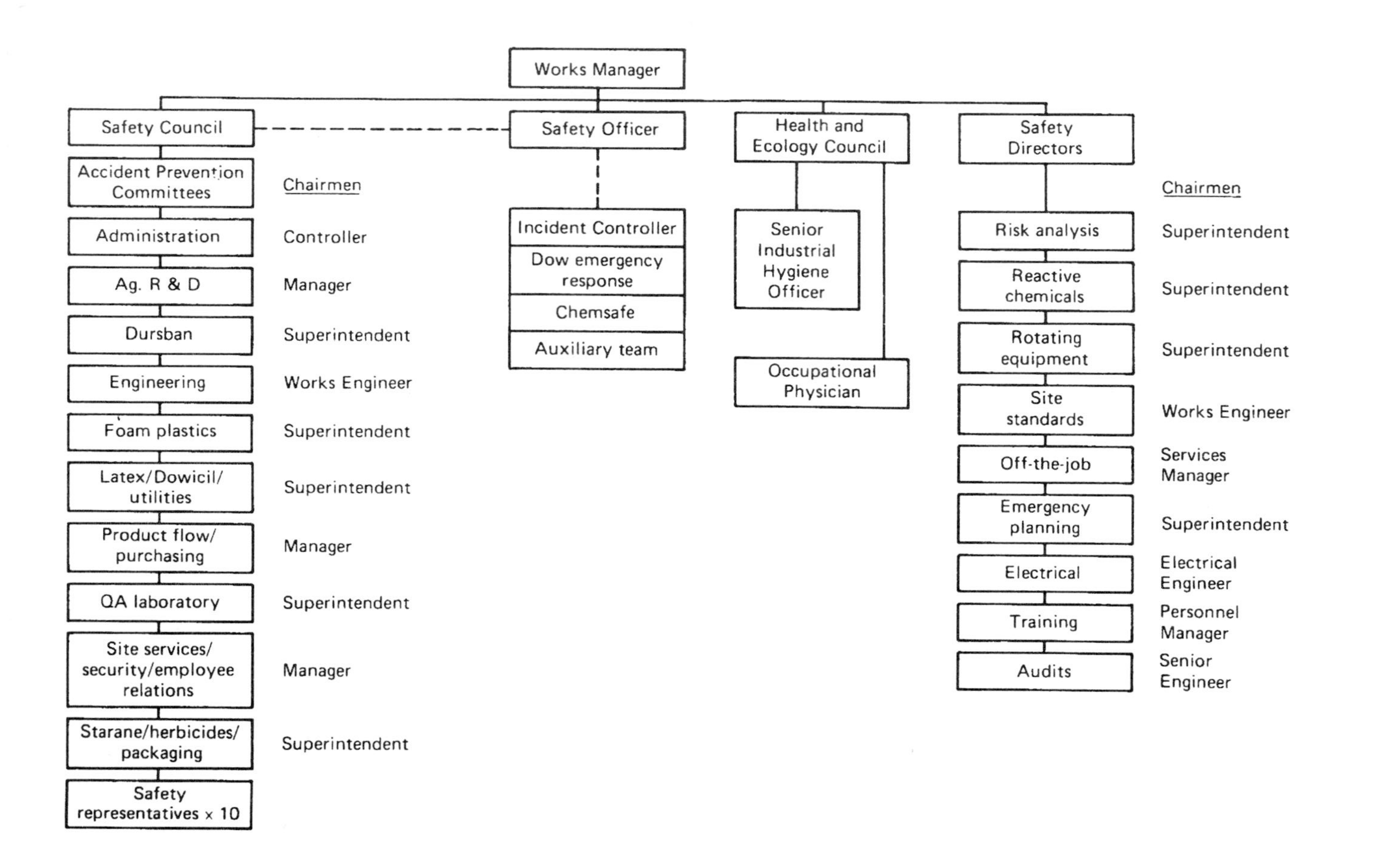

Fig. 19.3 Dow Chemical Company's site safety organisation at King's Lynn

1. *safety meetings* involving all employees for a minimum of 10 hours per year for process, craft and service personnel and 3 hours per year for office staff;
2. *safety inspections* (at least one per month);
3. *fire drills and simulated emergency trials* (two per employee per year);
4. *training* including job-related, firefighting, breathing apparatus, permit-to-work systems, first-aid, etc.;
5. *monitoring and enforcing safety standards and practices;*
6. *job safety analysis* with regular reviews and use in training.

Each APC should send a report or minutes to the safety council every month.

Each departmental head appoints one or more safety representatives who receive special training and are encouraged to take an active role in the safety programme.

19.1.10 Documentation

Most managements are well aware of the large amount of paperwork which must be kept and filed. Lees[9] lists some of the principal subjects on which documentation is needed for a chemical plant, with details of the documents themselves. The subject areas only are listed in Table 19.2.

A secure and efficient central filing system is required, with arrangements to ensure that working copies of documents are available to those needing them. Duplicate copies should be kept on microfiche or computer discs at a different, secure and fireproof location in case the only existing copies in the site office are destroyed by fire or explosion (as happened at Flixborough).

19.2 HS programmes and their elements

The following matters, most of which are discussed elsewhere in this book, should be considered when preparing a complete HS programme:

Table 19.2 HS-related subject areas on which documentation must be kept

Company systems	Fire protection
Standards, codes and legal requirements	Plant operation
Organisation	Training
Process design	Safety equipment
Plant layout	Hazard identification and assessment
Mechanical design	Security
Services design	Plant maintenance
Electrical, civil, structural design	Plant inspection
Plant buildings	Emergency planning
Control and instrumentation	Environmental control
Effluents, waste disposal, noise	Medical

1. The involvement of the safety department and of relevant line management at the planning stage and the analysis of health and safety factors in new projects [15].
2. The HS performance criteria required of new plant, machinery and equipment [12].
3. Evaluation of toxic properties of process materials and the precautions required in their use [7].
4. Instructions for the use of machines, for maintaining safety systems and for controlling health hazards [7].
5. Specific training for operatives, particularly those whose activities affect the safety of other workers.
6. Arrangements for medical examination and biological monitoring [7.7].
7. The provision of personal protective clothing and equipment, in consultation with those who have to use them [22].
8. Permit-to-work systems [18].
9. Emergency [20.3] and first-aid procedures [7.9].
10. Procedures for visitors and contractors.
11. Relevant instructions at all levels.

The main elements in the programme fall under the following headings, each of which is then discussed briefly (apart from (3) which is discussed in Chapter 17):

1. information and communications
2. training
3. hardware and protective systems
4. inspections and audits
5. accident reporting
6. measuring safety performance.

19.2.1 Information, communications and consultation

Personnel at all levels should know precisely where to get the information and decisions needed for their work. Management should ensure that all such information is correct, adequate and relevant for those who receive it. Most enterprises have developed their own jargon and abbreviations. To ensure that these are generally understood, it is recommended that a glossary of jargon, technical terms and abbreviations used within an enterprise be included in a handbook issued to all new entrants.

The problem of communications within an organisation varies with its size. In a small company where everyone knows everyone else and the boss is on first-name terms with the staff, communication is seldom a real problem. Person-to-person communication is the accepted norm.

In a large concern the directors are only names to most employees and are often separated from them by many layers of management. Internal communications then require special care to ensure that they are effective. Procedures and channels of communications between different departments should be established as close as possible to the working level. It is then important to know who is responsible for providing the various items of information which a particular group or individual will need. Permits for maintenance personnel to work on process plant provide examples of such procedures [18.3].

Communication has many pitfalls, including:

- language and translation problems [18.9.6, 19.3 and 23.4.1];
- inundation with too much information, much of it irrelevant to the needs of the person and the job [3.3.3];
- problems of communication between different shifts working on the same plant [20.2];
- ambiguity and poor self-expression;
- lack of explanation, and essential details;
- inadequate consultation;
- information lost or distorted through being relayed through too many intermediaries between its originators and those who need it.

A common fault of management is to instruct employees to achieve a certain result without explaining the reasons for it or how it is to be achieved. In many cases the only way to ensure that unusual instructions apply and are properly understood is to present them first as a draft to the recipients and then discuss it with them. Only when management is sure that the instructions apply and are understood are they presented in their final form. This, of course, takes rather longer than the simpler course of issuing a straight edict. The results, however, are far more satisfactory and pay for the extra trouble taken.

Consultation plays a major role in HS matters and requires a type of manager who is ready and willing to understand other people's point of view. Although consultation is easiest on an informal basis, formal channels of consultation between management and employees often have to be established. There is, however, a danger of consultation being carried too far and placing too great a strain on managers.

Information is liable to be lost or distorted during transfer when it is passed by word of mouth through several intermediaries. If the originators of the information do not have the time to discuss it with all its recipients personally, they should take pains to brief thoroughly those whose task it will be to explain it.

19.2.2 Training, codes and standards

Training methods are discussed in Chapter 21. Management is responsible for organising the training of its staff and workforce, particularly new employees. This must be appropriate to their work and its hazards and to the safety content of their job descriptions. Management should keep individual training records, review training annually, and ensure that every individual receives sufficient training, instruction and supervision to carry out his or her work without risk to health and safety. Special attention should be given to 'one-off' jobs which should be reviewed in advance jointly by supervisory staff and those who will do the work. Specialist training should be provided for safety representatives, for those joining firefighting and first-aid teams and for those who will have special duties in an emergency. Management needs to monitor the quality of the training given.

Codes and standards are discussed in Chapter 2. Management must ensure that it is familiar with and has copies of all relevant codes and standards and that these are incorporated into the appropriate training programmes and manuals. It must be alert to the issue of new and revised standards, particularly those affecting health and health monitoring, and ensure that steps are taken to comply with them. Management must ensure that records are kept of all inspections made to check compliance with standards. It should maintain a system of internal company standards for areas not covered by national or other published standards or where existing standards are inadequate.

19.2.3 Safety audits and measuring safety performance

Safety audits, surveys and inspections are valuable management tools so long as their recommendations are followed. They are used for assessing the strengths and vulnerability of management systems and technical features in HS programmes. The following definitions are given in the Chemical Industry Safety and Health Council (CISHEC) publication *Safety Audits*[14].

- A *safety audit* examines and assesses in detail the standards of all facets of a particular activity. It extends from complex technical operations and emergency procedures to clearance certificates, job descriptions, housekeeping and attitudes . . . An audit might cover a company-wide problem or a total works situation (say, its emergency procedures or effluent systems) or simply a single plant activity.
- A *safety survey* is a detailed examination of a narrower field such as a specific procedure or a particular plant.
- A *safety inspection* is a scheduled inspection of a unit carried out by its own personnel.
- A *safety tour* is an unscheduled tour of a unit carried out by an outsider such as the works manager or a safety representative.

The type of audit discussed here covers all aspects of a process site, plant or unit in regular production. It should be carried out by a team whose members are not involved in the plant or activity being audited. The expertise of the team should be compatible with the type of audit. Audits which delve into technical matters such as pressure vessel inspection require appropriate specialists. It is beneficial to include line managers of other plants or units in an audit team as well as one previous auditor of the same unit. Self-auditing, e.g. by the management or safety committee of the same plant, is, however, a valuable complementary exercise.

Audits are carried out in a formal way using a carefully drawn up checklist of items and descriptive standards for each item. Usually there are four standards for each item examined: poor, fair, good and excellent. Suitable checklists and standards for the chemical industry are available from BCISC[14] and Dow Chemical Company[15], while Lees[9] gives extensive references on safety audits and hazard identification.

Dow[15] recommends four features in a safety audit.

1. *Pre-audit survey questionnaires* which are handled by line management.
2. *Establishment of standards of performance* through interviews with management and examination of documents and records.
3. *Employee perception and implementation* through interviews with employees to assess their knowledge, involvement in and perception of the HS programme. These are supplemented by observations of employees at work by knowledgeable observers to assess compliance with site rules and standards.
4. *Inspection/observation of work environment* to identify work hazards from unsafe design, lack of protective features and exposure to materials in or evolving from the process.

Dow also recommends that advance warning of an audit be given. The object of the audit is to correct faults rather than find them, and the announcement that an audit will be made results in the elimination of some obvious faults and hazards. Auditors need to develop their powers of observation and perception in spotting disguised hazards. Dupont has published a programme[16] which helps to develop such skills.

A pre-audit meeting with the management of the plant, etc. being audited is advisable to ensure that the audit activities go smoothly and that personnel are available to be interviewed at mutually convenient times. Auditors should not hesitate to ask to see documents (e.g. operating instructions) to verify verbal statements. Both collective and individual interviews should be used, the former encouraging group dynamics while the latter often reveal information which would not be given in front of colleagues. The anonymity of persons interviewed must be respected.

A line manager or supervisor of the plant, etc. under audit should be asked to accompany the auditors inspecting it. He should be informed of all corrections and improvements required by the auditors so that he can start taking the necessary steps before the audit report is submitted to management. The main object of the inspection should be to determine whether the layout, design and condition of equipment and protective features are up to standard and to ensure that the protective features will work in an emergency.

The audit team should give a verbal report to management on completion of the audit followed by a clear and concise written report within two weeks. This should highlight features in urgent need of improvement.

Dow's outline of a site or plant audit falls under six main headings:

I PROGRAM FUNDAMENTALS
II FUNCTIONAL ASSISTANCE TO LINE
III PHYSICAL FACILITIES
IV MAINTENANCE
V GENERAL COMMENT
VI SPECIAL SYSTEMS

There are an average of about eight sub-headings for each main heading with an average of about six items to be audited for each sub-heading. As examples, under the sub-heading *Accident Investigation* which comes under main heading I, there are five audit items — Injuries, Losses, Potentially severe incidents, Cause determination and follow-up, and Analysis and trends. Under the sub-heading *Building construction* which comes under main heading III there are also five audit items — Type (open/closed structure), Materials (combustible walls, floors, roof), Structural members (fireproofing), Explosion relief, Smoke and heat venting.

As an example of the standards for activities audited, the following are given for *Supervisor safety training*.

Poor All supervisors have not received basic safety training
Fair All supervisors have received basic and some specialised training
Good Annual training required on some phase of safety and loss-prevention programme with documentation
Excellent In addition, specialised sessions conducted on specific operational problems.

It is clear that a thorough safety audit is a demanding exercise which requires meticulous planning. The need for such audits depends very much on the hazards of the processes used, on the worst consequences which could result, and on the actual HS record of the site, plant or unit. An independent safety audit is the most thorough method of measuring safety performance.

A list by Williams[17] of subjects for which audits are recommended and their frequency is given in Table 19.3. Most of these, however, come under the definitions given here of safety surveys and inspections rather than full audits and not all relate exclusively to safety.

19.2.4 Accident reporting and investigation

Employers in the UK process industries are legally bound under RIDDOR[18] to report many types of accident and most industrial diseases which

Table 19.3 Some safety audit activities

Activity	*Description*	*Interval* (months)
Plant safety review	Adequacy of operations, equipment and building safety	12
Job safety analysis	Standard operating procedures, to be updated where necessary	12
Operator review	Check for deviation from standard operating procedures and work habits	6
Supervisors' safety meetings	Education, training, drills, follow-up	3
Management development seminar	Development of management competence	1
Supervisory training	Training of foremen for supervisory role	1
Safety committee	Motivational safety suggestions	1
Plant managers' meeting	Communication, education, training, innovation, follow-up,	1
Foremen's meeting	Communication (vertical, horizontal), motivation, education, training	
Critical incident technique	Observation of unsafe acts, conditions. Reports of near-misses	1[a]
Central plant safety committee	Safety policy	As needed
Safety review committee	Review of safety of new processes and/or equipment	As needed
Works safety procedures review	Review of works safety procedures	As needed

[a] Plus continuous observation as needed

occur on their premises to the area office of HSE. They are also obliged to keep written records of these events. The investigation of accident causes is an important part of any HS programme and is frequently required to meet claims under common law.

Accidental injuries and industrial diseases in the UK which were sustained in the course of work and which keep an employee off work for more than three days should be reported by or on behalf of the employee, to the Department of Social Security (DSS). This serves several purposes.

1. It enables the victim to claim industrial injury benefit while off work. This is significantly more than unemployment benefit and is tax free.

2. It may form the basis for a claim for a disability pension under National Insurance.

3. It may form the basis for a lawsuit against the employer.

On receipt of the victim's report, DSS will send the employer two copies of a report form which gives some details of the victim's report of the accident. The employer is required to supply further details by completing both copies of the report form and returning them to DSS.

Managements must establish procedures to:

- ensure that they are informed promptly when an accident happens, and whether anyone was injured;
- provide first-aid and call for an ambulance, medical assistance, the fire brigade and the police where necessary;
- investigate and report on the causes of the accident, and delegate appropriate persons to carry this out;
- keep an adequate record of all accidents which are reported to HSE under RIDDOR and/or form the basis of a claim by an accident victim on DSS.

Managers' first action on hearing of an accident should be to ensure that these things have been done. In the event of a serious injury, management should inform HSE, the victim's family and the employer's insurers immediately, and should obtain details of the injuries as soon as possible. In the case of a fatality, the police and coroner's officers should also be informed.

Once any victims have been removed for treatment and everything necessary has been done to prevent further injury and loss, the site of the accident should be isolated and nothing should be disturbed. The supervisor of the area where the accident occurred should make an immediate examination. In the case of serious accidents, further examination by appropriate specialists such as safety advisers, chemists and engineers should be made. Photographs and samples should be taken and tests made which might shed light on the accident cause or prove useful in future training, taking care not to destroy evidence in the process and to preserve samples of materials involved for further tests.

Most serious accidents in the UK are investigated by HSE and many of their accident reports are published. This should not prevent the enterprise which had the accident from making its own investigation first, providing it does not thereby destroy evidence or otherwise prejudice the official investigation.

The investigation should start as soon as possible while witnesses' memories and evidence are fresh. Where a workers' safety representative requests it, a joint investigation should be carried out. Injured persons should, whenever possible, be interviewed to obtain their versions of events. The immediate investigation should be concerned with assembling information which enables the causes of the accident to be established rather than who was to blame. Since most accidents, especially major ones, have more than one physical cause, the investigator should not be content with finding a single cause, but look also for contributory and sub-causes. Once the technical causes are clear, management needs to discover the personal and organisational factors which allowed the accident to happen. Prompt steps should be taken to remove physical causes and correct organisational weaknesses to prevent a repetition of the accident, and to ensure that all its lessons are properly learnt. The scope of accident investigation is greatly widened when all accidents, whether causing injury or not, are investigated as part of a damage-control[19] programme.

Accident reports should contain the following features:[20]

- title
- contents list (unless the accident was minor and the report short)
- summary
- findings (the information gathered during the investigation)
- conclusions (based on the findings)
- recommendations
- appendices (where necessary, with tables, photographs, etc.).

The investigation of accident causes without apportioning blame is easier said than done, since most people will know who or which department was responsible for different possible causes. Thus in the case of a burst pressure vessel, the cause might lie with its design, its construction, its maintenance, the way it was operated, or with something very unusual for which nobody could be held responsible. It is then hardly surprising that these causes tend to be regarded as implications of responsibility by the persons or departments concerned. This makes the task of the investigators more difficult, especially when the accident was a serious one. Some of the pitfalls in investigating major accidents are listed in the conclusions of a paper[21] which this writer gave on the subject.

1. Accidents do not just 'happen', they are *caused*. The investigator's duty is to find the cause.
2. Frequently a major accident has been preceded by a number of minor incidents with a common cause. It is worth checking for these, both by interviewing personnel and by studying log books and instrument records.
3. Beware of bias in 'data' supplied by the parties concerned. Keep an open mind and investigate every source of data thoroughly. Expect some contradiction in statements.
4. Account for all the evidence. The theory with the least assumptions is the one most likely to be correct. Where alternative theories are plausible, estimate and quote their relative probabilities (e.g. from an event-tree analysis [14.10]).
5. Most hypotheses put forward will be seen by some persons as threats to their reputations and by others as self-justification. They may react in

ways which reduce the chances of reaching an objective conclusion. So keep your favourite hypothesis to yourself until you have enough evidence to nail it conclusively.

6. Remember that people are sensitive to criticism, especially after an accident. Their confidence in your discretion will often improve the chance of the truth being discovered. (This is not to suggest that you should assist in covering it up, although you may have the difficult task of persuading them that it should be revealed!)

Lees[9] gives several references on accident investigation and shows a useful procedure used by the Safety in Mines Research Establishment for examining faulty equipment.

19.2.5 Accident records, statistics and analysis

When considering any record, management should ask a few practical questions about it such as:

1. Where is it kept?
2. Who is responsible for keeping it?
3. Is it catalogued, with cross-references?
4. Is it kept in its original paper form or on computer discs or transferred onto microfilm or fiche?
5. Is it kept because the law says it should be kept? If so, for how long? Are there other copies and who keeps them?
6. Who has access to it?
7. What is its value and why is it being kept?
8. For how long will it be kept before being discarded?
9. What does it cost to keep it?

Of these questions, (7) is often the most difficult to answer. This writer finds there is a sort of 'Peter Principle'[22] about records which says 'You never need them until you have thrown them away'. It is truly remarkable how often following a major accident, records which might have shed valuable light on it are missing.

Most managers keep records which are important to them in their own offices, but there is a limit to how much paper can be kept there. Decisions are constantly having to be taken whether to discard a record entirely, make a microfiche before discarding it, or consign it to a central archive (if there is one). These decisions should be based on a common system within the organisation and not left to the whim of the individual or the moment.

Apart from legal obligations to keep them, most accident reports with details of investigations and the causes found are worth keeping in their entirety for possible future study. New facts may come to light which put a totally different complexion on the causation. When a new accident occurs under similar circumstances, it is important to know whether the causes were the same and if so, why the lessons of the first one were not applied. The same applies to medical, occupational and accident records of employees and to records of monitored health hazards such as noise and harmful substances in the working environment. These may contain useful data for epidemiologists in correlating diseases with particular substances or conditions. They may also provide data for assessing the need and effectiveness of protective measures, e.g. ventilation equipment. Fire records provide similar evidence about fire protection. Records of instrument and equipment failures and plant engineering inspections are essential to maintenance and safe operation. They also provide useful data on which to judge the reliability of instruments and equipment in particular service conditions. This may influence future designs and decisions on equipment purchasing.

Petersen[11] recommends that the system of injury records set up should enable the line manager's safety performance to be judged. For this he recommends:

- that accident records should be kept by the supervisor (by department);
- that they should give some insight into how the accidents seem to be happening;
- that they are expressed eventually in terms of dollars by department (by supervisor);
- that they conform to any legal and insurance requirements.

It is certainly useful to classify accidents and accidental injuries, not only by cause but also by severity, and in the case of injury, by the parts of the body affected, the occupation of the injured person and where he or she was working. It is best here to follow some recognised method of classification, either the one recommended by the International Labour Office (ILO),[23] or that used by the HS authority of the country concerned. There is no need to use complicated classification systems in their entirety. A few selected headings appropriate to the industry or enterprise are usually enough, sometimes with one or two special additions.

The ILO classify accidental injuries in five different ways.[23]

1. *According to the degree of injury.* Fatal, permanent disablement, temporary disablement and minor injuries.
2. *According to the type of event causing the injury.* Here there are nine main categories (e.g. falls of persons, being struck by flying objects), each with several sub-categories.
3. *According to the agency*, of which there are seven main agencies (e.g. machines, means of transport), each with several sub-divisions.

4. *According to the nature of the injury*, of which there are 16 headings such as fractures, dislocations, contusions and sprains.
5. *According to the bodily location of the injury*, of which there are seven main headings (e.g. head, neck and trunk) and a number of sub-headings.

Classified injury figures are far more use to management than total injuries, which give little guidance to prevention or protection.

When accidents which include losses and/or near-misses as well as injuries are recorded, it is important to classify them by cause as well as by effect and cost. These are essential features of a total-loss-control programme[24] which should result in the elimination of many hazards before they cause serious injuries. Heinrich's famous dictum[3] should be remembered:

> For a group of similar accidents, for every one causing a major injury there will be 29 producing minor injuries and 300 near misses.

This is not, of course, a hard-and-fast ratio. The actual ratio depends mainly on the occupation. That of minor to major injuries is lower for steel erectors than for carpenters.

Accident statistics are more useful when making comparisons or looking for trends than mere lists of accidents, but one should beware of making comparisons unless the statistical significance of the figures is known. Boyle[25] gives a salutary diagram, shown here as Fig. 19.4, of steps needed when designing a project using statistical analysis. He concludes by advising a trial check of the proposed analyses.

> The main reason for doing this is that it enables an estimate to be made of the time the complete analyses will take, there being little point in collecting so much information that the analysis takes too long to be useful. An additional benefit of a trial run is that it often shows problems which were not identified earlier, and enables final refinements to be made to the formulation of the questions before the full analysis is commenced.

Two comparative methods of presenting injury statistics are as frequency rates (FR) and as incidence rates (IR). Unfortunately there are differences in the bases of frequency rates recommended by the ILO and used by the UK authorities. In the USA the ILO recommendations are followed. These are:

$$\text{FR (frequency rate)} = \frac{\text{Total number of accidental injuries (of stated severity)} \times 1\text{ million}}{\text{Total number of man-hours worked}}$$

$$\text{IR (incidence rate)} = \frac{\text{Total number accidental injuries per year} \times 1000}{\text{Average number of workers at risk}}$$

In the UK accidental injury frequency rates are more usually quoted per 100 000 hours worked except for fatal injury frequency rates, which are quoted per 100 million hours worked.

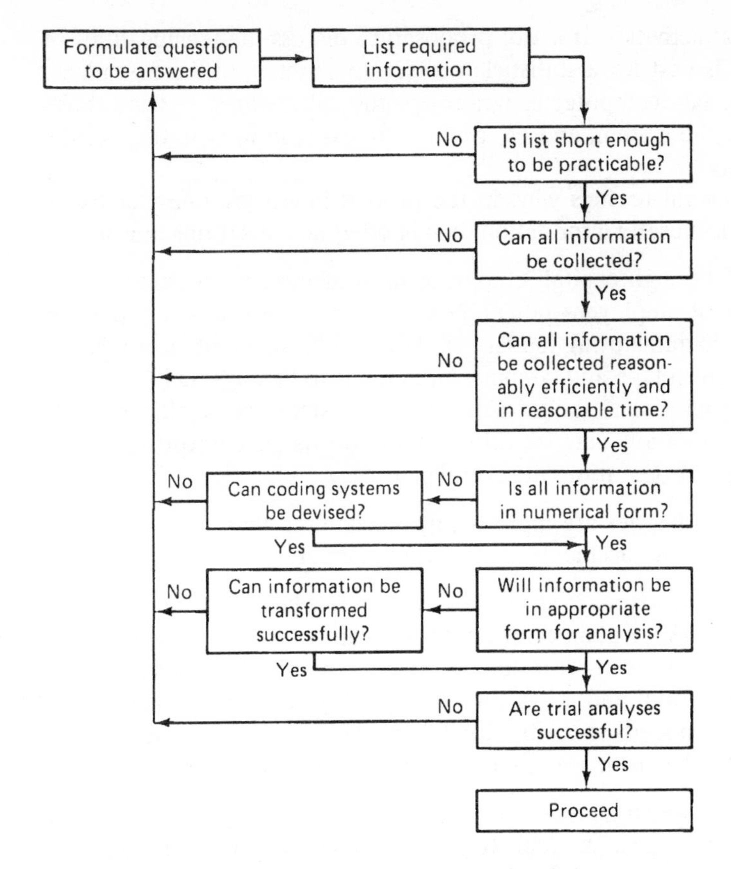

Fig 19.4 Steps in designing a project using statistical analysis

Another important indicator is the accident severity rate (ASR). In the USA this is given as:

$$ASR = \frac{\text{Total days charged} \times 1 \text{ million}}{\text{Employee-hours of exposure}}$$

In the UK it is given as:

$$ASR = \frac{\text{Man-hours lost} \times 100\,000}{\text{Man-hours worked}}$$

Some American companies like to combine their frequency and severity rates into a single measure, the frequency severity indicator, which is given as the square root of the product of their frequency and severity rates divided by 1000.

Several schemes have been proposed for presenting statistics of damage to plant, but none, according to HSE's APAU, have widespread support.[5]

The analysis of accident statistics requires the same statistical tools, especially analysis of variance, as those used for quality control and many other industrial activities. It is not proposed to discuss the technicalities of the subject. It is best for a statistician to set up a system in the first place, choose appropriate computer programs for the calculations, instruct those who will apply the system, and be available to assist in interpreting results where necessary.

There are several reasons why, in the process industries, injury statistics tend to supply less useful information than in other manufacturing industries.

- Because of the high capital: employee ratio of most process industries, the number of employees per enterprise of given size is lower than in most other manufacturing industries. As a result, the number of injuries per year of given type and circumstances tends to be too small for differences from one year to the next to have much statistical significance. Of course, the situation may be different in very large enterprises which carry out similar operations at several centres.
- The process industries are more highly mechanised than most others so that there are fewer accidents arising from manual work.
- The process industries are characterised by a few large accidents each of which may involve a number of employees, but which occur only rarely, perhaps once only or never in the lifetime of a plant. This leads to a high 'standard deviation' and a low level of significance in the annual injury statistics. It also leads to a situation where the emphasis is concentrated on preventing the last major accident rather than the next one.

Another factor which increases apparent accident rates in all industries is an improvement in accident reporting. Managements should be prepared for this and be careful not to blame supervisors for a rise in accident rates when they should be commending them for better reporting.

One benefit from the relatively small number of employees and hence accidental injuries in the process industries is that it allows more time and effort to be spent on investigating near-misses and non-injury accidents. Special attention should also be given to accidents during maintenance work and the operation of permit systems [18] and those caused by health hazards in the working environment [7.7].

19.3 Special management problems

Health and safety pose some difficult and often unexpected problems to management. Some which the writer has met are discussed here briefly.

19.3.1 Organisational blind spots

Whatever the technical cause of an accident, there is usually a blind spot in the organisation which allowed it to happen. The disasters discussed in Chapters 4 and 5 provide clear examples of such blind spots. As the subject has already been discussed in 3.3, little further needs be said on it here. In high-technology fields where major hazards exist, the need for truly independent monitoring of management decisions is recognised as perhaps the only way of avoiding blind spots. Unfortunately, political factors often intervene which undermine the independence, integrity and objectivity of the monitor. A former colleague, when introduced to a new consulting assignment of this nature, addressed those of us already working on it thus: 'Spare me the facts. Just tell me what are the politics of this assignment.' He showed a certain worldly wisdom in this question and was soon in a leading role. This does not only happen in consultancy but is found in all walks of life, including the management of large companies. By checking at the outset what the client or boss wants to hear, the resourceful consultant or junior can ensure that he does not upset him. This is fine so long as the ideas of the client or manager are sound. But when the consultant or junior merely reinforces firmly held misconceptions, the last state is worse than the first. Unfortunately, the management which most needs independent advice is often the one most wedded to its own misconceptions, which are difficult to remove without a damaging confrontation.

Of the factors which contribute to blind spots in the minds of otherwise competent people, mental stress is perhaps the most important and is discussed next.

19.3.2 Stress — a contributory factor

As Atherley has pointed out,[26] individuals and even entire organisations are sometimes unable to cope with the mental stresses generated within industry. This leads us to consider briefly the following psychological danger points which Lord Ennals in a foreword to Kearns's readable book[27] considered were 'totally ignored' by the Robens Report.

1. *Overpromotion* — stress, anxiety and often breakdown can follow from responsibilities beyond a person's capacities.
2. *Underwork* — sometimes agreeable for a time but leading to dissatisfaction, doubts about the worker's capacities, demoralisation and frequent spells of absence for minor complaints.
3. *Job definition* — it is essential that the employee should know the requirements of the job and to whom he or she is responsible. Uncertainties over these issues can become a major strain. The resulting stress can be taken by seniors as implying inadequacy for the job, when the real cause of trouble is lack of clarity by higher management.
4. *Lack of effective consultation and communication* — an all too common fault of management. Sudden unexplained changes in policy, take-overs and fears of redundancy can cause stress disorders. There is a correlation between the level of morale of employees and the quality of concern of senior management for the people they employ.

5. *Lack of financial security* — is especially felt by manual workers who have no job security, sick pay or pension schemes, which clerical and professional groups largely take for granted. This has obvious deleterious effects upon the psychological and physical health of workers.

Further consideration of these factors lies outside the scope of this book.

19.3.3 Changes in hazard awareness

Many serious hazards, especially to health, have only been recognised belatedly after they have been present in industry for many years. Examples are alpha naphthylamine, asbestos, vinyl chloride monomer, benzene and beryllium. In some cases the increased awareness of the hazard has led to a total ban on the use of the offending substance. Here managements are faced with a clear-cut choice: either to change to the use of a safer substance, or, where this is impractical, to shut down the operation altogether. In other cases the increased hazard awareness has led to very much tighter physical controls. In the case of vinyl chloride monomer (VCM) the controls needed are so stringent that the normal decision would probably have been to ban its use. The material is, however, so widely used and the investment in its production and use is so high that industry responded to the challenge by devising control methods to cope with the situation. Major companies such as ICI and Shell, which are normally in intense competition and take pains to conceal their research from each other, co-operated to develop the required analytical and control techniques.

While large companies generally have the financial and technical resources to improve control techniques in line with greater awareness of existing hazards, the problem can be a very serious one for smaller companies. It is also very much more difficult and costly to apply tighter controls to an existing plant which was designed for lower control standards than it is to apply them during the design of a new plant.

19.3.4 Problems in growth and decline

Technical management tends to be at its weakest both during the rapid expansion of a company or its use of a particular process and also during periods of falling sales and declining production.

In the first case the reason is that the technical expertise is very thinly stretched. This appears to have been the case at Flixborough [4], where the plant was only one of eight or nine similar ones which were designed and built within a few years for operation by DSM and its subsidiary companies. The top management in most cases were staff seconded from the parent company. This needs to be borne in mind when considering the organisational weaknesses which allowed the disaster to happen.

In the second case the organisation or one branch of it may be running down its staffing levels and economising as far as possible in preparation for final closure. All process plants are profitable only above certain production levels. Below a certain throughput the operation can only run at a loss. Sometimes a conscious decision is then taken to shut down the plant for good, sometimes to put it into 'mothballs'. Often a conscious decision is deferred in the Micawberish hope that something will turn up to rescue the situation. Something of this sort appears to have been the case with Union Carbide India at the time of the Bhopal tragedy [5.3]. Perhaps managements of large companies ought to consider creating and training special shutdown teams for controlling the closure of very hazardous plants.

Top management needs to be specially alert to the risks involved in the growth and declining phases of a major hazard project. In the first case the main need is for more experienced staff. In the second case both personnel and money are needed.

19.3.5 Problems of subsidiary companies

Of the five major accidents discussed in Chapters 4 and 5, three occurred on the site of an operating subsidiary of a foreign transnational company whose expertise had been developed and whose technical headquarters lay in the country of the parent company. In one case (Seveso), the accident occurred in a subsidiary of a subsidiary. In such cases there is a tendency for the parent company to supply only sufficient information to its subsidiary as it considers necessary to operate and maintain the plant, and the staff of the subsidiary company are chosen with these objects in mind. If the foreign subsidiary is acquired as a result of a take-over, there is often a brain drain from it to the parent company. It is obviously cheaper to concentrate higher technical effort and expertise at one centre than at several, and modern communications should allow top technical experts in the parent company to deal quickly with problems beyond the skills of the subsidiary company. There is often a gap between the competence of the staffs of parent and the subsidiary company. This is sometimes exacerbated by hierarchical structures, chauvinism and differences in language and standards. These problems are accentuated when the subsidiary company is in a Third World country [23]. Staff of the parent company also seldom feel so involved with problems of the foreign subsidiary as they do with those closer at hand. There is no simple solution to this problem although decentralisation of technical expertise and a less hierarchical organisation should help.

19.3.6 Funding difficulties

A company may experience real problems in providing adequate funds to meet HS objectives when it is operating at a loss and all expenditure is being cut to the bone. There is therefore much to recommend the creation of a special central trust fund for HS expenditure (including loss prevention) for the various projects of a major company and its subsidiaries. This would be quite separate from their individual operating budgets and be a charge on company reserves and current profits. Its purpose would be to ensure that money needed for essential HS expenditure is available when a project or company is in economic difficulties and the integrity of the plant and safety of its workers and the public are most at risk. Items covered by the fund should include the inspection and maintenance of protective systems (such as the vent gas scrubber at Bhopal), general maintenance of plant with significant hazard potential, and the salaries and wages of personnel with essential HS functions. There is evidence that some large companies have adopted similar measures as part of their internal insurance schemes. Governments should give practical encouragement.

19.3.7 Contractors

The safety role of contractors has been discussed in design, during commissioning and during work on site [18.9.6, 21.4]. Accidents during construction[28] lie outside the scope of this book. Management should ensure that the purchasing department of their company issues appropriate safety rules, terms and conditions as part of any contract involving work on site. Every contractor should attend a safety induction course before starting work unless the work is of such short duration that this is impracticable. In that case he or she should be accompanied while on site by an experienced employee of the operating company. All contractors' powered equipment and accessories should have a valid safety certificate issued by an independent authority. Contractors' ladders and steps should be inspected and registered by the appropriate department (e.g. the main stores) before being used on site.

19.3.8 Informing the public

Management should take the initiative in issuing warnings of special hazards and news of emergencies [20.3.2]. If it fails to do this, rumours will grow and the chances of erroneous reports will increase. It is recommended that a senior manager be appointed as the company's sole authoritative source of information in an emergency, assisted where possible by an experienced press officer. Other employees should be instructed not to comment but to refer inquiries to the company spokesman.

19.4 Computers and safety

The present phenomenal growth in the use of computers and robots is comparable to the Industrial Revolution which started over 200 years ago and seems unstoppable. A robot is seen here as a computer with the equivalent of arms, legs, hands, fingers, eyes and ears. The robot may not actually move but sit, as it were, permanently at a desk and control a large process plant.

There have been many failures in the use of computers, partly because those who had bought them had failed to analyse their own objectives, and partly because employees who had to work with them wanted to prove that they would not work.

Properties in which computer-controlled machines and robots outclass human workers are given in Table 19.4.[2]

Robots are chosen for jobs which are dull, dirty and dangerous (the three Ds), and hot, heavy and hazardous (the three Hs). There is more concern, however, over the employment of 'smart robots' equipped with 'expert systems' which are capable of doing the work of highly trained specialist staff. An expert system is defined as:[29]

> A computer system which reflects the decision making processes of a human specialist. It embodies organised knowledge concerning a defined area of experience and is intended to operate as a skilful, cost-effective consultant. An expert system comprises a knowledge base, inference engine, explanation programme, knowledge refining programme and natural language processor (Fig. 19.5).

Barnwell and Ertl[30] describe an expert system computer program as a 'shell' into which knowledge is placed by a human expert, when it becomes a 'loaded shell'. They believe that the use of expert systems in the process industries will become commonplace. Typical design applications include pipe sizing, control valve selection, materials of construction selection and steam/condensate system design. Plant management applications include capital cost estimating, reactor simulation and optimisation and utility usage strategy. (Shells could probably be loaded for most subjects discussed in this book.) The power and effectiveness of the plant manager or design engineer is greatly enhanced when he has several such loaded shells at his disposal, while fewer human professionals of his kind will be needed. One danger which many fear in this is that decisions may be taken on the basis of knowledge encapsulated in chips and about which the engineer or manager responsible for the decision is largely ignorant. The expert system

Table 19.4 Properties for which robots are preferred to human workers

Property	*Human*	*Robot*
1. Obedience	Variable, often sensitive to abuse	Good, only does what it is told
2. Dependence on oxygen supply	Complete	Seldom required
3. Resistance to radiation	Poor, needs special protection	Much improved
4. Temperature range	Narrow	Wider
5. Resistance to toxic hazards	Poor, needs special protection	Improved
6. Mechanical power	Low and limited for long periods	Limited only by design
7. Performance of repetitive work	Fair and easily impaired	Good if properly maintained
8. Information searching	Fair	Excellent
9. Disposability	Expensive	Easy

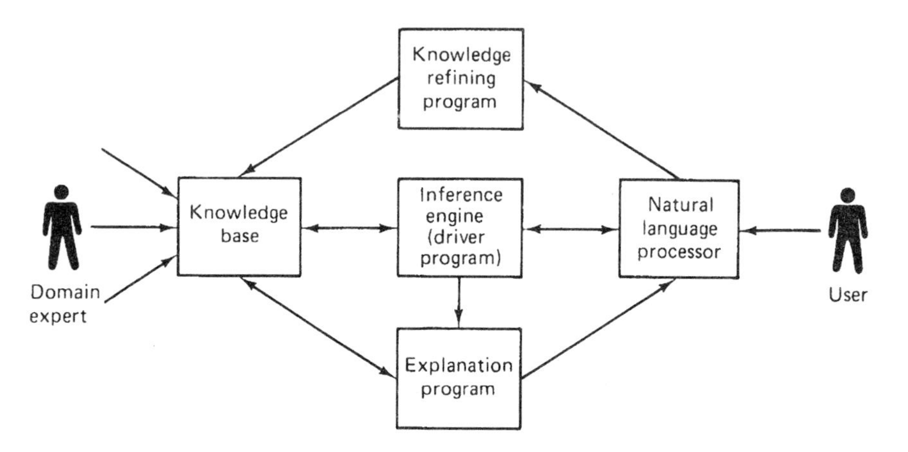

Fig 19.5 Components of an expert system

is a marketable commodity, and while it may solve problems, the basis on which it does so is valuable information which its owner is probably anxious to protect.

Because of their ability to improve the flow and speed the processing of information of all kinds, computers are widely used for all types of office duties in the process industries, as well as for computer-aided design (CAD) and computer-aided manufacture (CAM). The latter generally involve the use of 'programmable electronic systems' which may be used for the on-line control of process plant in 'real time'. (That is, the computer functions simultaneously with the plant in response to signals from it.)

Management must appreciate that computer technology is highly specialised, and its vocabulary often difficult for non-experts to understand. Even words such as *reliability* and *hazard* do not mean quite the same to computer specialists as to other professionals, and communication problems thus easily arise. Managers and safety personnel may therefore need training in the use of computers, particularly in the operation and maintenance of PES systems used to control and protect process plant.

Computers do not prevent the 'pollution of information' (such as leads to the organisational misconceptions discussed in 3.3). Unfortunately they can greatly magnify the effects of human errors while making them harder to trace. Computerised crime such as the theft of valuable information from electronic data banks, and the interference with computer software is also growing and very difficult to trace. According to Miller:[31]

> A programmer with less than a month's training can break the more elaborate procedures currently being used in large data banks within five hours,

and

> The programmer, for instance, could insert a secret 'door' in the monitor programme that would enable unauthorised people to bypass protective devices, or could 'bug' a machine in such a sophisticated manner that it might remain unnoticed for an extensive period.

Computer criminals include 'crackers' and 'crashers':[29]

> A cracker is a hacker who specialises in overcoming software protection systems. A crasher is a hacker who deliberately attempts to cause a serious interference with the operation of the computer system.

Unwittingly I bought an expert system for £13. It is a chess-playing program against which I can play, using the word-processing computer with which I wrote this book. Its skill can even be downgraded to match its human adversary. I was once a fair chess player, having played in a local team, yet now I can hardly beat the program at its second-lowest skill level. So I have no doubt it is an expert system!

Chatting one day with a seven-year-old friend Adam, whose mother is a computer consultant, and has a computer and chess-playing program similar to mine, I told him how hard I found it to beat the program. 'Why, it's simple,' said Adam, 'I can beat my mum's computer any time I want to.' Intrigued, I coaxed from him the secret of his success. 'I simply get the computer to change all my white pawns to queens, and then surround the black king with them.' I did not think the program allowed me to do that, and on checking found that it did not. Adam had 'fixed it'. He has a brilliant future open to him as a software cracker!

References

1. Drucker, P. F., *The Practice of Management*, Heinemann, London (1955)
2. King, R. W. and Majid, J., *Industrial Hazard and Safety Handbook* Butterworths, London (1980)
3. Heinrich, W. R., *Industrial Accident Prevention*, 4th edn, McGraw-Hill, New York (1968)
4. Lord Robens's Committee, *Health and Safety at Work*, HMSO, London (1972)
5. The Accident Prevention Advisory Unit of HM Factory Inspectorate, *Managing Safety*, HMSO, London (1981)
6. HSE review of work of The Accident Prevention Advisory Unit of HM Factory Inspectorate, *Effective policies for health and safety*, HMSO, London (1980)
7. Dow Chemical Company Corporate Safety and Loss Prevention, *Minimum Requirements*, 3rd edn, Dow Chemical Company, Midland, Mich. (1984)
8. The Council for Science and Society, *Superstar Technologies*, Barry Rose (Publishers) Ltd, London (1976)
9. Lees, F. P., *Loss Prevention in the Process Industries*, Butterworths, London (1996)
10. Powell, P. I., Hale, M., Martin. J. and Simon, M., *2000 Accidents*, National Institute of Industrial Psychology, London (1971)
11. Petersen, D. C., *Techniques of Safety Management*, McGraw-Hill, New York (1971)
12. Hearn, R. W., 'Management responsibilities for safety and health', in *Industrial Safety Handbook*, edited by Handley, W., 2nd edn, McGraw-Hill, Maidenhead (1977)
13. 'Health & Safety Policy and Principles of the Dow Chemical Company King's Lynn, 1st July 1986'
14. Chemical Industry Safety and Health Council, *Safety Audits*, London (1973)
15. Dow Chemical Company, *Guidelines for safety and loss prevention audits*, Dow Chemical Company, Midland, Mich. (1980)
16. E. I. DuPont, 'Safety training observation program' in *Safety Training Course for Supervisors*, DuPont Wilmington, Delaware
17. Williams, D., 'Safety audits', in *Major Loss Prevention in the Process Industries*, The Institution of Chemical Engineers, Rugby (1971)

18. S.I. 1985, No. 2023, *The Reporting of Injuries, Diseases and Dangerous Occurrences Regulations* (*RIDDOR*), HMSO, London
19. Bird, F. E. and Germain, G. L., *Damage Control*, American Management Association, New York (1966)
20. Adrian, E. W., 'Accident investigation and reporting' in *Safety at Work*, edited by Ridley, J., 3rd edn, Butterworths (1990)
21. King, R. W. and Taylor, M., 'Post accident investigations — in the aftermath of a catastrophe', in *Eurochem Conference — Chemical Engineering in a Hostile World*, Clapp & Poliak Europe Ltd, London (1977)
22. Laurence, J. and Hull, R., *The Peter Principle*, Souvenir Press, London (1969)
23. The International Labour Office, *International Recommendations on Labour Statistics*, Geneva (1976)
24. Fletcher, J. A. and Douglas, H. M., *Total Loss Control*, Associated Business Programmes, London (1971)
25. Boyle, A. J., 'Records and statistics' in *Safety at Work*, edited by Ridley, J., 3rd edn, Butterworths (1990)
26. Atherley, G. R. C., 'People and safety — stress and its role in serious accidents', in *Chemical Engineering in a Hostile World*, Clapp & Poliak Europe Ltd, London (1977)
27. Ennals, D., 'Foreword' to Kearns, J. L., *Stress in Industry*, Priory Press, London (1973)
28. King, R. W. and Hudson, R., *Construction Hazard and Safety Handbook*, Butterworths, London (1985)
29. Longley, D. and Shain, M., *Data and Computer Security — dictionary of standards, concepts and terms*, Macmillan, London, (1987)
30. Barnwell, J. and Ertl, B., 'Expert systems and the chemical engineer', *The Chemical Engineer*, No. 440, p. 41 (September 1987)
31. Miller, A., 'Personal privacy in the computer age', *Michigan Law Reporter*, **67** (6) 1090–1246 (1969)

20 Plant commissioning, operation and emergency planning

The term 'commissioning' as used here includes all activities needed to bring a newly constructed plant into regular production. These activities, which require a great deal of planning, organisation and training, are usually carried out in four stages. These correspond to specific points in the hand-over of the plant by the contractor to the owner.

- Inspections, cold trials and preparations made before mechanical completion.
- Pre-commissioning, in which steam and other utilities are introduced, further preparations and hot-running trials are carried out, and any necessary engineering corrections are made.
- First start-up. This begins with immediate preparations for the introduction of process materials, and continues until the plant is operating with all systems working at or near design throughput. It may typically last for a month. It usually includes test runs to prove licensors' and contractors' guarantees of yield, throughput, product quality, utility consumptions, etc.
- Post-commissioning. This term refers to the period which starts when the plant first comes on-stream and ends when it has settled down to regular production. It may take the best part of a year during which adjustments and modifications are made, faults are corrected, performance is brought up to scratch and test runs are carried out.

Operation [20.2] includes normal start-up, normal and emergency shutdown and most activities carried out by the operating or production department. Emergency planning [20.3] covers both controllable emergencies which lead to a safe plant shutdown (e.g. failure of electricity, cooling water or an internal plant function) and uncontrollable ones caused or accompanied by fire, explosion or toxic release. There is no sharp dividing line between these two types of emergency. What starts as an apparently controllable emergency sometimes escalates into a more serious one requiring speedy evacuation of plant personnel as well as plant shutdown. In planning for emergencies, local authorities and their fire and medical services will be involved as well as HSE.

The subjects of this chapter are so extensive that it is possible to deal with them only generally, with special emphasis on safety. The importance of starting to plan and train for commissioning, operation and emergencies *while the plant is still under design* cannot be overstressed. These exercises also usually reveal design shortcomings which, if not corrected, would hinder first start-up or cause hazards in operation.

20.1 Commissioning

The following critical dates correspond to points in the four stages of commissioning listed earlier:

1. the date of mechanical completion when the plant or unit [11.1] is provisionally accepted by the owner;
2. the date when process materials are first introduced into the plant or unit;
3. the date when the plant or unit is first brought on-stream;
4. the date on which test runs which meet performance guarantees are completed to everyone's satisfaction and the plant or unit is fully accepted by the owner.

Up to (1) the plant is normally under the control of the contractor, and from then it is under the owner's control.

The utilities (power, steam, etc.) and other offsites (storage, effluent treatment, etc.) on which the plant or unit depends should, if possible, have been commissioned before (2), although sometimes they have to be started up at the same time as the plant. Regular supplies of raw and auxiliary process materials must be available before (2) as well as proper means of handling and disposing of the products, which initially may be unsaleable.

The hazards and safety precautions needed undergo an abrupt change when point (2) is reached. The constructional hazards which previously predominated now give way to the hazards of the process materials, which may be flammable, toxic or highly reactive.

On large and complex plants some units may be still under construction while others are already in operation. The above critical dates are then staggered, to allow effort and personnel to be deployed most effectively. Where different units of a plant, or different plants in a new integrated complex, can be operated independently by introducing or withdrawing semi-processed material, it is generally best to get the last unit of the plant completed and working first, using bought-in materials. This enables markets for the product to be built up and problems (including product quality) solved before the front end of the plant is working. As an example, on a polymerisation plant, the powder-handling unit comprising drying, blending, pelletising and bagging operations might be commissioned first with imported wet polymer powder, and the equipment run-in before starting up the polymerisation unit. This is not then held back by difficulties with the powder-handling section.

Temporary hook-ups involving pumps, pipes, hoses, drums, tanks and hoppers, etc. are often needed for circulation of cleaning chemicals and other fluids during pre-commissioning, and for the transfer of process materials when the commissioning of a plant or unit is staggered. Such once-only operations, especially the use of hoses, tend to introduce hazards and need careful checking for safety.

Staged commissioning makes it necessary to erect temporary fencing to segregate areas where construction and welding are still proceeding from those in which process materials are being handled. Flammable and toxic process materials must be prevented from entering pipes and drains in areas where construction is still being carried out.

20.1.1 Roles, organisation and planning

Commissioning involves the personnel of several organisations with different interests, e.g. the operating company (and often its transnational parent), the main contractor, the process licensor, suppliers of packaged units and specialised equipment, sub-contractors and independent inspectors working on behalf of government, insurers and the plant owner. Many different but overlapping activities have to be crammed into a short time-span. Unless the roles and responsibilities of these organisations and their personnel have been carefully worked out, understood and agreed in advance, conflicts are likely to arise. From the date of mechanical completion and provisional acceptance by the owner, commissioning is best controlled by a single person, usually the owner's commissioning manager. In cases where the owner has no suitably experienced staff, he may arrange to have the contractor start up the plant, hand it over in a fully operational state, and train his personnel. Few contractors, however, have teams of experienced operators capable of starting up a complex plant. Owners thus generally need to recruit and train their operating and maintenance personnel well before start-up.

The owner should set up a task force at the start of design to ensure that its basis is safe and realistic, vet design work, carry out HAZOP and other safety studies, prepare operating and maintenance manuals, permit forms, checklists and other documentation and advise management of progress and problems. The task force should be led by the commissioning manager, preferably an experienced production specialist who will later be the plant's production manager. The task force should include the plant's project manager, process design manager and maintenance manager. Other experts including safety specialists should be available to assist the task force.

After completion of design, the task force leader moves to site to head the commissioning team where its members familiarise themselves with the plant, check that the contractor's work complies with design drawings and standards [20.1.2], complete the extensive documentation needed for commissioning, and arrange and negotiate any modifications found necessary for safe operation and maintenance. They will be responsible for recruitment, training, liaison with site contractors, plant acceptance testing, pre-commissioning and first start-up. The commissioning team has three main sections, each with its own leader, as well as experienced shift leaders for start-up. See Fig. 20.1.

The commissioning team should include the key personnel who will later operate and maintain the plant, together with extra experienced personnel. If these cannot all be released temporarily from other plants, some may have to be recruited on a short-term basis. Panic recruiting can result in square pegs in round holes. The writer recalls one fitter on a commissioning team who seemed strangely ill at ease with spanners. Enquiry revealed that he had previously been a tailor's fitter!

The human problems can be quite as serious as the technical ones and confrontation is often necessary to resolve them. There has to be clear agreement between the members of a commissioning team about who will do what and when. Group discussion is needed to allot tasks and ensure that each member is able and willing to play his or her part, especially when this involves working long hours on shift and coping with the unexpected. Good training [21] of personnel at all levels is also vital to successful commissioning. The commissioning manager should visit and become familiar with other plants which use the same process. For a large modern plant controlled from a central control station, where process information is shown on video display units, there is a growing trend to use a specially designed simulator to train operators [21.3.5].

20.1.2 Mechanical completion and provisional acceptance

The site contractor is responsible for building the plant in accordance with the terms and specifications of the contract. The owner's construction

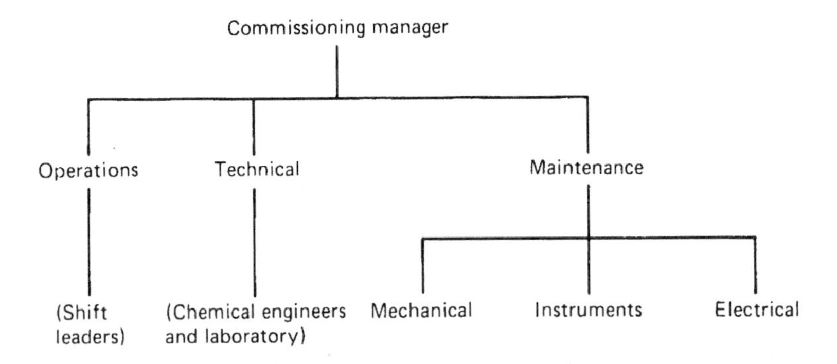

Fig. 20.1 Structure of a commissioning team

organisation, reinforced by maintenance personnel from the start-up team, checks the contractor's work as it proceeds. Good communications are essential and prompt and effective feed-back from site to the design office is necessary.

Electrical and compressed air systems are usually commissioned before mechanical completion of the plant or unit.

On mechanical completion the contractor carries out tests, inspections and checks such as those outlined below in conjunction with the owner's commissioning team, which carries out a complete inspection following a planned procedure. Each item is checked against a systematic checklist and the completed checksheets are incorporated into a dossier. Typical instructions include:

- Check that there is safe and unrestricted access for operators, lubrication and other workers to valves, switches, gauges, pump glands, sample points and other items where they may need to work during operation. Look out for tripping hazards and other booby traps and valves with inaccessible handwheels. Check that guard rails are securely fitted to the sides and floor openings of platforms and stairs, that vertical ladders are enclosed, and that non-slip flooring is used.
- Check that there is unrestricted fire brigade access and adequate water supplies.
- Check that no process fluids can enter the unit via feed and product lines, drains, vent and blowdown lines. Check that spectacle plates, line blinds and blank flanges are in position at the battery limits and that these can readily be turned or removed when first start-up begins.
- Check that pipework conforms to P&ID [15.10.4] specifications and is not unduly strained at joints.[1]
- Check that all pipes and equipment connected with unrestrained expansion bellows are designed and tested to withstand the thrust forces generated by the bellows at their maximum operating pressure.
- Check that screwed plugs are used only for air, water and nitrogen at pressures below 7 bar g. and on pipes not larger than 1½ inches NB.
- Check that welds have been inspected as per contract and found acceptable.
- Check that all gaskets, bolts, studs and other fastenings used in joints on pipes and equipment are suitable for the service conditions and that the fastenings have been correctly tensioned.
- Check that pipes are properly supported for both ambient and operating temperatures and pressures and for any likely vibration.[1]

- Check that spring-hangers are in sound condition and correctly set with retaining pins in position.
- Check that there are adequate valved vents and drains on high and low points on pipework, including temporary ones installed solely for pre-commissioning, and that they are in working order.
- Check that there are no drain holes on relief valve discharge lines inside buildings, nor on those handling flammable and toxic fluids when there is risk of a dangerous discharge to the atmosphere.
- Check that there are no small-bore pipe projections or equipment that can be accidentally broken off.
- Check that all small pipe branches, especially drains, are sufficiently clear of pipe supports.
- Check that drain lines lead to the correct drainage system and do not discharge over paved areas.
- Check completion of insulation and cladding except for flanged joints, for which properly fitting insulating covers should be ready to install after pre-commissioning.
- Check that painting, colour coding and numbering of pipes and equipment are complete.
- Check that valves can be easily identified, preferably by numbers marked on them or on adjacent pipework [20.2.1].
- Inspect internal parts of vessels, drums and heaters.
- Check that pipework is clean internally and clean it (chemically) where necessary.
- Pressure test process and utility equipment and pipes.
- Test, tag and check installation of relief valves.
- Test and commission electrical systems, check motors for rotation and run them uncoupled.
- Exercise motor control centres adequately.
- Flush, clean and fill lubricating systems and bearings with correct lubricants.
- Install pump seals and packings and suction strainers.
- Align and couple shafts of rotary machines and test-run them cold.
- Remove transport restrictions from instruments, and clean, pressure test, lubricate, adjust and calibrate them.
- Check proper interconnection and functioning of instrument loops including alarms and trips.

- Pressure test, check and commission pneumatic and hydraulic instrument systems and stroke and check control valves. (It is now possible to pre-commission, check and adjust the control system of a complex plant before any process materials are introduced with the aid of a process simulator.)[2]

All outstanding tests and faults brought to light are recorded onto 'punch cards' which are divided into two types:

1. faults and omissions,
2. Site alterations (which are generally to facilitate operation and maintenance and frequently for safety).

On completion of this work the plant unit is provisionally accepted by the owner.

20.1.3 Pre-commissioning

Pre-commissioning follows mechanical completion on which it is dependent and is carried out by the commissioning team with assistance from the contractor. All utility, effluent and other auxiliary systems which do not involve toxic or flammable materials are now commissioned, and all process systems and equipment are prepared for start-up. Typical tasks during pre-commissioning are:

- Blow out steam pipework with steam, commission steam and condensate systems, check steam traps.
- Commission all non-hazardous utility systems systems which are not yet working (e.g. cooling and process water, inert gas).
- Check, test and commission effluent and effluent-treatment systems.
- Check and test vent and pressure relief disposal systems.
- Dry out furnace linings.
- Water-wash and/or gas-purge piping systems, and drain water from low points.
- Dry piping and equipment as required.
- Install flow elements and orifice plates.
- Run steam turbines uncoupled and later coupled, where possible, to compressors and other driven equipment.
- Where possible, circulate suitable safe fluids through process systems with drivers coupled to pumps and other machines; bring these up to normal operating temperature and align them correctly.
- Remove temporary drain and vent valves and close openings as specified on piping drawings.
- Install non-hazardous solid catalysts, drying agents, adsorbents in process vessels, with supports and retaining grids etc.

Detailed reports are made on all work done during pre-commissioning including repairs, modifications and improvisations. A copy of these should be kept by the works maintenance department. Table 20.1, based on Pearson,[3] lists questions to be asked before first start-up.

20.1.4 First start-up and post-commissioning

These two phases are here treated as one. The dividing line between them is the final acceptance of the plant by the owner at the end of start-up. By

Table 20.1 Questions to be asked before process fluids are admitted to a new plant

1. Have the following been removed from the new plant area?
   (a) Contractors' huts, tarpaulins, etc.;
   (b) Non-flameproof equipment;
   (c) Rags, paper, wood, rubbish, dry grass, weeds.
2. Are the following in place and ready for use?
   (a) Perimeter fence and gate(s);
   (b) Security gateman and cabin;
   (c) Hazard and safety notices (no smoking, matches, general vehicles, etc.);
   (d) Nitrogen and/or other inert gas purge systems;
   (e) Oil/water separators and effluent-treatment plant;
   (f) Fire alarm system;
   (g) Fire main, hydrants, hoses, monitors and foam system and supplies;
   (h) Fire extinguishers;
   (i) Eye-wash bottles, first-aid boxes and kit;
   (k) Emergency personal showers;
   (l) Water sprinklers and deluge systems;
   (m) Steam hoses;
   (n) Gas detectors;
   (o) Plant lighting (normal and emergency);
   (p) Pressure relief, flare and/or blowdown system;
   (q) Instrument air and electrical supplies and back-up;
   (r) General utilities, water, power, steam, fuel, etc.
3. Have the following been informed of the plant start-up and its consequences?
   (a) Construction personnel (including restrictions on smoking, welding, etc.);
   (b) Fire services;
   (c) Local authorities;
   (d) Neighbouring plants;
   (e) Records section.
4. Other questions
   (a) Are operating and maintenance personnel properly trained and organised?
   (b) Have drains been flushed and checked free of obstructions?
   (c) Can isolation blinds, etc. at battery limits be readily turned or removed?
   (d) Have shift fire and first-aid teams been nominated and trained?
   (e) Has all welding been done?
   (f) Is there an effective permit system [18] for further engineering work, particularly welding, which may be required?

then most of the temporary and visiting personnel present during commissioning will have left. There may, however, still be many outstanding problems. It is usually left to the regular operating, technical and maintenance personnel to cope with them, although there is often a need for additional experienced staff for 'trouble shooting' during post-commissioning.

First start-up is usually the owner's responsibility. It is the acid test when both aspirations are realised and errors and omissions are relentlessly exposed.

First start-up should not begin until pre-commissioning is complete and all questions listed in Table 20.1 can be answered 'Yes'. The first steps in first start-up are usually to:

- introduce any hazardous solid catalysts, adsorbents, etc. into the plant;
- remove all line blinds and turn all spectacle plates at the battery limits on pipelines carrying hazardous materials into and out of the plant or unit;
- purge the plant or unit with inert gas and leak test any joints broken and remade since the last leak test [10.3.2];
- line up valves in the plant or unit as detailed in the operating manual for the initial introduction of process fluids.

The start-up now depends entirely on the process and should follow the procedures laid down in the operating manual. First start-up (like regular operation) consists very largely of opening and closing valves in the right order and when certain pre-conditions are met. The special hazard of mistaken valve identity is discussed later [20.2.1].

Batch and semi-batch processes can usually be started up in a more relaxed manner than continuous ones, since the former usually allow the process to be interrupted to take stock of the situation. Most continuous units consist of a number of stages. It is important to ensure that the products from any stage have reached their design composition, temperature and pressure before being admitted to the next one. Where the stage is one of physical separation only, such as fractional distillation, the process piping should allow its products to be mixed and recycled to the inlet of the stage. This enables the stage to be brought on-stream on total recycle until its products reach the required composition, before entering the next stage. Where the stage involves chemical reaction, such a recycle may be impossible. It is thus sometimes necessary to provide for withdrawal and disposal of off-specification reactor products until these have reached the required composition. Once the start-up of a continuous process gets under way, its operation becomes a very dynamic affair which often requires both quick and correct reactions from the operators.

At the same time, start-up cannot be rushed. Sufficient time must be allowed in heating up equipment or allowing operations such as distillation, crystallisation and reaction to settle down to a steady state. Catalytic reactors can be specially temperamental and sometimes seem to be governed more by black magic than by science. Two examples from personal experience are given. In one, several weeks were spent trying unsuccessfully to start a stirred semi-batch reactor in which gases were to react with solid particles immersed in some of the liquid product. Only then was it discovered that the stirrer was too high in the reactor and not reaching the liquid at the start of the reaction. In another a fixed-bed gas-phase catalytic reactor failed to show any reaction at all for several weeks. After an expert on the process had visited the site, tried several remedies without success, and left, the reaction suddenly started when everyone had lost hope and the process was about to be written off. From that time the reaction never failed to start and the reason why it failed earlier was never discovered.

One hazard of starting up exothermic reactors, both semi-batch and continuous, is that the reaction may fail to start as intended when the concentration of one of the reactants is low. Unless this failure to react is recognised and the flow of one of the reactants is halted until it is cured, the reaction may start suddenly when the concentrations of reactants is high, and become uncontrollable. This is a common hazard of autocatalytic reactions. Sometimes it can be cured by adding reaction products to the reactor before introducing feed materials.

Eventually the whole unit is, hopefully, operating at reduced throughput under instrument control. Controllers are tuned, plant samples are taken and analysed, readings are recorded, and, where appropriate, extrapolated to design conditions. From instrument readings, mass balances are made over sections of the plant and discrepancies investigated. It is advisable to bring the whole plant or unit up to its design throughput as soon as practicable in order to check performance of individual items and of the plant or unit as a whole against design expectations.

Careful monitoring of signs of future trouble is needed during first start-up. These include leaks, vibration, unusual noises, corrosion, differential thermal expansion, cavitation, steam and water hammer. First start-up is a time when damage readily occurs. Furnace tubes may be overheated and warp, compressor and turbine rotors may be overstressed and their shafts bent, and vessel linings and mechanical seals may be damaged by hard particles left inside equipment.

First start-up can proceed smoothly like clockwork, or it can be long drawn-out and beset by many problems. Some plants (such as the two discussed in 13.9) could not be started up at all and had to be written off and scrapped. Others such as the Nypro works rebuilt after the 1974 explosion were scrapped soon after commissioning because they were not economically viable in the changed economic climate. Well-established processes usually give fewest problems. Plants employing brand-new technology and which have been designed on the basis of laboratory and pilot plant experiments tend to give most trouble, particularly when the process fluids pose severe and unusual corrosion or fouling problems.

Troubles in first start-up are frequently attributed to equipment failure, when in fact the wrong equipment had been chosen in the first place. As an example, a high-speed centrifugal pump was installed to handle a polyethylene slurry from a polymerisation reactor. The high shear forces shredded the polymer particles into a mass of fibres resembling asbestos which quickly blocked the pump casings and discharge lines. A similar problem was found when the flow of slurry was throttled by a control valve. Both problems were solved by changing to a slow-running 'Mono' single-rotor screw pump with a synthetic rubber lining, which was fitted with a variable-speed drive. This both pumped the slurry gently and controlled the flow, thus eliminating the control valve.

Similar troubles arise when the equipment is suitable for the normal process fluid in contact with it but cannot cope when the properties of the fluid change significantly. Thus a centrifugal compressor may surge or overheat if the properties of the gas change. A settler separating a light hydrocarbon from water will not work if traces of a strong emulsifying agent are present.

The short residence times in much of the equipment of modern plant often demands a very rapid response which only a finely tuned control instrument can give. Thus first start-up can reveal many a 'chicken and egg' situation which the process designer had not foreseen, although experienced operators often develop special knacks to deal with them. The writer once designed a pilot cracking plant which only a certain green-fingered Norman could start up. When asked what his secret was, he would only say, 'It's all quite simple and I just don't understand what you are doing wrong.' Eventually the plant had to be modified so that less-skilled operators could start it up.

There is an increasing tendency to use computers to assist in commissioning, even when it is not intended to employ them during normal operation. Simulators are used to tune control instruments [17.8.3] and to train operators [21.3.5]. Computers are sometimes linked to the plant instruments during commissioning and used to log data, for simple calculations such as mass, heat and energy balances, and for reconciling discrepancies between instrument readings and suggesting where the fault lies. They may also be used for additional alarms to alert operators of possibly dangerous changes in process variables at an early stage. A further use for computers is to forecast changes in the consumptions of steam and other utilities during commissioning. This can be of considerable help in planning.

It not infrequently happens that a plant being started up has to be shut down, isolated, made safe for welding and modified and the whole process of commissioning begun again. Unfortunately such modifications may nullify HAZOP studies on the original design on which much time and money had been spent, and a further HAZOP study has to be done hurriedly on the modified design.

Generally a process plant reaches its highest state of integrity at the end of the post-commissioning phase, after which it slowly deteriorates. Successive overhauls partly restore its integrity but seldom to the original level.

20.2 Plant operation

The plant manager is responsible for the safe and successful operation of process plant around the clock. For this he must ensure that there are proper job descriptions for the operators and supervisors working under him, that their numbers are adequate and that they are well selected and trained. Many of these will be working on shift, which causes special problems of stress, rest and home-life to which he must respond. He also needs to ensure that fairly standardised procedures are adopted, even when there appears to be more than one safe way of carrying out a particular operation. The main reason for this is that if different shift teams develop their individual operating methods, with different orders in which things are done, continuity is lost at shift change. Jobs which should have been done are left undone and the chances of accidents increase.

These procedures should be incorporated into specific operator training for the plant (including simulator programs [21.3.5]). They also form an essential part of the plant-operating manual which should be available in draft form when the plant is commissioned, even though it may have to be modified in the light of commissioning experience. Many instructions in the operating manual fall into the category of 'Safe systems of work'. These should be highlighted in the text according to their importance to safety.

The plant manager needs to motivate his supervisors and operators and monitor their performance, both during and outside his own normal working day. Above all, he needs to set a good example in giving clear instructions, listening to problems and complaints, following agreed safety guidelines and promptly investigating dangerous incidents whether or not they caused loss or injury. He and his senior day supervisors should occasionally visit the plant at night to audit the standards of safe operation and compliance with laid-down procedures, and he should encourage or reprimand where this is called for.

Before discussing the plant-operating manual, it is appropriate to say something about valve identification and operation, since a high proportion of the operations performed on process plants consists of opening, closing and adjusting valves, which must be done at the right time and often in a particular order.

20.2.1 Identification and operation of hand-operated valves

Mistakes in valve identification, in the order in which they are operated, and failures to pass correct information about valve settings from shift to shift are a potent cause of accidents.

Over 40 years ago while working as a shift supervisor in an oil refinery, this writer mistakenly caused the wrong valve to be opened on a high-octane gasoline plant so that over 100 000 gallons went to the wrong tank. The loss cost the company several times the writer's annual salary. The mistake would probably not have occurred if the valves had been clearly identified. Without valve numbers it is also difficult for shift operators to leave clear reports of the settings of their valves to the following shift.

Clear identification of valves, both on the plant itself and in operating manuals, instructions and P&IDs should be a cardinal point of good operating procedure. Unfortunately this is not always appreciated, and contractors do not usually number hand-operated valves unless the contract requires them to do so. It is thus often left to the operating company to provide valve numbering. This may involve redrawing the P&IDs since there is often not room on the original ones to add valve numbers.

Valve numbers can most simply follow as an extension to the pipe numbers as shown on the P&ID. The valve number is then a composite of the pipe number plus a serial-plus-type number which is unique for that particular valve. Since tags attached to valves have an unfortunate tendency to disappear, and valves are sometimes changed during maintenance, it is recommended that the composite valve number (with an arrow pointing to the valve) be stencilled on the pipe (or pipe insulation cladding) next to the valve connected to it.

The correct sequence of valve operation is sometimes vital, e.g. on batch processes, on start-up and shutdown of continuous ones, when taking samples and when draining water from vessels containing a gas or liquefied gas under pressure. The expansion or evaporation of gas passing through the valve may cause ice to form in it which prevents it being closed. At least one major disaster started in this way (Fézin, 1966). There should be two valves of different sizes in series on such drain lines, the larger one closest to the vessel. To drain water, the larger valve is opened first and the smaller one then 'cracked open' to allow the water to drain under supervision. Should gas escape and the second valve freeze, there should then be no difficulty in closing the first one across which there is little pressure drop.

On plants whose operations are controlled or assisted by PES programs [17.9], the correct sequence of valve operations should be written into the program so that if valves have to be opened or closed manually, the operator receives a 'prompt' on the VDU to remind him of the correct operating order.

Clear information about the settings of all hand-valves must be available and passed on at shift change. This can be given on a log sheet with a list of valve numbers in the first vertical column, and subsequent vertical columns for each shift to mark whether they left the valve open, closed or partly open. Another method which gives a better visual impression is to mount a P&ID showing all valves on a panel, with a small captive screw with a distinctively coloured slot inserted at the centre of each valve symbol to show the state of the valve. Every time a valve is opened or closed the corresponding screw is turned. (The writer first saw this done many years ago but has rarely met it again.) A permanent record of the state of the valves at each shift change can then be made by a Polaroid camera mounted in front of the panel.

20.2.2 The plant-operating manual: general

The manual is probably best drafted during plant design at the same time that the HAZOP study is made, by a small team which includes the commissioning manager (destined later to become the plant manager), the process designer or (design manager) and a safety specialist. A technical writer and an illustrator may also be needed. The question of who writes the manual is, however, less important than that it be written. The team should go through all the motions of testing and prestart-up activities, start-up, operation at normal and reduced throughput, and shutdown, as an imaginative round-table paper exercise with the aid of suitable 'models' (e.g. PFDs, material and energy balances and P&IDs). It should choose between alternative methods of operation before finalising the draft. The order in which operations (including valve opening and closing) are carried should be stated, all valves being referred to by their number. The times required for various operations should be given. Attention should be drawn to operations which have to be completed or brought to a particular state before others are started.

The manual should contain not only detailed operating instructions for all likely situations but also essential information about the processes, plant, equipment, hazards and protective features, as well as P&IDs and other illustrations. The manual may be of loose-leaf type to enable revisions to particular sections to be incorporated. It is suggested that personnel of shift leader status and higher should be given their personal copies of the complete manual, and that a copy be available (e.g. in the control room) for all operating personnel to refer to. Those below shift leader status should be provided with personal copies of those parts of the manual needed for their work, i.e. the operating instructions for the units on which they are working.

So far as safety is concerned, the depth and detail needed in the manual depend largely on the degree of hazard of the plant and the possible consequences of failing to follow the procedures laid down. (Sufficient detail is always needed for any plant to be operated economically.) The manual should be an important part of the training for personnel new to the plant. They should be examined on their understanding of it before they play an active operating role.

The operating manual should always be kept up to date, but revisions should only be made with the signed approval of the plant manager, after considering their implications for safety. If modifications to the plant become necessary during commissioning, the manual should be revised accordingly.

The operating procedures drafted before a plant is commissioned can never be perfect. Thus they should be carefully assessed and revised after first start-up.

Operators often find improvements and short-cuts. These should be recorded and discussed with the plant manager, and, if possible, with others involved in writing the manual, before any revisions are made. There may be sound (perhaps safety) reasons against a suggested short-cut which an operator with limited technical education fails to appreciate. (In Iran the writer encountered uneducated operators who believed that the performance of a distillation column depended on the behaviour of a benevolent *jinnee* inside it!) Warnings with explanations about the dangers of short-cuts should be written into the manual.

The normal operation of a plant may be continuous, batch or a blend of the two. Thus a plant consisting of two continuous distillation columns and a continuous reactor may contain a cyclic adsorption dryer with two drying vessels which operate alternately, one always being on continuous drying service while the other is being purged and regenerated. Although batch processes are usually simpler and easier to control than continuous ones, they involve more operational steps and valve changes, which need to be detailed in the manual.

20.2.3 Suggested contents for a plant-operating manual

The following chapter headings are suggested:

1. Introduction, safety and general
2. Units, symbols and nomenclature
3. Plant description
4. Start-up on commissioning or after major overhaul
5. Normal plant start-up
6. Normal plant operation
7. Normal plant shutdown
8. Emergency shutdown
9. Emptying and purging plant.

The following additional information should be included (e.g. in appendices):

- requirements, properties and hazards of raw materials, chemicals, catalysts and other ancillary materials;
- detailed equipment, instrument and valve lists and performance details such as characteristic curves of centrifugal pumps and compressors;
- operational details of special packaged units;
- lists and details of special safety items and protective systems such as showers, first-aid kit, personal protective equipment, monitoring equipment for harmful substances, pressure-relief devices and disposal systems, firefighting equipment, sprinklers, fire-water systems;
- sampling methods and programme, on-stream analysers and tests (e.g. S. G.) made in plant.

The scope of each chapter is described briefly below.

Chapter 1 should explain the purpose and scope of the manual, the purpose and general principles of the plant, with specifications or typical analyses of the main feed materials and products. It should state the hazards of the process and materials processed, their possible consequences, and describe the protective features used. It should explain and stress the need to follow the procedures given in the manual, especially those categorised as 'Safe systems of work'. For these, checklists should be prepared and issued to operators against which to tick off each step as they take it. It should cover communications with other plants, departments, managers and emergency services, giving their telephone numbers. It should discuss plant staffing, job functions and responsibilities, log books and plant hand-over by outgoing to incoming shifts and arrangements for overlap at shift change (which is especially important for shift supervisors).

Chapter 2 should explain the units of measurement and technical terms used and also the drawing symbols employed for equipment and instrument items. In the absence of company standards, the drawing symbols and abbreviations given in *The Piping Guide*[1] are recommended. These include those of the Instrument Society of America[4] which are widely used internationally.

Chapter 3 should describe the various process, storage and utility units of which the plant is composed, referring to drawings and lists included in the manual and drawing attention to critical and safety features.

Chapter 4 should list and describe the steps required in the initial start-up [20.1.4] in a logical sequence so that each operation follows smoothly from the preceding one. It should list the various problems and hazards which may be encountered, with their symptoms, and explain how to diagnose and resolve them. It should give instructions for the disposal of solid wastes, liquid effluents, vent gases and processed materials leaving the unit, and the steps to be taken to ensure that downstream units are ready to receive the latter. Special attention should be given to the start-up of reactors and how to recognise whether a reaction is proceeding properly or not.

Chapter 5 should repeat those parts of Chapter 4 which apply to a normal start-up (i.e. after a short shutdown). This and the following chapters should be included in the standard operating instructions issued to all operators.

Chapter 6 should include guidelines on:

- pump changes and gland adjustments;
- record keeping;
- checking relief valves, alarms, controls, the temperature of pump glands and bearings, the current consumption of electric motors, and other items essential to safety;
- lubrication;
- Checking and reconciling instrument readings and recognising faults in instrument control loops;

- recognising and dealing with fault conditions, especially when several alarms are triggered in sequence;[5]
- the scope of 'running' maintenance and the conditions under which it may be done;
- relief of excess pressure;
- maintenance of throughput and product quality when throughput, feed composition, product grade, catalyst activity, heat transfer, weather and other process conditions change;
- action required in the event of leaks, instrument failures, excessive vibration and other danger signals.

The ranges of 'safe' operating conditions of equipment and process units and their limits should be discussed. While maximum and minimum operating temperatures, pressures and shaft speeds are easily appreciated, regions of unstable conditions often exist within these limits. Certain concentrations of process materials (e.g. caused by mixing them in the wrong order or proportions) may lead to runaway chemical reactions [8.2] or rapid corrosion [13]. While the plant and its controls should have been designed to give a wide range of safe operating conditions, deterioration in performance and reduced integrity which occur with age can narrow this range considerably. Guidance should be given on the steps to be taken in upset conditions and how to decide when an unscheduled plant shutdown is necessary.[5] The use of condition-monitoring equipment for corrosion, vibration, lubricant contamination, etc. should be explained.

Chapter 7 should give the procedure for normal plant shutdown as a result of shortage of feedstock, full product storage or plant inspection and maintenance. It is assumed here that the shutdown is brief and that equipment will not have to be opened. The final state of the plant including services and the approximate quantities of process materials to be left in each part of it after shutdown should be stated, together with all the steps needed to reach this state. Special requirements on shutdown, such as leaving parts of the plant filled with nitrogen under slight positive pressure, or at a temperature higher than ambient with some means of heating, should be stated.

Chapter 8 should give the procedure for rapid shutdown in emergencies such as fire or failure of electrical power, cooling water, steam or instrument air. Besides total shutdown, it should deal with partial emergency shutdowns which may follow the failure of an important plant item such as a compressor or turbine.

If the plant presents major hazards, the actions required of operators in an emergency shutdown should be as few and simple as possible, so that they can escape almost immediately. There is usually an interval (only 45 seconds at Flixborough) between the first warning of an explosion or other event likely to kill or seriously injure personnel still remaining on the plant, and the event itself. This suggests that the plant and its instrumentation should be designed so that by actuating a special switch, it shuts itself down automatically without creating further hazards. This is most readily achieved when PES control is used. The switch might, for example, do the following:

- shut off the air and/or electric power supply to all control instruments and manual valve loading stations in the control room, thus causing all control valves to take up their 'fail-safe' positions;
- shut off the normal electric power supply, thus stopping electrically powered pumps and other motors, and shutting off electric heaters;
- close remotely operated isolation valves on feed, product and service lines at the plant battery limits.

Such drastic action may itself lead to other hazards which should have been considered and provided for when the plant was designed. This means that the design team (and the HAZOP study team) should have systematically considered the consequences of automatic emergency action for each of the possible events leading to it. For example, the loss of power to the reflux pump of a distillation column with a steam-heated reboiler could lead to a temporary pressure rise in the column and the opening of pressure relief valves, even if the steam supply was shut off when the pump stopped. Another example is the sudden loss of stirring and cooling on a reactor in which an exothermic reaction is taking place. Valves cannot be closed too fast on liquid pipelines without risk of liquid hammer and rupturing the pipe or valve.

It is therefore often necessary to provide emergency supplies of electricity (e.g. from a stand-by diesel-driven electricity generator) and compressed air (e.g. from an air reservoir located on the plant) which come into operation automatically when the normal supply fails, in order to operate certain valves and motors for a limited time at the start of an emergency. Such operations should again be as automatic as possible and should ideally require no action on the part of an operator. Emergency lighting (often from batteries) is also essential and this should switch itself on automatically when needed.

A tricky problem can arise over the starting of an emergency electrical generator which takes over the power supply automatically if the normal supply fails. In this case there may be an interval of a minute or more with no power between the failure of the main supply and the availability of emergency power. There will then be a peak power demand as some motors start on full load. Provided this temporary loss of power can be tolerated by the process, the instrumentation which controls automatic plant shutdown could contain a time-delay which applies only to power failure, so that the shutdown is only triggered if the emergency supply fails to start and continue functioning within a short given period. But if at the time of the power failure the steam and fuel supplies remained normal, the dangers mentioned earlier of excess pressure and run-away reactions would be accentuated. This problem can sometimes be reduced by the use of steam-turbines instead of electricity for critical pumps such as column reflux.

Most plant emergencies which result from a service failure are handled quite safely without any need for plant operators to abandon their posts. For these 'safe' emergency shutdowns it is usually possible to devise a modified shutdown procedure which falls between a normal and a major emergency shutdown.

The manual should explain the consequences of failure of each service in turn, unit by unit, and advise on the most appropriate action to take. It should also deal with the consequences of failure of critical equipment such as compressors and turbines. Thus in the case of electrical power and/or compressed air failure, operators may be able to use emergency power and air supplies to empty excess process materials (especially flammable ones) from the plant to a storage unit as well as shutting manual isolation valves at critical locations. They may even feel compelled to do this in the case of a major emergency, so as not to endanger other lives. (One brave operator at the time of the Flixborough disaster went through the burning wreckage to close manual isolation valves on a liquid ammonia storage sphere when the lines to the sphere had been damaged and the sphere itself had been displaced by the explosion.) Positive action may be required on the part of operators in many fire situations, including small or incipient plant fires which may be dealt with by portable extinguishers, hose reels and steam snuffing hoses, fires on storage tanks which may require the contents of a burning tank to be pumped out, and fires under LPG storage spheres. The last may require the sphere to be depressurised and sometimes water to be pumped into its base, e.g. to prevent the escape of LPG from a leaking joint on the pipework below it. Planning for such fire emergencies should be done jointly with those responsible for firefighting. It must be clearly understood who will have overall control and what the operators' responsibilities will be.

Chapter 9 should describe the steps (discussed in 18.6) needed to empty, clean, and purge each unit and section of the plant after a normal or emergency shutdown in preparation for inspection and/or maintenance. It should state the operators' responsibilities when plant and equipment is handed over before and after inspection/maintenance, and it should explain the permit system used.

20.3 Planning for major emergencies

Under the CIMAH Regulations [2.5.5], all manufacturers who operate major hazard installations must prepare, test and and rehearse emergency plans to the satisfaction of HSE and their local authorities, who are also obliged to prepare emergency plans. Though separate, these plans should be prepared in conjunction and dovetail together. The subject is dealt with by Lees,[5] who gives many references to pre-1980 literature, and in CIA and

HSE booklets.[6,7] More recent papers about on-site[8] and off-site arrangements[9] and sites with toxic hazards[10] include further references. Training for emergencies is discussed in 20.3.9 and 21.3.4. Although the main emphasis is on major hazard installations,[11] a major emergency (defined as one which involves several departments, serious injuries, loss of life and/or extensive damage[6]) can happen on many other installations.

Emergency plans by their very nature deal with incidents which have already started or taken place, and do not prevent them from happening in the first place. Only the main points of such plans are discussed here. The first step in preparing one is to check the safety of the installation as regards its design, operation and maintenance.

The main objectives of an emergency plan are to:

- rescue victims and treat them;
- safeguard others, arranging for their escape or evacuation where necessary;
- contain the incident and control it with minimum damage;
- identify the dead;
- inform relatives of casualties;
- provide reliable information to the news media;
- preserve relevant records, equipment and samples which may be needed as evidence for subsequent investigations.

To achieve these objectives the plan should make the best possible use of works and outside services and personnel. Unlike most plans which start from a clearly defined situation, an emergency plan may have to start from any one of several abnormal ones. Emergency plans should therefore be simple and flexible. They should be tested as they are made, and updated later as necessary. Simplicity increases the chances that the plan can be followed and flexibility allows for adaptation to a changing situation.

While simple, the plan must be appropriate to the emergency which arises. Unfortunately this has not always been the case. A careful objective examination of the main risks is needed to ensure that it does. The most unlikely scenarios such as an aircraft crash or a severe earthquake (in the UK) on the site of the installation can usually be ignored. Those involving the explosion or release of hazardous substances present on the installation must, however, be considered, as must the effects of toxic and ecotoxic substances which may be formed as a result of fire or abnormal plant conditions (e.g. dioxin and the combustion products of polyurethane foams). The 'domino' effects of incidents on neighbouring installations should also be considered. Thus the disruption of cooling water supply to a neighbouring steel plant by the Flixborough explosion in 1974 would have produced a further major emergency had not the fire brigades been able to quickly improvise an emergency water supply.

Nypro's emergency plan at the time which called for all plant and laboratory personnel to assemble in the control room in the event of fire is an example of an inappropriate plan, since it completely failed to cater for the consequences of a large escape of hot flammable liquid under pressure in the plant. A simplified plan of the works is shown in Fig. 20.2. All those already in the control room were killed, and one man who had followed the plan and run from a safe place about 200 metres north of the control room was killed just outside it. Those in the adjacent laboratory who saw the initial release of vapour from the cyclohexane oxidation unit ignored the plan, escaped by running to the north west and survived, despite being caught by the blast of the explosion in the open.

Two lessons might be learnt from this organisational misconception [3.3].

1. A bad emergency plan is worse than none at all because it inhibits independent thought and gives people an illusory impression that they will be safe by following the plan.

2. Persons whose lives and safety may depend on the plan should be consulted on how the plan will affect them, if possible while it is being prepared. They should be encouraged to think out for themselves and make notes on the various possible emergency situations, and how they would react to them. Those responsible for preparing or revising the plan should study these notes and discuss them with their writers before the plan is finalised. By taking such ideas into account when the plan is being made, mistakes and shortcomings may be avoided and greater confidence created. There is a clear role here for safety committees and representatives [19.1.9].

20.3.1 Identification of major hazard situations and assessment of risk

Both subjects are dealt with elsewhere in this book, and in the references quoted. Potentially major hazard installations include:

1. Those where more than minimum quantities (in most cases a few tons) of toxic and flammable materials, particularly gases and vapours, may be released on loss of containment;

2. Those containing more than minimum quantities of the following:
   (a) Any gas contained at pressures of 100 bar g. or higher;
   (b) Liquid oxygen;
   (c) Explosive, unstable and highly reactive materials and mixtures of them.

The minimum quantities (not always identical) are given in the NIHHS and CIMAH Regulations and in the reports of HSC's advisory Committee on Major Hazards.[11]

In assessing the risk of major hazards to people and the spread of damage, the following must be considered:

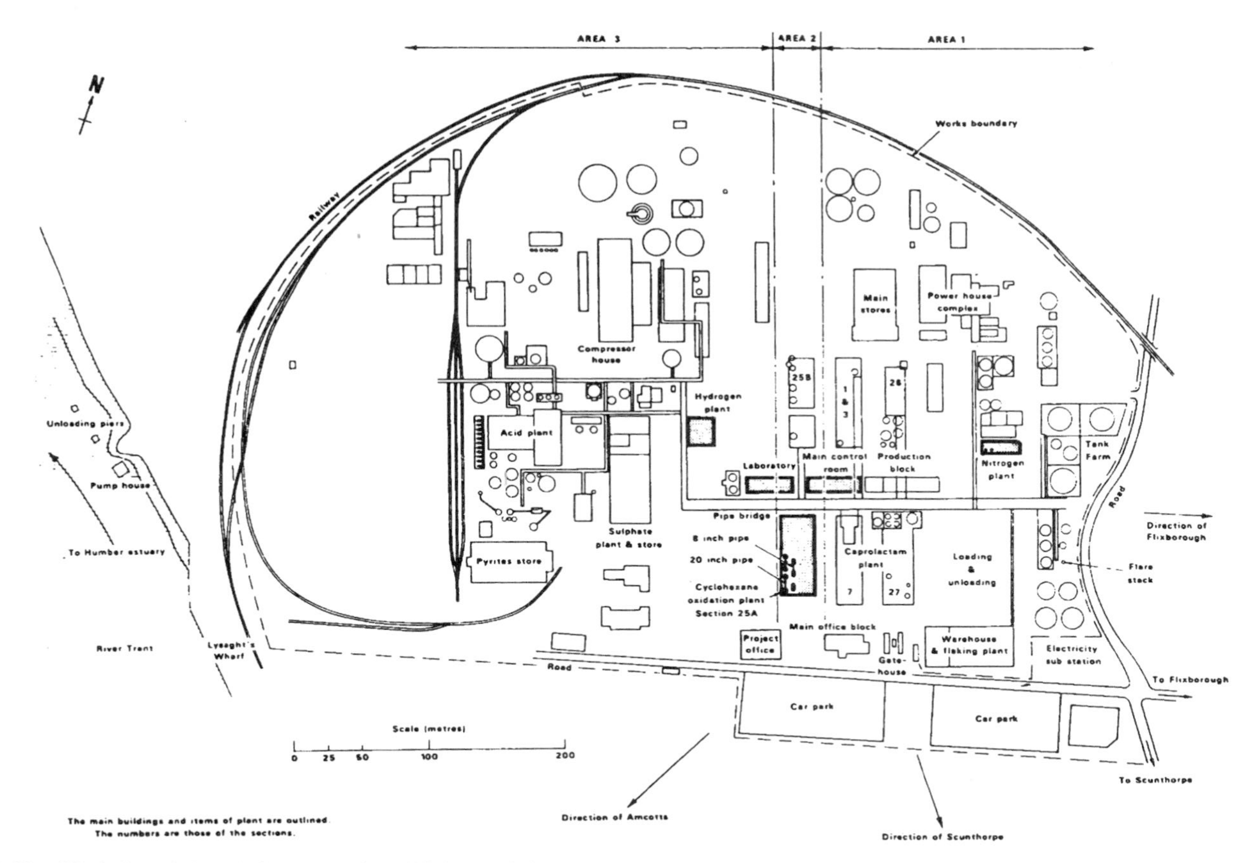

Fig 20.2 Simplified plan of Nypro's former works at Flixborough (courtesy the Controller, HMSO)

1. the type of incident expected (fire, explosion, toxic release) and its probable duration;
2. the location of the incident in relation to neighbouring plants, storage areas and built-up areas;
3. prevailing winds;
4. areas most likely to be affected;
5. population densities in the areas possibly affected;
6. possible damage or contamination of drains, crops, water supplies;
7. possible 'domino effects', i.e. an explosion in one area causing the release of flammable or toxic materials elsewhere;
8. possible effects of collapse of buildings and structures;
9. presence of radioactive sources;
10. topography and physical features of site.

Models may be used to assess the spread and dispersion of emissions of flammable and toxic materials in various circumstances. Since airborne toxic materials are harmful at much lower concentrations than merely flammable ones, their effects extend to much greater distances.

Having assessed the risk, one should examine the adequacy of existing resources (first, from the works and second, from the local fire, ambulance and hospital services) to handle the most serious foreseeable emergency, and then in collaboration with outside services decide what further provision or action is needed.

Typically questions raised at this point relate to:

- adequacy of works firefighting resources before arrival of outside fire service;
- adequacy of drains for fire water;
- ease of isolation of plant and equipment; should more isolation valves be motorised for remote operation?
- adequacy of fire protection of instrument and electrical cables, structural steel and tanks;
- adequacy of relief valves for fire conditions;
- adequacy of plant alarm, evacuation assembly and roll-call arrangements for the scenarios postulated, including the presence of contractors, drivers and visitors;
- alternative methods of communication if the usual one fails or proves inadequate.

20.3.2 Liaison with outside authorities

Outside authorities must be prepared to deal with the effects of major accidents both inside and outside the installations in which they originate.

Resources include police, ambulance and hospital and other local authority services and district inspectors of the HSE. They may also include the services of neighbouring firms under mutual-aid schemes.

In a major emergency the police will co-ordinate the activities of the various emergency services, and a senior police officer will be designated as the Incident Controller.

The fire service, while maintaining a brigade adequate for normal requirements, may not be able to cope with serious incidents in isolated locations. This should be discussed between the works management and the chief fire officer. On being called to an incident, the senior fire officer present will take charge of firefighting, rescue and salvage, and the brigade will also assist in dealing with the escape of toxic materials.

The HSE need to see and approve in advance the plans for dealing with emergencies and satisfy themselves that they are practical, sufficiently detailed and rehearsed. The local HSE inspector should be advised promptly of a major emergency once outside services and key personnel have been informed.

Liaison with outside services should ensure that:

- there is a properly co-ordinated plan which will be effectively controlled;
- works procedures are in harmony with plans developed by outside authorities;
- outside services understand the nature of the risks and have appropriate knowledge, equipment and materials to deal with them;
- the equipment of the works and outside services is compatible;
- the personnel of the works and outside services who are likely to have to co-operate in an emergency already know each other;
- the appropriate type and number of outside services reach the scene promptly.

20.3.3 Works organisation for major emergencies

No universal organisation suitable for all circumstances can be given but the following basic features generally apply. Nominated persons should be trained to fill and deputise for two key roles in a major emergency — *works incident controller and works main controller*. Their duties are spelt out in more detail in the references quoted earlier. They will normally be day staff (e.g. plant or works managers) but senior shift staff should be nominated and trained to deputise for them to ensure that trained persons are available for these roles at any time. If a major emergency arises when the day staff nominee is away, his shift deputy will act for him until he arrives and can take over.

Other personnel who may be needed in addition to the works fire, first-aid, engineering and security personnel include:

- persons with special knowledge of the plants, processes and their hazards;
- checkers for assembly points;
- a mobile analytical team for monitoring the environment for harmful gases and vapours, etc.;
- a public relations officer or team;
- additional incident controllers, telephonists and assistants for the main key staff.

All personnel liable to be needed in an emergency should carry means of quick identification to avoid delays at police check points, etc.

The works incident controller will proceed to the scene of the incident, assess the scale of the emergency, take responsibility and activate the major emergency procedure if this has not already been done. He should wear a distinctive hat and/or jacket, have a portable two-way radio-telephone and an assistant with him. In the absence of the works main controller or pending his arrival, the incident controller will normally direct shutting down and evacuation of affected plant and ensure that outside services and key personnel have been called in.

The works main controller will go to the emergency control centre as soon as he is aware of the emergency and take over from whoever is deputising for him. His duties include:

- calling in outside services and key personnel;
- informing other organisations in the neighbourhood;
- establishing communications and liaising with the works incident controller;
- exercising operational control over those parts of the works outside the affected area, directing the shutdown of plants, and the evacuation of personnel as necessary;
- ensuring that casualties are attended to;
- ensuring that relatives are informed;
- liaising with police and fire services and HSE and advising them as necessary, particularly on the possible effects of the emergency on areas outside the works;
- controlling traffic movement in the works;
- recording or arranging for a chronological record of the emergency to be made (e.g. on a long-playing tape recorder kept in the emergency control centre);
- arranging for personnel to be relieved and provided with food and drinks;

- contacting the local meteorological office for early warning of weather changes;
- issuing statements as required to news media;
- liaising with his company's head office;
- controlling the clean-up and rehabilitation of affected areas after the emergency.

Two 'musts' for the works main controller are good communications and reliable up-to-the-minute information.

Checkers on each shift should be nominated to proceed immediately to each assembly point once a major emergency is declared with a list of names of all persons known to be on the works at the time. Their duties are to record the names of all reporting to the check point and relay these to the emergency control centre. If time clerks are present they are an obvious choice, but otherwise shift workers who have no specific emergency duties should be nominated and trained as checkers.

A **mobile analytical team** is recommended by Essery[8] of ICI. Its main purpose is to provide rapid analyses of the atmosphere for harmful substances at various locations and relay them by radio-telephone to the emergency control centre, so that the works main controller has reliable information on the extent and spread of toxic and flammable materials.

The team would normally be laboratory personnel but their organisation, equipment and means of transport depend on the size and layout of the works and its hazards.

20.3.4 Emergency control centre

At least one pre-arranged emergency control centre should be established and provided with adequate means of communication with areas inside and outside the works, as well as maps, site plans and relevant data and equipment to assist those manning the centre, not forgetting toilet facilities. The centre should be reasonably close to the scenes of possible incidents, yet sufficiently far from them and well enough protected to be able to function in a major emergency. It should be close to a road to allow for ready access by persons and equipment, etc. needed. It will be manned by the works main controller, other designated key personnel, and senior officers of the outside services. To cater for the possibility that a single emergency control centre becomes inoperable as a result of the incident (e.g. through being downwind of a toxic or flammable gas release), an alternative emergency control centre at a different location (usually on the opposite side of the works) should also be established. The police will, if necessary, assist in the establishment of an emergency control centre. A list of suggested equipment for an emergency control centre is given in Appendix 5 of the CIA booklet.[6]

20.3.5 Assembly points

Two or more clearly marked assembly points should be chosen in safe places on different sides of and well away from the areas at risk. Their purpose is twofold:

1. to enable the names of those known to have been present on the plant to be checked as safe and not in need of rescue;
2. to enable those assembled there to be recalled to duty if and when it is safe to do so.

Employees should proceed to the assembly point if escape is imperative or if evacuation proves necessary. They should know their location and have simple instructions on which one to make for, depending on the wind direction and the apparent site of the incident. Generally they should be in a building whose risk of collapse resulting from an explosion is realistically low and where there is adequate protection from the hazard of flying glass fragments. If toxic gases or vapours are liable to be released in an emergency, the assembly room should have tight-fitting doors and windows and no other openings to the outside atmosphere. If the latter is heavily contaminated, those in the room should be able to survive in it until the danger passes, or until they are rescued or able to leave, wearing escape breathing apparatus. Unless the risks of a heavier-than-air gas or vapour emission or of flooding can be ignored, the assembly point should not be below ground-level. Where possible, the assembly point should be at or near a manned gate in the works perimeter fence. Access to the assembly point should be possible at all times, if necessary by breaking a special seal on a door lock. Every assembly point should be provided with emergency lighting (torches or batteries) and means of communication with the emergency centre. Every assembly point should be manned by a checker [20.3.3] as soon as possible after a major emergency starts.

20.3.6 Raising the alarm and declaring a major emergency

Every works should be provided with a sounding alarm, duplicated where necessary, so that it can be heard everywhere in the works. Very large works may be divided into zones, each provided with an independent alarm. It is best to have an alarm which gives at least two distinctive sounds, one for 'alert', the other for the real emergency. Electronic devices which produce a pulsed tone which varies between two frequencies are recommended by Hirst.[12] The alarm should be actuated by an electrical signalling system with enough call points spread over the works (or zone) for the 'alert' to be raised by anyone without going far. The alert may also be triggered by suitable fire and/or gas detectors. Only a limited number of senior personnel (including those on shift) should be authorised to sound the major emergency alarm. Once this is raised, the police/fire services, key works personnel and neighbouring firms must be informed immediately.

If more than one type of major emergency (e.g. release of flammable or toxic gases) is possible on the same site, consideration should be given to providing distinctive alarm signals for each, although their numbers should be strictly limited and all employees carefully instructed in their meaning in order to avoid confusion.

The Major Emergency may announce itself to most of those present before warning can be given. Even so, the alarm should, if possible, be sounded.

20.3.7 Evacuation, searching and accounting for personnel

The possible hazards to personnel while escaping from their workplaces to assembly points should be considered, and special readily accessible protective equipment as well as advice and training may have to be provided for this. In some circumstances it is safest for personnel not to attempt to escape at once, but to remain where they are until the danger has passed or until they can be rescued. The safest place much depends on the type of incident. Provided there is no nearby explosion, people are less at risk from toxic gases and thermal radiation in a closed building than in the open. Buildings, however, are more vulnerable to explosion damage than the human body,[13] so it seems that people may be safer in the open if there is an explosion. The draught of a large fire and the blast of an explosion speed the dispersion of airborne toxic materials.

Escape-type breathing apparatus [22.7] may be needed for toxic hazards, and skin and eye protection for thermal radiation. Hand torches and mobile spotlights will be required at night if normal lighting fails, while alternative escape routes will be needed if the normal one is impassable.

The importance of accounting for personnel following a major incident, and the consequent need for them to report at once to the checker at the assembly point on making good their escape or evacuating, require emphasis. If because of injuries or other reasons a person escaping cannot report at the assembly point, he or those helping him should ensure that his escape is reported to the main controller in some other way, e.g. via the security guard at the works gate or the police.

Escape may be spontaneous in the area where a severe incident arises but evacuation will be more controlled in areas further from it. The emergency plan should include measures which ensure that personnel in these areas are quickly warned if and when they should shut down operations and evacuate.

A good security system whereby the names of all visitors, contractors' men, drivers, etc., with their times of entering and leaving and their locations on site are recorded at the main gate-house is essential. Despite such measures, there may still be uncertainty, particularly at shift change, in knowing precisely who was present on the site at the time of an emergency.

The following measures are required to account for personnel.

1. The incident controller should arrange as soon as possible, subject to his discretion that he is not thereby seriously endangering other lives, for a search to be made to locate and rescue casualties. A further search should be undertaken by the local authority fire brigade on arrival, advised as necessary on particular risks by the incident controller.
2. Nominated personnel should record the names and works numbers of casualties taken to hospitals, mortuaries, etc. and the addresses of these places. The names and addresses of fatal casualties should be reported to the police.
3. Nominated works personnel (checkers) should record the names and departments of those reporting at assembly points, and advise the emergency control centre, with special emphasis on persons feared to be missing.
4. A responsible person at the emergency control centre should collate the lists, check them against the nominal roll of those believed to be on site, and inform the police of any thought to be missing. Where missing persons might reasonably have been in the affected area, the incident controller and senior fire officer should be informed and a further search made.

Those responsible for drawing up emergency plans should consider the use of electronic aids in searching for and locating casualties on a disaster site. One is the Breitling Emergency wrist-watch (made by Breitling Montres SA, PO Box 1132, CH-2540, Grenchen, Switzerland). This contains a miniature radio transmitter which emits an uninterrupted signal for 20 days or more which can be detected at a radius of up to 20 km by a receiver tuned in to its frequency. The signal may be switched on manually or automatically. By issuing all persons entering a major hazard site with such devices and equipping the incident controller and/or the search team with locating radio receivers, the task of searching for survivors should be simplified. Such devices must, however, comply with electrical safety requirements [10.5.3] so that they do not introduce ignition hazards into zones where flammable gases may be present. Another device, used by fire brigades to locate casualties after a disaster, is the thermal image camera.[14]

20.3.8 Post-emergency duties

Many other special duties which mostly lie outside the scope of this book follow any major emergency. These include comforting and attending to the needs of relatives, ministering to the injured and dying, investigating and publicising the causes of the incident, public relations and rehabilitation of affected areas and property.

20.3.9 Training and rehearsals

Training for emergencies has a threefold aspect: that of works personnel, that of outside service personnel, and joint training exercises involving both.

Works personnel are generally not used to the types of incident discussed here and all require training and rehearsal. A Millbank film *Rescue Team Alert*, an I. Chem. E. video training module and a computer-based training program which simulates process plant emergencies are available. Rediffusion Simulation Ltd offer tailor-made 'emergency response trainers' for oil and gas plant.[2] Special training is needed both individually and as team members for key personnel, including part-time works teams for fire-fighting, first-aid, atmospheric monitoring and emergency engineering operations [21.4]. All on-site emergency procedures (mostly shutdown, isolation and evacuation) for every process and storage unit should be rehearsed regularly, where possible by doing the real thing but where more appropriate by simulation. All employees liable to be involved in emergencies should receive initial and refresher training.

Professional emergency service personnel may be assumed to have adequate general training, but they need to become familiar with the hazards, geography, facilities and special problems of the works, and with key works personnel.

Joint exercises involving works and emergency service personnel are essential and probably form the most important part of emergency training. Due to the cost and interruptions to production caused by full-blown exercises, these should be complemented by 'table top' and 'control post' exercises. In the former it is possible to 'go through the motions' of the parts played by personnel of different organisations following a particular hypothetical event, several of which may be treated in a single session, with the various personnel sitting round a table with a tape recorder running. The more realistic 'control post' exercises are designed to test communications, with key personnel working in the locations they would use in an emergency. These should allow the different professional emergency services to test their own roles and their co-ordination with other organisations. They should also prove the accuracy of telephone numbers and other plan details, and the availability of special equipment and materials needed in various emergencies. Periodic full-scale practices are also needed to increase confidence and ensure that nothing important has been ignored or forgotten. All exercises need careful preparation. Results should be studied and the lessons learned should be circulated and discussed.

References

1. Sherwood, D. S. and Whistance, D. J., *The Piping Guide*, Chapman and Hall, London (1979)
2. Rediffusion Simulation Ltd, *Simulation in the Oil and Gas Industry* (brochure), Manor Royal, Crawley
3. Pearson, L., 'When it's time for startup', *Hydrocarbon Processing*, **58**(8), 116 (1977)
4. ANSI/ISA S5.1–1984, *Instrumentation Symbols and Identification* (available from London Information (Rowse Muir) Ltd, Index House, Ascot, Berks SL5 7EU)
5. Lees, F. P., *Loss Prevention in the Process Industries*, Butterworths, London (1996)

6. Chemical Industry Safety and Health Council of the Chemical Industries Association Ltd, *Major Emergencies*, 2nd edn, London (1976)
7. HSE, *Control of Industrial Major Accident Hazard Regulations — Further Advice on Emergency Planning*, Booklet HS(G)25, HMSO, London
8. Essery, G. L., 'Planning for the worst — on-site arrangements', paper given at Major Hazards Summer School, Cambridge, organised by IBC Technical Services Ltd, London (1986)
9. Cooney, W. D. C., 'The role of public services — off-site emergency arrangements', *ibid*
10. Lynskey, P., *The Development of an Effective Emergency Procedure for a Major Hazard Site*, European Federation of Chemical Engineering, Publication Series No.42, available from the Institution of Chemical Engineers, Rugby
11. HSC Advisory Committee on Major Hazards, *First, Second and Third Reports*, HMSO, London (1976, 1979 and 1984)
12. Hirst, R., *Underdown's Practical Fire Precautions*, Gower Technical, Aldershot (1989)
13. Roberts, A. F., 'Vapour Cloud Explosions and BLEVES', paper given at Major Hazards Summer School, Cambridge, organised by IBC Technical Services Ltd, London (1986)
14. Treliving, L. 'Thermal image cameras: the new lifesavers', *Fire International*, **92**, 81 (April-May 1985)

21 Safety training

The need for safety training of all ranks in the process industries is clear from most chapters of this book. That of supervisors is specially important because of their many responsibilities for safety.

Safety training requires:

- a properly prepared programme and allocation of adequate time and resources;
- careful analysis of the jobs to be trained for and their hazards;
- good instructors who are thoroughly familiar with their subjects;
- appropriate training media and methods — a good programme usually employs a combination of several different ones.

Training films, aids and sources of information on process safety are available from many organisations. Safety training should form part of normal job training. Psychological factors are important in training, which is most effective when it satisfies three important human motivators in the trainees[1] [19.1.7]:

- recognition of his effort and achievement;
- acceptance by members of his group;
- maintenance of his self-respect.

Adult trainees should not be treated as children. Three important laws of learning which apply in training are *primacy*, *recency* and *frequency*.[1] Primacy means learning the right way of doing something from the start. This is more difficult if the worker has been doing it the wrong way and has first to break a bad habit. Recency means that we tend to remember what we have just learnt better than what we learnt, say, last year or the year before, and frequency means that we need regular reminding or revision of what we have learnt to prevent it being forgotten.

Studies in the USA[2] have shown that most people forget 90% of what they were told, 80% of what they were shown but only 35% of what they were both told and shown three days earlier. They are most likely to remember instruction when they perform a task and describe the task while doing it.

21.1 Training aims and framework

The (UK) Chemical and Allied Industries Training Review Council (CAITREC) states some essential features of training in the process industries.[3]

1. All new employees should receive Induction Training which includes relevant Health and Safety information.

2. All new employees, and all people changing job in-company, should receive job-specific training based on an analysis of training needs, including relevant Health and Safety aspects.

3. All employees should receive health and safety information and training.

4. All training programmes should specify:
   - the name of the person responsible for ensuring delivery of the programme,
   - the objectives, i.e. what the trainee will be able to do on completion of the programme,
   - the trainers, i.e. who will carry out each part of the instruction,
   - the time and place of training.

5. All training programmes should:
   - build upon the trainee's existing skills, knowledge and experience,
   - specify the standard of performance required after training.

6. Trainee job-performance should be monitored during and after completion of training.

7. After training, the trainee's manager/supervisor should assess his performance against standards set for his job.

8. All training should be recorded to show:
   - the date of training,
   - the training that has yet to be completed,
   - the time taken to reach acceptable performance.

9. All training programmes should be monitored and revised as necessary.

21.1.1 Training needs

Training is most needed:

1. for new employees
2. when new equipment and processes are introduced
3. when procedures have been revised
4. when new information has to be imparted
5. where performance and morale need to be improved.

Indicators of the need for training are:

- above-average accident and injury rates for the type of work
- high labour turnover
- excessive waste, poor yields and poor-quality products
- works, factory or plant expansion.

21.1.2 Training objectives and levels

Training objectives must be set, based on:

1. detailed descriptions of the jobs being trained for, including possible hazards;
2. the initial levels of knowledge and skills of trainees;
3. the levels to be reached during training.

Training courses for skilled jobs (such as plant operation and maintenance) should be graded, e.g. as basic, intermediate, advanced and supervisory. New and inexperienced employees should be required to take the basic course and pass its test before being assigned to work involving risk to themselves or others. They should have a stipulated amount of appropriate work experience before starting more advanced training. Similar considerations apply at all levels. Awards such as a badges or ties showing the level reached by successful trainees (e.g. bronze, silver and gold like those for life-saving and dancing) offer useful incentives as do cash bonuses or steps in the wage scale. It is best to engage new trainees for skilled jobs on a probationary basis. They should be required to pass an appropriate test before being taken on the permanent pay-roll.

Jobs to be trained for should be analysed and broken down into steps. The time and method required to teach each step and test the trainee must be assessed when planning a training programme. While training should not be so narrowly based that its logic remains unclear, it should keep to its subject.

Management should ensure that jobs which are critical to safety are performed only by workers who have successfully taken the relevant training course and test, or preferably one a stage more advanced. The latter is of great advantage in ensuring a pool of trained personnel when production facilities are expanded.

21.1.3 Induction of new workers

It is important to include a positive health and safety (HS) message in the information imparted to every new employee before he starts work. He may, however, absorb only part of it because of the newness of his surroundings and the many other things he had to take in at the same time. The message should therefore be reviewed and amplified soon afterwards, i.e. about two weeks later. It should include the following.

1. Management's interest in preventing accidents. This should be illustrated by the company's safety policy, programme and record.
2. To prevent accidents, certain rules and procedures must be followed. A company safety-rule booklet should be given to the new worker and explained. All its rules should be logical and have been discussed and agreed with workers' representatives. They should be enforceable, e.g. by suspension of offenders. The rule booklet should include the following items where applicable:
   - smoking rules,
   - permits-to-work,
   - first-aid and its organisation,
   - personal protective clothing and equipment (including its issue, safekeeping, inspection and maintenance),
   - work clothing (including the above provisions),
   - raising a fire alarm, firefighting and its organisation,
   - electrical equipment,
   - 'housekeeping',
   - emergency and evacuation procedures,
   - procedures for reporting accidents and injuries and getting medical attention.
3. Every employee should report to his supervisor unsafe conditions which he encounters in his work.
4. No employee is expected to undertake a job until he has learned how to do it and is authorised to do it by his supervisor.
5. No employee should undertake a job which appears to him to be unsafe.
6. If an employee suffers an injury, even a slight one, he is required to report it at once.

Training should be discussed, including essential training provided by the company. The new entrant should also be encouraged to take any good and appropriate training courses run externally.

This message should be given by someone with proven ability (and, where necessary, training) as an instructor. He or she may be a member of the personnel department or preferably, the manager of the worker's future department, or a safety professional. The message should be carefully prepared and presented and may be supplemented by a brief safety video or film which reinforces the points discussed. New workers should be encouraged to ask questions which should be answered in a friendly way. Any safety rules which are a condition of employment should be enforced from the start.

A checklist of safety topics should be discussed with the new entrant and each should be ticked off as it is discussed. On completion the form should be signed by both the person giving and the person receiving the indoctrination. It should then be attached to the employee's record to confirm that the safety indoctrination has taken place.

If the new employee is examined medically, the doctor and/or nurse should explain the work of the medical department, encourage him or her to use its services and stress the importance of reporting all injuries, sicknesses, skin complaints, dizziness or irritability.

21.1.4 Initial safety training

On reporting for work to his supervisor, the new employee is usually entrusted to an experienced worker who will first familiarise him with the welfare facilities, the plant, its safety features and controls, the work done and any special terms and jargon used. For about the first two weeks the new employee's work is usually limited to assisting experienced workers in non-hazardous routine tasks. When he is familiar with his new surroundings, his supervisor, who should know the HS message given to him on his first day, should meet him and review it in detail. At the same time, the supervisor should get to know him personally, assess his experience, and discuss with him specific safety aspects and potential hazards of the department and the job for which he will be trained. He should discuss any departmental safety rules, fire protection, the use of any personal protective equipment needed for the job, location of emergency showers and eye-wash units and the department's safety programme. The supervisor should record the points covered on the employees' safety orientation checklist. He should discuss what training the new entrant will next receive and what work he will be doing during training, and make proper arrangements for both.

21.2 On-the-job training

Most on-the-job training should be to a definite programme and given by the supervisor or by an experienced worker with proven training ability nominated by the supervisor. The programme organiser is responsible for reviewing and, where necessary, developing the training methods used.

The first source to be considered is the plant documentation. This should include operating and maintenance manuals [20.2.2], process material data sheets, permit systems [18], operating and maintenance reports, emergency procedures [20.3] and sometimes departmental safety rules (as distinct from those covering the entire works). The documentation should contain information on potential hazards, how to recognise them and how to control them mechanically or procedurally. The training organiser has to select what is relevant to the trainee's educational level and needs. For the average new worker who may not understand the meaning of the terms 'flash point' and 'occupational exposure limit', a set of data sheets giving detailed properties of all process materials in a plant would be confusing. The training organiser might therefore short-list the more hazardous process materials using the criteria shown in Table 21.1,[1] and provide abridged safety data sheets for their safe handling as shown in Table 21.2.[1] Trainees should, however, have the right to see and obtain copies of more detailed material data sheets.

Table 21.1 Criteria for selection of hazardous process materials for chemical safety data sheets used for training new process workers[1]

Health	1. Materials absorbed by the skin
	2. Dusts with OEL ≤ 1 mg/m^3
	3. Vapours and gases with OEL ≤ 1000 ppm
	4. Corrosive and irritating materials
	5. Others recommended by the industrial hygienist
Flammability	6. Materials with flash-point < 38°C
Reactivity	7. Materials readily capable of detonation, explosive decomposition or reaction at ambient temperature and pressure
	8. Materials capable of detonation or explosive decomposition under a strong initiating source or when heated while confined
	9. Materials which react explosively with water
	10. Other unstable materials capable of reacting violently, especially with water and materials which may form explosive mixtures with water

Table 21.2 Contents of chemical safety data sheet for training new process workers[1]

1. Name and description of material (including chemical name)
2. Nature of hazard (toxic, flammable, reactive, corrosive, etc.)
3. Prompt first-aid procedure for exposed personnel
4. Protective equipment needed
5. Handling precautions
6. Prompt spill-control method

Other information where appropriate on:

7. Engineering control method
8. Fire-extinguishing agent
9. Incompatibility with other chemicals
10. Unusual fire and explosion hazards
11. Simple description of symptoms expected in exposed persons

A trainee operator for a particular unit of a plant must know and understand the operating and emergency procedures for that unit, but needs only a general understanding of others which affect it. The trainer should present the information in small doses, checking that each is learnt and understood before proceeding.

Three job training tools which are applied systematically in the USA are Job Safety Analysis, Job Instruction Training and 'Over the Shoulder Coaching'.

21.2.1 Job safety analysis (JSA)[1,4]

This is a technique for identifying potential hazards in each step of a job and eliminating them by specifying a safe procedure or changing an existing one, or by the use of particular equipment or tools. Jobs usually selected for JSA are those in which:

- accidents have frequently occurred;
- there have been disabling injuries;
- there is a high potential for severe injury or damage;
- the work is new, resulting from a change in equipment, process or procedure.

Typical jobs for which JSA may be used in the process industries are breaking a flanged pipe joint, clearing a blockage in a pipe, taking a sample of a process fluid under pressure, starting a compressor or adding chemicals to a reactor. Broadly defined jobs such as building a plant and narrowly defined ones such as pressing a button or tightening a screw are not suitable for JSA.

The trainee should participate in and study safety analyses of jobs which will form part of his or her work. A JSA is best done by the line supervisor for that job with its work crew. The job is first broken down into a sequence of steps each starting with an action word such as 'remove' or 'open' with a description of what is being done. An experienced and co-operative worker is briefed on the purpose of the exercise and asked to do the job while the other participants watch and record each step, where possible with the help of a video camera. The participants then attempt by a brainstorming type of approach to identify all hazards and potential accidents associated with each step, regardless of their probability and without at that time attempting to devise means of controlling them. It helps here to consider various types of injury and look for exposures which could cause them.

The participants then review the hazards and develop corrective action to control them, and define a safe procedure for the job. This may involve a change from the previous one used. Changes in tools or equipment may also be required, or improved guarding or interlocks may be necessary. The analysis is recorded on a form (Fig. 21.1) with three columns, the first show-

<table>
<tr><td></td><td colspan="3">JOB:</td><td>DATE:</td></tr>
<tr><td>JOB SAFETY ANALYSIS
TRAINING GUIDE</td><td>TITLE OF MAN WHO DOES JOB:</td><td colspan="2">FOREMAN/SUPR:</td><td>ANALYSIS BY:</td></tr>
<tr><td colspan="2">DEPARTMENT:</td><td colspan="2">SECTION:</td><td>REVIEWED BY:</td></tr>
<tr><td colspan="4">REQUIRED AND/OR RECOMMENDED
PERSONAL PROTECTIVE EQUIPMENT:</td><td>APPROVED BY:</td></tr>
</table>

SEQUENCE OF BASIC JOB STEPS	POTENTIAL ACCIDENTS OR HAZARDS	RECOMMENDED SAFE JOB PROCEDURE
Break the job down into its basic steps, e.g., what is done first, what is done next, and so on. You can do this by 1) observing the job, 2) discussing it with the operator, 3) drawing on your knowledge of the job, or 4) a combination of the three. Record the job steps in their normal order of occurrence. Describe what is done, not the details of how it is done. Usually three or four words are sufficient to describe each basic job step. For example, the first basic job step in using a pressurized water fire extinguisher would be: 1) Remove the extinguisher from the wall bracket.	For each job step, ask yourself what accidents could happen to the man doing the job step. You can get the answers by 1) observing the job, 2) discussing it with the operator, 3) recalling past accidents, or 4) a combination of the three. Ask yourself: can he be struck by or contacted by anything; can he strike against or come in contact with anything; can he be caught in, on, or between anything; can he fall; can he overexert; is he exposed to anything injurious such as gas, radiation, welding rays, etc.? for example, acid burns, fumes.	For each potential accident or hazard, ask yourself how should the man do the job step to avoid the potential accident, or what should he do or not do to avoid the accident. You can get your answers by 1) observing the job for leads, 2) discussing precautions with experienced job operators, 3) drawing on your experience, or 4) a combination of the three. Be sure to describe specifically the precautions a man must take. Don't leave out important details. Number each separate recommended precaution with the same number you gave the potential accident (see center column) that the precaution seeks to avoid. Use simple do or don't statements to explain recommended precautions as if you were talking to the man. For example: "Lift with your legs, not your back." Avoid such generalities as "Be careful," "Be alert," "Take caution," etc.

Fig. 21.1 Job safety analysis form (courtesy National Safety Council, Chicago)

ing the basic steps of the job, the second the hazards of potential accidents associated with each step and the third the recommended safe job procedure and any other corrective action which needs to be taken.

An incidental benefit of JSA is that it often develops new ideas for saving material and labour which more than pay for the time and effort required to do it.

21.2.2 Job instruction training (JIT)[1]

Where a simulator is available for the job this may be used until the trainee has mastered it. Otherwise most JIT is done on or at the controls of real plant or equipment, often under normal operating conditions. JIT can be applied to a sequence of operations which, when started, often dictate their own pace.

The instructor may be the trainee's supervisor or a worker with experience of the job and the method of instruction, or a special instructor. He first has to decide on the speed of teaching and what skills the learner should acquire in the time available. He needs to analyse the job and check the following:

- every step in the standard procedure being taught;
- any health risks and adequate provision for them;
- any personal protective equipment needed;
- the safety of methods used for handling materials;
- opportunities for trainee error and their consequences;
- any equipment hazards present and provisions against them;
- protection against fire and explosion hazards;
- potential emergency situations and arrangements to control them;
- adequate emergency shutdown procedure.

He must make sure that all equipment and materials are ready and that each trainee can be placed in such a position while observing the instructor's demonstration that he will see the task as he would if he were doing it himself. (He should not be placed opposite the instructor but could sit or stand beside him and view his actions in a mirror facing them both.)

JIT is broken down into four parts:

1. preparation
2. presentation
3. application
4. testing.

During preparation the instructor describes the job and the way it should be done, discusses the more important points about it and tries to find out what each trainee knows about it already. The instructor should follow a format such as that used by the American National Safety Council (Fig. 21.2). In stating what must not be done, he should explain why.

The instructor then illustrates and, if possible, demonstrates each step of the operation, referring frequently to the standard procedure of which the trainee should have a copy. He discusses it with the trainees to find out how much he has absorbed and, if necessary, repeats all or part of the demonstration.

He then checks what the trainee has learnt by getting him to explain how the process is operated and describe what goes on in each step. He next asks the trainee to operate the process and describe the key points of each step as he does it. The instructor watches, corrects and explains as needed. The trainee then repeats the cycle with the instructor watching him until both are confident that he can do it safely on his own.

The trainee is then invited to carry out the operation on his own after being told from whom to get help if needed. He should be impressed with

HOW TO GET READY TO INSTRUCT

Have a Timetable—
how much skill you expect him to have, by what date

Break Down the Job—*
list important steps. pick out the key points. (Safety is always a key point.)

Have Everything Ready—
the right equipment, materials and supplies.

Have the Workplace Properly Arranged—
just as the trainee will be expected to keep it.

*Use JSA, Job Safety Analysis breakdown to locate hazards.

JOB INSTRUCTION TRAINING (JIT)

HOW TO INSTRUCT

1. Prepare
Put trainee at ease.
Define the job and find out what he already knows about it.
Get him interested in learning job.
Place in correct position.

2. Present
Tell, show, and illustrate one IMPORTANT STEP at a time.
Stress each KEY POINT.*

3. Try Out Performance
Have him do the job—coach him.
Have him explain each key point to you as he does the job again.
Make sure he understands.
Continue until YOU know HE knows.

4. Follow-Up
Put him on his own.
Designate to whom he goes for help.
Check frequently. Encourage questions.
Taper off extra coaching and close follow-up.

*Safety is always a key point.

SAFETY TRAINING INSTITUTE
NATIONAL SAFETY COUNCIL

Fig. 21.2 Format for Job Instruction Training (courtesy National Safety Council, Chicago)

the dictum 'if you are not sure — don't do it'. The supervisor must check frequently until certain that the trainee has mastered the operation, and then occasionally to ensure that he is able to cope on his own in unusual situations.

21.2.3 Over-the-shoulder coaching[4]

This is a flexible and direct method of training which allows the trainee to develop and apply his skills under the guidance of a skilled and safe operator who has the time, patience and desire to help him. As with JIT, it can be applied in conjunction with simulators of complex plant [21.3.5]. The coach now sits in a room separated from the trainee's by a window with 'one-way vision', so that he can observe the trainee's actions. They should be able to communicate with each other by telephone. The coach feeds the trainee with various simulated plant situations on VDU screens or verbally and the trainee has to respond to them.

The coach should keep a careful record, e.g. on a chart, of the progress of each of his trainees. Because of its personal nature the method can be very effective.

21.3 Training media and methods

A well-designed training programme may use several different media and methods.

21.3.1 Printed media

Books and notes are easy to produce, survive minor mishaps such as spills of hot coffee and are invaluable for reference and records. Each trainee can be given a personal copy which he or she can study almost anywhere. Whatever other methods are used, the booklet or manual is usually an essential accompaniment.

The written word is, however, quickly forgotten. Books, moreover, are passive aids to learning. While the student may respond to the book, the book cannot detect any difficulties which the student is having.

Written training material may include exercises to which answers are available, but there is nothing to stop the student from cheating or losing interest unless the book is used in conjunction with personal instruction. Printed instructions are most useful for tasks which do not require rapid hand and eye co-ordination, although even for these an illustrated pocket book can be a helpful supplement to practice.

21.3.2 Programmed instruction

This method was first used with special printed texts (developed for correspondence courses), and later as computer programs [21.3.5]. It depends on breaking the subject down into many small parts and then concentrating on each in turn. The student is given encouragement whenever he demonstrates that he has mastered a task, and is then allowed to progress to the next one. If he cannot master a task the first time, he will have the option of further study in which the subject matter is presented differently, usually in even smaller pieces. It provides a more thorough training than the mere reading of a book.

21.3.3 Personal instruction

This can take a variety of forms ranging from individual coaching and counselling to a lecture to a large audience. Instructors are more versatile than books and can sense and probe students' learning problems. The spoken word alone is not, however, a good means of communication because it is easily misheard, misunderstood and forgotten. To overcome these drawbacks the speaker may articulate slowly to allow students to take notes, which is a distraction.

The teacher's voice is best reinforced by demonstrations or visual aids, such as a chalkboard, flip-chart, overhead slide projector, slide and film strip projector or three-dimensional models, as well as printed notes and illustrations. Personal instruction is generally used in conjunction with a printed text. Conferences, discussions, case studies, role-playing exercises, demonstrations, drills and panels are forms of personal instruction which are discussed later [21.3.7/8].

21.3.4 Films and video cassettes

Film and video programmes can be powerful training tools. Several excellent films on specific safety topics are available for hire or purchase. They require considerable skill and planning to produce, with a carefully prepared script, often professional actors, proper lighting and subsequent editing. Although a film can be stopped for discussion, most projectors do not allow single frames to be viewed as 'stills' and it takes time to wind a film forwards or backwards if one wants to project only part of it.

Low-budget films and videos produced in-house with works personnel as actors can be effective if the training programme is well designed and there are enthusiastic and skilful amateur actors and a film maker on the staff who are willing to co-operate. Otherwise it may be possible to involve a small commercial film unit or a film/video teacher at a local art college or evening class.

Video cassettes are largely replacing film as an instructional medium. They can be used with conventional TV sets and video players for small audiences and with video projection screens for larger ones. Video training cassettes produced in-house with amateur video cameras can be indexed to allow a particular 'rush' to be selected and viewed with far less delay than in the case of a film. Most video projectors can be stopped to allow a single shot to be examined, although the resolution may not be good.

Video discs are a newer development which allow very rapid transfer from one part of the disc to another during viewing. At present all video sequences have to be shot first on tape and then transferred to discs, which adds to their cost.

21.3.5 Computers[5]

Computers were introduced as a training aid to extend the availability of training when teachers were in short supply. They are used in three principal ways: computer-assisted learning (CAL), computer-managed learning (CML) and for keeping student records. CAL developed as a means of programmed instruction, using a computer terminal or desk-top computer with a screen, keyboard and sometimes a printer and a 'mouse', at which one or two students study from a program 'written' on 'floppy' disc. It is not difficult to make a program which can carry out a restricted silent dialogue with a student on a selected subject. A text similar to a page in a programmed instruction book is displayed on the screen. This is followed by a display of questions, usually with a multiple choice of answers from which the student select by pressing a key. In other cases the answers are not displayed and the student has to type his own. The CAL program checks and evaluates the answers, records the student's marks and proceeds to the next step if the answer was correct, or returns to remedial work if it was wrong. At the end of a lesson the student's marks are added, and if a printer is available a record of the lesson can be made for the student to retain. Once the program has been developed and polished, the computer can act as an individual tutor of unlimited patience to a number of students, although it cannot entirely replace a teacher. It can also be used for drill and practice programs where the student has to answer a number of questions drawn at random from an 'item bank'.

In another form of CAL known as 'revelatory', the program contains a hidden model of some real-life situation and allows the student to develop a feeling for its behaviour under various circumstances. This form of CAL has been developed into sophisticated and expensive multi-media training simulators. Special tailor-made simulators are now used to train operators of complex PES-controlled oil and petrochemical plant.[6] In these the operator sits at a control station provided with VDU screens which give him detailed information about the state of the plant, and a keyboard on which he issues instructions to the control system.

The simulator has a nearly identical control station at which the operators sits observed by a trainer who can feed him with various plant scenarios to which he has to respond. Such simulators are proving themselves in reducing start-up times, improving operator performance, and contributing to safer operation.[6] Simulators have also been installed inside plant control rooms as an aid to operation, where they can be used to try out the effect of a particular operating strategy before applying it to the real plant [20.2.2].

CML is used to assist the examination and assessment of students and for the general administration of a teaching establishment. Computers are also used to keep students' records and report on their work.

21.3.6 Interactive video

Portable equipment which combines an audio-video system with a computer, keyboard, mouse, graphics facility and a colour monitor is now proving popular for training operators for complex chemical plant, as well as for communications, exhibition display and other purposes. An example is Ivan Berg's 'Take Two' system[7] (marketed in the UK by Quadrant Network) which uses video tape or discs controlled by a computer program. It allows computer text and graphics to be superimposed on the video picture, or the video sequence to be interrupted while the computer program is run.

Such systems lend themselves well to courses of programmed instruction. These can be made from bought or in-house-shot video discs which are combined with computer programs authored in-house. The effect of seeing and hearing a scene on the video screen while assimilating its lesson, on which the students are then interrogated by the computer, makes it a more powerful training tool than either computer or video alone. The complete program naturally requires more effort to produce. Examples of its use are:

- instruction and examination of trainee operators on the operating manual for a hazardous plant;
- training operators and maintenance fitters on permit systems and plant isolation, hand-over and maintenance procedures;
- warning and instructing operators on changes in procedures such as those brought about by plant modifications;
- instructing operators on the diagnosis and treatment of plant fault conditions as revealed by alarm signals;
- training personnel on procedures and special duties in major emergencies.

21.3.7 Conferences

Conferences are widely used for teaching management subjects and for solving problems common to a special group such as plant supervisors whose contacts with each other are normally limited. Good leaders who can draw out information and opinions from participants and sum up conclusions are vital. A conference called to discuss a particular problem can yield good returns in educating its participants. It is important that its leader and members know its scope and limitations well in advance. The leader, while keeping speakers within the terms of reference of the conference, must not succumb to the temptation of trying to steer the discussion in the way he or she wants it to go. A closely controlled 'conference' is one in name only.

The safety professional is often faced with the need to call a conference of production supervisors and others to discuss problems such as putting a policy or directive into practice, on-the-job training, plant hand-over at shift change, the high incidence of a certain type of accident, the company's permit-to-work system and first-aid. If a purpose of the conference is to recommend action and it does this, its members should know what becomes of their recommendations. The best conferences are usually on matters which only affect those present, when the conclusions drawn are mainly for their own guidance.

Safety professionals and managers at all levels should develop their skills at leading conferences. This involves the following sequence.

1. the leader states the problem;
2. he or she tries to break this down into segments to keep the discussion orderly;
3. he encourages free discussion;
4. he makes sure that all significant points have been properly understood and that members have given sufficient thought to them;
5. he notes any conclusions reached;
6. he states the final conclusions which truly represent the group's findings, any agreed action items and the names of those responsible for implementing them.
7. he ensures that a concise report of the conference with its conclusions and action items is made and copies are distributed to all conference members and others concerned.

21.3.8 Fire training

It cannot be repeated too often that the humble fire extinguisher is the first line of defence against disaster. Nearly all fires are small when they start and can be readily put out by the prompt use of first-aid equipment [10]. However, an extinguisher is only as good as the person who uses it. Everyone, without exception, should have hands-on experience with the equipment which is appropriate to the risks in their area. For office workers this could be repeated perhaps once a year, but plant operatives will need more frequent training. Extinguishers which have been withdrawn from the plant for checking and refilling can be used for training, thus providing a useful confirmation of the safe maximum exposure to plant conditions. Realistic fires should be used, and operatives should be asked to tackle the largest fire that they could reasonably be expected to extinguish. Escape drills will also be needed, particularly in office buildings.

Exercises should be planned which involve the local fire brigade. On a large site a visit could be made to a different plant each month, with the operation of jets, monitors, deluge systems and foam-making equipment as appropriate. The opportunity can be taken to provide lunch for everyone concerned, so that there can be informal discussions between the plant management, safety personnel, works fire brigade, and the visiting officers and firefighters.

Voluntary members of a works fire team should have regular training, perhaps once a week, as paid overtime. Full use should be made of all the training facilities provided by the local authority and the Home Office. Teams should also be encouraged to take part in the annual competitions of the IFPA and BFSA. If a full-time works fire brigade is employed, then everyone should have adequate professional training. All members should be encouraged to sit the examinations of the Institution of Fire Engineers. If an officer has passed the IFE graduateship examination then by the Home Office reciprocity agreement he or she will have the same academic standing as a Station Officer in a local authority brigade. Degrees in Fire Engineering are now available from a number of universities: senior safety personnel should be encouraged to attend these courses.

One problem with the members of a works fire brigade is that they do not put out many fires. However, it should be emphasised to them and to top management that their main function is to prevent fires. They should be engaged in an almost continuous fire-prevention survey in addition to the responsibility of maintaining all of the detection and firefighting equipment. When they are called to a fire, it will be their problem for the vital few minutes before the local authority brigade arrives. Effective use of this time may very substantially reduce the total losses from the fire. Their firefighting role will then by taken over, unless they are operating some equipment which is unique to the site. They can then become fully engaged in damage limitation (salvage) work, and in providing advice on the plant and hazardous processes.

On a site where fire, safety and security operations have been combined in one department, it may be effective to have combined patrols. Four-wheel drive vehicles can be converted into surprisingly well-equipped fire and rescue appliances, and can be sent out on regular fire/security patrols around the site. The personnel involved are then continuously up-dated on new developments.

21.3.9 Other methods

Other methods of group training include discussions, case studies, role-playing, drills, demonstrations and panels. All are useful provided they encourage participation by the trainees.

Discussions

These should be held formally as mini-conferences, to exchange ideas and standardise procedures and techniques. They should allow students to participate and pool their knowledge.

Role-playing

Incidents based on real situations are re-enacted by selected members of the group, playing roles and making their own decisions. These are then discussed by the group and its instructor to highlight behaviour patterns.

Drills

These consist of repetition of the task and its various components under guidance to develop important and fundamental skills. They are important to ensure the safe performance of tasks critical to safety such as firefighting and first-aid, starting pumps, taking samples and lubricating moving machinery. The safe method must be well established before a drill is carried out and the limits of its applicability should be made clear and stressed.

Demonstrations

The operation is demonstrated by the instructor as in job instruction training [21.2.2].

Panels

These are planned sessions in which two or more 'experts' in turn answer questions from a selected audience (such as trainees) on various aspects of an assigned subject. A panel benefits from having a good leader. It sometimes reveals differences in the views of experts which would not otherwise come to light.

21.4 Training for special safety responsibilities

Groups of personnel with special responsibilities include:

- professional managers, engineers, chemists, etc.;
- health, safety and security professionals;
- supervisors;
- process plant operators;
- skilled engineering tradesmen — electricians, crane, truck and mobile plant drivers, welders and others;
- trained volunteer workers with special duties in emergencies — part-time firemen, first-aiders, personnel 'checkers' and telephonists;
- contractors;
- safety representatives.

Training within industry in the UK is now largely in the hands of non-statutory training organisations (NSTOs) which are run by the industries themselves. Those for the chemical and petroleum industries which have high hazard potential are:

1. The Chemical and Allied Industries Training Review Council (CAITREC).[3] This is closely linked with the Chemical Industries Association.[8]
2. The Petroleum Training Federation (PTF), which runs several courses by arrangement on its members' sites.[9]

While all personnel need safety training which is often specialised, only that of supervisors [21.4.1] and process plant operators [21.4.2] is discussed in any detail here. In the UK the Chemical Industries Association has established 11 self-help regional training organisations within the chemical industry. Its training department runs short residential HS courses in the UK for managers, supervisors, safety representatives and other groups. It co-operates with the City and Guilds of London Institute in the training, assessment and certification of skilled craftsmen, process workers and laboratory technicians and has developed computer-based training packages in several of these fields.

Much of managers' safety training consists of self-study in which this book can play a part. Lees's list[10] of topics in the safety training of managers is given in Table 21.3.

It should not be forgotten that degree courses, for both first and second degrees, are now available in a number of universities. These cover different aspects of safety management and both fire and safety engineering. Some are in modular form, so that a course can be selected which is best suited to individual needs. Also, it may be possible to complete one module at a time so that a course is taken part-time over several years.

While all employees should have the most basic instruction in first-aid and fire duties, most companies in addition need volunteer workers trained to higher levels for special emergency duties [21.3.8]. This need depends on the extent of both the company's and the local authority's full-time fire, medical and ambulance services and the speed with which they will respond. Thus a petrochemical works might have a medical centre with a trained nurse available during office hours only, and a fire appliance with a professional skeleton crew available for 24 hours a day. The works would need trained voluntary first-aiders able to give artificial respiration and to cope with fractures and other injuries outside office hours, and trained voluntary firemen to make up fire crews for the fire appliance at any time. The training of these volunteers is usually arranged through the company's or the local authority's fire and medical services [21.3.8].

It is important that adequate numbers of trained volunteers for these special tasks are present while work is being done. To ensure this, managements should provide proper recognition and incentives.

Before contractors are employed in works under the control of the operating company, it is essential that they and their personnel should have received appropriate HS training and be familiar with the relevant works' safety rules and procedures. These conditions should be stated in the contract [18.9.6]. Dow Chemical Company, for instance, insists that every

Table 21.3 Some topics in safety training of managers (by kind permission of Professor Lees)

Managerial responsibility for safety and loss prevention
Legal requirements
Principles of safety and loss prevention
Company safety policy, organisation and arrangements
Hazards of the particular chemicals and processes
Accidents and accident prevention, statistics and case studies
Pressure systems
Trip systems
Principle of independent assessment
Plant maintenance and modification procedures, including permits-to-work and authorisation of modifications
Fire prevention and protection
Emergency planning arrangements
Training of personnel
Information feedback
Good housekeeping
Sources of information on safety and loss prevention including both people and literature
Case histories

contractor must attend a safety induction course before starting work unless the job is of such short duration that it would not be practicable to do so. In this case the contractor would be treated as a visitor and accompanied at all times by a Dow employee while on the site. Table 21.4 gives the headings of Dow's safety rules and procedures[11] for contractors working on their sites.

To ensure that its own safety guidelines for contractors are enforced on its numerous sites, Dow lists the following questions to be asked by its executives:

- does the site have a safety and loss-prevention manual for contractors?
- is safety referred to in the purchase contract?

Table 21.4 Headings of Dow's safety rules and procedures for contractors

Accident procedure	Cranes, hoists, lifting machines and lifting tackle
Emergency plan	Power tools and equipment
Assembly points	Welding, burning and cutting
Working on site	Cutting into drums, tanks and vessels
Chemicals	Ladders
Asbestos	Scaffolds
Housekeeping	Roofwork
Machinery safety	Piling operations
Tags	Right of search
Excavations	

- are site safety rules, emergency procedures and special hazard procedures a part of each job specification?
- are pre-job meetings held and if so what is their content?
- does Dow receive a written outline from each contractor of an accident-prevention programme prior to bid acceptance?
- how are 'service' contractors handled?
- are field audits done to check for violations?
- who performs these field audits?
- how is contractor's performance recorded?
- are site safety rules translated into local languages used by construction personnel?
- is there a policy of restricting access of non-essential people to plant while it is being started up?
- are contractors permitted to use plant air as a source of breathing air for respirators, or under hoods of sandblasters?

21.4.1 Supervisors

Supervisors have many duties which are closely linked with safety such as:

1. establishing methods of work
2. instructing people how to do jobs
3. assigning people to jobs
4. supervising people at work
5. maintaining equipment and the workplace

The company should recognise the key role of supervisors in safety and ensure that they understand and accept it. This often requires patient discussion with the supervisors, e.g. by the safety professional. There is also a special need for supervisors' safety training, the details of which depend on the manufacturing operation and the work of the supervisors' departments or teams. The American National Safety Council's (NSC) twelve-hour 'Key man development course'[12] for which instructors' course notes and visual aids are available forms a useful foundation for developing a safety training course for supervisors and for more specialised training later.

The instructor should be a company safety professional, a division manager or a general supervisor. He should know the supervisors' work from first-hand experience and have proven talent for training. He needs the time to study the course material available and adapt it to his company's situation, using examples and problems drawn from it. The twelve training sessions of the NSC course have the following headings:

1. safety and the supervisor
2. know your accident problems
3. human relations
4. maintaining interest in safety
5. instructing for safety
6. industrial hygiene
7. personal protective equipment
8. industrial housekeeping
9. materials handling and storage
10. guarding machines and mechanisms
11. hand and portable power tools
12. fire protection

The NSC recommends that the course be formally organised, where possible in company time, in weekly one-hour sessions. Ideally, it should be opened by a senior manager or director of the company. Records of student attendance and performance should be kept and some form of certificate or award presented to candidates who complete the course satisfactorily, if possible at a dinner to mark its successful completion.

Any of the training methods and media discussed earlier may be used for more specialised safety training of supervisors and for updating those who have taken the basic course previously.

21.4.2 Process plant operators

Mistakes made by process plant operators can have disastrous consequences, and while every effort should be made to design plants which reduce the consequences of human error, high levels of skill are needed for many operating jobs. Crossman[13] suggested that operators should be responsible, conscientious, reliable and trustworthy. Lees[10] classifies the work of process operators as (1) simple tasks, (2) vigilance tasks, (3) emergency behaviour, (4) complex tasks and (5) control tasks, and discusses the problems of operator error and fault diagnosis. He gives lists (reproduced here with his permission as Tables 21.5 and 21.6) of topics which should feature in the general and safety training of operators and all workers.

The question is often raised as to how much theory a process plant operator needs to know. A higher technical education is no substitute for specific training in process tasks. While it should enable the operator to adapt readily to different types of plant, he may become bored with the routine and constant shift work. A few years spent as a process plant operator is, however, a useful background for future managers and designers.

In the UK, the revised City and Guilds training scheme[14] in which CAITREC are closely involved provides a substantial technical education for process plant operators, particularly in the chemical, petroleum, pharmaceutical, food and other related industries. It is offered at several technical colleges for part-time and block-release training and is suitable for workers of all ages, being intended to supplement the training and experience gained in their employment. The scheme consists of three courses:

Part I Basic knowledge, including science and communication
Part II Industrial science and process calculations
Part III Physical chemistry of processes

Safety aspects are considered at all stages and form an integral part of the scheme. The competence of process operators is assessed based on standards set for the common elements of ten operating tasks — start-up, running and shutdown of continuous plant, batch plant operation, materials handling, filling and packing, preparing plant for maintenance, hand-over, emergency procedures and other routines. Special emphasis is placed on fault-finding ability. Assessment is by a combination of internal assessors nominated by the site, and external assessors appointed by CIA and the City and Guilds of London Institute. 'Certificates of Process Operations Competence' with three levels are awarded.

- Level 1 applies to simple operations with basic instrumentation, simple equipment, a limited number of operator/equipment interactions and a small number of parameters under the operator's control.

Table 21.5 Topics in the training of process operators

Process goals, economics, constraints and priorities	
Process flow diagram	
Unit operations	
Process reactions, thermal effects	
Control systems	
Process materials, quality, yield	
Process effluents and wastes	
Plant equipment	
Instrumentation	
Equipment identification	
Equipment manipulation	
Operating procedures	
Equipment maintenance and cleaning	
Use of tools	
Permit systems	
Failure of equipment and services	
Fault administration	Alarm monitoring
	Fault diagnosis
	Malfunction detection
Emergency procedures	
Fire fighting	
Malpractices	
Communications, record-keeping, reporting	

Table 21.6 Some topics in safety training of workers

Topics
Workers' responsibility for safety
Legal background, particularly in the UK's Health and Safety at Work etc. Act 1974 and the Factories Act 1961
Company safety policy, organisation and arrangements, in particular general safety rules, safety personnel, safety representatives and safety committees
Hazards of the particular chemicals and processes
Fire/explosion hard (flammable mixture, ignition source). Ignition sources and precautions, including electrical area classification, static electricity, welding, smoking. Fire spread, fire doors. Action on discovering fire or unignited leakage
Toxic hazard. Action on discovering toxic release
Emergency arrangements, including alarm raising, alarm signals, escape routes, assembly points
Protective clothing, equipment use and location
Fire-fighting methods, equipment use and location
First-aid methods, equipment use and location
Lifting and handling
Security, restricted areas
Accident reporting
Case histories
Permit systems
Good housekeeping
Health, medical aspects

- Level 2 applies to operations of intermediate complexity.
- Level 3 applies to complex operations with a high degree of instrumentation, many operator/equipment interactions and parameters within the operator's control and a significant frequency of problems.

The assessments include the candidate's knowledge of process technology as given in the revised City and Guilds course 060 on process plant operation.[14]

To conclude this chapter, special attention should be drawn to the excellent safety training films, videos and other visual aids which are available from a number of sources. While some cater for all workers in the process industries, others cater for special groups — operators, maintenance personnel, contractors and others. Films such as *Nobody's fault* and *Is there anything I've forgotten*? leave a lasting impression. To gain maximum benefit from such films they should be shown twice to the same audience, as many of the finer points tend to be missed on first viewing. After the first showing, a discussion session led by an instructor should be held to highlight the lessons to be learnt and to answer questions. Some time after the second showing it is useful to subject viewers to a written examination in which they answer questions related to the film. Their answers provide a good indication not only of what they have learnt and understood but also of their ability to communicate in writing.

There is no excuse today for lack of safety training in the process industries when there is such an abundance of good training material available.

References

1. Kubias, F. O., 'Tools and techniques for chemical safety training' in *Safety and Accident Prevention in Chemical Operations*, edited by Fawcett, H. H. and Wood, W. S., 2nd edn, Wiley-Interscience, New York (1982)
2. Bird, F. E. and O'Shell, H. E., 'Incident recall' in *National Safety News*, **100** No. 4, 58–62 (1969)
3. Chemical and Allied Industries Training Review Council, *Guidelines for good training*, Chemical Industries Association Ltd, London
4. National Safety Council, *Accident prevention manual for industrial operations*, Chicago (frequently revised)
5. Rushby, N. J., *An Introduction to Educational Computing*, Croom Helm, London (1979)
6. Pathe, D. G., 'Simulator a key to successful plant start-up', *Oil and Gas Journal*, 7 April (1986)
7. 'Take five mark two', *Audio Visual* (April 1987)
8. *CIA Training bulletin*, No. 5, November 1987, Chemical Industries Association Ltd, London
9. *PTF Course information brochure*, Petroleum Training Federation, London
10. Lees, F. P., *Loss Prevention in the Process Industries*, Butterworths, London (1996)
11. Dow Chemical Company Ltd, *Rules and procedures for contractors' personnel*, King's Lynn, Norfolk (1987)
12. National Safety Council, *Key man development course*, Chicago
13. Crossman, E. R. F. W. and Cooke, J. E., 'Manual control of slow-response systems', in *International Congress on Human Factors in Electronics*, Long Beach, California (1962)
14. City and Guilds scheme pamphlet, *060 Process plant operation*, City and Guilds of London Institute, 76 Portland Place, London (1988)

22 Personal protection in the working environment

22.1 Introduction

Personal protective equipment (PPE) is the most fundamental but primitive of all safety gear. We all wear protective equipment such as shoes to protect our feet, clothes to protect our bodies and gloves to protect our hands. Yet in the employment or work environment PPE should only be used as a last resort when all other forms of protection are not reasonably practicable.

There is a hierarchy of preventative measures which should be used in the work environment starting with elimination, then substitution, safety by safeguarding and if none of these other measures are at all practicable then and only then should safety by using PPE be considered. The COSH[1] Regulation 7 says that hazardous substances should be reasonably and practically controlled or prevented, by using methods other than PPE.

EC member states (including the UK) have all brought in regulations governing PPE at work. Within the UK this is referred to as the Personal Protective Equipment at Work Regulations 1992.[2] These regulations came fully into force on 1 January 1993. The UK regulations in turn refer to the Control of Lead at Work Regulations, the Ionising Radiation Regulations, the Control of Asbestos at Work Regulations, the Noise at Work Regulations, and the Construction (Head Protection) Regulations 1988.

Under the PPE Regulation 4 all employers shall ensure that suitable PPE is provided to their employees who may be exposed to risk to health or safety while at work except where and to the extent that such risk has been adequately controlled by other means which are equally or more effective.

In layman's terms PPE should only be used to control hazards as a 'last resort'. For example employers would be expected to provide warm working environments rather than supplying workers with clothing to keep them warm, noisy machines should be made quiet rather than supplying workers with ear defenders, slippery floors should be cleaned and rendered less slippery rather than supplying workers with special footwear or safety harnesses. An employer would be expected to carry out adequate ventilation rather than supply gas masks.

When it is not reasonably practicable to carry out risk reduction in any other way then PPE is a legal requirement. Indeed it is common practice throughout the process industry to see people using PPE in the form of protective clothing such as aprons, clothing for adverse weather conditions, gloves, safety footwear, safety helmets, high-visibility waistcoats, eye protectors, life jackets, respirators, under-water breathing apparatus and safety harnesses.

Both by law and common sense the PPE supplied must be appropriate for the hazard and risk being experienced. For this reason it would be expected that a hazard and risk assessment should be undertaken. The hazards that could be experienced (Fig. 22.1) include cold, noise, fumes, cuts, stabbing etc., and these can be related to various parts of the body such as ears, eyes, hands, feet etc. It will often happen, within the process industries, that the same part of the body, e.g. the face or nose or eyes, needs protecting for a number of reasons, e.g. to protect the head from falling obstacles a hard hat would be worn and this hard hat might have goggles attached to it, as well as ear defenders and respiratory protection. It is common sense that the various pieces of PPE should be compatible with each other.

It is also essential that the PPE supplied should be compatible with the wearer. It is obvious that different people are different sizes and so equipment should either be supplied in different sizes or should be made readily adjustable. The equipment should fit the wearer correctly and comfortably. Above all the protective equipment supplied should be practical and effective in adequately controlling the risks involved without itself increasing the overall risk. It should be realised that wearing heavy or cumbersome protective equipment will reduce the effective and practical way a person can move, and it would be expected that production would decrease and the wearing of PPE may necessitate more frequent rest periods.

22.2 Assessment of PPE

Under the Management of Health and Safety at Work Regulations 1992,[3] management have an obligation to carry out a hazard and risk assessment and it follows that they must also assess the type of PPE they are supplying to their workers. The large employer within the process industry will in all probability have qualified experienced personnel capable of carrying out this function within their own staff, but the smaller organisation can always seek expert help and advice either from the suppliers of equipment or from competent consultants. Once potential hazards are known there may be several types of PPE that would be suitable. Risk at the workplace and which parts of the body are endangered are the two key elements to be considered.

			Risks																				
			The PPE at Work Regulations 1992 apply except where the Construction (Head) Protection Regulations 1989 apply										The CLW, IRR, CAW, COSHH and NAW Regulations[1] will each apply to the appropriate hazard										
			Mechanical					Thermal															
			Falls from a height	Blows, cuts, impact, crushing	Stabs, cuts, grazes	Vibration	Slipping, falling over	Scalds, heat, fire	Cold	Immersion	Non-ionising radiation	Electrical	Noise	Ionising radiation	Dust fibre	Fume	Vapours	Splashes, spurts	Gases, vapours	Harmful bacteria	Harmful viruses	Fungi	Non-micro biological antigens
Parts of the body	Head	Cranium																					
		Ears																					
		Eyes																					
		Respiratory tract																					
		Face																					
		Whole head																					
	Upper limbs	Hands																					
		Arms (parts)																					
	Lower limbs	Foot																					
		Legs (parts)																					
	Various	Skin																					
		Trunk/abdomen																					
		Whole body																					

(1) The Control of Lead at Work Regulations 1980, The Ionising Radiations Regulations 1995, The Control of Asbestos at Work Regulations 1987, The Control of Substances Hazardous to Health Regulations 1988, The Noise at Work Regulations 1989.

Fig. 22.1 Specimen risk survey table for the use of PPE

Having sensibly assessed the type of equipment needed, for example by using the risk survey table Fig. 22.1, and bought equipment from reputable manufacturers, it naturally follows that workers should receive information, instruction and training on the appropriate ways to wear and use PPE.

22.3 Maintenance and replacement of PPE

The employer is obliged to maintain the equipment provided, or replace equipment that becomes worn or defunct. Hard hats, being made of plastic, will deteriorate over time. Their age or life expectancy will be advised by the maker so that the employer can budget and arrange to have them replaced at the end of their life.

Some PPE is for one-off use, e.g. paper boiler suits, disposable gloves, or disposable respiratory protective equipment such as face masks.

Some equipment will have a life expectancy of a few years. If this is the case then employers should arrange for it to be adequately cleaned and sterilised so as to reduce cross-infection between users.

Non-disposable equipment must be stored in adequate accommodation to protect it from deterioration, damage, or harmful effects such as damp, sunlight, fungal attacks or general abrasion.

22.4 Use of PPE

The employee or the user of equipment should use it in accordance with the employer's instructions which in turn would be based on the manufacturer's instructions. In other words there is an obligation that, having been supplied with PPE, workers must use at properly.

22.5 Standards for PPE

The HSE has produced guidance on the Personal Protective Equipment at Work Regulations[4] and this lists many British and CEN or EN standards. The *British Standard Year Book*[5] also deals with many pieces of PPE. The American National Standard Practice for Respiratory Protective Equipment is ANSI Z88.2-1980.

22.6 Types of PPE

22.6.1 Body protection

Within the process industries there are many and varied hazards, calling for many types of body protection. Figure 22.2 describes some types of work wear.

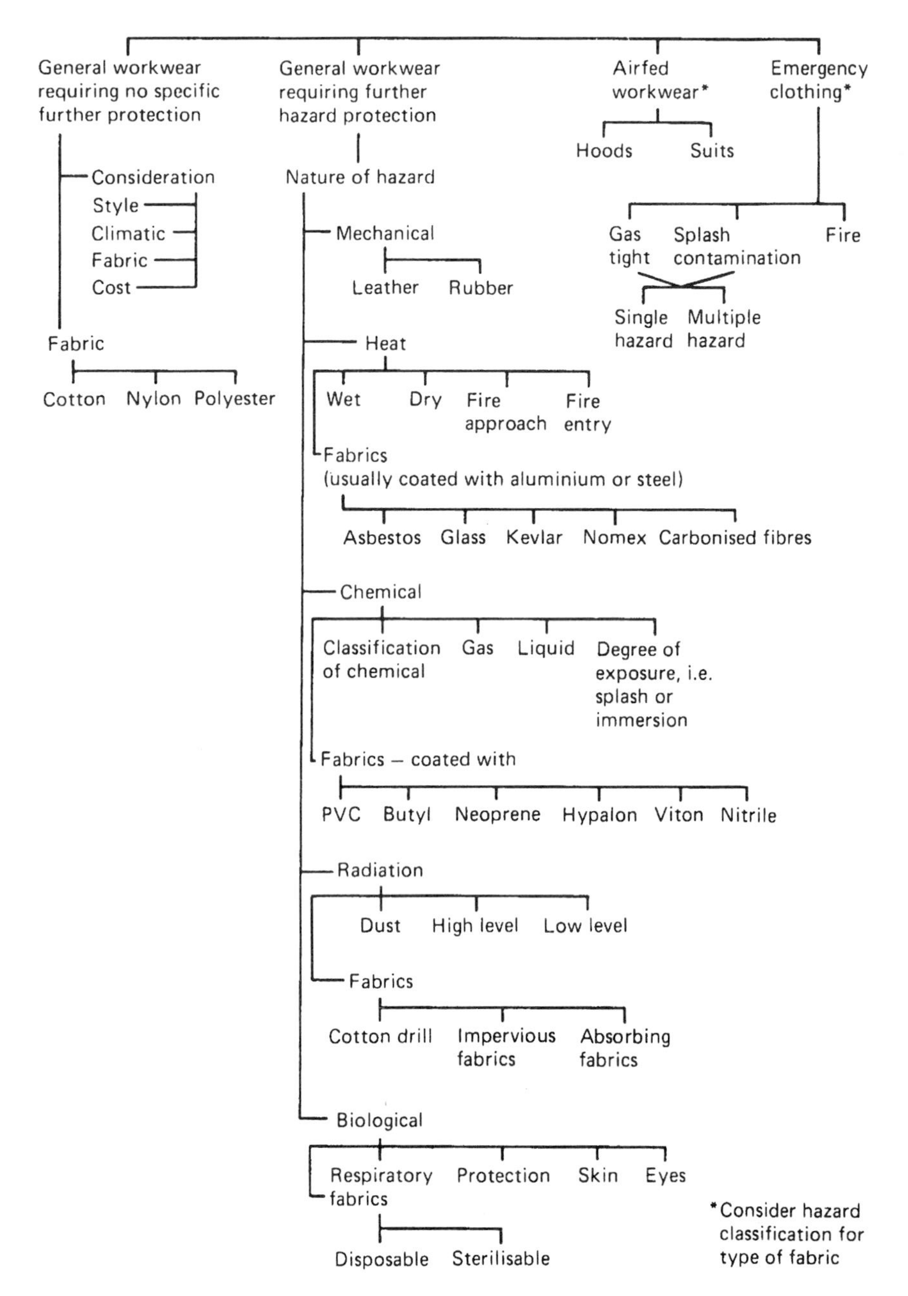

Fig. 22.2 Types of workwear

Body protection will be necessary to protect the worker from excess heat, or excess cold, or could be needed for protection from sparks or flame. Flame-retardant cotton or woollen substances will be essential when working near molten metal or where electrical equipment may be liable to arc. Such suits should fasten up to the neck and safety helmets should have rear flaps to cover the back of the neck. Gauntlets of heat-resistant leather should be worn for hand and wrist protection.

In some parts of the process industry, particularly food and pharmaceuticals, clean working clothes are needed partly to prevent contamination of the product by the worker as well as to protect the wearer from the product being handled.

It is obvious that clothing that has been contaminated whether by asbestos or tobacco dust, etc., should not be worn outside the factory premises or taken into workers' homes. Adequate laundry services should be provided for the equipment and full showering and changing facilities for the employees.

It is clear that the operators in process plant often have to work outdoors, and in these circumstances foul weather clothing should be light-weight and wind and rain repellent, with a flexibility at low temperatures and the facility for removable cold weather clothing to be worn underneath.

22.6.2 Hand and arm protection

Hands and arms can be injured by cuts and bruises, extremes of ambient temperature (hot or cold), skin irritation and dermatitis by contact with toxic or corrosive liquids, or they can be burnt with molten metal, hot steam or electric shock.

To safeguard workers there will be purpose-made gloves, supplied by manufacturers specialising in such products, capable of protecting them from the hazards listed above. But it must be remembered that all types of gloves will reduce the dexterity and effectiveness of the grasp or grip of the workers. It should also be realised that manual dexterity is lost when hands get cold, and so gloves will need to be provided for outdoor activities as well as work in cold rooms or refrigerated sections of a factory.

22.6.3 Head protection

There are four widely used types of head protection:

- industrial safety helmets which can protect against falling objects or impact with fixed objects;
- caps, helmets etc. which can protect the scalp and hair from entanglement or can protect the head from contamination with toxic or nuisance dusts, or protect the product from dirt carried within the hair;
- industrial scalp protectors such as Bump Caps can protect process workers from striking their heads against fixed objects when working in confined spaces;
- crash helmets, cycling helmets and climbing helmets which are intended to protect wearers should they be involved in a fall.

The head protection should be compatible with the work to be done. Most head protections come with internal harnesses and it is essential that these are adjusted by the workers to fit their own heads. For outside workers in the process industry the use of chin straps is recommended, because gusts of wind and falling debris can dislodge head protection and there is a tendency for workers to hold onto their hats, rather than look after their own safety by using handrails. Chin straps securely fit the hat to the wearer's head and can increase comfort. Should a head protector be struck by a missile it should be discarded and a new one supplied. The interior of a hat should be kept clean and sweat bands should be regularly cleaned and/or regularly replaced.

22.6.4 Foot protection

The safety boot or shoe is the most common type of safety footwear, and would normally have a steel toe cap. For the fashion-conscious they can have the appearance of trainers or 'Doc Martins'. For some parts of the process industry it will be necessary to have a steel sole, to prevent penetration of soft soles by sharp objects.

For use in slippery or wet conditions special materials will be necessary to prevent slipping and protect the feet from corrosive or toxic materials.

Anti-static footwear prevents the build-up of static electricity on the wearer and it can also reduce the danger of igniting flammable atmospheres.

Wellington boots will protect the wearer from water and other liquid contamination, and can be supplied with steel toe caps and steel mid soles. The material of the boot can be selected from rubber, polyurethane and PVC, depending upon the chemical resistance necessary.

Foundry boots will have to be made from heat-resistant substances which will not readily melt or burn, and they should be made so that they can easily be removed should there be a danger of splashing. Thermal footwear can be provided for working in extra cold or extra hot conditions.

Safety footwear should be properly maintained and checked at regular intervals for worn or deteriorated components. It is also essential to maintain foot hygiene and so regularly worn shoes should from time to time be sterilised and cleaned on the inside.

22.6.5 Vision and eye protection

Eye protection needs to safeguard against the hazards of impact by missiles or walking into fixed objects, splashes from toxic or chemical substances, or molten metal, dust, gases, welding arcs and non-ionised radiation, and in some processes laser light.

The Display Screen Equipment Regulations[6] call management attention to hazards arising from using computer or word processor screens, and recommend that appropriate eye tests and safety spectacles should be supplied.

Eye protection can be provided either as specialist optical glasses which provide protection from splashes and penetration, etc. or eye goggles or visor shields worn over or external to optical spectacles.

22.6.6 Noise and hearing protection

Noise can be hazardous by impairing human hearing, by interfering with communication and by reducing morale and general awareness. It is discussed in detail in a UK code of practice[7] and in several books[8] including my earlier one on construction hazard.[9] Apart from construction and maintenance activities, excessive noise in the process industries is mostly found in places such as boiler houses, compressor rooms and where fluids under pressure are discharged to the atmosphere or flared.

Most jobs in the process industries require good hearing which can be tested speedily. Hearing records should be kept. Periodic retesting of workers' hearing may reveal excessive exposure to noise at work.

The (UK) Noise at Work Regulations 1989

These regulations, which result from an EC directive, aim at protecting the hearing of workers from damage by noise. Their first requirement on employers is to assess the problem in noisy areas, e.g. where people have to shout or have difficulty in being understood by someone about 2 m away. The assessment should be made and recorded by a competent person. Noise levels should be reduced as far as practicable to below 90 dB(A), e.g. by fitting silencers and enclosing noisy machinery in acoustic hoods. Noise emissions of new machinery should be checked on purchase.

Managements should restrict the number of workers in zones with noise levels above 85 dB(A) to a minimum, inform them of the risk, and provide workers in zones with noise levels between 85 and 90 dB(A) with ear protectors if they ask for them.

In zones where noise levels exceed 90 dB(A), warning notices should be displayed and all working in them should be provided with ear protectors and trained in their use. Employees are obliged to wear ear protectors when in such zones. Employers are obliged to maintain machines, etc., in these zones to prevent noise from increasing and to maintain the ear protectors provided. Actions indicated at 90 dB(A) also apply where peak sound pressures may exceed 200 Pa.

Makers and suppliers of machines are obliged to provide information on the noise they are likely to generate.

Hearing protection

Ear-plugs (inserts) are mainly for workers continuously exposed to noise levels between 90 and 100 dB(A). Ear-muffs give protection to noise levels up to 120 dB(A). For still higher noise levels acoustic helmets are available. Since ear protectors cause a communication problem, care is needed when selecting them not to accentuate it by overprotecting the wearer.

Ear-plugs are worn in the ear canal, sealing the entrance to the ear. Some are conformable and allow the plug to be compressed before insertion in the ear, where it expands to give a comfortable fit. Others which are premoulded to a predetermined shape are available in a range of sizes. Some have a valve system which, it is claimed, absorbs high noise levels while transmitting speech sounds. There are also semi-inserts attached to a headband which keeps them in the right position.

Disadvantages of ear-plugs are:

- it is difficult for a supervisor to check whether they are being used;
- there is a hygiene problem when a worker decides to fit ear-plugs when his hands are dirty;
- the conformable type tend to be displaced if the wearer moves his jaws sharply.

Ear muffs consist of two hard cups which fit over the ears, foamed plastic or rubber cushions which fill any gaps between the cup and the head, and a semi-rigid headband which keeps the cups and cushions in contact with the head. This can be worn over the head, behind the neck or under the chin. The cups may also be attached to some safety helmets by adjustable side-arms. The cushions are liable to degrade from mechanical abuse or sweat from the wearer and therefore need regular inspection and replacement.

Ear muffs are of two types, circumaural and superaural. The former, which enclose the ears, are commoner and more effective except where spectacles with normal side-arms are worn. The latter, which are lighter, seal against the ears themselves and are less affected by spectacle frames.

22.7 Breathing and respiratory protection

Breathing and respiratory protective equipment (RPE) should only be used as a final resort to achieve health and safety. Wherever reasonably practicable adequate general ventilation should clear the workplace of toxic, flammable, corrosive or other harmful fumes, dusts or gases. If general ventilation cannot be reasonably maintained, then adequate local exhaust ventilation (LEV) should be undertaken so that dust, fumes or particles will be captured where generated and transported via ductwork to decontaminating plant.

22.7.1 Nominal protection factor

There are numerous types of RPE and the recommendations for their selection, use and maintenance are contained within BS 4275.[10] Breathing apparatus or RPE is graded according to its nominal protection factor (NPF). The required NPF for any given work site can be calculated by dividing:

$$\frac{\textbf{the time-weighted average}}{\textbf{the exposure limit}}$$

The time-weighted average is found by monitoring the work shop using a sampling device. The exposure limit is the allowable limit within the work shop and can be found by consulting HSE Guidance Document EH42 *Monitoring Strategies for Toxic Substances*.[11]

22.7.2 Disposable RPE

It is estimated that some 10 million disposable respirators are bought in the UK each year and as many as 40 000 people regularly wear them each year. Their NPF factor is 10 and they are made to BS 20918 and BS 60116. Such disposable RPE does not need cleaning or any other maintenance procedure but is thrown away after each shift.

22.7.3 Half masks or ori nasal masks

The NPF factor for these is 10 and they are made to BS 2091. It is estimated that one-third of a million such masks are bought in the UK each year and they can have a life expectancy of between six months and five years. It is estimated that 500 000 employees could be using them. Such masks are more comfortable to wear than disposable masks but both have approximately the same protection factor. Half masks are usually made from rubber and must be properly maintained. They are fitted with renewable or removable filters, which should be changed at each shift, and the outer filter should be renewed at least once a month. Half masks should be cleaned each day, and there is thus a cost involved in maintaining them.

22.7.4 High-efficiency dust respirators

These respirators have an NPF factor of 1000 and are manufactured to BS 4555. They are designed for protection against the higher levels of toxic particulate material. The masks can be adapted for respirators or breathing apparatus and hence it is difficult to estimate how many are sold or in use. Their life expectancy is between one and five years and the filters are likely to last two months.

22.7.5 Ventilated visors and helmet respirators

Helmets with hoods are provided with a small transportable fan which sucks in air, filters it and then delivers it to the breathing zone. The NPF factor is 100.

22.7.6 Positive pressure powered respirators (high efficiency)

Such respirators are made to BS 4558, and have an NPF of 2500. The face mask used for this equipment is similar to the high-efficiency respirator but air is supplied via an electric fan.

22.7.7 Breathing apparatus

Breathing apparatus with a trailing compressed air line supplying clean fresh air to the user has an NPF factor of 2000 and is supplied to BS 4667.

It has a full face mask with compressed air fed through a line with valves attached to a belt, and has a life expectancy of five years. The air should be cleaned by a filtration unit from a compressor; it would not be acceptable just to use the compressed air mains in a factory unless it was adequately filtered.

22.7.8 Self-contained breathing apparatus

Self-contained breathing apparatus is used intermittently, often for rescue purposes. A high-efficiency face mask is supplied with clean, fresh air from air cylinders worn on the operator's back. The NPF figure is 2000 and they are made to BS 4667. Such self-contained breathing apparatus will need adequate maintenance and cleaning. It should also have warning systems to indicate when the cylinder is running empty. Extensive training is needed for operators using self-contained breathing apparatus and it is rarely used in normal work.

22.7.9 Blasting helmets

Blasting helmets are used when operators are carrying out blast cleaning of structures, castings etc. A full protective suit made in rubberised canvas is donned by the operator, and then an independent blasting helmet is applied over the head and fixed to the full suit. External clean air is supplied via a compressor with a filter, or from a compressed air supply, again with a suitable filter. Work inside a full blasting suit is very difficult, work efficiency will be low, fatigue will be high and such suits should only be used when all other precautions cannot be reasonably applied.

22.8 Other personal hazards

The process industries, while having hazards which are largely peculiar to them, are not exempt from common ones which affect workers throughout industry. Some of those not treated elsewhere in this book are discussed briefly here.

22.8.1 Manual handling

Of all accidents reported to the HSE, 34% are attributed to manual handling, and the process industry has a track record no better than the average. The UK record is echoed across the Continent and for this reason the Manual Handling Regulations 1992[12] were enacted throughout the EU. There is now a statutory duty within Europe for each employer, so far as reasonably practicable, to avoid the need for employees to undertake any manual handling operations at work which involve a risk of being injured. It is now incumbent upon employers to arrange for the elimination of manual handling and substitute mechanical means where reasonably practicable.

Where it is not reasonably practicable to avoid manual handling then the employer has to make a suitable and sufficient assessment of all such operations and for this purpose the Manual Handling Regulations have published Schedule 1 which is reproduced as Fig. 22.3.

In most cases employers will be able to carry out the assessment themselves; if not they can delegate it to organisations which have the necessary skill and expertise. Such experts will have an understanding of the legal requirements of the Regulations, the nature of handling operations and a basic understanding of human capabilities. For the more high risk activities it is probable that the skills of safety professionals, industrial engineers and ergonomic experts will be required.

For many handling jobs, productivity as well as health and safety can be optimised by applying an ergonomic approach where consideration is given to the task as well as the load moved, the working environment and the individual's capability. An ergonomic approach would involve fitting the task to the operator rather than selecting a worker to carry out a physically demanding operation.

Employers have a responsibility to those employed to reduce the risk of injury arising from manual handling operations to the lowest level reasonably practicable. In practice this means that employers should consider supplying mechanical aids, maintaining those aids and making them readily available. They should give adequate support and seating to people who are undertaking manual handling operations, make loads to be moved as light as possible and, where the centre of gravity can vary, this should be indicated on the outside of packages.

It should be realised that the provision of PPE (such as gloves, aprons, overalls, gaiters and footwear) may protect a worker from one type of injury, but at the same time the restrictions caused by wearing it may make manual handling more risky.

A fully ergonomic approach to safety in the workplace will include the environment, and ventilation, lighting, rest periods, the nature and condition of flooring, and working at the appropriate height or at different levels. There are well tried ergonomic principles such as the positioning of feet and keeping the back straight when lifting, avoiding jerking, moving your feet before your torso, etc., but it is recommended that guidance is drawn from technical books devoted to ergonomics, or from properly qualified experts.

The HSE have produced guidance on the Manual Handling Regulations 1992[13] but it should be realised that these are only guidelines and not limits. Inexperienced employers can persuade people to carry out work their bodies cannot cope with and this in the past has resulted in a high incidence of injury.

22.8.2 Vibration

Vibration can be both a safety and health problem, causing:

- fatigue and failure in metal parts;
- fixtures to work loose;

– More detailed assessment, where necessary:

Questions to consider: (If the answer to a question is 'Yes' place a tick against it and then consider the level of risk)	Yes	**Level of risk:** (Tick as appropriate) Low	Med	High	**Possible remedial action:** (Make rough notes in this column in preparation for completing Section D)
The tasks – do they involve:					
• holding loads away from trunk?					
• twisting?					
• stooping?					
• reaching upwards?					
• large vertical movement?					
• long carrying distances?					
• strenuous pushing or pulling?					
• unpredictable movement of loads?					
• repetitive handling?					
• insufficient rest or recovery?					
• a workrate imposed by a process?					
The Loads – are they:					
• heavy?					
• bulky/unwieldy?					
• difficult to grasp?					
• unstable/ unpredictable?					
• intrinsically harmful (eg sharp/hot?)					
The working environment – are there:					
• constraints on posture?					
• poor floors?					
• variations in levels?					
• hot/cold/humid conditions?					
• strong air movements?					
• poor lighting conditions?					
Individual capability – does the job:					
• require unusual capability?					
• hazard those with a health problem?					
• hazard those who are pregnant?					
• call for special information/training?					
Other factors – Is movement or posture hindered by clothing or personal protective equipment?					

Fig 22.3 Manual handling risk assessment

- objects resting on high surfaces to move and fall;
- inability to read instruments and instructions,
- ill-health such as 'white fingers' among persons whose hands or bodies are exposed to excessive vibration.

Like noise, much unwanted vibration can be avoided by careful design and initial choice of machines, etc. and by improved mounting. Again like noise, it can be surveyed by specialists with suitable instruments. Draft UK standards for human exposure to vibration are available.[14]

22.8.3 Electromagnetic radiation

Other forms of radiation sometimes used in the process industries present hazards if not properly controlled. The main ones are listed in Table 22.1, with typical operations in which they occur and the risks to which they give rise.

Table 22.1 Hazards of electromagnetic radiation

Form of radiation	*Use*	*Health risk*
Microwaves	Heat treatment	Deep burns
Infra-red radiation	Drying and heat treatment	Eyes
Ultra-violet radiation	Welding	Skin and eyes
Lasers	Measurement, cutting	Eyes
X-rays	Metal inspection	Whole body
Gamma-rays	Level measurement	Whole body

References

1. S.I. 1988, No. 1657, *Control of Substances Hazardous to Health Regulations*, 1988, HMSO, London
2. S.I. 1992, No. 2966, *Personal Protective Equipment at Work Regulations 1992*, HMSO, London
3. S.I.1992, No. 2051, *Management of Health and Safety at Work Regulations 1992*, HMSO, London
4. HSE guidance on the PPE at Work Regulations
5. *British Standard Year Book*
6. S.I. 1992, No. 2792, *Display Screen Equipment Regulations*, HMSO, London
7. Old edition ref 14
8. Old edition ref 15
9. Old edition ref 23
10. BSI, BS 4275 [p. 11]
11. HSE, *GN EH 42, Monitoring Strategies for Toxic Substances*, HMSO, London
12. S.I. 1992, No. 2793, *Manual Handling Regulations 1992*, HMSO, London
13. HSE, Guidance on the Manual Handling Regulations
14. Old edition ref 24

23 Hazards in the transfer of technology (TT)

The risks of accidents and ill-health generally increase when technologies cross national frontiers, particularly when the recipient is a 'Third World' or 'developing' country (DC). The world's worst industrial disasters have occurred in DCs although the installations where they happened were only of medium size for their type. The release of toxic methyl isocyanate vapour in Bhopal, India, in 1984 caused over 2000 deaths and 200 000 injuries, while two weeks earlier the LPG fire and explosions at Mexico City caused 650 deaths and several thousand injuries. Four out of five of the disasters discussed in Chapters 4 and 5 happened on plants operated by a subsidiary of a foreign company which had developed the technology and had some responsibility for its use by the subsidiary.

There is no simple panacea. Even adjacent countries such as Burma and Thailand, Israel and Jordan, India and Pakistan have very different legal and political systems and cultures. For the industrialist who wishes or intends to introduce a certain new technology into a particular DC, one of the first steps should be to visit the country and find out the ground rules. However, the International Labour Office (ILO) code of practice[1] on HS in TT to DCs, abstracted here in 23.2, covers many common denominators to the problems.

23.1 Definitions and historical introduction

Most of us are used to thinking of technology in terms of complete processes and artefacts, including both hardware and software. But when we consider the hazards of TT, it is often useful to think of technology in the limited sense of 'know-how' which is created and retained in people's minds. This corresponds to the definition given in the *Concise Oxford Dictionary* as the 'science of the industrial arts'. We all use a great deal of (technologically-based) hardware in our daily lives which few of us understand and fewer still could service, let alone create. The hardware is assembled from a multiplicity of components, the construction of each of which depends on the specialised know-how of a small group of people.

The importance of this 'know-how' element struck the writer forcibly when working in Uganda in 1976 for a UN agency on the rehabilitation of industries which had collapsed after the exodus of expatriate technical personnel [23.6].

Technology, however, unlike mathematics and pure science, has little significance without the hardware associated with it and the goods it helps to create. In the limited sense technology consists of patents, written instructions, drawings, computer programs, models, manuals, reports and training. To translate it into hardware and use it effectively requires capital, workers, training and a range of supporting infrastructure and services. Research and development are needed to create new technology. Organisation, discipline, standards and quality control are required to apply it.

The development of technologies and their transfer from one region of the world to another have gone on for thousands of years. Countries of the Western World can trace an almost continuous lineage of their technologies from the ancient Egyptians, 6000 years ago, through Greece, Rome, the Arab world and Spain, with implants from China (including gunpowder and porcelain) and other countries. People in other parts of the world, particularly in Africa, the Arctic, Australasia, and the Americas, lived until recently cut off from this mainstream of technological development. This may have led to the present uneven spread of industrial development and the concepts of 'developed' and 'developing' countries, although the latter term is largely a euphemism.

The USA, Canada, Western and most of Eastern Europe, the former USSR, Japan and Australia have well-developed industries and broad infrastructures. The DCs have poorer populations, predominantly agricultural societies, and relatively undeveloped industries and infrastructures. They include areas with extremes of climate which hinder most human activities.

At first a country lacking a new technology may import its products until its market for them becomes sufficient to justify importing the technology and setting up indigenous production facilities. With the growth of international trade and multinational enterprises, many new technology-based plants have been built in DCs. Sometimes the products are primarily made for export. Reasons for this include cheaper labour, raw materials or energy, or freedom from controls on HS and environmental pollution which the same industries would face in their country of origin.

In Third World countries with expanding populations, low life expectancies, widespread poverty and unemployment, fatal accident frequency rates in industry are higher than in industrialised countries, although they are only one of many reasons for premature death. Such statistics for 22 countries (chosen solely because they had common data bases) are given in Table 23.1.

Regulations in DCs are generally less stringent than in more industrialised ones and standards such as occupational exposure limits for airborne toxic substances are often lacking. Regulatory agencies in DCs are often thought of as impotent and ineffective and seldom have the equipment and trained staff needed to monitor the observance of even those national limits that exist. Where prosecutions of manufacturing firms for the death or incapacity of workers from occupational exposure are successful, the penalties are far lower than in a developed country. In short, life is cheaper in DCs than in industrialised ones.

Workers in poorer countries are rarely worse off as a result of the import of hazardous industries nor do their representatives always oppose it. If the choice is between starvation and early death or a job with a low but certain wage, accompanied by a one-in-ten risk of cancer after 20 years, a rational

Table 23.1 Fertility, mortality and fatal accident frequency rates for 22 countries at different stages of development

A = Annual rate of natural increase per 1000, 1980–1985[2]
B = Male life expectancy at birth, years 1980–1985[2]
C = Fatal accident frequency rate in manufacturing industry per 100 000 worker-years, 5-year average from 1981 to 1986[3]

Country	*A*	*B*	*C*
Developing countries			
Egypt	28	57	17
Togo	29	49	60
Zimbabwe	35	54	13
Panama	23	69	15
Peru	26	57	23
Bahrein	28	66	25
Intermediate development			
Cyprus	12	72	5
Hungary	–2	66	12
Ireland	8	70	27
Korea (South)	13	63	18
Spain	5	73	12
Yugoslavia	6	68	6
Developed countries			
Austria	0	70	7
Canada	7	72	8
Czechoslovakia	2	67	6
Finland	3	70	4
France	4	71	6
German Federal Republic	–1	71	9
Hong Kong	9	74	4
Netherlands	4	73	2
Switzerland	2	74	9
United Kingdom	1	71	2

worker would choose the latter, as was pointed out by delegates at a 1981 International Labour Office (ILO) symposium on safe TT to DCs.[4] (In some DCs, as Table 23.1 shows, most workers would not expect to live for another 20 years in any case.) Strongest opposition to the uncontrolled transfer of hazardous technologies came from workers' representatives from industrialised countries, where tighter health controls were already in force.

23.2 The ILO code of practice[1]

Of the various UN agencies and other international bodies [23.2.12], the ILO is specially concerned with the HS problems (which in this context include working conditions) of TT. In 1988, it published a code of practice on the subject.[1] This is specially addressed to designers, importers, exporters and users of technology, to national authorities responsible for HS aspects of imported technology and to contractors involved in installing and operating new technological hardware. Selected parts are abstracted here and passages quoted directly from the code are shown in parentheses.

Much of the code stems from bitter experience and its full implementation will be an uphill task. Like other UN agencies, the ILO has no powers of compulsion over its member states, and depends on voluntary ratification of its conventions, recommendations and codes. This limits their effects unless their provisions are incorporated into strictly enforced national legislation. Important subjects which are not dealt with directly in the code are:

- economic factors which underlie many of the hazards of TT;
- the need for adequate funding to control these hazards.

Several examples of the problems which led to the code are given later in this chapter.

23.2.1 General provisions

Methods of TT covered by the code include:

1. the use of experts;
2. the supply of machinery and equipment directly or under a contract which also provides for the TT;
3. the acquisition of patented technology through a licence agreement;
4. the use of turnkey contractors to set up and commission the plant;
5. the direct import of technologies by foreign companies.

The following basic principles apply:

1. the technology-exporting country should furnish the recipient country with all standards, national regulations and other relevant information about the operation and development of the technology and why it is used;

2. the recipient country should compile from other sources all available HS information on the proposed technology;
3. the competent authorities in the recipient country should use the information thus collected to judge the safety and suitability of the proposed technology;
4. HS information compiled by the recipient country should be made public so that all concerned can deal expeditiously with initial proposals for the TT;
5. the technology-exporting country should not export technology involving processes, equipment or substances which are prohibited in its own territory because of their potential to cause serious risks to HS;
6. imported technology should be subject to HS standards, regulations, practices or guidelines which are no less stringent than those applied to the same technology in the exporting country.

23.2.2 General factors to consider in TT

1. Any necessary modifications should be made to the original technology to take adequate account of the differences between the receiving and the supplying country.
2. Technology should not be selected for transfer on purely economic or technical data, but only after careful study of all factors affecting HS.
3. Operatives and maintenance staff in the recipient country who will have to operate and maintain the processes, plant and equipment must be properly trained so that they can work safely.
4. Proper maintenance and repair facilities should be available to or within the DC.

23.2.3 Decisions to be made before TT

1. Technology-receiving countries should draw up lists of technologies whose import should be (a) prohibited and (b) subject to restrictions. In doing this, note should be taken of substances which are (a) prohibited or restricted in industrialised countries and (b) subject to stringent HS precautions.
2. Before importing any technology, the recipient country should ensure that its HS and social insurance infrastructure are sufficiently developed to provide the necessary medical surveillance, treatment and compensation for any resulting occupational injuries and diseases.
3. HS implications should be considered in the following particulars of choosing and implementing a foreign technology:

- alternative technologies which serve the same purpose (with a view to selecting the safest);
- pre-investment studies including feasibility and environmental impact;
- process and manufacturing studies;
- layout, design and engineering studies and machinery specifications;
- equipment selection, plant construction and start-up;
- personnel selection needed for the imported technology;
- technical assistance needed for training, commissioning and various management aspects.

23.2.4 HS-related standards, risk appraisal, consultants, checklists and regulations

1. The suppliers of technology, plant and equipment should clearly inform the purchaser which HS-related technical standards have been used in the design. This information should also be given to the relevant workers' organisation (where there is one). Internationally recognised standards should be used wherever possible.
2. Qualified HS experts should be associated with the design work.
3. The location and design of the plant should be subjected to independent risk appraisal.
4. HS checklists should be used as an aid to risk appraisal for all relevant aspects of the project, including plant location, design and the materials and chemicals used in the manufacture.
5. Regulations adopted should refer to and comply with ILO Conventions and Regulations and codes of practice and/or national statutes and regional directives issues by inter-governmental agencies.

23.2.5 Major hazard installations requiring special HS attention

1. Major hazard installations are defined as those which by the nature or quantity of dangerous substances present could cause a major accident in one of the following categories:
   - release in tonnage quantities of toxic gases which are lethal or harmful at considerable distances from the point of release;
   - release in kilogram quantities of extremely toxic substances which are lethal or harmful at considerable distances from the point of release;
   - release in tonnage quantities of flammable liquids or gases which form a large cloud which in turn burns or explodes;
   - the presence of unstable or highly reactive materials which may explode.

2. The supplier of a technology which requires the storage, processing or production of dangerous substances should inform the technology-receiving country whether the technology involves activities which are classified as a major hazard in the supplier's or any other country.

The code gives design principles for major hazard installations and lists the information which the technology supplier should provide to the recipient country and action required by the recipient country. These matters are dealt with in more detail in ILO's manual on major hazard control,[5] and throughout this book.

23.2.6 Administrative and institutional arrangements

1. Legal standards governing HS and working conditions in TT should be linked with existing HS legislation and be enforced by a competent authority.
2. Licence agreements for TT should state whether the legal standards and regulations of the licensee's or licensor's country should apply, but in general the more stringent of the two should apply.
3. The validity of such licence agreements should be subject to approval by the HS authorities of the technology-receiving country.
4. Such licence agreements should cover appropriate HS aspects including training of national personnel.
5. New techniques incorporated into renewed agreements should comply with all relevant HS rules and regulations.
6. The granting of patents should stipulate that the technology-receiving country must be kept fully informed on all relevant HS provisions and means of hazard assessment and control used in the production of the patented item. The same should apply to the granting of trademarks.
7. Each country should establish the necessary institutional arrangements to ensure that HS aspects are considered in TT.

These institutional arrangements should include the preparation and harmonisation of national standards and regulations.

23.2.7 Training and education

1. Training programmes should be specifically adapted to the needs of technology-receiving countries.
2. Cultural aspects have a strong influence on attitudes towards risk. They should therefore be recognised in their entirety and taken into account by training organisers and trainers prior to training.
3. Trainers must be trained to the required level of expertise.

4. Workers should not pay for their training within industry.
5. The training of designers should include consideration of factors in technology-receiving countries such as climate which may affect design.
6. 'The training of engineering students from developing countries who study at universities and colleges in industrialised countries should emphasise the adaption of technology to local conditions. To promote this training DCs should be given the chance to contribute to the curricula of universities and colleges in industrialised countries.'
7. 'The understanding of the problems related to TT should be promoted by means of training material and special publications, and other measures such as courses, discussions and seminars. These promotion efforts should be directed at policy-makers, industrial planners, management in private and public enterprises, supervisors and foremen, workers and trade union officials, the staff of labour, medical and factory inspectorates, occupational hygienists, economists, engineers, chemists, safety officers, vocational and safety trainers, and agricultural and other workers.'

23.2.8 Collection and use of information

1. Technology suppliers should provide all HS information relevant to the technology to the authorities in the recipient countries and to the users. This should be updated periodically and when changes are made to the technology.
2. Such information should be drafted in an agreed language which is understood by the users of the technology. It should take into account all factors which influence the use of the technology in the recipient country and be supported by case studies and experience gained in the application of the technology.
3. The technology supplier should be consulted before the user makes any modification or adaption of the technology. This should be specified in the technology supply contract.
4. All available information and expertise should be shared with national tripartite safety councils and similar non-governmental organisations in the user country.
5. Multinational enterprises should make available information on relevant HS standards which they observe in other countries, particularly special hazards and related protective measures associated with new products and processes. They should co-operate with international organisations in preparing international HS standards.
6. Technology-receiving countries should be encouraged to exchange HS information including their field experience, the successful adaptation, modification of imported technologies, and also to exchange technical personnel.

23.2.9 Action at company level

1. The technology selected should take HS aspects fully into account and the influences which local climate and cultural factors have on them.
2. During planning studies the technology supplier should consult the recipient country to obtain all necessary information required for design and should provide to that country all information needed for proper planning.
3. Planning should include studies of similar existing technologies to note (among other things) the effect of the technology on the environment and social system of the country of origin.
4. A competent representative of the technology-receiving country who will be involved in the operation of the process and plant using the technology should be present during the design of the process and plant.
5. The technology supplier should develop as part of the project documentation a safety specifications book containing specific information about the safe operation of the process and plants. This should include details of all hazard analyses provided by the licensor and the technical codes and standards used during design and construction. (The ILO manual includes a checklist to be used in preparing the safety specifications book.)
6. Personnel selected to operate the new technology should understand it well, and be professionally and technically qualified and motivated to work in the DC.
7. Technical advisers (generally from the technology supplier) should be employed for long enough for full TT, including responsibility for management, HS and working conditions, to local personnel.
8. Job descriptions of the technical advisers should detail their duties and responsibilities in HS matters.
9. Top management of the enterprise should formulate a written HS policy and ensure that a safety manual highlighting operation and maintenance is written and made available with the policy to all within the enterprise.
10. Where technology is received in 'package' form, personnel of the recipient country should be trained to fully understand all HS aspects rather than merely following instructions mechanically.
11. Occupational hygiene standards should be adapted to local conditions.
12. In packaging chemicals and other materials used in the process, conditions of transit and the handling and storage conditions in the recipient country should be considered.
13. The directors of the enterprise should maintain an adequate HS programme (similar to that discussed in Chapter 19 of this book).

14. Safety instructions and other notices should be in the languages of the workers employed, and easily understandable symbols should be used. Texts should be displayed in a durable form and protected against damage.
15. Employers should equip their workers with personal protective clothing and equipment which fits their physique and suits the prevailing climatic conditions.

23.2.10 Action at the national level

1. A technology-receiving country should develop the necessary occupational HS infrastructure to deal adequately with all the problems related to HS involved in TT. (This may include the setting up of a special national standards body where none existed before.)
2. National negotiators for TT should have been adequately trained in HS requirements in order to ensure the inclusion of these matters in the TT process.
3. 'Where policies for the progressive take-over of foreign enterprises by national interests are adopted by the national authorities, care should be taken that the resulting mixed or national enterprises have the full background knowledge, information, experience and competence, including staff skills, to deal with safety and health and working conditions aspects, as well as the ability to handle all emergencies.'

Other measures include the promotion of consultancy services, the sponsorship and publication of technical journals and the formation and development of professional organisations within DCs.

23.2.11 Action at the regional level

This includes co-operation between national organisations in different countries, the establishment of regional technology centres and the pooling and interchange of technical expertise at the regional level to assist in the diagnosis, identification and solution of relevant HS problems.

23.2.12 Role of international organisations

1. The ILO should continue its efforts in the dissemination of relevant technical information, in the promotion of its exchange, in the development of training and training materials, in strengthening existing national HS institutions and in assisting DCs through its technical co-operation programme.
2. The work of other international organisations in the HS field should include the provision of technical information, the maintenance of lists of suitable consultants, the provision of advice and assistance to DCs, especially technical assistance in the development of hazard-control

systems, and international standard-setting activities such as the preparation of relevant conventions and codes of practice.

3. Projects financed by international agencies should include HS requirements in their guidelines. The various international organisations concerned with TT should co-operate more in HS aspects of their work.

Such organisations include the United Nations Industrial Development Organisation (UNIDO), the United Nations Conference on Trade and Development (UNCTAD), the World Bank and various regional development banks. The World Health Organisation (WHO) and the United Nations Environment Programme (UNEP) already have a joint International Programme with the ILO on Chemical Safety (IPCS).

23.3 Examples of the spread of hazardous technologies

Many products and intermediates have been belatedly found to cause disease at low levels of exposure. Several years have sometimes elapsed before the hazards were recognised and appropriate measures introduced to protect workers and the public. Examples are asbestos, benzidine, benzene, vinyl chloride, cadmium and beryllium. Increased awareness of other health risks such as arsenic, chromium, mercury and lead has also resulted in tighter controls in industrialised countries. It is thus not surprising that some companies have closed down some of their more hazardous manufacturing operations in their home countries and set them up in poorer ones where controls are less stringent (see para. 5 of 23.2.1). This has even been encouraged by the governments of some developing countries, which, to attract foreign investment, set up 'export processing zones' with special concessions for foreign firms, including exemption from certain industrial safety legislation.

The following examples of such spread of hazardous technologies are due to Castleman.[6]

23.3.1 Asbestos textiles

The USA was long a world leader in the manufacture of asbestos textiles, although most of the asbestos used was mined in Quebec. Although general recognition of the health hazards of asbestos is comparatively recent, some American insurance companies selling workers' compensation insurance appear to have recognised the risk as far back as 1918. By 1965, the high incidence of cancer among asbestos workers was well established. In 1972 the Occupational Safety and Health Administration (OSHA) set a temporary standard of 5 million fibres/m^3 in workroom air. This was reduced in 1976 to 500 000 fibres/m^3 and further later.

The industry faced considerable expense and difficulty in meeting even the 1976 standard of 500 000 fibres/m^3. The US industry declined while imports from Mexico, Taiwan and Brazil rose from almost nothing to 4.5 million pounds in 1976.

23.3.2 Arsenic

Arsenic is used to make pesticides, herbicides, wood preservatives (still in some countries) and glass. It is present in the ores of many metals, particularly copper, and unless special extraction equipment is installed, much of the arsenic in the ore is discharged into the atmosphere during smelting as arsenious oxide. Research in the USA from 1950 showed that deaths from lung cancer in counties where copper, lead and zinc ores were smelted were significantly higher than elsewhere and that airborne arsenic was the principal cause. The situation became so serious that in 1975 OSHA proposed to reduce the limit for airborne arsenic in the working atmosphere from 500 mg/m^3 to 4 mg/m^3.

The high cost of complying with the new limits caused US smelters to restrict their use of ores of high arsenic content and switch where possible to domestic ores of lower content. Copper ores and concentrates from several countries (Peru, Mexico, the Philippines and Namibia) which were once imported in large quantities into the USA have high arsenic contents. Partly because of this, new export-oriented smelters were constructed in these countries, where there was little regulation of the arsenic and other pollutants discharged to the atmosphere. A plant producing arsenic was shut down in the USA and new ones were built in Mexico and Peru.

23.3.3 Benzidine dyes

A useful range of cheap dyes based on benzidine or 4–4′-diaminodiphenyl was invented in the last century and widely used in textiles. Later an unusually high incidence of bladder cancer was found among benzidine workers. A retrospective study of benzidine workers in one US company showed that 17 out of 76 had developed bladder cancer. The manufacture of benzidine dyes was subsequently banned in a number of industrialised countries, many of which, however, still import and use them. The principal sources of benzidine dyes in the mid-1970s (when their manufacture in the USA virtually ceased) were Romania, Poland, India and France. A large new export-based plant came into operation in South Korea. The dyes themselves are apparently safe provided their free benzidine content is controlled at less than 20 ppm.

23.4 Problems of culture, communication and language[7]

Established cultural norms and attitudes vary widely throughout the world. Unless properly understood, they can aggravate the HS problems of importing new technologies. Thus the colour green, which is associated with 'safe' in the West, can have different associations in the Orient. A shake of the head which means '*No*' in the West means '*Yes*' in parts of Asia. In some parts of the world the new day starts at sunset. While these cultural differences can be bridged by care and local knowledge, attitudes fostered by religion can present greater difficulties.

In societies where events are thought to be controlled by divine fate rather than human action, attempts to implement safety practices may be met with complete incomprehension or (worse) by suspicion that the implementer is attempting to interfere with the Divine Will. The wearing of safety hats contravenes the tenets of one religion. Islam has strictly enforced periods of fasting, mourning and celebration during which its followers tend to become more accident prone. Road accidents are more numerous during the month of Ramadan in Moslem countries, when no food or drink is taken between sunrise and sunset.

Excessive politeness and fear of 'losing face' are widespread attitudes in several societies and tend to inhibit the free flow of information. In some friendly South-east Asian countries, many people, often to their own cost, are too polite to give '*No*' for an answer, while fear of losing face may inhibit the speaker from admitting that anything is amiss, and hence enabling corrective action to be taken.

The concept of maintenance is lacking in some pre-industrial societies. The writer has seen several factories under local management in two African countries which ground to a halt due to lack of maintenance and failure to purchase spares.

23.4.1 Language [18.9.6]

Language can be a major barrier to communications. We in the English-speaking world, which has developed a wide technical vocabulary and has a relatively simple grammar, are fortunate in the widespread use of English as a common language. Yet within departments of many English-speaking enterprises special meanings have developed for familiar words and entirely new ones have been coined which are not to be found in any dictionary. The meaning of words often depends on their context, their intonation, and the gestures that accompany them, and can only be grasped after lengthy initiation.

Technical comprehension and communication are thus harder for those brought up in many other languages, particularly those of small countries which are little spoken outside their borders. Even Spanish, the principal language of most of South and Central America, has a limited technical vocabulary. Several Peruvian engineers have told the writer that they find it more ambiguous than English, even for non-technical discussions. Attempts to translate technical terms can be quite hilarious. Thus the meaning of a Peruvian term used for 'heat-exchanger' was given in the writer's Spanish dictionary as 'brothel keeper'.

To overcome problems in communicating safety messages in Third World countries, some experts recommend the use of pictures and visual aids, without written words. The pictures of workers should be ones with which the locals can identify, and not copies of posters showing workers in Western clothing. 'Match-stick' figures seem to be universally acceptable.[7]

23.5 Problems of standards in developing countries[8]

The role of standards in enabling industries to develop efficiently and safely has been discussed earlier [21]. Standards which we take for granted are most appreciated when they are lacking. They play important roles in all technologies and in the cultural infrastructures to which they relate. Until the new technology arrives, there is often little need for particular standards, but once it comes, it may do so in a 'standards vacuum' [23.2.1, para. 6].

Sometimes the problem is partly solved by creating a special 'multinational enclave' in the developing country around the industrial enterprise where the new technology is being introduced. In this enclave, a set of standards borrowed from the country of origin of the new technology is enforced, while the rest of the country carries on as before. Problems then arise when the enterprise is taken over by the developing country. Only a handful of people who held responsible positions when the enterprise was under foreign control may be aware of the standards in use and appreciate their importance. This is not a new twentieth century problem but is as old as history and was encountered in Britain when the Romans left.

A frequent question for countries receiving new technologies is whether they should wait for international standards to be agreed and published before creating their own, or whether to adopt or adapt standards from the country from which they received the technology. In the first case the country may have no effective standard for many years. In the second it risks dependence on the technology-exporting country for the supply of plant, equipment and/or materials. In this situation it is probably best to solve the problem at company level first, by adopting the most technically satisfactory, if restrictive, foreign standard. The creation of a national standard can then be postponed, perhaps until there is an adequate international one on which it can be based.

The following examples of standard-related problems in developing countries are taken from first-hand experience.

23.5.1 An LPG storage and distribution depot

One of several LPG (propane) depots built in the 1980s in a DC included pressure storage spheres, a road tank-car filling station and a cylinder-filling shed with equipment for filling several sizes of portable cylinder. In Europe it would have rated as a Major Hazard Installation. The depot included a 'stenching unit' which was designed to inject a chemical with a powerful and offensive odour into the incoming propane to enable customers who purchased it in rented cylinders to detect leaks. The contract called for the pressure-relief system to be designed to API standards (RP 520), while the LPG was to be stenched to meet the UK specification for 'commercial propane' (BS 4250). These standards had no status in the country where the depots were built.

Fluids passing the relief valves should have been led through pressure 'knockout' vessels to separate liquid propane before the gas passed to a high-level vent pointing upwards, well away from possible sources of ignition. No knockout vessels were provided and any fluid discharged would have passed through short vertical lines terminating with inverted U bends, whose open ends pointed downwards over a road used by LPG road tank-cars.

An even more serious mistake was that other pressure-relief lines discharged directly into a large shed for cylinder filling, where up to 100 people would normally be working. This was discovered when relief valves discharged propane into the shed while the plant was being commissioned.

The plant was far from leak-tight when liquid propane was first fed into the storage spheres and the stenching unit was started. So much stenched LPG entered the atmosphere that the plant manager was inundated with complaints from neighbouring residents. The stenching unit was therefore shut down and commissioning was continued!

23.5.2 A national institute for occupational safety and health

In the outskirts of a large Middle East city, a modern four-storey purpose-built building had been donated to the nation by an international agency to house a new national institute for occupational safety and health, whose functions included advising industry on fire protection. It had laboratories on two floors in which flammable organic liquids and LPG were used, and about 100 offices, with a staff of about 500. The building had two stairwells at opposite ends, one landing in the main foyer at the front of the building, the other leading to double doors which opened onto waste ground at the rear. These doors had only conventional locks. 'Panic' latches (Fig. 23.1) which would have allowed the doors to be locked to outsiders but easily opened from the inside had not been fitted. The two stairwells were required as alternate fire exits, which under most countries' fire regulations should have been enclosed.

As a result of break-ins, the security chief of the enterprise had ordered these doors to be kept locked. This made the lowest flight of stairs unusable. The cleaning contractors, who previously had to take their daily harvest of waste paper several kilometres to the city refuse dump, now found a handy dumping space within the building itself. When this writer inspected the stairwell the bottom flight was filled with waste paper and laboratory rubbish.

Situations such as these which cause nightmares to safety advisers from industrialised countries are all too common in DCs where shortage of foreign currency inhibits the import of appropriate safety equipment.

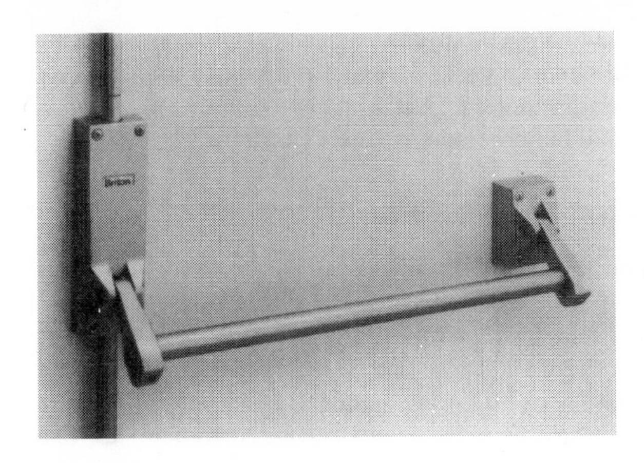

Fig. 23.1 Panic bolt (courtesy Newman Tonks Engineering Ltd)

23.6 Uganda 1976

This writer had the unusual experience of spending several months in Uganda in 1976, the fifth year of Amin's rule, as a consultant employed through UNIDO to study and advise on the rehabilitation of the country's non-agricultural industries. These had been shattered not by war but through government policies — part of the growing pains of national independence. Although HS did not feature in our terms of reference, the gradual breakdown of imported technology caused serious accidents and illness. The fragility of technological implants in an isolated and primarily agricultural society became only too apparent.

The causes of the industrial collapse were twofold.

1. The recent exodus of most technical and managerial staff of Indian and British origin, coupled with an acute shortage of trained and experienced Africans.

2. Restrictions in the supply of imported machinery, spare parts, fuel oil and raw materials, most of which came overland through Kenya. The reasons for this were partly political and partly economic.

By 1970 Uganda had several industrial projects based on imported Western technologies, mostly owned by subsidiaries of British and Kenyan–Indian companies. Much of the trade and many of the technical jobs in the country (such as electrical and instrument maintenance) were done by descendants of Indian workers brought in 60 years earlier to build a railway from the port of Mombasa to the capital, Kampala, later extended west to the copper mines at Kilembe. The success of this Indian minority and their discouragement of would-be African sons-in-law caused pent-up frustrations among Africans which encouraged Amin to expel them.

By the time our team arrived, several of the largest and most technically advanced factories had been forced to close down, while others were still running precariously at well below their rated capacities. Other less mechanised factories which had been specially designed for ease of maintenance were, however, still running smoothly. Of the many industries we studied, only a few which had special HS problems are discussed here.

23.6.1 Cement and brick production

Uganda had two cement plants, one at Tororo in the east of the country making a slow-setting cement used to construct the Nile dam and hydroelectric station at Jinga, and a larger one built later at Hima near the Kilembe copper mines in the west. These had been effectively nationalised by Amin. Their management in 1971 was mainly British, with Indians in many skilled jobs. In the next two years most British and Indian staff had left, leaving only a handful of trained Africans.

Early casualties of both works were the electrostatic precipitators which removed fine dust from the flue gases. At Tororo a layer of fine cement dust quickly settled on the flat roof of the raw materials yard. Not appreciating the hazard, nobody bothered to remove it. When rain came, the dust set hard and gradually the roof became increasingly thicker until the weight was too much for the supports, and it collapsed early in 1973, killing several workers.

The lesson failed to reach or was not learnt at Hima, where the main factory roof collapsed from the same cause in August 1973, killing more workers, damaging the gantry crane and halting production for several months.

Both works were shut down during our visit in 1976. They were in desperate need of maintenance and spares of all kinds, short of fuel (oil and charcoal from Kenya) and subject to long and frequent power cuts. There was little likelihood of restarting the Tororo plant. We visited Hima during an unsuccessful attempt to start the plant, which lay at the centre of a cloud of choking dust. Workers with inflamed eyes wearing strips of cloth over their mouths and noses were manhandling hot, dusty process materials because of the breakdown of mechanical handling equipment. None of the temperature-measuring instruments were working. After spending several hours at the works the writer became quite ill and had to return quickly to Kampala for medical treatment (thus missing the opportunity of visiting Kilembe Mines which had similar problems on a larger scale, and the Murchison Falls).

In contrast to the continuous and mechanised cement works, the brick and tile industry which used local clays, and coffee husks as fuel, was in a relatively healthy state. A semi-continuous Hoffmann kiln designed by Swiss engineers had been built almost entirely from local materials on a deposit of ball clay near Kampala in 1958. All equipment had been selected for ease of maintenance. The only imported machine was a Morondo extruder which formed the clay into the required shapes before these were air-dried and loaded daily into a fresh section of the kiln. It produced a range of hollow bricks, blocks and floor sections, roofing tiles, ridges, channels and grilles, and was working almost at full capacity.

23.6.2 Glass bottles

A small glass factory making bottles (for beer and soft drinks) and tumblers had been set up by an Indian group in 1968. This had an oil-heated tank furnace and used local sand, the fuel and other raw materials being imported. In 1972 the company was nationalised, when the Indian owners and staff left. Little maintenance was done during the next three years. The annealing section first broke down, as a result of which some 25% of the beer bottles made burst after filling, causing injuries. Shortly before our arrival the brickwork of the furnace collapsed, spilling its entire contents of molten glass onto the floor of the building, causing casualties and secondary damage.

23.6.3 Bottled beverages

During the 1950s and 1960s many middle-class Ugandans developed a taste for bottled beer and aerated beverages, in preference to the traditional and more nutritious plantain wine known as pombé, which is fermented in earthenware jars and drunk while still fermenting through hollow bamboo canes. Two foreign-owned breweries (one of which was nationalised in 1972) and a bottling plant for non-alcoholic beverages were hence set up in the 1950s and 1960s. Most of their raw materials, malt, hops, sugar and flavouring materials were imported. Production rose steadily until 1973 after which it fell rapidly for several reasons. These included shortage of foreign exchange to buy raw materials, lack of maintenance and shortage of spare parts, loss of skilled expatriate personnel, and poor quality bottles whose breakage was accentuated by a shortage of bottle openers (in place of which teeth were often used and broken). The run-down of the breweries and bottling plant was accompanied by many accidents and injuries, including the death of at least one man who entered an apparently empty vat containing carbon dioxide gas.

23.6.4 Industrial gases

A factory set up as a subsidiary of a British company to produce industrial oxygen, nitrogen, acetylene, argon, nitrous oxide and hydrogen was still operating precariously in 1976 but suffering severely from lack of maintenance which affected its safety. The following examples were noted:

1. holes in the floor which (it was said) could not be repaired due to lack of cement;
2. gas cylinders not painted with their distinguishing colours due to lack of paint, leading to mistakes in identifying their contents;
3. missing nuts and bolts (in short supply) on flanged joints;
4. lack of rubber cushions for unloading gas cylinders from trucks;
5. shortage of protective caps on cylinders;
6. use of an improvised intercooler on a gas compressor to replace one which had failed — this was thought to be about to burst at any time.

Throughout the country there was an acute shortage of welding goggles. At one factory making hand hoes from scrap-iron, gas welders were working with no form of eye protection. Several appeared to be suffering from cataract and one was nearly blind.

23.7 Important lessons for technology importers

The interests of sellers and buyers of technology are usually different. Only those of the importing country which will have to live with the imported technology are considered here. Consultants engaged to carry out objective feasibility studies on projects in DCs, more often than not in the writer's experience, find themselves under pressure from one of the possible parties to the TT. The following important lessons are among those pointed out briefly in the ILO code of practice [23.2].

1. A common mistake is to opt for the latest and most sophisticated technology, regardless of the level of technical development and education in the country and of possible difficulties in being able to get spare parts and special process materials. In Burma, for example, a country which has been very isolated for the last 40 years, there was general praise for a large Chinese-built textile mill which was designed for ease of maintenance (like the brick factory in Uganda [23.6.1]).
2. The siting of major hazard installations well away from populated centres needs special vigilance, and strict measures are required to ensure that unofficial settlements like Brazil's *favelas* (Fig. 23.2) do not spring up on their doorsteps. The scale of the disasters at Bhopal and Mexico City in 1984 was caused by the close proximity of major hazards to population centres.
3. Where hazardous materials are used, whether highly flammable like LPG or highly toxic like chlorine, the quantities stored and their inventories in process should be restricted to the bare minimum.
4. Nationals of the technology-importing country should be trained so that they can take charge of, operate and maintain the installation from the earliest possible moment after commissioning.
5. Good maintenance, whilst important for all plant and storage installations, is even more so for its protective systems.
6. The dangers of a hazardous plant are particularly high when it is being started up, and when the manufacturing operation is being run down or in economic difficulties.
7. Pay special attention to the effects of climate when that in the technology-importing country differs significantly from that in the country where the technology was developed and is now used. These effects are

Fig 23.2 A typical Brazilian favela: '. . . a jumble of rude shacks made of wood, cardboard, scraps of metal . . .' (Popperfoto)

far-reaching and easily overlooked. Remember, for example, that in a hot climate human work may be impossible near furnaces, etc. although this may present few problems in a cold one; materials thought of as liquids in a temperate climate may freeze or deposit solids in cold ones; bacteria and fungi which are dormant in a cold climate multiply fast and cause serious problems in hot and humid ones.

References

1. International Labour Office, *Safety, health and working conditions in the transfer of technology to developing countries — An ILO code of practice*, Geneva (1988)
2. *United Nations Demographic Yearbook 1986*, New York (1986)
3. International Labour Office, *ILO Statistical Yearbook, 1988*, Geneva

4. International Labour Office, *Inter-regional tripartite symposium on occupational safety, health and working conditions specifications in relation to transfer of technology to the developing countries*, Geneva (1981)
5. International Labour Office, *Major Hazard Control – A practical manual*, Geneva (1988)
6. Castleman, B. I., 'The export of hazardous factories to developing nations' (paper reported in reference 4)
7. Brown, D. H., 'Safety problems faced by UK firms working abroad', *Industrial Safety Data File*, United Trade Press, London (1984)
8. King, R. W., 'The role of standards in the safe transfer of technology to developing countries' (paper reported in reference 4)

List of abbreviations

Abbreviations given here do not include:

- letters used only for quantities in equations,
- those in common English usage,
- symbols for SI and derived units,
- symbols for items shown in figures accompanying the text.

ABPI	Association of British Pharmaceutical Industry
ABS	acrylonitrile-butadiene-styrene (copolymer)
abs.	absolute (pressure)
a.c.	alternating current
ACGIH	American Conference of Governmental Industrial Hygienists
ACOP	Approved Code of Practice
ACTS	Advisory Committee on Toxic Substances (HSCs)
AD	Appointed Doctor (in UK)
AFFF	aqueous film forming foam
AIT	autoignition temperature
AMAO	assumed maximum area of operation
AMOCO	American Oil Company
ANSI	American National Standards Institute
APAU	Accident Prevention Advisory Unit (of HMFI)
APC	accident prevention committee
API	American Petroleum Institute
ARC	accelerating rate calorimetry
ASME	American Society of Mechanical Engineers
ASR	accident severity rate
ASTM	American Society for Testing and Materials

BA breathing apparatus
BASEEFA British Approvals Service for Electrical Equipment in Flammable Atmospheres
BASF Badishe Anilin und Soda Fabrik
BLEVE boiling liquid expanding vapour explosion
BP boiling point
BS British Standard

CAA Clean Air Act
CAD computer-aided design
CAITREC Chemical and Allied Industries Training Review Council (UK)
CAL computer-assisted learning
CAM computer-aided manufacture
CBI Confederation of British Industry
CEN European Committee for Standardisation
CENELEC European Committee for Electrotechnical Standardisation
CERCLA Comprehensive Environmental Response, Compensation and Liability Act
CFC chlorofluorocarbon
CHETAH chemical thermodynamic and energy hazard evaluation
CHP cumene hydroperoxide
CIA Chemical Industries Association (UK)
CIMAH Control of Industrial Major Accident Hazards (UK Reg.)
CISHEC Chemical Industry Safety and Health Council (UK)
CL confidence level
CM condition monitoring
CML computer-managed learning
COP code of practice
COSHH Control of Substances Hazardous to Health (UK Reg.)
CPLR Classification, Packaging and Labelling (of Dangerous Substances) Regulations (UK)
CRUNCH dispersion model for continuous releases of a denser-than-air vapour into the atmosphere
CSWIP Certification Scheme for Welding Inspection Personnel (UK)

dB(A) decibel, 'A' scale
d.c. direct current
DC developing country
DENZ computer program for the calculation of the dispersion of dense toxic or explosive gases in the atmosphere
DDT 2, 2-bis [p-chlorophenyl] 1, 1, 1-trichloroethane
DEn Department of Energy
DHSS Department of Health and Social Security (pre-1989) (UK)
DIN Deutsches Institut für Normune
d.p. differential pressure (cell)
DSC differential scanning calorimetry

DSM	Dutch State Mines
DSS	Department of Social Security (UK)
DTA	differential thermal analysis
DTI	Department of Trade and Industry (UK)
EDNA	ethylenediamine dinitramine
EEC	European Economic Community (also EC)
EEMUA	Engineering Equipment Material Users Association
EH	Environmental Hygiene (series of HSE Guidance Notes)
EHH	extra high hazard
EIS	environmental impact statement
ELH	extra low hazard
EMA	Employment Medical Adviser (UK)
EMAS	Employment Medical Advisory Service (UK)
e.m.f.	electro-motive force
EML	estimated maximum loss
EO	ethylene oxide
EPA	Environmental Protection Agency
EPCRA	Emergency Planning and Community Right to Know Act
F&E	fire and explosion (in Chapter 11)
F&EI	fire and explosion index (in Chapter 11)
FAFR	fatal accident frequency rate
FIFR	fatal injury frequency rate
FIRTO	Fire Insurers Research and Testing Organisation
FMEA	failure modes and effect analysis
FOC	Fire Offices' Committee (UK)
FPA	Fire Protection Association (UK)
FR	frequency rate
FRP	fibre-reinforced plastic
FWPCA	Federal Water Pollution Control Act
g.	gauge (pressure)
GLC	gas–liquid chromatography
GRP	general-purpose synthetic rubber (styrene–butadiene)
GS	general series (of HSE Guidance/Technical Notes)
HAZAN	hazard analysis (fault tree type)
HAZOP	hazard and operability (study)
HC	hydrocarbon
HD	high density (polyethylene)
HMFI	Her Majesty's Factory Inspectorate
HS	health and safety (in a general sense)
HSC	Health and Safety Commission (UK)
HSE	Health and Safety Executive (UK)
HSWA	Health and Safety at Work etc. Act

Hz	Hertz (cycles per second)
IATA	International Air Transport Association
ICC	Interstate Commerce Commission
I. Chem. E.	Institution of Chemical Engineers (UK)
ICI	Imperial Chemical Industries
IEC	International Electrotechnical Committee
ILO	International Labour Office (or Organisation)
IMCO	Intergovernmental Maritime Consultative Organisation
IOI	International Oil Insurers (London)
IOSH	Institution of Occupational Safety and Health
IP	Institute of Petroleum
IPCS	International Programme for Chemical Safety
IR	infra-red
IR	incidence rate
IRI	Industrial Risk Insurers (Chicago)
ISO	International Standards Organisation
ISPEMA	Industrial Safety (Protective Equipment) Manufacturers Association
JIT	job instruction training
JSA	job safety analysis
JT	Joule-Thomson
LC_{50}	lethal concentration (for 50% of rat population)
LD_{50}	lethal dose (for 50% of rat population)
LEV	local exhaust ventilation
LL	Lower Limit (of flammability)
LNG	liquefied natural gas
LPC	Loss Prevention Council
LPG	liquefied petroleum gas
MAWP	maximum allowable working pressure
MDHS	methods for the determination of hazardous substances
MEL	maximum exposure limit
MF	material factor (in Chapter 11)
MHAU	Major Hazards Assessment Unit (of HSE)
MIC	methyl isocyanate
MP	member of parliament
MP	melting point
MPPD	maximum probable property damage
MSDS	material safety data sheet
MTBF	mean time between failures
MTTF	mean time to failure
NADOR	Notification of Accidents and Dangerous Occurrences Regulations (UK)

NCB	National Coal Board (now British Coal)
NDT	non-destructive testing
NEPA	National Environmental Policy Act
NFPA	National Fire Protection Association
NIG	National industry group
NIHHS	Notification of Installations Handling Hazardous Substances (UK Reg.)
NIOSH	National Institute for Occupational Safety and Health (USA)
NJAC	National Joint Advisory Council (UK)
NPF	nominal protection factor
NSTO	non-statutory training organisation
NTSB	National Transportation Safety Board
OEL	occupational exposure limit
OES	occupational exposure standard
OFCE	open flammable cloud explosion
OIM	Oil installations manager
OM	organisational misconception
OPSO	office of pipeline safety operations
OSHA	Occupational Safety and Health Administration (USA)
P	proportional control (single mode)
P&ID	piping and instrumentation diagram
P&V	pressure and vent/vacuum (valve)
PCN	personnel certification in NDT
PE	programmable electronics
PES	programmable electronic system
PETN	pentaerythritol tetranitrate
PFD	process flow diagram
PI	proportional plus integral control (two mode)
PID	proportional plus integral plus derivative control (three mode)
PPC/E	personal protective clothing/equipment
ppm	parts per million
psi	pounds per square inch
PSV	pressure safety valve
PVC	polyvinylchloride
QC	Queen's Counsel (leading UK barrister)
RAM	random-access memory
RCD	residual current device
RCRA	Resource Conservation and Recovery Act
RDX	cyclotrimethylamine trinitramine
RIDDOR	Reporting of Injuries, Diseases, Dangerous Occurrences Regulations

ROM	read-only memory
RoSPA	Royal Society for the Prevention of Accidents
RPE	respiratory protective equipment
SCC	Stress-corrosion cracking
S.G.	specific gravity
S.I.	Statutory Instrument
SI	Système International d'Unités
SIC	standard industrial classification (UK)
SOAP	spectrometric oil analysis procedures
SRD	Safety and Reliability Directorate (UK)
St	Stoichiometric
SWT	School of Welding Technology (Welding Institute's) (UK)
TCDD	2,3,7,8, tetrachlorodibenzoparadioxin ('dioxin')
TCP	2,3,5 trichlorophenol
tetryl	tetranitroaniline
TIG	tungsten inert gas (welding)
TLV	threshold limit value (USA)
TNT	trinitro toluene
TSR	temporary safe refuge
TT	technology transfer
TWA	time-weighted average
UCC	Union Carbide Corporation
UCIL	Union Carbide India Ltd
UFD	utility flow diagram
UL	Underwriters' laboratory
UL	Upper limit (of flammability)
UKOOA	UK Offshore Operators Association
UNCTAD	United Nations Conference on Trade and Development
UNEP	United Nations Environment Programme
UNIDO	United Nations Industrial Development Organisation
UP&ID	utility piping and instrument diagram
UPS	uninterruptible power supply
UV	ultra-violet
VCM	vinyl chloride monomer
VDU	visual display unit
VGS	vent gas scrubber
WHO	World Health Organisation

Index